BIOFERTILIZERS TECHNOLOGY

Prof. Dr. S. Kannaiyan
Former Vice-Chancellor (TNAU),
AL-85, 4th Street, 11th Main Road,
Annanagar,
Chennai-600 040.

Dr. K. Kumar
Professor,
Department of Agrl. Microbiology,
Tamil Nadu Agricultural University,
Coimbatore 641 003.

Dr. K. Govindarajan
Professor,
Department of Agrl. Microbiology,
Tamil Nadu Agricultural University,
Coimbatore 641 003

New Delhi • Jodhpur

Published by:

Scientific Publishers (India)
5 A, New Pali Road, P.O. Box 91
Jodhpur 342 001 (India)

Branch Office
Scientific Publishers (India)
4806/24, Ansari Road, Daryaganj
New Delhi - 110 002 (India)

E-mail: info@scientificpub.com
Website: www.scientificpub.com

Print : 2025

ISBN: 978-81-72336-55-4
eISBN : 978-93-86347-42-8

Printed in India

FOREWORD

The major challenge in the 21st Century is to sustain the increase in productivity forever decreasing per capita land and water resources. So far, the mankind has successfully managed to provide the basic necessities with the increase in production of food, fuel and fibre. Their basic weapon has been non – renewable fossil fuel for energy intensive agricultural management. But the impact has began to show its teeth. There is urgent need for environment friendly, economically viable but sustainable technologies to satisfy the requirement of huge population of this planet. Sustainable agriculture requires the management of resources in a way to fulfil changing human needs without damaging or deteriorating the quality of environment and conserving vital natural sources.

Lowlands comprise of 87% of the 145 M ha of world rice area. Lowland rice – based cropping system are characterized by soil flooding during most of the rice growing season. Rainfall distribution, availability of irrigation water and prevailing temperatures determine when rice or other crops are grown. Nitrogen is the most required nutrient in low-land rice based cropping system. Reducing fertilizer N use in these cropping systems, while maintaining or enhancing crop output, is desirable from both environmental and economic perspectives. This may be possible by enriching the soil N through biological nitrogen fixation (BNF), minimizing soil N losses and by improved recycling of N through plant residues.

Favourable conditions for BNF is one of the reasons for the relatively stable yield of rice in the rice based cropping system. The reduced soil conditions favour heterotrophic nitrogen fixation. The flood water is the site for photodependent nitrogen fixation by free-living cyanobacteria, photosynthetic bacteria and symbiotic cyanobacteria with *Azolla*. The rhizosphere of rice also provides favourable conditions for microaerophilic bacterial N_2 fixation. In addition, the inclusion of legumes for grain, forage and green manure production might increase the total N output from the cropping system and further add to soil N through BNF. Though, BNF by individual systems in different crops has been investigated in detail, the BNF in a cropping system has not been studied well. As a result, the relation between the different N_2 fixing system, especially indigenous ones, are not fully understood and it is not clear if their activities are independent or related. Moreover, because of technological and socioeconomic limiting factors, the agronomic potential of BNF is still largely underutilized in the rice based cropping system. Methodological progress and comprehensive evaluations of BNF in rice fields are still needed to develop and test agricultural practices that take advantage of BNF in this important agroecosystem.

In this context, this book entitled **'Biofertilizers Technology'** gives a clear insight into the potentials and prospects various nitrogen fixing systems, both conventional and novel, that could be exploited for the maximum advantage of the diverse crops grown in the rice based cropping systems. Due importance has been given for other microbial systems which are involved in making available phosphorus and micronutients to the crops raised in this major cropping system. Any technology, which has commercial implications, could be sustained only through appropriate and viable quality control programmes. It is interesting to note that attempts have been made to throw much light on the recent approaches in the quality control of biofertilizers through molecular biological and biotechnological methods. I am sure, that, the readers of the book will get a new dimension of thinking on the beneficial role of biofertilizers in a cropping system as a whole. I congratulate and compliment the editors of the book, Prof. Dr. S. Kannaiyan, Dr. K. Kumar, and Dr. K. Govindarajan and all the scientists who have contributed research papers, for their sincere efforts to bring out this most useful and productive volume. I am proud to note, that, this book is an outcome of the National Workshop on 'Recent Developments in Biofertilizer for Rice Based Cropping System' held at Tamil Nadu Agricultural University, Coimbatore as per the mandate of the National Agricultural Technology Project sponsored by Indian Council of Agricultural Research, New Delhi. I sincerely hope, that, this piece of work will have a long lasting impact in the field of BNF and biofertilizers technology.

New Delhi

Dr. Panjab Singh
Secretary, DARE &
Director General,
Indian Council of Agrl. Research

PREFACE

In India, the compulsion of increasing crop yield per hectare is simply a matter of necessity to achieve production target of atleast 240 million tonnes to feed over 1 billion people in the ever deteriorating land : man ratio. Our nation is fourth largest user of chemical fertilizers (18.7 million tonnes of NPK nutrients) in the world and its soils are still being depleted of their inherent nutrients reserve as a result of wide gap between additions and removals. One tonne of produce removes from soil on an average 32 kg nitrogen, 12 kg phosphorus and 58 kg potash. At present, India produces about 210 million metric tonnes of food grains for its growing population of more than one billion people. In the recent plan, the target is to produce an additional 30 million tonnes food grains and 5 million tonnes of pulses and oil seeds. In order to meet the food requirement of ever increasing population, the present nitrogen fertilizer requirement for crop production is estimated to be about 11.4 million tonnes as against the supply of 3.9 million tonnes. At present, this vast gap cannot be filled-up merely through the production of synthetic nitrogenous fertilizers due to scarcity of high cost raw materials such as fossil fuels. Biological nitrogen fixation is the key to sustain agricultural productivity and application of biofertilizers in the field is the viable alternative.

Biofertilizer technology enables the use of living organisms for the nourishment of the plants either by fixation of atmospheric nitrogen through the process of BNF or by solubilization of mineral nutrients like phosphorus. Biofertilizers are apparently environment and farmers friendly renewable sources of non – bulky and low cost organic agro – input. While *Rhizobium*, Blue Green Algae (BGA) and *Azolla* are crop specific, bio – inoculants like *Azotobacter*, *Azospirillum*, Phosphorus Solubilizing Bacteria (PSB), Vesicular Arbusular Mycorrhiza (VAM) etc. could be regarded as broad spectrum biofertilizers.

Rice based cropping system is the major cropping system practised in India which includes the rotation of crops involving rice, pulses, oil seeds, cotton, sugar cane, green manures etc., Rice – rice is the most dominant cropping system under irrigated conditions in South and Eastern India, while rice – wheat, rice – groundnut, rice – legumes, rice – mustard and rice – potato are some of the predominant rice based cropping systems with 200 per cent cropping intensity in different North Indian States. Various rice based cropping systems with 300 per cent cropping intensity have been reported from different parts of India, ranging from rice – rice – rice to rice followed by different cereals, pulses, oil seeds, vegetables and fibre crops. Rice – rice – pulse (green gram / black gram) is the predominant cropping system of major rice growing areas of Tamil Nadu.

Rice based cropping systems may include lowland rice and upland crop. So far, research had been focussing attention on individual crops disregarding the fact that each crop was only a component of a cropping system. Hence, the nutrient management practices for the individual crops were based on the responses of individual crops to the major nutrients without considering the system as a whole. Under these circumstances, there is urgent need for developing integrated nutrient management practices for the cropping system as a whole rather than for individuals crops. Integrated use of organic manures, biofertilizers and crop residues in rice based cropping system not only could substitute a portion of the costly fertilizer N but also could help to maintain fertility in intensively cultivated rice soils. The practice of using green manures, biofertilizers, including pulses in the cropping system and incorporating their haulms integrated with inorganic fertilizers has received a great deal of attention for efficient and economic management of N for rice based cropping system.

This book entitled **'Biofertilizers Technology'** is the outcome of the discussions and deliberations of the National Workshop on 'Recent Developments in Biofertilizers for Rice Based Cropping System' held at Tamil Nadu Agricultural University, Coimbatore during August 16 -18, 2001 to develop a strong, workable and compatible package of nutrient management through organic and inorganic sources exclusively for rice based cropping system. Leading scientists and experts from different parts of India working on the basic and applied aspects of various biofertilizers used in rice based cropping system have shared their valuable experiences and useful informations in this area of research in their research papers. This book is first of its kind in focussing the application of biofertilizers technology in the major cropping system of India *viz.*, rice based cropping system. We hope, that, the book will be a very good resource material for people in different walks of life *viz.*, agricultural policy makers, crop management scientists, researchers, students, extension workers and literate farmers. We profusely thank all the authors, who have richly contributed research papers to this very productive volume. We gratefully acknowledge the financial assistance extended by Indian Council of Agricultural Research, New Delhi under National Agricultural Technology Project-Team of Excellence on Biofertilizers for Rice Based Cropping System, for conducting the National Workshop from which the focal theme and the idea of publishing this book emanated.

Coimbatore – 3
20.09.2002

S. Kannaiyan
K. Kumar
K. Govindarajan

LIST OF CONTRIBUTORS

S. P. Adhikary
P.G. Department of Botany,
Utkal University,
Bhubaneswar 751004, Orissa

N. Anand
Director, Centre for Advanced Studies in Botany, University of Madras,
Guindy Campus, Chennai – 600 025.

Anjali Pal
Department of Soil Science,
Indira Gandhi Agricultural University,
Raipur- 492 012 (Chhattisgarh)

Anurag
Department of Soil Science,
Indira Gandhi Agricultural University,
Raipur- 492 012 (Chhattisgarh).

M. Asia Mariam
Dept. of Biotechnology,Centre for Plant Molecular Biology,
Tamil Nadu Agricultural University,
Coimbatore-3.

M. Bharathi
Associate Professor, Dept. of Biotechnology, Centre for Plant Molecular Biology, Tamil Nadu Agricultural University, Coimbatore-3.

R.N. Bisoyi
Regional Director,
Regional Biofertilizers Develp. Centre,
A-156, Sahidnagar,
Bhubaneswar – 751 007 (Orissa).

K. Amutha
W/o M. Tamilmani, Scientist 'C',
No. 974, Scientist Hostel-3, DRDO Complex,
CV Raman Nagar, Bangalore – 560 093.

R Anandham
Department of Agricultural Microbiology,
Tamil Nadu Agricultural University,
Coimbatore - 641 003

S. Anthoni Raj
Dean, Anbil Dharmalingam Agrl. College & Res. Inst., Trichy – 620 009.

K. Arulmozhiselvan
Associate Professor, Dept. of Soil Science & Agricultural Chemistry, Tamil Nadu Agricultural University, Coimbatore-3.

D.J. Bagyaraj
Professor Emeritus, Department of Agricultural Microbiology,
University of Agricultural Sciences (UAS),
GKVK, Bangalore-560 065.

Binu Koshy
Azolla and Cyanobacterial Research Lab.,
Department of Agricultural Microbiology,
University of Agricultural Sciences,
GKVK, Bangalore-560 065.

D. Balachander
Asst. Professor, Dept. of Agrl. Microbiology,
Tamil Nadu Agrl. University,
Coimbatore-641003

A. Balasubramanian
Professor & Head,
Department of Central Farm,
Tamil Nadu Agricultural University,
Coimbatore-3.

Chhaya D Saboji
Bioconversion Technology (AICRP on Res.),
Dept. of Agrl. Microbiology,
University of Agrl. Sciences,
Dharwad-580 005, Karnataka.

P.K. Chhonkar
Head, Division of Soil Science and Agrl. Chemistry, Indian Agricultural Res. Institute, New Delhi – 110 012.

V. Gayathri
Centre for Advanced Studies in Botany,
University of Madras,
Guindy Campus, Chennai – 600 025.

G.S. Geeta
Associate Professor, Bioconversion Technology (AICRP on Res.),
Dept. of Agrl. Microbiology,
University of Agrl. Sciences,
Dharwad-580 005, Karnataka.

S. G. Gholve
Department of Agronomy, MPKV,
Rahuri (M .S.)

G. Gopalaswamy
Professor, Dept. of Biotechnology,
Centre for Plant Molecular Biology,
Tamil Nadu Agricultural University,
Coimbatore -641 003.

M. Govindan
Associate Professor,
Regional Agricultural Research Station,
Pilicode - 671 353, Kerala.

K. Govindarajan
Professor, Dept. of Agrl. Microbiology,
Tamil Nadu Agricultural University,
Coimbatore 641 003.

S. Gunasekaran
Professor, Dept. of Agrl. Microbiology,
Tamil Nadu Agrl. University,
Coimbatore-641003

T.R. Guruprasad
Associate Professor,
Agricultural Research Station,
University of Agricultural Sciences (B),
Kankannady,
Mangalore – 575 002, Karnataka

P.S. Herle
Associate Professor,
Agricultural Research Station,
University of Agricultural Sciences (B),
Kankannady,
Mangalore – 575 002, Karnataka.

N. A. Janardhana Gowda
Associate Professor and Head,
Agricultural Research Station,
University of Agricultural Sciences (B),
Kankannady,
Mangalore – 575 002, Karnataka.

B. Jeberlin Prabina
Ph.D. Scholar,
Department of Agrl. Microbiology,
Tamil Nadu Agricultural University,
Coimbatore 641 003

K. Jeyaram
Dept. of Biotechnology, Centre for Plant Molecular Biology, Tamil Nadu Agricultural University, Coimbatore-3.

P. Kamalasanan Pillai
Vivekananda Kendra, Natural Resource Development Project,
Vivekanadapuram – 629 702, Kanyakumari

S. K. Kamble
Department of Agronomy,
MPKV,
Rahuri (M .S.)

I. Kannan
Department of Agricultural Microbiology,
Tamil Nadu Agricultural University,
Coimbatore - 641 003

S. Kannaiyan
Former Vice-Chancellor (TNAU),
AL-85, 4th Street, 11th Main Road,
Annanagar, Chennai-600 040.

K. Kavitha
Department of Agrl. Microbiology,
Tamil Nadu Agricultural University,
Coimbatore 641 003

A.S. Kumaraswamy
Professor, Agricultural Research
Station, University of Agricultural
Sciences (B), Kankannady, Mangalore
– 575 002, Karnataka, India

N. Lakpale
Department of Soil Science,
Indira Gandhi Agricultural University,
Raipur- 492 012 (Chhattisgarh).

M. Madhaiyan
Dept. of Agricultural Microbiology,
Tamil Nadu Agricultural University,
Coimbatore - 641 003.

K. Manjappa
Assistant Professor of Agronomy,
Agricultural Research Station (Paddy),
Banavasi Road, Sirsi - 581 401 (Kar.)

S K. Mishra
Dept. of Microbiology, FBS & H,
Rajendra Agricultural University,
Pusa (Samastipur) – 848 125 (Bihar)

V.S.R. Murugesan
Centre for Advanced Studies in
Botany, University of Madras,
Guindy Campus, Chennai – 600 025.

P. Nagarajan
Professor of Agrl. Microbiology,
Horticultural Research Station,
Periyakulam- 625604.

P. Pandiyarajan
Professor of Agrl. Microbiology,
Anbil Dharmalingam Agricultural
College & Research Institute,
TNAU, Trichy - 620 009

B.D. Kaushik
Head, Division of Microbiology,
Indian Agricultural Research Institute,
New Delhi 110 012.

K. Kumar
Professor, Department of Agrl. Microbiology,
Tamil Nadu Agricultural University,
Coimbatore 641 003.

K. Kumutha
Asst. Professor, Department of Agricultural
Microbiology, Tamil Nadu Agrl. University,
Coimbatore-641 003.

K. Lenin
Centre for Crop Nitrogen Fixation,
School of Life and Environmental Sciences,
University of Nottingham,
Nottingham NG7 2RD, UK.

P. Malliga
NFMC, Department of Microbiology,
Bharathidasan University,
Tiruchirapalli-620 024

A. Manjunath
Professor, Dept. of Agricultural Microbiology,
University of Agrl. Sciences, (UAS),
GKVK, Bangalore-560 065.

K. Mohana Sundaram
Professor & Head, Dept. of Millets,
Tamil Nadu Agrl. University,
Coimbatore-641003.

L. Nabachandra Singh
Dept. of Agronomy, College of Agriculture,
Central Agricultural University, Iroisemba,
Imphal-795 001,Manipur (India).

H. Nayak
P.G. Department of Botany,
Utkal University,
Bhubaneswar -751004, Orissa.

J. Prabakaran
Professor & Head,
Dept. of Agrl. Microbiology,
Agrl. College and Research Institute,
Madurai-625 104

P. C. Prabu
Dept. of Soil Sci. and Agri. Chemistry,
Tamil Nadu Agricultural University,
Coimbatore-3

D. Pragatheswari
Department of Agricultural Microbiology,
Tamil Nadu Agricultural University,
Coimbatore - 641 003.

A. N. Prasad
Dept. of Microbiology, FBS & H,
Rajendra Agricultural University, Pusa
(Samastipur) – 848 125, Bihar.

K. Premalatha
Department of Agricultural Microbiology,
Agricultural College and Research Institute,
Madurai – 625 104.

S. Premalatha
Vivekananda Kendra,
Natural Resource Development Project
Vivekanadapuram – 629 702

S. Rajamony
Vivekananda Kendra,
Natural Resource Development Project,
Vivekanadapuram – 629 702, Kanyakumari

M. Rangarajan
Professor (Rtd.),
Dept. of Agrl. Microbiology,
Tamil Nadu Agrl. University,
Coimbatore-641 003

J. K. Sahu
P.G. Department of Botany,
Utkal University,
Bhubaneswar 751004, Orissa.

K. C. Samal
Regional Research and Technology
Transfer Station, Orissa Uniiversity of
Agriculture and Technology, Chiplima,
Orissa-768 025.

P. Santhanakrishnan
Controller of Examinations,
Tamil Nadu Agricultural University,
Coimbatore - 641 003.

V.S. Saravanan
Dept. of Agricultural Microbiology,
Tamil Nadu Agricultural University,
Coimbatore-641 003.

G. G. Selvakumar
Dept. of Agrl. Microbiology,
Tamil Nadu Agricultural University,
Coimbatore - 641 003.

G. Selvi
Centre for Advanced Studies in
Botany, University of Madras,
Guindy Campus, Chennai – 600 025.

J. Sempavalan
Department of Agricultural Microbiology,
Tamil Nadu Agrl. University,
Coimbatore-641003

P. M. Shanmugam
Department of Agronomy,
Tamil Nadu Agricultural University,
Coimbatore-3

S.N. Shinde
Department of Agronomy, MPKV,
Rahuri (M .S.)

M. K. Shivaprakash
Associate Professor,
Azolla and Cyanobacterial Research
Laboratory, Dept. of Agricultural
Microbiology, University of Agri. Sci.,
GKVK, Bangalore-560 065.

A. Singh
Department of Soil Science,
Indira Gandhi Agricultural University,
Raipur- 492 012 (Chhattisgarh)

R. Sridar
Professor, Dept. of Agri. Microbiology,
Tamil Nadu Agricultural University,
Coimbatore - 641 003.

P. Subramanian
Professor, Dept. of Environmental Sciences,
Tamil Nadu Agricultural University,
Coimbatore - 641 003.

M.D. Sundaram
Professor & Head, Department of Agricultural Microbiology, Faculty of Agriculture, Annamalai University, Annamalai Nagar- 608 002.

S. Sundaravarathan
Asst. Professor,
Dept. of Agrl. Microbiology,
PAJANCOA, Serumavilangai,
Nedungadu Post, Karaikal – 609 603.

M.Thangaraju
Professor,
Department of Agrl. Microbiology,
Tamil Nadu Agricultural University,
Coimbatore 641 003

A. Vaishampayan
Photobiological Nitrogen Fixation Research Laboratory, Department of Genetics and Plant Breeding, Institute of Agricultural Sciences, Bananas Hindu University, Varanasi-221005

K. Vijila
Associate Professor (Microbiology),
National Pulses Research Centre,
Vamban Colony
P.O. Pudukkottai – 622 303, TN

P. Yasotha
Department of Agricultural Microbiology,
Tamil Nadu Agricultural University,
Coimbatore-3

Sp. Sundaram
Professor,
Department of Agricultural Microbiology,
Tamil Nadu Agricultural University,
Coimbatore - 641 003.

K. Tedia
Department of Soil Science,
Indira Gandhi Agricultural University,
Raipur- 492 012 (Chhattisgarh)

V. Udayasuriyan
Professor, Dept. of Biotechnology,
Centre for Plant Molecular Biology,
Tamil Nadu Agricultural University,
Coimbatore-3.

R. Venkitaswamy,
Professor of Agronomy,
TNAU Agricultural Research Station,
Aliyarnagar – 642 101.

C. K. Yamini Varma
Regional Agricultural Research Station,
Pilicode - 671 353, Kerala.

CONTENTS

Foreword *iii*

Preface *v*

List of Contributors *vii*

SECTION – I
CYANOBACTERIAL BIOFERTILIZER

1. Biological fertilizers for sustainable production in rice based cropping system 1

   — S. Kannaiyan

2. Recent phycotechnological advances concerning bio-N fertilization of rice 27

   — A. Vaishampayan

3. Certain recent developments in using cyanobacterial biofertilizers 41

   — V. Gayathri, S.R. Murugesan, G. Selvi, N. Anand

4. Cyanobacterial biofertilizer technology 53

   — B.D. Kaushik

5. Growth response and nitrogen fixation of rice field cyanobacteria to pesticides 60

   — J.K. Sahu and S.P. Adhikary

6. Growth, nitrogen fixation and extracellular amino acids of cyanobacteria from rice fields at different temperatures 84

   — H. Nayak and S.P. Adhikary

7. Effect of phyto-extracts harbouring insecticidal property on growth and nitrogen fixation of cyanobacteria 93

   — M.N. Jha, A.N. Prasad and S.K. Mishra

8. Effect of physico-chemical parameters on akinete differentiation and germination and its application in biomass production in *Anabaena* sp. 102

   — M. K. Shivaprakash, Vidya Kulkarni and Binu Koshy

9. Cyanobacterial biofertilizers for problem soils 111

— *G. Gopalaswamy*

10. Assessment of growth and biomass production by cyanobacteria isolated from different soil environments 122

— *G. Selvakumar, G. Gopalaswamy and S. Kannaiyan*

11. Characterisation of potential cyanobacterial strains for biofertilizer and their effect on paddy. 126

— *Chhaya D. Saboji and G.S. Geeta*

12. Potential of coir waste based BGA fertilizer for paddy cultivation 130

— *P. Malliga*

13. A new approach for inoculating cyanobacteria for rice cultivated in sodic soils 141

— *P. Pandiyarajan and P. Nagarajan*

SECTION - II
AZOLLA BIOFERTILIZER

14. Mineralization pattern of nitrogen fixing green manures *Sesbania rostrata* and *Azolla microphylla* in wetland rice soil ecosystem and their effect on rice 149

— *S. Sundaravarathan and S. Kannaiyan*

15. Estimation of biological nitrogen fixation by ^{15}N isotope technique 169

— *K. Arulmozhiselvan*

16. Effect on biofertilizers with graded levels of nitrogen on lowland rice and its carryover effect on succeeding crop of rapeseed 182

— *L.N. Singh*

17. Effect of integrated nutrient management in rice – wheat cropping system in western Maharashtra 137

— *S.G. Gholve, S.K. Kamble and S.N. Shinde*

18 Impact of *Azolla* biofertilizer on rice yield in Hirakud command areas of Orissa 193

— *K.C. Samal*

19. Effect of *Azolla* biofertilizer on the growth and yield of rice in Coastal Karnataka 198

— *N.A. Janardhana Gowda, A.S. Kumaraswamy T.R. Guruprasad and P.S. Herle*

20. Performance of cyanobacterial and *Azolla* bioinoculants in rice in Coimbatore and Erode Districts of Tamil Nadu 202

– *P. Yasotha, K. Kumar and S. Kannaiyan*

21. A new method for intercropping of *Azolla* in coconut gardens 205

— *R. Venkitaswamy*

22. *Azolla* - a sustainable feed substitute for livestock with special emphasis on silpauline based production and feeding technique 209

— *P. Kamalasanan Pillai, S. Rajamony and S. Premalatha*

SECTION - III
AZOSPIRILLUM BIOFERTILIZER

23. Studies on *Azospirillum* associated with rice varieties 221

— *K. Govindarajan and K. Kavitha*

24. Effect of *Azospirillum* inoculation on rice grown in acidic soils of Kerala 225

— *M. Govindan and C.K. Yamini Varma*

25. *Azospirillum* – A biofertilizer for rainfed transplanted rice 232

— *K. Manjappa*

26. Response of winter rice (Rabi) to inorganics and biofertilizers applications 238

— *P.M. Shanmugam, A. Balasubramanian and P.C. Prabu*

SECTION - IV
RHIZOBIUM BIOFERTILIZER

27. Symbiotic nitrogen fixation in legumes associated with rice cropping systems 245

— *J. Prabakaran and M. Rangarajan*

28. Induction of desiccation tolerance in greengram rhizobia 253

— *K. Kumutha, J. Sempavalan and K. Vijila*

29. Formulation of stress tolerant bio-inoculants for rainfed cropping system 256

— *S.B. Gupta, K. Tedia, A. Singh, Anurag, N. Lakpale, Anjali Pal and P.K. Chhonkar*

30. Field evaluation of liquid and carrier based rhizobial inoculants in redgram 265

— *S. Gunasekaran, D. Balachandar and K. Mohana Sundaram*

31. Studies on synergism between *Rhizobium*, plant growth promoting Rhizobacteria (PGPR) and phosphate solubilizing bacteria in blackgram 269

— *S. Gunasekaran, D. Balachandar and K. Mohana Sundaram*

SECTION - V
ENDOPHYTIC DIAZOTROPHS

32. Harnessing stem nodule nitrogen fixing system for lowland rice 275
— M.D. Sundaram

33. Induction of lateral rootlet formation, systemic translocation and endophytic colonization of *Azorhizobium caulinodans* ORS 571-IK-SK-M_4 by its combined inoculation with Agrobacterium rhizogenes. 280
— I. Kannan and S. Kannaiyan

34. Effect of seed inoculation of *Azorhizobium caulinodans, Azospirillum brasilense* and *Pseudomonas fluorescens* on their survival in the spermosphere and rhizosphere of cereal crops. 289
— K. Amutha and S. Kannaiyan

35. Induction of nodulation in rice using *Azorhizobium caulinodans* under pot cultures conditions 293
— K Amutha and S. Kannaiyan

36. Endophytic nitrogen fixation by *Gluconacetobacter diazotrophicus* in sugarcane 297
— M. Thangaraju

37. Occurrence of pink-pigmented facultative methylotrophic bacteria in tropical plants 305
— M. Madhaiyan, Sp. Sundaram and S. Kannaiyan

38. Studies on pink-pigmented facultative methylotrophic bacteria on *Glycine max* 314
— M. Madhaiyan, Sp. Sundaram and S. Kannaiyan

39. Effect of pink- pigmented facultative methylotrophic bacterial isolates on the cotton seed quality 320
— M. Madhaiyan, Sp. Sundaram and S. Kannaiyan

SECTION - VI
MICROBIAL SOLUBILIZATION & MOBILIZATION OF NUTRIENTS

40. A new mixed bioinoculum –'azophos' 329
— K. Premalatha, P. Subramanian and S. Anthoni Raj

41. Soil solution phosphorus status and myco-rrhizal inoculation efficiency of selected tropical grain legumes in an alfisol 334
— D. Pragatheswari, A. Manjunath, M. Madhaiyan and K. Kumutha

42. Effect of inoculation with vesicular arbuscular mycorrhizal (VAM) fungus (*Glomus mosseae*) on selected tropical grain legumes 342
— D. Pragatheswari, A. Manjunath and M. Madhaiyan

43. Screening of AM fungi for efficiency on growth and biochemical constituents of mulberry 347

— *K. Kumutha, Sp. Sundaram, J. Sempavalan and P. Santhanakrishnan*

44. Effect of insoluble phosphate and dual inoculation on soybean 354

— *K. Kumutha, J.Sempavalan and P. Santhanakrishnan*

45. Development of pellet formulation of sulphur oxidizing bacterium (SOB) for groundnut 359

— *R. Anandham and R. Sridar*

46. Use of sulphur bacteria for increased yield and oil content of groundnut 365

— *R. Anandham and R. Sridar*

47. Solubilization of silicate and concurrent release of phosphorus and potassium in rice ecosystem 372

— *S. Anthoni Raj*

48. Isolation and characterization of zinc sol-ubilizing bacteria (ZSB) from paddy soil 379

— *V.S. Saravanan and S. Anthoni Raj*

SECTION - VII
OTHER BIOINOCULANTS

49. Cloning of new *Bacillus thuringiensis* genes for management of rice pests 389

— *V. Udayasuriyan, K. Lenin, M. Asia Mariam and M. Bharathi*

50. Studies on the effect of *Fusarium* culture filtrate on paddy seedling development 394

— *K. Jeyaram, K. Govindarajan and S. Kannaiyan*

SECTION - VIII
QUALITY CONTROL OF BIOFERTILIZERS

51. Quality control and constraints in biofertilizer production technology 401

— *D.J. Bagyaraj*

52. Current advances in production and quality control of rice based biofertiliser 409

— *R.N. Bisoyi*

53. Importance of quality control in biofertilizers 421

— *M. Thangaraju*

54. Molecular techniques for checking genetic purity of strains in the cyanobacterial inoculum 430

— *K. Kumar*

55. Evaluation of nitrogen fixing potential, ammonia excretion and protein content of the cyanobacterial isolates under laboratory conditions for developing quality control standards for cyanobacterial biofertilizers 441

— *B. Jeberlin Prabina, K. Kumar and S. Kannaiyan*

56. Growth pattern and chlorophyll content of the cyanobacterial strains for their utilization in the quality control of cyanobacterial biofertilizers. 446-450

— *B. Jeberlin Prabina , K. Kumar and S. Kannaiyan*

SECTION – I

CYANOBACTERIAL BIOFERTILIZER

CHAPTER - 1

BIOLOGICAL FERTILIZERS FOR SUSTAINABLE PRODUCTION IN RICE BASED CROPPING SYSTEM

S. Kannaiyan

ABSTRACT

Rice based cropping system is the major cropping system practised in India which includes the rotation of crops involving rice, pulses, oil seeds, cotton, sugar cane, green manures *etc.* The rice based cropping system offers lot of scope for the effective utilization of a wide range of biofertilizers such as *Azolla*, blue green algae (BGA), *Azospirillum*, *Rhizobium*, *Azorhizobium caulinodans*, *Acetobacter diazotrophicus* and other heterotrophic N_2 fixing bacteria, which help to increase the yield by reducing the cost of cultivation. It thus has dual advantages of being sustainable without endangering the environment and being highly cost effective. It has been proved beyond doubt that the application of nitrogenous biofertilizers such as *Azolla*, BGA, *Azospirillum*, *Acetobacter*, *S. rostrata etc.* is economically rewarding and environmentally safe. An integrated package of biofertilizers, N_2 fixing green manures, organic manures and inorganic fertilizers will be considered as a holistic approach for Integrated Plant Nutrient Supply (IPNS) to maintain soil fertility and sustainability in rice based cropping system. Since these biological sources represent a great diversity in chemical, physical and biological characteristics, their efficient use depends upon the particular agro-ecological environment and local availability. Therefore, it is very much essential to develop a strong, workable and compatible package of nutrient management through organic and inorganic sources for rice based cropping system based on scientific facts, local conditions and economic viability. In the context of global concern over the adverse effects of fertilizers and pesticides on long term soil fertility, soil productivity and environmental safety, new strategies should be developed for increasing crop production taking into account a more realistic, inexpensive and pragmatic programmes to utilize the biofertilizers which not only increase the crop productivity but also improve the long term soil fertility and sustainability.

INTRODUCTION

Rice (*Oryza sativa*) is the major staple food crop of the tropical world and rice- based cropping patterns are prevalent in much of the tropics. Rice is grown on 145 M ha, 87% of which is in wetland (lowland) cultures, with the remaining grown on dry lands. The annual cultivation period for rice is determined by the distribution of rainfall, availability of irrigation water, and temperatures. Upland crops are typically grown during the nonflooded period, particularly in the transition season either at the end of the rains or before the onset of heavy rainfall. Cool season crops, such as wheat and winter legumes, are grown during the non-flooded season in area where cool temperatures limit rice production.

Nitrogen is the input required in largest quantities for lowland rice production. Soil N and biological nitrogen fixation (BNF) by associated organisms are major sources of N for lowland rice. More than 50% of the N used by flooded rice receiving fertilizer N is derived from the combination of mineralization of soil organic N (Broadbent, 1984) and BNF by free living and rice plant – associated bacteria (Roger and Ladha, 1992). The remaining N requirement is normally met with fertilizer. It is estimated that the world must produce 60% more rice in the year 2030 than it is produced at present to meet the demand created by increasing population and rising income. To achieve food security through sustainable agriculture, the requirement for fixed nitrogen must be increasingly met by BNF rather than by industrial nitrogen fixation.

The flooded rice soil ecosystem is favourable for biological nitrogen fixation. The major biological nitrogen fixing systems available in rice eco-system include cyanobacteria that inhabit in flood water and heterotrophic bacteria in the rhizosphere (Ladha *et al.*, 1997). It was observed that cyanobacterial inoculation increased the rice yields by an average of 337 kg grain / ha / crop (Roger and Watanabe, 1986). Heterotrophic nitrogen fixation ranges from 11-16 kg N / ha and contributes 16-12 per cent of total nitrogen requirement (Shresta and Ladha, 1996). Aquatic plant like the water fern *Azolla* and semi- aquatic legumes such as *Sesbania* are known to fix N_2 by symbiotic means and contribute nitrogen for rice crop (Kannaiyan, 1992). *Sesbania* spp. have the potential to produce biomass of 6-8 t / ha which is roughly equivalent to application of urea at 100-200 kg N / ha (Ladha *et al.*, 1993). Rice cropping system provides favourable aquatic environment for growth, multiplication and nitrogen fixation by diazotrophs utilizing the solar energy. Thus, these potential group of nitrogen fixing system may be exploited for harnessing atmospheric nitrogen so as to increase the productivity of crops in the rice based cropping system.

BIOFERTILIZERS APPLICATION IN RICE BASED CROPPING SYSTEM

Azolla

Azolla is a free-floating water fern, which in symbiotic association with the cyanobacterium - *Anabaena azollae* contributes substantial amount of biologically fixed nitrogen to the rice crop. The nitrogen fixing water fern, *Azolla* has been used as a biofertilizer for rice crop in India, China, Vietnam, Thailand, Philippines, Korea, Sri Lanka, Bangladesh, Pakistan, Nepal, Burma, Indonesia, Brazil and West Africa (Kannaiyan, 1987). *Azolla* is used as biofertilizer for rice crop and it could contribute 40-60 kg N ha^{-1} per crop (Kannaiyan, 1995). The important factor in using *Azolla* as a biofertilizer for rice crop is its quick decomposition in soil and efficient availability of its nitrogen to rice plants (Kannaiyan, 1990).

Table 1. Performance of *Azolla* Biofertilizer in the Demonstration Trials Conducted in the Farmer's Holdings in Rice Growing Areas of Tamil Nadu.

Name of the district	No. of trials conducted	Season	Variety	Average grain yield (t ha^{-1})		Per cent increase over control
				Uninoculated	Inoculated	
Coimbatore	97	Rabi	IR 50 / IR 20 / ADT 36	5.60	6.20	10.2
Erode	124	Rabi	ADT 36 / Entry 2	6.50	7.08	8.9
Salem	49	Rabi	J 13	8.80	9.70	10.2
Thanjavur	47	Rabi	IR 20	5.00	6.33	26.6
Tirunelveli and Tuticorin	30	Rabi	IR 50 / IR 20	6.51	7.40	13.7

Srinivasan (1980) reported that *Azolla* as green manure could reduce the nitrogen input by 25-30% and increased the yield by 35.8% in rice (Kannaiyan, 1981). The results of the demonstration trials conducted to test the performance of *Azolla* biofertilizer in the farmer's holdings in the rice growing areas of Tamil Nadu revealed that an additional grain yield of rice ranging from 8.9-26.6% could be obtained due to the application of *Azolla* biofetilizer (Table 1). The use of *Azolla* in rice field has considerably increased the soil microbial as well as soil enzyme activity (Kumar, 1994 and Kannaiyan, 1995), which in turn maintains the soil fertility. Thangaraju and Kannaiyan (1990) reported that *A. microphylla* and *A. filiculoides* coupled with USG application significantly increased the grain yield of rice. Dual cropping of *Azolla* is a most effective, ecofriendly practice which play a major role in sustainable rice farming (Kannaiyan, 1995). The N use efficiency of nitrogen fixing green manure plants *viz., Azolla* and *Sesbania rostrata* by rice crop could be enhanced by application of neem cake at 250 kg ha^{-1} during incorporation in lowland rice ecosystem (Kumar *et al.*, 1995). Rice production could be significantly increased with *Azolla* hybrids in combination with 75 kg N ha^{-1} Urea Super Granule (Gopalaswamy and Kannaiyan, 1998).

Cyanobacteria

The agronomic potential of cyanobacterial N_2 fixation in rice fields was first recognized in India during 1939 by De who attributed the natural fertility of tropical rice fields to N_2 fixing blue green algae. Singh (1961) and Venkataraman (1972) extensively studied the role of cyanobacteria in rice cultivation in different rice growing areas of India. Inoculation of BGA is highly suitable during dry season. Application of dried soil based blue-green algal flakes at the rate of 10 kg per ha is recommended for rice crop. At the time of inoculation of BGA, 500 g of single super phosphate may be mixed with 10 kg of soil based BGA which encourages better growth and establishment in rice field. A composite starter culture consisting of more than one species or genera is more preferable. Fast growing species, which can tolerate certain amount of extremities, are more suitable than highly susceptible ones. The algal species should have a fair amount of tolerance to agricultural chemicals such as pesticides and weedicides. The algal inoculant is to be applied 10 days after transplantation of rice crop. Continuous application of blue green algae to the rice fields for at least 3 or 4 consecutive seasons sustains the high crop yield at reduced levels of commercial nitrogen fertilizers input in ensuing cropping seasons (Kannaiyan, 1993). The demonstration trials conducted to test the performance of BGA biofertilizer in the farmer's holdings in rice growing areas of Tamil Nadu indicated the increase in grain yield of rice to the tune of 9.5 to 13% over uninoculated control due to the application of BGA biofertilizer (Table 2).

Table 2. Performance of BGA Biofertilizer in the Demonstration Trials Conducted in the Farmer's Holdings in Rice Growing Areas of Tamil Nadu, India.

Name of the district	No. of trials conducted	Season	Variety	Average grain yield ($t\ ha^{-1}$)		Per cent increase over control
				Uninoculated	Inoculated	
Coimbatore	172	Kharif	ADT 36	5.30	5.98	13.0
Erode	138	Kharif	IR50 / ADT36	6.29	7.10	12.9
Tirunelveli and Tuticorin	138	Kharif	ASD16	6.00	6.57	9.5

Recently, several efficient cyanobacterial cultures were isolated from the acid soils of Tamil Nadu. The predominant genera are *Anabaena* (25%). *Westiellopsis* (17.5%), *Nostoc* (15.0%) *Oscillatoria* and *Phormidium* (25.0%) and the rest were unicellular algae *Microcystis* (5%). The acid tolerant cyanobacterial culture *Westiellopsis* AT-TGK-4A7 performed better both under normal and acid pH levels like 4,5 and 6. The study on the effect of inoculation of acid tolerant cyanobacterial cultures on rice variety ASD 20 revealed that the inoculation of composite cultures of the acid tolerant cyanobacteria comprising *Anabaena*-AT-TGK-5C10, *Nostoc*-AT-TGK-5A4, *Oscillatoria* AT-TGK-5C9, *Westiellopsis* AT-TGK-4A7 and *Westiellopsis*- AT-TGK- 5A9 was found to be better in increasing the growth and yield of ASD 20 besides excreting maximum ammonia in the flood water (Tamil Selvam, 1998).

The algal symbiont *A. azollae* is known to colonize and immobilize well in polyvinyl foam (Kannaiyan *et al.*, 1994). Kannaiyan *et al.* (1997) have shown ammonia excretion by *A. azollae* (AS-DS) and *Nostoc muscorum* (DOH) immobilized in polyvinyl foam at different foam density. Mahesh (1994) has shown the induction of rice seedlings (Var. ADT 36) growth due to the inoculation of PU foam immobilized *A. azollae* and *A. variabilis*. Similarly, the free-living as well as symbiotic cyanobacterial cultures under immobilized state in polyurethane foam excreted significantly higher amount of ammonia (Uma and Kannaiyan, 1998). The inoculation of immobilized algal symbiont would increase the growth of rice seedlings by virtue of their continuous photoproduction of ammonia in flood water system (Samal and Kannaiyan 1998). A carrier based immobilized cyanobacterial inoculant was prepared to maintain better quality (Kannaiyan and Kumar, 2000). The polyurethane foam, sugarcane waste and paper waste immobilized cyanobacteria were used as a carrier for these immobilized cyanobacteria .

Azotobacter

Azotobacter is an aerobic, free living, heterotrophic N_2 - fixer. Of the several species of *Azotobacter, A. chroococcum* happens to be the dominant inhabitant in arable soils capable of fixing N_2 (2-15 mg N_2 fixed / g of carbon source) in culture media (Mishustin and Shilnikova, 1969, 1971). The numbers of *A. chroococcum* in Indian soils rarely exceeds 10^5/g soil due to lack of organic matter and the presence of antagonistic microorganisms in soil. The bacterium produces an anti-fungal antibiotic which inhibits the growth of several pathogenic fungi in the root region thereby preventing seedling mortality to a certain extent (Subba Rao, 2001). Seed germination and plant stand are improved in plants upon inoculation with improved strains of *A. chroococcum* capable of elaborating growth substances (Shende *et al.*, 1977) eventhough quantitative assessment of these physiological attributes *in situ* in soil has not been made. Those experiments were followed up by many Indian scientists with Indian isolates *A. chroococcum* (Sundara Rao *et al.*, 1963) who found that vegetables such as brinjal, tomato, cabbage and onion benefited largely to variable extents from 2 to 50 per cent over uninoculated control. There is a host varietal dependence and internal colonization of roots by improved strains of *A. chroococcum* (Subba Rao, 2001). Notwithstanding all these limitations, one of the best selling microbial products in India happens to be *A. chroococcum* because the farmers can see improvement in plant vigour in the early stages of growth, which however, begins to diminish to the level of uninoculated controls as the plants advance in age. Incidentally, India and probably Egypt are two countries, which advocate the use of *Azotobacter*. However, there have been no attempts to monitor the fate of the added inoculant in the rhizosphere. Virtual absence of quality control measures provides room for many shady operators of this product in India. Increase in yield due to *Azotobacter* inoculation could be obtained in crops like rice, wheat, *Sorghum*, maize, pearl millet, cotton, vegetables and sesame. In irrigated wheat, significant yield increases to *Azotobacter* inoculation were obtained in 342 out of 411 trials conducted on farmers fields and the response ranged from 34 to 247 kg/ha (Kundu and Barman, 1998).

Azospirillum

Azospirillum group of bacteria is an associative micro aerophilic N_2-fixing bacteria that colonizes the root mass of cereals, millets and grasses and fixes N in loose association with plants. Field demonstrations conducted on diverse soils and climatic conditions revealed sizeable gains in productivity of different crops consequent to use of *Azospirillum* biofertilizers. *Azospirillum* has shown positive interaction with applied N in several cereal crops with an average response equivalent to 15-30 kg /ha of applied N. The association of *Azospirillum* with many crop plants in India was reported by Subba Rao *et al.* (1993). The isolation of *Azospirillum* was made for the first time in Indian soils by Lakshmi Kumari *et al.* (1976). The performance of *Azospirillum* in pot experiments (Subba Rao *et al.*, 1972) and field trials (Subba Rao, 1979) was reported. Several experiments were conducted at different locations in India under All India Co-ordinated Project on Biological Nitrogen Fixation, and found positive response in *Sorghum* and pearl millet. Nayak and Rao (1977) and Charyulu and Rao (1979) have reported the association of *Azospirillum* with rice roots. Field experiments conducted in different places in India have indicated the positive response of rice cultivars to *Azospirillum* inoculation (Subba Rao, 1979 and Rao *et al.*, 1983). Experiments conducted at IRRI, Philippines showed that inoculation with *Azospirillum lipoferum* promoted early tillering and also the growth of rice and significantly increased filling rate of grain and the grain weight per plant at harvest (Watanabe and Lin, 1984 and Nayak *et al.*, 1986). In winter rice, such reduction in fertilizer N was possible in 31 out of 55 trials. The method of inoculation of *Azospirillum* is also considered very important for effective colonization of *Azospirillum* bacteria. Inoculation of *Azospirillum* by seed treatment, seedling root dipping and soil application proved effective than individual soil application (Gopalaswamy *et al.*, 1989). At lower and higher levels of inorganic nitrogen, the inoculation effect was not significant (Table 3).

Table 3. Effect of *Azospirillum* inoculation on the grain yield of Rice (ADT 36) under graded levels of Nitrogen

Treatment	Grain yield (t/ha)			
	1985	1986	1987	Mean
0 kg N/ha	3.1 d	3.0 d	3.1 c	3.1 d
0 kg N/ha + *Azospirillum*	3.1 d	3.4 d	3.5 d	3.3 d
25 kg N/ha	3.6 d	3.4 d	3.5 d	3.5 d
25 kg N/ha + *Azospirillum*	3.8 bc	4.0 c	3.9 cd	3.9 c
50 kg N/ha	4.1 ab	4.3 bc	3.7 cd	4.0 bc
50 kg N/ha + *Azospirillum*	4.4 a	4.5 ab	4.2 bc	4.4 b
75 kg N/ha	3.6 cd	4.5 ab	3.9 cd	4.0 d
75 kg N/ha + *Azospirillum*	4.4 a	4.9 a	5.0 a	4.7 a
100 kg N/ha	4.3 ab	4.7 ab	4.2 bc	4.3 b
100 kg N/ha + *Azospirillum*	4.0 ab	4.7 ab	4.5 ab	4.4 ab
CD (P= 0.05)	0.5	0.4	0.5	0.4

In a column, means followed by the same letter are not significantly different at 5% level by DMRT

Rhizobium

The root nodulating bacteria was first isolated and named as *Bacillus radicola* that was changed later to *Rhizobium* (Buchanan, 1926). *Rhizobium* is one of the important nitrogen-fixing bacteria, which helps the legume to maintain the soil fertility by way of their symbiotic association and nitrogen fixation. A legume plant having effective root nodules not only can meet its own nitrogen requirement but can also enrich the soil nitrogen content and thus improving soil fertility and sustainability. The N_2 fixation by different legumes in associations with specific species of *Rhizobium* under ideal and controlled conditions is estimated to be 100-300 kg ha^{-1} in alfalfa, 102-150 kg ha^{-1} in clover, 37-196 kg ha^{-1} in cluster bean, 112-152 kg ha^{-1} in peanut, 68-200 kg ha^{-1} in pigeon pea, 49-130 kg ha^{-1} in soybean, 36-63 kg ha^{-1} in chickpea, 46 kg ha^{-1} in pea and 53-85 kg ha^{-1} in cowpea (Lee and Wani, 1989). *Rhizobium* inoculant consisting of the strains from blackgram and peanut performed better over the single strain in green gram recording the maximum grain yield, whereas it was not consistent in blackgram under field conditions (Oblisami *et al.*, 1976).

Inoculation success of *Rhizobium* in legumes at field levels has been documented from time to time (Table 4). Remarkable increases in peanut pod yields have been achieved with *Rhizobium* inoculation (Nambiar, 1985; Staphorst *et al.*, 1975). It has been reported that peanut has been found to nodulate freely and effectively with a wide range of cowpea group *Rhizobium* strains. It has also been indicated that up to 80 per cent of nitrogen requirement of the plants could be obtained through dinitrogen fixation (Staphorst *et al.,* 1975). Responses to *Rhizobium* inoculation have been demonstrated with principal grain legumes such as red gram (*Cajanus cajan*), gram (*Cicer aritinum*) and green gram (*Vigna mungo*) at different parts of India (Subba Rao, 1976). Similarly, field trials conducted on soybean over six years also indicated increased grain yield due to rhizobial inoculations (Table 5).

Table 4. Frontline Demonstration Trials on the Use of *Rhizobium* in Peanut Conducted by Tamil Nadu Agricultural University

Year	No. of trials conducted	Pod yield (kg ha $^{-1}$)		% increase over control
		Inoculated	Control	
1991-92	6	3378	2825	19.57
1992-93	2	3810	3100	22.90
1993-94	3	1787	1480	20.74
1994-95	2	1910	1460	30.80
1995-96	4	2235	1811	23.41
1997-98	7	1678	1403	19.60
1998-99	11	1947	1698	14.66
1999-2000	8	1803	1533	17.61

Table 5. Frontline Demonstration Trials on the Use of *Rhizobium* in Soybean Conducted by Tamil Nadu Agricultural University

Year	No. of trials conducted	Grain yield (kg ha $^{-1}$)		% increase over control
		Inoculated	Control	
1991-92	3	1688	1519	11.12
1992-93	4	1860	1600	17.12
1993-94	4	2091	1613	29.63
1994-95	5	2003	1613	24.18
1995-96	3	1907	1618	17.87
1997-98	6	1296	1151	12.57

Inoculation of appropriate strains of rhizobia to legumes is known to enhance the yield. But both success and failure at field experiments have been documented (Subba Rao, 1979; Subba Rao *et al.,* 1977). Colonization of the rhizosphere by *Rhizobium* sp. is considered to be the most vital factor for efficient N_2 fixation. But *Rhizobium* sp. grows more slowly than many indigenous species of the rhizobia, presumably is at a competitive disadv-antage. Competition among *Rhizobium* for nodulation of their legume host is an important aspect of the ecology of the root nodule bacteria. This problem is highlighted where an inoculation with highly effective strain does not always result in yield increases.

Studies were conducted to study the survival and multiplication of bacteria introduced into soil (Acea *et al.,* 1988). The population of *R. leguminosarum* biovar *phaseoli, R. meliloti, Agrobacterium tumifaciens, Micrococcus flavus, Corynebacterium* sp. and *Pseudomonas* sp. declined because of their susceptibility to predation, starvation or possibly antibiotic producing or lytic microorganisms. Bacteriocin producing *Rhizobium* strains have been reported to inhibit the growth of non-bacteriocinogenic strains (Ahlawat and Dadarwal, 1996).

Failure to obtain the desired response is due to (i) the presence of native ineffective strains which could not be displaced by the introduced effective strains (ii) the presence of antagonists of rhizobia which minimize the numbers of rhizobia in the rhizosphere (iii) the soil conditions such as acidity, alkalinity, application of pesticides, herbicides *etc* and (iv) the presence of effective native rhizobial strains in large numbers (Subba Rao, 1979).

Leguminous green manures

Green manures can be fitted into rice farming systems in either the pre-rice or post-rice phase (Garrity and Finn, 1988). Pre-rice green manures are cultivated for 6 to 10 weeks before the establishment of the rice crop and incorporated into the soil during land preparation. Post-rice green manures are relay-established into the growing rice (Liu, 1988) or sown after rice harvest (Chen, 1988), when the period available for growth is usually longer than that of pre-rice legumes. Post-rice green manures have historically played a greater role than pre-rice species and are exemplified by *Astragalus sinicus* for the cool tropics (Chen, 1988) and *Indigofera tinctoria, Calopogonium* spp.(Ali and Narciso, 1993), *Crotalaria juncea, Clitoria ternatea, Desmanthus virgatus* and *Macroptilium atropurpureum* in the warm tropics. Prototype species for the pre-rice niche include flood-tolerant legume species such as *Sesbania* and *Aeschynomene* spp.(Ladha *et al.,* 1992) or the more or less drought-tolerant legumes *Crotalaria* and *Tephrosia* spp. (Abrol and Palaniappan, 1988).

Flood-tolerant legumes have been most prominent in recent research, a development that was triggered by the recognition of their high N_2-fixing capacity and the world-wide spread of the stem-nodulating *Sesbania rostrata* (Dreyfus *et al.,* 1985; Kalidurai and Kannaiyan, 1991). *S. rostrata* produces nodules in both roots and stems and can grow in flooded as well as dry soil condition. The stem nodulation caused by the new genus *Azorhizobium caulinodans* (Dreyfus *et al.,* 1988) is unique and interesting because of its ability to fix nitrogen under free-living condition (Balasubramani *et al.,* 1992). An attempt was made to develop *S.rostrata* mutants by gamma irradiation and screened for increased growth, biomass production, nodulation and nitrogen fixation. Gamma irradiation of *S. rostrata* seeds at 50,55,60,65 and 70 kR reduced the plant height compared to the wild parent. Gamma irradiation significantly increased the number of compound leaves plant^{-1}, branches plant^{-1}, total biomass, stem nodulation and stem nodule nitrogenase activity than the wild parent. However, gamma irradiation had no significant influence on root nodulation and root nodule nitrogenase activity of *S. rostrata* (Sundaravarathan and Kannaiyan, 1997).

Endophytic association of *A. caulinodans* in non-legumes

Associations between non-legumes and diazotrophic bacteria have been studied for many years with the expectation that these bacteria will provide combined nitrogen to the plant for enhanced crop production (Sloger and Van Berkum, 1992). One approach is to seek naturally occurring endophytic diazotrophs in cereals and to optimise endophytic nitrogen fixation. Currently, this approach is being exploited with sugar cane (Boddey *et al.*, 1995). An alternative approach is to determine whether rhizobial strains can colonize non-legumes internally. Previous observations suggest that some strains of rhizobia are able to induce, at low frequency, nodule-like structures on the roots of wheat, rice, maize and oilseed rape without the need for any enzyme treatment (Cocking *et al.*, 1990). These rhizobia all enter their host plant root system intercellularly, between adjacent plant cells (Cocking *et al.*, 1994). Rhizobia that are capable of symbiotic nitrogen fixation with the non-legume *Parasponia* sp. also gain entry into the roots of *Parasponia* through ruptures in the root epidermis (Webster *et al.*, 1995). A small percentage of lateral roots of inoculated wheat and rice plants also revealed that rhizobia were within and between cells of the root cortex. Among the rhizobia used in the previous work, *Azorhizobium caulinodans*, which induces root and stem nodules on the tropical legume *Sesbania rostrata*, is especially interesting, since in addition to forming nodules on emerging lateral roots (Tien *et al.*, 1983; Ndoye *et al.*, 1994), it is able to fix nitrogen in the free-living state (Kitts and Ludwing, 1994).

Preliminary studies on the feasibility of extending N_2-fixation to non-legumes revealed the formation of nodule-like structures in roots of rice, maize and Sorghum when inoculated with *A. caulinodans* together with 2,4-D at 0.5 and 3.0 ppm (Amutha and Kannaiyan, 1999). Nodule like structures were not formed when either *A. caulinodans* or 2,4-D alone were applied. Further research revealed the stimulation of lateral roots formation in rice, maize, pearl millet (*Pennisetum typhoides*), sugarcane (*Saccharum officinalis*), and a grass (*Leptochloa* sp.) (Amutha and Kannaiyan, 2000). Recently it has been found that inoculation of *A. caulinodans* alone could induce nodule-like structures in rice and baby corn (Buvana and Kannaiyan, 1999).

Studies of plant-microbe interactions have benefited from the use of genetically modified microorganisms harbouring reporter genes under the control of constitutively expressed promoters. Stimulation of lateral root formation by naringenin in rice has been reported (Kumar *et al.*, 1996). By using the bacterial strains carrying a marker gene, it is now possible to detect both visually and enzymatically the internal colonization of inoculated diazotrophs (Gopalaswamy *et al.*, 2000). Recently, it has been clearly established that the diazotroph, *Azorhizobium caulinodans* - ORS 571 colonizes the xylem vessels of the rice roots (Gopalaswamy *et al.*, 2000). Once the bacterium is established in xylem with adequate colonization, it might be capable of fixing nitrogen.

Acetobacter diazotrophicus

More recently, a new species of nitrogen fixing bacterium, *Acetobacter diazotrophicus* was isolated from sugarcane in different parts of the world. Isolation of *A. diazotrophicus* from the root, leaf, bud and stem samples of sugarcane cultivars planted in 10 different locations of Tamil Nadu province of India was carried out by Jayakumar and Thangaraju (1996). Inoculation of the acid tolerant and sucrose loving bacterium to sugarcane enhances the cane yield by 17 to 20 t/ha with a 50 per cent reduction in the recommended chemical nitrogen (Muthukumarasamy *et al.*, 1994). The results of the field experiments showed that inoculation of *A. diazotrophicus* to sugar cane enhances the cane yield and sugar yield (Table 6) than *Azospirillum* inoculation (Jayakumar, 1996). Combined inoculation of *A. diazotrophicus* with *Azotobacter*, VAM and 75 per cent nitrogen recorded the highest cane yield (Panneerselvam, 1997 and Thangaraju, 2000).

Table 6. Response of sugarcane to *Acetobacter diazotrophicus* inoculation

Treatment	Cane yield		Sugar yield
	($t\ ha^{-1}$)	% increase over uninoculated control	Sugar yield ($t\ ha^{-1}$)
100% fertilizer N	102.67	-	11.01
75% fertilizer N	95.50	-6.98	9.44
A.diazotrophicus + 100% fertilizer N	118.33	15.25	13.05
A.diazotrophicus + 75% fertilizer N	114.67	11.69	12.44
Azospirillum + 100% fertilizer N	116.67	13.64	12.81
Azospirillum + 75% fertilizer N	113.00	10.06	12.08

The key bacteria recognized so far in N_2-fixing sugar cane are *Acetobacter diazotrophicus* (Cavalcante and Dobereiner, 1988; Reis *et al.*, 1994) and *Herbaspirillun* spp. (Baldani *et al.*, 1986). Dobereiner *et al.* (1993) suggested that this organism is an obligate endophyte and that in sugarcane, it could be capable of conducting a complementary metabolism with *A. diazotrophicus* by consuming organic acids formed by the latter from sugar, thus maintaining homeostasis for pH within the plant tissues.

Phosphobacterial biofertilizer

Several soil bacteria and fungi notably species of *Pseudomonas, Bacillus, Penicillium* and *Aspergillus etc.* secrete organic acids and lower the pH in their vicinity to bring about solubilization of bound phosphates in soil (Sundara Rao and Sinha, 1963). Inoculation of composite culture of phosphate solubilizing microorganism comprising of *Pseudomonas striata* and *Aspergillus awamori* recorded increase in the yield of cotton than the single strain inoculation (Kundu and Gaur, 1980a). Increase in the yield of wheat and potato were demonstrated due to inoculation of peat based cultures of *Bacillus polymyxa* and *Pseudomonas striata* (Kundu and Gaur, 1980b). Saving of 50 per cent of recommended level of P_2O_5 is possible in sugarcane by inoculation with phosphobacteria as the cane yield and sugar yield of 50 per cent P_2O_5 and phosphobacteria treatments are on par with 100 per cent P_2O_5 application (Kathiresan *et al.*, 1995). However, Kumaraswamy *et al.* (1992) observed that the positive influence of phosphobacteria was conspicuous even at 0 and 75 per cent of phosphorus levels in sugarcane.

Phosphate solubilizing microorganisms (PSM) inoculation along with rock phosphate has shown increased benefit in rice in terms of yield. *Aspergillus awamorii* seed inoculation to cowpea resulted in increased yield either with rock phosphate or with a combination of single super phosphate and rock phosphate (Nagaraju and Nanjundappa, 1996). Inoculation of PSM to sunflower along with rock phosphate increased the grain yield (Andhani Gowda, *et al.*, 1998). Significant increase in grain yield of wheat and paddy was obtained in field inoculations with *Pseudomonas striata* and rock phosphate application to soil (Table 7).

Table 7. Effect of phosphate solubilizing microorganisms on the yield of wheat and paddy

Treatment	Yield (kg ha^{-1})	
	Wheat	Paddy
Control	4975	4866
Bacillus polymyxa - No P	5000	4966
B. polymyxa with MRP*	5000	5300
Pseudomonas striata - No P	5066	5083
P. striata with MRP	5466	5200
Aspergillus awamori - No P	5166	4783
A. awamori with MRP	4966	5000
Super phosphate at 50 kg P_2O_5/ha	5500	5083
Super phosphate at 100 kgP_2O_5/ha	5633	5600
MRP at 100 kg P_2O_5/ha	5250	5166

*MRP – Mussoorie rock phosphate

Vesicular Arbuscular Mycorrhiza

Mycorrhiza is the mutualistic association between plant roots and fungal mycelia. Vesicular Arbuscular Mycorrhiza (VAM) and endotrophic mycorrhizae formed by aseptate phycomycetous fungi are associated as an obligate symbiont with majority of agricultural crops growing under broad ecological range. Many graminaceous plants and legumes are highly susceptible to VAM colonization. VAM endophytes are not host specific although evidence is showing that certain endophytes may form preferential association with certain host plants (Mosse, 1977). The transfer of nutrients mainly phosphorus and also zinc and sulphur from the soil to the cells of the root cortex is mediated by intracellular obligate fungal endosymbiont of the genera *Glomus, Gigaspora, Acaulospora, Scleroscystis* and *Endogone* which posses vesicles for storage of nutrients and arbuscules for funneling these nutrients into the root system (Powell and Bagyaraj, 1984).

MAXIMISING N USE EFFICIENCY IN RICE BASED CROPPING SYSTEM

Neem cake and N use efficiency

Neem cake is widely used as organic fertilizer for several crops. But it seems to have good nitrification inhibition capacity, due to the presence of alkaloids solanin, meliantrool, azadiractin, nimbin and nimbidin which possess a furan ring system responsible for inhibition (Jotwani and Srivastava, 1981). When neem cake is mixed with urea, the rate of inhibition increase with increasing proportion of neem cake. The nitrification inhibition period of neem cake is around 10-25 days. It retains ammonium N for longer period and higher NH_4^+ -N contents are noticed in the soil with neem cake coated urea application (Subbiah and Kothandaraman, 1980).

The application of nitrification inhibitors along with urea in low land rice could inhibit nitrification and suppress the transformation of ammoniacal N and thereby accumulating, more ammoniacal N for effective utilization by rice crop (Manickam *et al.,* 1976; De Datta, 1986). The increase in rice yields due to the application of DCD mixed urea and neem cake mixed urea was reported (Sharma and Sinha, 1977). In the experiments conducted at Tamil Nadu Agricultural University, Coimbatore, the grain and straw yields of rice were enhanced significantly due to the application of neem cake and DCD. Of the two nitrification inhibitors, neem cake recorded maximum total N content in the plant, grain and straw yield when added during the incorporation of N_2 fixing green manure plants, *viz., Sesbania rostrata* and *Azolla microphylla* compared to DCD. Among different treatments with different combinations of green manures, fertilizer N and nitrification inhibitors, *S. rostrata* incorporation (20 t ha^{-1}) with fertilizer N (75 kg ha^{-1}) and neem cake (250 kg ha^{-1}) registered the highest nitrogen accumulation, grain and straw yield in IR 64 rice. It has been apparently shown that both the nitrification inhibitors neem cake and DCD have influenced the grain yield of rice considerably. The beneficial effect on IR 64 rice, as a result of the addition of nitrification inhibitors during the incorporation of *A. microphylla* and *S. rostrata* was attributed to the significant biological and chemical changes brought by the inhibitors in rice soil which eventually influenced the availability of nitrogen to rice crop. Similar effect of nitrification inhibitors in increasing the yield of lowland rice and reducing the N loss have been well documented (Manickam *et al.,* 1976; Kamire and Sonar, 1979; De Datta *et al.,*

1983; Kumar *et al.,* 1992 and 1995). Recently, Kannaiyan (1996) more clearly elucidated the inhibitory action of neem cake on the N loss mechanism in rice soil eventually increasing the N use efficiency of *Azolla* biofertilizer and inorganic fertilizer N by rice crop (Fig. 1). Application of neem cake along with *S.rostrata* incorporation resulted in significant reduction in the accumulation of nitrate N in flooded water and the population of nitrifying bacteria in rice soil (Sundaravarathan and Kannaiyan, 1997). Influence of *S.rostrata* resulted in increased evolution of CO_2 and CH_4 than application of fertilizer N alone (Sundaravathan and Kannaiyan, 1998).

Urea super granules (USG) and N use efficiency

Deep placement of fertilizer in reduced soil has been considered the most efficient method to increase fertilizer N efficiency in lowland rice and is becoming increasingly relevant since the introduction of modified urea materials like urea super granules (USG). The large size urea granule commonly referred to as urea super granule, can be placed in the soil by hand or with mechanical operator. The advantage of deep placement of urea super granules is that various losses like ammonia, volatilization, denitrification, run off and leaching can be minimised (Mikkelsen *et al.,* 1978; Savant *et al.,* 1983 and Kannaiyan, 1995). An additional advantage is that this mode of fertilization does not seem to interfere with dinitrogen fixation in flood water by *Azolla* and blue green algae (Roger *et al.*, 1980).

Several years of agronomic data from the International Network on Soil Fertility and Fertilizer Evaluation For Rice (INSFFER) have shown that deep placement is generally more efficient than the traditional split application of urea (De Datta 1986). Sharma (1995) has documented the positive influence of USG on rice soil based on the field experiments conducted at different locations.

When the nitrogen fixing green manures like *Azolla* and *Sesbania rostrata,* biofertilizer like blue green algae are used, the application of neem cake coated urea and deep placement of USG proved better (Kannaiyan, 1993). The application of different forms of urea along with *Azolla* inoculation has also been evaluated in the field experiments conducted during early and late wet seasons of 1984-85 at Tamil Nadu Agricultural University (Latha *et al.,* 1988). Significant increase in grain yield was obtained over 75 kg N ha^{-1} as prilled urea with *Azolla* as dual crop, due to the application of 75 kg N ha^{-1} as urea super granule or neem coated urea along with *Azolla* inoculation as dual crop in IR 50 rice during early wet season. The increased yield obtained due to the application of urea super granule and neem coated urea with *Azolla* as dual crop may be attributed to the continuous availability of nitrogen for rice crop. From our results, it is evident that the nitrogen fixing green manure plants *viz., Azolla* and *Sesbania rostrata,* biofertilizer like blue green algae could be effectively utilized by incorporation with different forms of urea *viz.,* neem coated urea, urea super granule *etc.* to increase the rice production under low land system (Kumar *et al.,* 1995).

Fig. 1. ROLE OF NEEM CAKE ON EFFICIENT UTILIZATION OF NITROGEN IN RICE CROP (Kannaiyan, 1996)

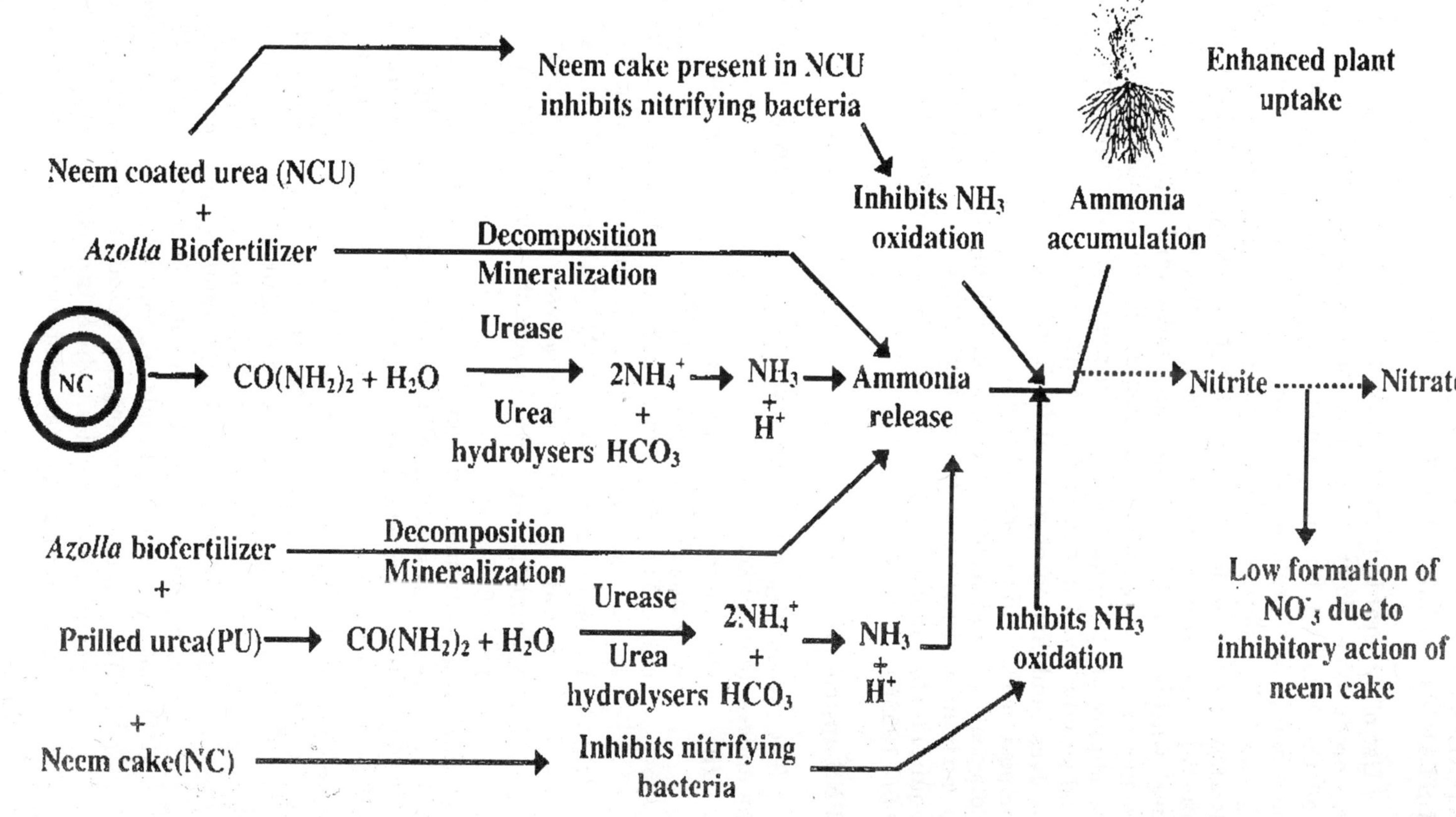

INTEGRATION OF ORGANICS AND INORGANICS FOR SUSTAINABLE CROP PRODUCTION

Use of mineral fertilizer is the quickest and surest way of increasing the crop production. Major changes in the pricing policies, after the 90's disrupted the balanced fertilizer application. Nitrogen fertilizer dominated for sometimes, and thereby affecting the balance use of major nutrients for crops. Alternatively, therefore, search has been made to exploit the potential alternate sources of plant nutrients harmoniously, by integrating the use of minerals, fertilizers, organics, biofertilizers and recyclable urban wastes, rural wastes, industrial wastes and farm yard mannure to increase crop production and at the same time sustain the soil fertility. The basic concept is the supply of the required plant nutrients, for sustaining the desired crop productivity. The Integrated Plant Nutrient Supply (IPNS) system, primarily involves combination of mineral fertilizers, organic manures, crop residues, compost or N_2-fixing microbial systems in relation to the system of land use, ecological, soil and economic conditions. The approach focuses mainly, at low to medium external input taking into account, the soil fertility level, crop requirements and management for a targeted yield per unit area (Kannaiyan *et al.,* 1998).

IPNS approach

The present day management practices particularly for IPNS is based on the understanding like:

(i) Organic or biological sources of plant nutrients complemented with mineral fertilizers in fulfilling the nutrient requirements of crops. The magnitudes of the contributions vary, very widely to sources, combination of sources and agro ecological conditions.

(ii) In many instances of IPNS, where combined application of mineral and organic sources were involved, it resulted in synergistic effects by increasing significantly the fertilizer use efficiency.

(iii) The humus fraction, which constitute the residual organic sources substantially improves the soil physical conditions. Better soil structure, could provide better environment for root development and aeration. Soil aggregation, is quite often improved, which is attributed to the actions of several compounds, particularly the polysaccharides and fulvic acid components of the organic sources. It not only improves the water holding capacity of the soil, but also prevents soil erosions etc. Application of organic resources play a substantial role in reducing the plant parasitic nematodes and the soil-borne phytopathogens, besides supplying the nutrients to crop plants. The application of organic resources also trigger the microbial population dynamics, which directly reflects on the transformation of nutrients in compounds. Long term fertilizer practices in India and China during the past one century showed that integrated use of organic manures and fertilizers, in addition to maintaining higher productivity, provide stability in crop production. The effect has been more pronounced in acid soils than in normal soils.

(iv) Under high input production system, the productivity curve shows a plateau and it cannot be further increased with any further increment of mineral fertilizer alone. But, addition of organic sources could again increase yields due to increased soil productivity and fertilizer use efficiency. Again, results of long term experiments, conducted in several agro ecological location, in our country revealed that responses was higher for farm yard manure along with NPK for rice, wheat, maize, soyabean and several other crops.

(v) The cost of cultivation is held within the reach of the average farmer under IPNS, since the crop requirement is partly met from cheaper organic sources like farm yard manure, green manures, legumes and biofertilizer. The potential N_2 fixing biological systems like *Rhizobium*-legume symbiosis, for increasing the grain legumes production, *Azolla – Anabaena* symbiosis, for increasing rice production, associative symbiosis with *Azospirillum* in cereal crops, *Acetobacter diazotrophicus* in strong association with sugarcane roots are potential biological agents contributing substantial amounts of nitrogen to the crops. Besides the nitrogen fixing biological systems, phosphorus solubilizing bacteria such as *Bacillus* and *Pseudomonas* and the Vesicular Arbscular Mycorrhizae (VAM) play a triggering role in phosphorus mobilization to crop plants.

IPNS models and packages

Based on the experiences gained in the IPNS, FAO has taken the pioneering initiative with the following conceptual programmes like identifying the main organic and biological sources which vary from country to country. For example, in Bangladesh, Pakistan, Sudan and Zambia animal wastes and the biological nitrogen fixing *Rhizobium* are the major natural resources. In Srilanka and India, recycling of rice straw and legume crop residues are the major sources while in Guinea and Zaire, it is the crop residues. FAO has undertaken development of IPNS model systems and field trial networks to generate informations for technology transfer programmes in Asia.

It is high time, that IPNS packages, should be developed more in the form of alternative options and it should be within the overall limits of recommended nutrient application rates. It also leads sometimes to a negative trend by loading all available components in a haphazard manner.

Biomass production in fallows or cultivation of leguminous green manure crop like *Sesbania* and Kolingi (*Tephrosia purpurea*) on the field bunds and borders can be attempted, to put it in more appropriate areas where limited farm lands are available.

- Computer models for an IPNS calculations programmes of plant nutrients should be developed based on general agronomic assumption.
- Site specific packages based on farming systems, farm size, availability of inputs and socio-economic factors should be developed.
- IPNS should be developed for cropping system, rather than for individual crops.
- IPNS practices for major agro-ecological zones and for various categories of farms should be developed.
- Forests and pastures are to be encouraged, in non-cultivable areas, for transport of biomass to the nearby farm lands.

Soil health management through IPNS

The basic concept of integrated plant nutrient supply is the promotion and maintenance of soil fertility for sustaining crop productivity through optimizing all possible sources like organic, inorganic and biological in an integrated manner appropriate to each farming situation in its ecological, soil and economic possibilities. The principal aim of this IPNS is efficient and judicious use of all the major sources of plant nutrients in an integrated manner, so as to get maximum economic yield without any deleterious effect on physico-chemical and biological properties of the soil. The IPNS system mainly concentrates on the following objectives.

1. To achieve potential yield of crops
2. To optimize and to sustain the resource base
3. To minimize the cost of cultivation
4. To demonstrate the need for soil health management
5. To inculcate the principle of integrated nutrient management

The major components of integrated plant nutrient supply system are fertilizers, FYM/ compost, green manures, crop residues / recyclable wastes and biofertilizers. These components possess great diversity in terms of chemical and physical properties, nutrient release efficiencies, positional availability, crop specificity and farmer's acceptability. Therefore, the combination of different components to ensure optimum nutrient supply to a prediction system may depend on land use, ecological, social and economic conditions.

Besides these components, IPNS considers some factors for increasing the available nutrients in the soil. Those factors are,

1. Management of physical condition of soil,
2. Reclamation of problem soils,
3. Water management, *etc.*

Soil health management is fundamental to sustainable intensification of agriculture. The IPNS approach affords the opportunity to include stem-nodulating legumes such as *Sesbania rostrata* in the farming system and to incorporate *Azolla*, or blue-green algae and other sources of symbiotic and non-symbiotic nitrogen fixation in rice based cropping system.

CONCLUSION

Biological nitrogen fixation is an important process in rice farming systems because it is an inexpensive source of nitrogen for increasing the productivity of crops. The nitrogen fixing biological systems such as *Azolla*, cyanobacteria and legume green manures were able to adopt well under wetland rice field ecological conditions and fixes considerable amounts of nitrogen. The biomass of these biological systems decomposes rapidly in rice soil and supply the nitrogen to rice crop. Besides nitrogen addition, these biofertilizers or biomanures also contributes significant amounts of phosphorus, potassium, sulphur, zinc, iron, molybdenum and other micronutrients. The organic acids released during the mineralization process of the biomass of biomanures would accelerate the phosphorus availability in rice soil. These biofertilizers or biomanures were known to influence the total soil microbial population such as bacteria, fungi, actinomycetes and also substantially activates the population dynamics of nitrogen fixing free-living bacteria like *Azotobacter, Azospirillum* and also the free-living symbiotic bacterium such as *Rhizobium*. It is evidently clear now that the application of biofertilizers triggers the accumulation of soil enzymes which directly reflects on soil fertility index. The use of biofertilizers would reduce the cost of chemical fertilizers involved in crop production. The effective utilization of biofertilizers for crops will not only provide economic benefits to the farmers but also improves and maintain the soil fertility and sustainability in rice soil ecosystem. It is estimated that 40% of the current global production depends on the use of N- fertilizer. Hence, it is suggested that the chemical fertilizer and organic nitrogen source could be integrated in the ratio of 80:20 for intensive crop production in the rice based cropping system.

REFERENCES

Abrol, I.P. and S.P. Palaniappan 1988. Green manure crops in irrigated and rainfed low land rice - based cropping system. In: **Sustainable Agriculture: Green Manure in Rice Farming.** The International Rice Research Institute, Los Banos, IRRI, Philippines. pp. 71-82.

Acea, M.J., C.R. Moore, and M. Alexander 1988. Survival and growth of bacteria introduced into soil. **Soil. Biol. Biochem., 20 (3):** 509 – 515.

Ahlawat, O.P. and K.R. Dadarwal 1996. Bacteriocin production by *Rhizobium* sp. Cicer and its role in nodule competence. **Indian. J. Microbiol., 36:** 17-23.

Ali, M. and J.H. Narciso 1993. The perception and reality of GM use in rice: An economic evaluation. **IRRI Social Science Division Paper Series.** The International Rice Research Institute, Los Banos, Philippines. p. 42.

Amutha, K. and S. Kannaiyan 1999. Effect of 2,4-D, NAA and inoculation with *Azorhizobium caulinodans* on induction of lateral rootlets and para nodules in rice. **Indian. J. Microbiol., 39:** 125-128.

Amutha, K. and S. Kannaiyan 2000. Effect of kinetin and inoculation of *Azorhizobium caulinodans* on induction of lateral rootlets and para nodules in rice. **J. Microb. World, 2(1):** 1-7.

Andhani Gowda. M.J., V. Ajay Kumar, V. Suresh Babu and K.S. Sreeramulu 1998. Productivity and economics of sunflower as influenced by sources of phosphatic fertilizers under varied moisture regimes. **Curr. Res., 27 :** 142-143.

Balasubramani, G., K. Kumar and S. Kannaiyan 1992. Studies on nitrogen fixing activity in stem nodulating *Sebania rostrata* In: **Biological Nitrogen Fixation and Biogas Technology** (eds.) S. Kannaiyan, K. Ramasamy, K. Ilamurugu K. Kumar. Tamil Nadu Agrl. University, Coimbatore. pp.84-89.

Baldani, V.L.D., M. A.De, B. Alvarez, I. J. Baldani and J. Dobereiner 1986. Establishment of inoculated *Azospirillum* spp. in the rhizosphere and in the roots of field grown wheat and *Sorghum*. **Plant Soil, 90 :** 35-46.

Boddey, R. M., O.C.de Oliveira, S. Urquiaga, V.M. Reis, F.L. de Olivares, I.J.Baldani, and J. Dobereiner 1995. Biological nitrogen fixation associated with sugarcane and rice: contributions and prospects for improvement. **Plant Soil, 174:** 195-209.

Broadbent, F. E. 1984. Plant use of soil nitrogen. In: **Nitrogen in Crop Production**. Ed. R.D Hauck. ASA – CSSA – SSSA, Madison, WI. pp. 171-182.

Buchanan, R.E. 1926. What name should be used for the organisms producing nodules on the roots of leguminous plants? **Proc. Dowa . Acad . Sci. 33:** 81-90.

Buvana, R. and S.Kannaiyan. 1999. Role of bioflavonoids on induction of nodule like structures in cereals and grass with inoculation of *Azorhizobium caulinodans* **Paper presented in State level seminar on Modern Trends in Plant science.** A.V.C. College, Mayiladuthurai, Tamil Nadu, India, p.6.

Cavalcante, V.A. and J. Dobereiner. 1988. A new acid tolerant nitrogen fixing bacterium associated with sugarcane. **Plant Soil, 108:** 23-31.

Charyulu, P. B. B. N. and V. R. Rao. 1979. Nitrogen fixation in some Indian rice soils, **Soil Science, 128:** 86-89.

Chen, L. 1988. Green manure cultivation and use for rice in china. In: **Sustainable Agriculture: Green Manure in Rice Farming.** The International Rice Research Institute, Los Banos. Philippines. pp 63-70.

Cocking, E. C., M.K. Al Mallah, E. Benson and M.R. Davey. 1990. Nodulation of non legumes by rhizobia In: **Nitrogen fixation: Achievements and Objectives.** (eds.) P.M. Gresshoff, E.C. Roth, G. Stacey and W.E. Newton. Chapman and Hall, New York, USA. pp. 813-823.

Cocking, E. C., G. Webster, C. A.Batchelor and M.R. Davey. 1994. Nodulation of non legume crops A new look. **Agro-Food - Industry Hi- tech** January / February 21-24.

De Datta, S.K. 1986. Improving nitrogen fertilizer efficiency in lowland rice in tropical Asia. **Fert. Res., 9:** 171-186.

De Datta, S.K., I.R.P. Fillery and E. I. Craswell. 1983. Results from recent studies on nitrogen fertilizer effiency in wetland rice. **Outlook on Agriculture. 12:** 125-134.

Dobereiner, J.V., V.M. Reis, M.A. Paula and F. Olivares. 1993. Endophytic diazotrophs in sugarcane, cereals and tuber plants. In: **New Horizons in Nitrogen**

Fixation. (eds.) R. Palacios, J. Mora and W.E. Newton. Kluwer Academic Publishers, Dordrecht, The Netherlands. pp. 671-676

Dreyfus, B.L., G. Rinando and Y. R. Dommergues. 1985. Observations on the use of *Sesbania rostrata* as green manure in paddy fields. **Mircen. J., 1:** 111-121.

Dreyfus, B.L., J.L.Garcia and M. Gillis. 1988. Characterization of *Azorhizobium caulinodans* gen, nov. sp. nov., a stem nodulating nitrogen fixing bacterium isolated from *Sesbania rostrata*. **Int. J. Syst. Bacteriol., 38:** 89-98.

Garrity, D.P. and J.C. Finn 1988. Farm - level management systems for green manure crops in Asian rice environments. In: **Sustainable Agriculture: Green Manure in Rice Farming.** The International Rice Research Institute, Los Banos, IRRI, Philippines. Pp. 111-129.

Gopalaswamy, G. and S. Kannaiyan 1998. Biomass production, nitrogen fixation and biochemical characterization of *Azolla* hybrids. **Indian J. Microbiol., 38:** 81-84.

Gopalaswamy, G, P. Vidhyasekaran and S. Chelliah 1989. Effect of *Azospirillum lipoferum* inoculation and inorganic N on wetland rice. **Oryza, 26** : 378-380.

Gopalaswamy, G., S. Kannaiyan, J.O' Callaghan, M.R. Davey and E.C. Cocking 2000. The xylem of rice (*Oryza sativa*) is colonized by *Azorhizobium caulinodans*. **Proc. R. Soc. Lond. B 267:** 103 - 107.

Jayakumar. P and M. Thangaraju 1996. Assessment of nitrogen fixation by *Acetobacter diazotrophicus* in sugarcane. **TNAU Newslett., 26** : 1-2.

Jayakumar. P. 1996. Isolation, characterization and assessment of nitrogen fixation by *Acetobacter diazotrophicus* in sugarcane. M.Sc. (Ag.) Thesis, Tamil Nadu Agric. Univ. Coimbatore, India.

Jotwani, M.G. and K.P. Srivastava 1981. Neem insecticide of the future: Chemistry, toxicology and future strategy. **Pesticide, 15**(2): 12-14.

Kalidurai, M., and S. Kannaiyan 1991. Studies on the biomass production nodulation and nitrogen fixation in stem nodulating *Sesbania rostrata*. In: **Biological nitrogen fixation associated with rice production.** (eds.) S.K. Dutta and C. Sloger, Oxford and IBH Publishing Co. Pvt. Ltd., New Delhi. pp.291 - 305.

Kamire, K.K. and K.R. Sonar 1979. Mineralization of urea-N with 'Neem' (*Azadirachta indica*) and 'Karanj' (*Pongamia glabra*) cakes. **J. Maharashtra Agric. Univ., 4**: 10-13.

Kannaiyan, S. 1981. *Azolla* biofertilizer for rice. **INSFER Training Seminar**, Intl. Rice Res. Inst., Manilla, Philippines, p.1-11.

Kannaiyan, S. 1987. *Azolla* biotechnology. **Tech. Bull.** Tamil Nadu Agric. Univ., Coimbatore, Tamil Nadu, India. pp.1-10.

Kannaiyan, S. 1990. Blue green algal biofertilizers. In: **Biotechnology of biofertilizers for rice crop**. (ed.) S. Kannaiyan, Tamil Nadu Agric. Univ. Publication, Coimbatore, Tamil Nadu, India. pp.225.

Kannaiyan, S. 1992. Studies on the factors influencing the growth and nitrogen fixation in *Azolla* In: **Biofertilizer Technology Transfer**. Ed. L.V. Gangawane, Associated Pub Company, New Delhi. pp. 239-254.

Kannaiyan, S. 1993. Algal biofertilizer technology for rice. **Ind. Phyco. Rev., 2** : 73-82.

Kannaiyan, S. 1994. Nitrogen Contribution by *Azolla* to rice crop. Proc. Indian Nat. Science Acad. B. 59: (3 & 4) p.309-314.

Kannaiyan, S. 1995. Biofertilizers for rice status and scope. In : **Rice Management Biotechnology**, (ed.) S. Kannaiyan, Assoc. Publ. Co., New Delhi, pp. 237-265.

Kannaiyan, S. 1996. Immobilization of cyanobacteria in solid matrix for ammonia production and development of immobilized substrate based cyanobacterial inoculants for rice crops. **Paper presented in X Southern Regional Conference on Microbiol Inoculants,** A.V.V.M. Sri Pushpam College, Poondi, India, p. 1

Kannaiyan, S. 1998. Integrated plant nutrient supply system. **Bulletin of Indian Institute of Soil Science, 2**: 10-13.

Kannaiyan, S. 2000. **Biofertilizer technology and quality control**. Publication Directorate, Tamil Nadu Agric. Univ., Coimbatore, India. p. 256.

Kannaiyan, S., S.J. Aruna, S. Merina Prem Kumari and D.O. Hall. 1997. Immobilized cyanobacteria as a biofertilizer for rice crops. **J. Appl. Phycol., 9:** 167-174.

Kannaiyan, S., S.K. Goyal, S.S. Goyal and D.W. Rains. 1988. Some factors influencing the growth and sporulation in *Azolla*. **Phykos, 27:** 197-205.

Kannaiyan, S., and K. Kumar 2000. PUF immobilized cyanobacterial biofertilizer for rice. In: **First International global Sustainable Biotech Congress Memori Journal** (ed.) Sudhir Meshram, Nagpur University, Nagpur, India, pp.12-18.

Kannaiyan, S., K.K. Rao and D.O. Hall 1994. Immobilization of *Anabaena azollae* from *Azolla filiculoides* in polyvinyl foam for ammonia production in photobioreactor system. **World J. Microbiol. Biotechnol., 10 :** 55-58.

Kathiresan, G., G. Manickam and P. Parameswaran. 1995. Efficiency of phosphobacteria addition on cane yield and quality. **Cooperative sugar, 26 :** 629-631.

Kitts, C.L. and Ludwing, R.A. 1994. *Azorhizobium caulinodans* respires with at least 4 terminal oxidases. **J. Bacteriol., 176:** 886-895.

Kumar, K. 1994. Studies on the development of *Azolla* mutants and their nitrogen fixing potential, microbial decomposition and biofertilizer value for rice. Ph.D Thesis, Tamil Nadu Agrl. Univ., Coimbatore, Tamil Nadu, India.

Kumar, K. G. Gopalaswamy and S. Kannaiyan. 1992. Effect of incorporation of nitrogen fixing plants *Azolla microphylla* and *Sesbania rostrata* with nitrification inhibitors - neem cake and dicyandiamide. In: **Biological Nitrogen Fixation and Biogas Technology**. (eds.) S. Kannaiyan, K. Ramasamy, K. Ilamurugu and K. Kumar, Tamil Nadu Agric. Univ. Coimbatore, (TN), India. pp. 49-53.

Kumar, K., S. Kannaiyan and E.C. Cocking. 1996. Studies on the colonization of *Azorhizobium caulinodans* in rice roots and their role in symbiotic nitrogen fixation. **Paper presented on 37th Annual Conference of the Association of Microbiologist of India**, Chennai, India. p. 118 (Abstr.).

Kumar, K., M Thangaraju and S. Kannaiyan. 1995. Use of nitrification inhibitors and nitrogen use efficiency in low land rice soil. In: **Rice Management Biotechnology** (ed.) S.Kannaiyan, Assoc. Publ. Co., New Delhi, pp.101-100.

Kumarasamy, K., N. Gurusamy and S. Rajasekaran. 1992. Efficacy of phospho-bacterium biofertilizer on the yield & quality of sugarcane. **Indian Sugar,** pp. 841-845.

Kundu, S and K.K. Barman. 1998. Biological Nitrogen Fixation. In: **50 years of natural resource management research.** (eds.) G.B Singh and B.R. Sharma, Indian Council of Agricultural Research, New Delhi. pp. 199-213

Kundu, B.S and A.C. Gaur. 1980a. Effect of nitrogen fixing and phosphate solubilizing microorganisms as single and composite inoculants on cotton, **Indian J. Microbiol., 20(3)** : 225-229.

Kundu, B.S and Gaur, A.C. 1980b. Establishment of nitrogen fixing and phosphate solubilizing bacteria in rhizosphere and their effect on yield and nutrient uptake of wheat crop. **Plant Soil, 57** : 223-230.

Ladha, J, K., R. P. Pareek and M. Becker. 1992. Stem - nodulating legume- *Rhizobium* symbiosis and its agronomic use in lowland rice. **Adv. Soil Sci., 20:** 147-192.

Ladha, J.K., F.J., deBrujin and K.A. Malik. 1997. Introduction: assessing oppurtunities for nitrogen fixation in rice – a frontier project. **Plant Soil 194:** 1-10.

Ladha, J. K., Tirol – Padre, C. K. Reddy and W. Ventura. 1993. Prospects and problems of biological nitrogen fixation in rice production: A critical assessment. In: **New horizons in nitrogen fixation**. (eds.) R. Placios, J. Mora and W.E. Newton. Kluwer Academic Publishers. The Netherlands. pp. 677-682.

Ladha, J.K., I. Watanabe and S. Saono 1988. Nitrogen fixation by leguminous green manure and practices for its enhancement in tropical low land rice. In: **Green manure in rice farming**, Inte. Rice Res. Instt., Los Banos, Philippines. pp. 165-183.

Lakshmi Kumari, M., S. K. Kavimandan and N. S. Subba Rao 1976. Occurrence of nitrogen fixing *Spirillum* in roots of rice, *Sorghum*, maize and other plants. **Indian J. Exp. Biol., 14** : 638 - 639.

Lee, K.K and S.P. Wani. 1989 . Significance of biological nitrogen fixation and organic manures in soil fertility management. In: **Soil fertility and fertilizer management - semi arid tropical India,** (ed.) IFDC Christianson Muscleshoals, Albama, USA, pp. 89-108.

Liu, C.C. 1988. Integrated use of green manure in rice fields in South China. In: **Sustainable Agriculture: Green Manure in Rice Farming**. IRRI, Los Banos, Philippines. pp. 319-331.

Manickam, T.S., A.Selvaseelan and K.K.Krishnamurthy. 1976. Nitrification of urea treated with nitrification inhibitors, coated fertilizers and slow acting fertilizers. **Fert. News, 21**(9):64-65.

Mikkelsen, D.S., S.K.De Dutta and W.N.Obcemae.1978. Ammonia volatilization losses from flooded rice soils. **Soil Sci. Soc. Am. J., 442:** 725-730.

Mishustin, E.N. and V.K. Shilnikova. 1969. Free living nitrogen fixing bacteria of the genus *Azotobacter*. In: **Soil Biology, Reviews of Research**. UNESCO Publications. pp. 72-124.

Mishustin, E. N., and V. K. Shilnikova. 1971. **Biological Fixation of Atmospheric Nitrogen**. Mac Millan Press Ltd., London, UK. P.415.

Mosse B 1977. Plant growth responses to vesicular arbuscular mycorrhizae. Responses of *Stylosanthes* and maize to inoculation in unsterilized soils. **New Phytol., 78** : 277-288.

Muthukumarasamy, R., G. Revathi and A. R. Solayappan. 1994. Biofertilizers-A supplement or substitute for chemical nitrogen for sugarcane crop. **Co-operative Sugar, 25** : 287-290.

Nagaraju, A. P. and G. Nanjundappa. 1996. Effect of sources and levels of phosphorus using P-solubilizing inoculant on cowpea. **Crop res., 12(3):** 387-389.

Nambiar, P.T.C. 1985. Response of groundnut (*Arachis hypogaea*) to *Rhizobium* inoculation in the field: Problems and prospects. **MIRCEN J. Appl. Biol. Biotech., 1:** 293-309.

Nayak, D. N. and V. R. Rao. 1977. Nitrogen fixation by *Azospirillum* sp. from rice roots. **Arch. Microbiol., 115:** 359.

Nayak, D. N., J. K. Ladha and I. Watanabe. 1986. The fate of marker *Azospirillum lipoferum* inoculated into rice and its effect on growth, yield and N_2 feeding by ^{15}N dilution techniques. **Biol. Fertil. Soils, 2:** 7-14.

Ndoye, I., F. De Billy, J. Vasse, B. Dreyfus and G. Trucjet 1994. Root nodulation of *Sesbania rostrata*. **J. Bacteriol., 176:** 106 - 1068.

Oblisami, G., K. Balaraman and T. Natarajan 1976. Effect of composite cultures of *Rhizobium* on the pulse crops. **Madras. Agric. J., 63:** 587-589.

Panneerselvam, P. 1997. Studies on *Acetobacter diazotrophicus* and its interaction with VA-mycorrhizal fungi in sugarcane. M.Sc. (Ag.) thesis, Tamil Nadu Agric. Univ., Coimbatore. India. p 140.

Powell, C. L. and D. J. Bagyaraj. 1984. **VA Mycorrhiza,** CRC Press, Florida, USA.

Rao, V. R., D. N. Nayak, P. B. D. Charyulu and T. K. Adhya. 1983. Yield response of rice to root inoculation with *Azospirillum*. **J. Agric. Sci., 100:** 689-691.

Reis, V. M., F. L. Olivares and J. Dobereiner. 1994 . Improved methodology for isolation of *Acetobacter diazotrophicus* and confirmation of its endophytic habitat. **World J. Microbiol. Biotechnol., 10** : 101-104.

Roger, P.A., and I. Watanabe 1986. Technologies for utilizing biological nitrogen fixation in wetland rice: Potentialities, current usage and limiting factors. **Fert. Res. 9:** 39-77.

Roger, P. A., and J.K. Ladha 1992. Biological N_2 fixation in wetland rice fields: Estimation and contribution to nitrogen balance. **Plant Soil 141:** 41-55.

Roger, P.A., S.A. Kulasooriya, A.C. Third and E.T. Craswell.1980. Deep placement of nitrogen fertilizer application compatible with algal nitrogen fixation in wetland rice soils. **Plant Soil, 57: 137-142.**

Samal, K. C. and S. Kannaiyan. 1998. Immobilization of algal symbionts of *Azolla* in polyurethane foam and its effect on growth and yield of rice crop. In: **Cyanobacterial Biofertilizer for rice crop** (ed.) S. Kannaiyan, Tamil Nadu Agricultural University, Coimbatore. pp. 191 - 199..

Savant, N.K., E.T. Craswell and R.E. Kiamond, 1983. Use of urea supergranules for wetland rice: A Review. **Fert. News, 28(8)** : 27-35.

Sharma, N.N. and U.P.Sinha. 1977. Neem cake blending of urea for raising efficiency of applied nitrogen to wheat. **Indian. J. Agron., 22:**106-107.

Sharma, S. K. 1995. Urea Super Granules - Its performance and prospects for rice production. In : **Rice Management Biotechnology**. (ed.) S. Kannaiyan, Assoc. Publ. Co., New Delhi, pp. 117-134.

Shende, S.T., R. Apte and R. Singh 1977. Influence of *Azotobacter* on germination of rice and cotton seeds. **Curr. Sci., 46** : 675.

Shresta, R.K., and J. K. Ladha 1996. Genotype variation in promotion of rice nitrogen fixation as determined by nitrogen ^{15}N dilution. **Soil Science. Soc. Am. J., 60:** 1815-1821.

Singh, R.N. 1961. The Role of Blue Green Algae in Nitrogen Economy of Indian Agriculture, ICAR Publications, New Delhi.

Sloger, C., and P. Van Berkum. 1992. Approach for enhancing nitrogen fixation in cereal crops. In: **Biological Nitrogen Fixation Associated with Rice Production.** (eds). S.K. Dutta and C. Sloger. Oxford and IBH Publishing Co., New Delhi, India. pp. 229 - 234.

Srinivasan, G. 1980. *Azolla* manuring and grain yield of rice. **Inte. Rice Res. Newslett., 6** : 25.

Staphorst, J. L., B. W.Strijdom and J. F. Otto. 1975. Nitrogen fixing ability of rhizobia which nodulate groundnuts in South African soils. **Phytophylactica, 7:** 133-135.

Subba Rao, N. S. 1976. Field reponse of legumes in India to inoculation and fertilizer application. In : **Symbiotic Nitrogen Fixation in Plants** (ed.) P.S. Nutman, Cambridge Univ. press, Cambridge, UK. pp. 255-268.

Subba Rao, N. S. 1979. Biofertilizers in Indian Agriculture Problems and prospects. Fert. News, 24: 84-90.

Subba Rao, N.S. 2001. An appraisal of biofertilizers in India. In: **Biotechnology of Biofertilizers**, (ed.) S. Kannaiyan, Narosa Pub. House, New Delhi (In Press).

Subba Rao, N.S., and Tilak, K.V.B.R. 1977. **Rhizobial cultures and their role in pulse production Souvenir Bulletin,** Directorate of Pulses Development , Lucknow, India. pp 31-34.

Subba Rao, N.S., W.V.B. Sundara Rao, A.N. Sen and Y.D. Gaur 1972. **Final Technical Report on *Rhizobium*, PL-480 Scheme,** Division of Microbiology, Indian Agri. Res. Inst., New Delhi.

Subba Rao, N. S., G. S. Venkataraman and S. Kannaiyan. 1993. **Biologial nitrogen fixation**, Indian Council Agric. Res., New Delhi, p. 112.

Subbiah, S. and G.V. Kothandaraman. 1980. Effect of neem cake blending of urea on ammonium and nitrate N content in soil under field conditions. **J. Indian. Soc. Soil. Sci., 28**:136-137.

Sundara Rao, W. V. B and M. K. Sinha. 1963. Phosphate dissolving organisms in soil and rhizosphere. **Indian J. Agric. Sci., 33:** 272-278.

Sundaravarathan, S. and S. Kannaiyan. 1997. Studies on the effect of gamma irradiation on the growth, biomass production, nodulation and nitrogen fixation by stem nodulating leguminous green manure *Sesbania rostrata.* **Presented in National Conference on Recent Trends of Researches in Microbiology and Plant Physiology in India**, held at University of Burdwan, Burdwan, West Bengal.

Sundaravarathan, S. and S. Kannaiyan. 1998. Decomposition of *Azolla microphylla* and *Sesbania rostrata* with fertilizer N and neem cake under submerged soil conditions Indian J. Microbiol., 38: 89-93.

Tamilselvam, B. 1998. Selection of fast growing and higher nitrogen fixing and acid tolerant cyanobacterial culture and their utility as biofertilizer for rice. **M.Sc. (Ag.) Thesis, Tamil Nadu Agricultural University, Coimbatore, Tamil Nadu, India.**

Thangaraju, M. 2000. **Prospects of biofertilizer usage in sugarcane.** In: Proc. of symp on current status of biofertilizer technology and its promotion in Karnataka, Tamil Nadu, Kerala and Pondicherry. Ed. R N Bisoyi.. Regional Biofertilizer Development Centre, Bangalore, India. pp 202-208.

Thangaraju, M. and S. Kannaiyan. 1990. Use of nitrification inhibitors with *Azolla* biofertilizer and fertilizer on nitrogen use efficiency in rice. In : **Biotechnology of Biofertilizers for Rice Crop** (ed.) S. Kannaiyan, Tamil Nadu Agrl. University, Coimbatore, Tamil Nadu , India. pp. 131-140

Tsien, H. C., B. L.Dreyfus and E. L. Schmidt. 1983. Initial stages in the morphogenesis on nitrogen - fixing stem nodules of *Sesbania rostrata*. **J. Bacteriol., 156:** 888-897.

Uma, D. and S. Kannaiyan. 1998. Effect of the insecticide carbofuran on ammonia excretion by cyanobacteria under immobilized state in polyurethane foam. In: **Cyanobacterial biofertilizer for rice crop** (ed.) S. Kannaiyan, Tamil Nadu Agricultural University, Coimbatore. pp. 160-169

Venkataraman, G.S. 1972. **Algal Biofertilizers and Rice Cultivation.** Today and Tomorrow's Printers and Publishers, New Delhi.

Watanabe, I and C. Lin 1984. Response of wetland rice to inoculation with *Azospirillum lipoferum* and *Pseudomonas* sp. **Soil Sci. Plant Nutr., 30:** 118-124.

Webster, G., M.R. Davey, and E.C. Cocking 1995. *Parasponia* with rhizobia: a neglected non-legume nitrogen - fixing symbiosis, **Ag. Biotech News Info., 7:** 119N - 124N.

CHAPTER - 2

RECENT PHYCOTECHNOLOGICAL ADVANCES CONCERNING BIO-N FERTILIZATION OF RICE

A. Vaishampayan

ABSTRACT

The flooded water and the surface of the soil provide the sites for aerobic phototrophic nitrogen fixation by free-living cyanobacteria and *Azolla-Anabaena* symbiotic N_2-fixing complex. The free-living cyanobacteria, majority of which are heterocystous and nitrogen-fixing, contribute on an average 20-30kg N ha^{-1}, while the value is up to 60 kg ha^{-1} in the case of *Azolla-Anabaena* system (the most beneficial cyanobacterial symbiosis from agronomic point of view). Synthesis and excretion of organic/growth promoting substances by the cyanobacteria are also on record. During the last 2-3 decades a great volume of work has been accomplished on the various important fundamental and applied aspects of both kinds of cyanobacterial biofertilizers (the free-living cyanobacteria, and the cyanobacterium *Anabaena azollae* in symbiotic association with the water fern *Azolla*), which include strain identification, isolation, purification and culture; laboratory analyses of their N_2-fixing activity and related physiology, biochemistry and energetics; identification of the structure and regulation of nitrogen-fixing (*nif*) genes and nitrogenase enzyme; symbiotic biology of *Azolla-Anabaena* mutualistic N_2-fixing complex; free-living cyanobacterial strain improvement through mutagenesis with respect to constitutive N_2 fixation and resistance to the non-congenial agronomic factors; preliminary meristem mutagenesis in *Azolla* for achieving reduced phosphate dependence, temperature tolerance and significant sporulation/ spore germination under controlled conditions; mass production biofertilizer technology of free-living and symbiotic (*Azolla-Anabaena*) cyanobacteria; interacting and agronomic effects of both kinds of cyanobacterial biofertilizer with rice; and economics of rice cultivation with the cyanobacterial biofertilizers. While the researches, so far, are an indicative of the strong potential of cyanobacterial biofertilizer technology in rice growing countries, these open up a vast area of more concerted basic, applied and extension work in future to make these self-renewable natural nitrogen resources further promising at field level in order to help reducing the requirement of inorganic N to be the bare minimum, if not zero.

INTRODUCTION

The flooded water and the surface of the soil provide the sites for aerobic phototrophic nitrogen fixation by free-living cyanobacteria, often known as blue-green algae (Vaishampayan, 1996; Whitton, 2000) and *Azolla-Anabaena* (fern-cyanobacterial) symbiotic N_2-fixing complex (Singh, 1979; Vaishampayan, 1994; Kundu and Ladha, 1995; Vaishampayan *et al.*, 1998a, 2000), and the reduced zone and the rhizosphere of the rice plants are the sites for heterotrophic nitrogen fixation (Matsuguchi, 1979). The present literature is an attempt to review some of the significantly important findings out of a host of work accomplished in the area of phycotechnological advances towards using cyanobacterial-based biofertilizers for increased rice production.

A variety of free-living cyanobacteria are now well identified as efficient components of cyanobacterial biofertilizers. Nevertheless, a few cyanobacterial species form symbiotic associations with plants (algae, i.e., diatoms; fungi, i.e., lichens; bryophytes, i.e., liverworts, hornworts and mosses; pteridophytes, i.e., *Azolla*; gymnosperms, i.e., cycads; and angiosperms, i.e., *Gunnera*), animals (marine sponges and echiruoid worms), non-photosynthetic protists (belonging to the group Glaucophyta), bacteria and hollow shafts of hairs of polar bear (Rai, 1990), of which the water-fern *Azolla*, enlodging the N_2-fixing cyanobacterium *Anabaena azollae* is another established major cyanobacterial biofertilizer. Both forms of these cyanobacterial biofertilizers are the key agents of phycotechnological applications in beneficial wetland (rice) agriculture. The present article focusses mainly on some genetic improvements made with respect to this group of biofertilizers, as it is hard to cover their entire biological and agricultural aspects under one roof.

The Free-living Cyanobacteria

Rice fields of the several rice growing countries, e.g., Japan, Thailand, China, Philippines, Bangladesh and India have been investigated to show the luxuriant presence of cyanobacteria (Venkataraman, 1981), the majority of which are capable of fixing the atmospheric nitrogen and their presence in the rice fields is considered to maintain the depleting nitrogen levels in the rice field soils (Venkataraman, 1993). These may be unicellular, filamentous, non-heterocystous and filamentous heterocystous (Table 1) and their role as a bio-N fertilizer in supporting enhanced grain yield in wet rice fields has been unequivocally demonstrated (Table 2).

Table 1. A list of so far known nitrogen-fixing cyanobacterial genera

Unicellular	Filamentous non-heterocystous	Filamentous heterocystous		
Aphanothece	*Lyngbya*	*Anabaena**	*Anabaenopsis*	*Aulosira*
Chroococcidiopsis	*LPP group*	*Calothrix**	*Camptylonema*	*Chlorogloea*
Dermocarpa	*Microcoleus chthonoplastes*	*Chlorogloe-opsis*	*Cylindrosper-mum*	*Fischerella**
*Gloeocapsa (Gloeotheca)**	*Myxosarcina*	*Gloeotrichia*	*Haplosiphon*	*Mastigo-cladus*
Myxosarcina	*Oscillatoria*	*Nodularia*	*Nostoc**	*Nostochopsis*
*Pleurocapsa group**	*Plectonema boryanum*	*Rivularia*	*Scytonema**	*Scytonemato psis*
Synechococcus	*Pseudoanabaena*	*Stigonema*	*Tolypothrix*	*Westiellopsis*
Xenococcus	*Schizothrix*	*Westiellopsis*	*Westiellopsis*	*Trichodesmi um*

* Some strains of these genera live symbiotically with other plants (after Venkataraman, 1993; Sinha and Hader, 1996a)

Table 2. Average rice grain yield with 10 kg ha^{-1} cyanobacterial biofertilizer applied in Experimental fields unsupplemented or supplemented with inorganic N fertilizer

N level*	Rice grain yield (kg ha^{-1})			
	With biofertilizer	Without biofertilizer	Per cent increase	Reference
0	2,541	2,079	22.2 %	Venkataraman (1981)
50 N	4834	4,176	15.75%	Vaishampayan *et al.* (1996)
100 N	5,485	5,112	7.29%	Venkataraman (1981)
100 N	5,654	5,079	11.32%	Vaishampayan *et al.* (1996)

*50 and 100 N represent the 50% and 100% of the recommended N levels.

The cyanobacteria have been critically examined to form the N_2-fixing apparatus, i.e., heterocysts, and synthesize active nitrogenase specifically in zero or least combined N supplied condition in laboratory (Stewart and Rowell, 1975; Singh *et al.*, 1978a,b; Vaishampayan and Singh, 1981a,b; Vaishampayan, 1982) and fields (Mikheeva *et al.*, 1990), except in *Anabaena* strain CA (ATCC 33047), where covalent modification mechanism of nitrogenase during inhibition by combined N is non-operative (Bottomley *et al.*, 1979), and nitrate reductase-less mutants of *Nostoc muscorum* (Singh and Sonie, 1977; Vaishampayan and Prasad, 1982). Reportedly, a highly suppressed *in situ* cyanobacterial nitrogenase activity is operative in wet fields at the recommended field dose of nitrogenous fertilizers between 20 and 100 days after rice transplantation (Singh, 1975). Nevertheless, fertilizer needs of the crops may be reduced when weeds are controlled with herbicides, as these exert a more perfect weed control than physical methods. However, there are impending dangers of the damage of "higher plant-type" O_2-evolving photosynthesis of the cyanobacteria (as non-target species) by the use of herbicides, as the latter is designed to be either photosynthetic inhibitors or uncouplers (Vaishampayan and Prasad, 1984; Prasad and Vaishampayan, 1994; Vaishampayan, 1998; Vaishampayan *et al.*, 1998c). Thus, photosynthesis turns out to be a disadvantage with respect to the viability of the N_2-fixing cyanobacteria in pesticide-treated modern agricultural fields. This paradoxical situation raises an essential need to use high doses of synthetic N fertilizers (Singh *et al.*, 1988), which do repress the N_2-fixing activity of the few surviving cyanobacterial cells (Vaishampayan *et al.*, 1998c). In fact this is one of the prime reasons why rate of cyanobacterial contribution of soil N is supposed to be less as compared to *Azolla-Anabaena* symbiosis or the other heterotrophic N_2 fixers (Prasad and Vaishampayan, 1994). Rather, under the situation of repressed nitrogenase, the surviving cyanobacterial cells are forced to consume a part of the applied synthetic nitrogenous fertilizers for maintaining their own survival. Since the practice of agro-chemicalization has to continue to maintain the high yielding crop varieties, a unified concept has evolved during the last couple of decades towards developing ammonia derepressible pesticide-resistant cyanobacterial mutants as a viable and efficient cyanobacterial biofertilizer in modern agro-chemicalized rice fields (Meeting, 1988; Reddy and Roger, 1988, Venkata-raman, 1993; Anand, 1998; Vaishampayan *et al.*, 1998c).

While some of the pesticides at doses recommended for field application were not found to affect the cyanobacteria (Goyal, 1989), the others proved to be extremely injurious (Vaishampayan *et al.*, 2000b,c), causing inhibition of either N_2 fixation/fixed N assimilation directly, or indirectly through inhibiting chlorophyll biosynthesis, photochemical generation of reducing power, photophosphorylation, fixed N assimilation (references cited in Table 3). Of the various N_2-fixing cyanobacterial mutants now available (Vaishampayan *et al.*, 1998b) two of the *Nostoc muscorum* mutants, resistant to the herbicide amitrole (3-amino-1,3,4-triazole) and the fungicide carbendazine (2-methoxycarbamoyl-benzamidazole), isolated by Mishra *et al.*, (1991) and Prasad *et al.* (1991), respectively, have been found to be nitrogenase-derepressible (constitutive for N_2 fixation, irrespective of the level of fixed N present in the exterior environment). The same have been tested with rice (Vaishampayan *et al.*, 1996) and found to support an increase in grain yield by 28-32% in three traditional rice varieties, i.e., "Saket-4", "Lanjhee" and "Adamchini", and 15-20% in three high yielding rice varieties, i.e., "Pant-4", "HUR-36" and 'Saryu-52".

THE *AZOLLA-ANABAENA* SYMBIOTIC N_2-FIXING COMPLEX

The ability of the endosymbiont *Anabaena azollae* to fix substantial atmospheric dinitrogen inside *Azolla*, has made the association useful in rice culture in countries like Thailand (Moore, 1969), Indonesia (Brotonogero *et al.*, 1982), India (Singh, 1979), China (Lumpkin, 1982), Philippines (Watanabe *et al.*, 1977), Vietnam (Tuan and Thuyet, 1979), Taiwan (Lee and Lin, 1981), Brazil (Fiore and Ruschel, 1982), Italy (Espinosa-Abarca *et al.*, 1985), Mexico (Ferrera-Cerrato and Romero, 1982) and West Africa (Reynaud, 1982). High yield increases in rice through *Azolla* application have been reported (see Singh, 1979; Kannaiyan, 1994; Vaishampayan *et al.*, 2000a). The problem, however, with this beneficial bio-material has been primarily with respect to its high thermo-sensitivity, greater phosphate dependence and lack of sporulation/spore germinated under the controlled condition. Had the spore been easily germinated these would have been the ideal starting population for the genetic improvement purpose. Yet, through the use of shoot/frond meristem mutagenesis in *Azolla-Anabaena* symbiotic N_2-fixing complex (Vaishampayan & Banerjee, 1995; Vaishampayan *et al.*, 1992, 1998c, 2000a), low P-requiring and up to 40^{0}C temperature tolerant *Azolla pinnata* could be isolated which have supported a still higher rice grain yield as compared to the natural *Azolla* (Table 4), leading to a grater N economy (saving of inorganic N) (Table 5), and a significantly higher incidence of sporulation and spore germination (Table 6).

Table 3. Biological effects of various pesticides studied in cyanobacteria

Pesticide	Organism	Effect	Sensitivity level (ppm)	Reference
Atrazine	*Nostoc muscorum*	Photosynthetic inhibitor	10.00	Mishra *et al.* (1991a)
BHC	*Anabaena* sp.	Inhibitory to N_2 fixation	50.00	Das (1977)
Blitox	*Nostoc muscorum*	do	10.00	Vaishampayan & Prasad (1982)
Carbaryl	do	Mutagenic	do	Vaishampayan (1985d)
2,4-D	do	Inhibitory to N_2 fixation	11.00	Lundkvist (1970)
do	*Cylindrospermum* sp.	do	do	do
do	*Nostoc punctiforme*	do	do	do
do	*Tolypothrix tenuis*	do	4.50	Hamdi *et al.* (1970)
Diquat	*Nostoc muscorum*	Mutagenic	10.0	Vaishampayan (1984g,h, 1985b,c)
Dithane	do	Photosynthetic inhibitor	10.00	Vaishampayan and Prasad (1981)
Diuron	do	Inhibitory to CO_2/N_2 fixation	2.34	Vaishampayan *et al.* (1984c, 1985a)
Eptam	*Calothrix brevissima*	Inhibitory to N_2 fixation	0.10	Ibrahim (1972)

Hg-fungicide	*Nostoc muscorum*	Inhibitory to CO_2/N_2 fixation	10.00	Prasad *et al.* (1986)
Lasso	do	Mutagenic	do	Singh *et al.* (1979)
Linuron	*Anabaena* sp.	Inhibitory to N_2 fixation	10.00	Das (1977)
Machete	*Nostoc muscorum*	Mutagenic	do	Singh and Vaishampayan (1978)
Malathion	Many species	Inhibitory to N_2 fixation	100.0	Da Silva *et al.* (1975)
MCPA	*Nostoc muscorum*	do	19.00	Lundkvist (1970)
do	*Nostoc punctiforme*	do	do	do
do	*Cylindrospermum* sp.	do	do	do
Molinate	*Tolypothrix tenuis*	do	25.00	Hamdi *et al.* (1970)
Monuron	*Nostoc muscorum*	Inhibitory to CO_2/N_2 fixation	0.34	Vaishampayan (1984a,b)
do	*Anabaena* sp.	do	do	Das (1977)
Paraquat	*Nostoc muscorum*	Mutagenic	10.0	Vaishampayan (1984 g,h, 1985b,c)
Propanil	*Tolypothrix tenuis*	Growth stimulatory	0.010	Ibrahim (1972)
do	*Calothrix brevissima*	do	do	do
do	*Anabaena cylindrica*	do	0.030	Wright *et al.* (1977)
do	*Nostoc entophytum*	do	do	do
do	*Gloeocapsa alpicola*	Growth inhibitory	0.005	Wright *et al.* (1977)
do	*Tolypothrix tenuis*	do	0.100	Hamdi *et al.* (1970)
do	*Calothrix brevissima*	do	do	Ibrahim (1972)
do	*Nostoc entophytum*	do	0.170	Wright *et al.* (1977)
do	*Anabaena variabilis*	do	5.000	do
do	*Tolypothrix tenuis*	Chlorophyll biosynthesis inhibitor	1.800	Hamdi et al. (1970)
do	*Anabaena variabilis*	Photosynthetic inhibitor	8.000	Wright *et al.* (1977)
do	*Tolypothrix tenius*	do	do	do
do	*Nostoc entophytum*	do	do	do
do	*Nostoc muscorum*	do	50.00	Vaishampayan *et al.* (1978)
Sandoz 6706	do	Phosphorylation inhibitor	10.00	Vaishampayan *et al.* (2000b)

Trilluralin	*Tolypothrix tenuis*	Inhibitory to N_2 fixation	25.00	Hamdi *et al.* (1970)

Table 4. Per cent increase in grain yield of Saket-4 (after Bhan, 2000) and HUR-36 (after Dey, 1999) rice varieties over respective controls on the application of wild type or mutant *Azolla*

Rice variety	Inorganic N level (kg ha^{-1})	Per cent increase in grain yield with	
		Wild type Azolla	Mutant Azolla
Saket-4	0	51.30	73.57
	40	24.37	32.92
	60	27.41	38.22
	80	6.02	13.21
HUR-36	0	39.04	60.88
	40	11.78	19.83
	60	13.40	24.08
	80	2.56	5.64

Table 5. Saving of inorganic N fertilizer due to the application of mutant *Azolla* to obtain the optimum grain yield in Saket-4 (after Bhan, 2000) and HUR-36 (after Dey, 1999) rice varieties

Rice variety	Treatment	Grain yield (q ha^{-1})	Saving of fertilizer N (kg ha^{-1})	Additional gain due to *Azolla* (q ha^{-1})
Saket-4	$N_{40} A_0$	16.49	20 kg ha^{-1}	4.02
	$N_{40} A_1$	20.51		5.43
	$N_{40} A_2$	21.92		
	$N_{60} A_0$	23.31		6.39
	$N_{60} A_1$	29.70		8.91
	$N_{60} A_2$	32.22		
	$N_{80} A_0$	32.02*		
	$N_{80} A_1$	32.89		1.87
	$N_{80} A_2$	35.12		4.10
HUR-36	$N_{40} A_0$	32.04	20 kg ha^{-1}	3.78
	$N_{40} A_1$	35.85		6.36
	$N_{40} A_2$	38.43		
	$N_{60} A_0$	39.77		
	$N_{60} A_1$	45.10		6.33

N_{60} A_2	49.35	9.58
N_{80} A_0	49.07*	
N_{80} A_1	50.33	1.26
N_{80} A_2	51.84	2.77

N_0 = 0 kg N ha^{-1}; N_1 = 40 kg N ha^{-1}; N_2 = 60 kg N ha^{-1}; N_3 = 80 kg N ha^{-1}

A_0 = Without *Azolla*; A_1 = Wild type *Azolla*; A_2 = Mutant *Azolla*

*The optimum grain yield obtainable with 80 kg inorganic fertilizer N ha^{-1} (without biofertilization); the same is obtainable with only 60 kg N ha^{-1} N when biofertilized with mutant *Azolla*, thus making possible a saving of 20 kg N ha^{-1} in both the rice varieties through mutant *Azolla* application.

Table 6. Per cent sporulation [sporocarps produced per 25 g fresh fronds (after Banerjee, 1994)] and germination [per 100 megasporocarps (after Dey, 1999)] in wild type and mutant *Azolla pinnata* whole plant spore bearing the N_2-fixing endosymbiotic cyanobacterium, *Anabaena azollae*) at graded P levels and varying temperatures (the values presented are the means of 30 random samples + standard errors)

Character	Temperature (°C)	Wild type			Mutant		
		P-0	P-10	P-20	P-0	P-10	P-20
Sporulation	25.0	0	65 ± 1.80	104 ± 0.70	0	116 ± 1.67	120 ± 1.70
	30.0	0	38 ± 0.74	77 ± 1.70	0	111 ± 0.70	116 ± 2.13
	35.0	0	12 ± 0.63	48 ± 0.95	0	106 ± 0.70	110 ± 1.67
	37.5	0	0	0	0	98 ± 1.12	105 ± 0.40
	40.0	0	0	0	0	71 ± 1.00	84 ± 1.58
Spore Germination	25.0	0	10.0 ± 0.81	30.0 ± 0.70	0	38.0 ± 1.38	39.6 ± 0.67
	30.0	0	2.0 ± 0.70	17 ± 0.83	0	25.8 ± 0.66	35.8 ± 1.28
	35.0	0	0	2 ± 0.66	0	19.6 ± 0.74	31.6 ± 0.74
	37.5	0	0	0	0	14.2 ± 0.79	22.4 ± 0.86
	40.0	0	0	0	0	5.2 ± 0.37	8.0 ± 0.32

P-0 (phosphate unsupplemented); P-10 (10 mg l^{-1} phosphate); P-20 (20 mg l^{-1} phosphate)

The production of nitrogen-fixing cyanobacteria for N-fertilization in rice-fields in relatively simple tubular reactors may be a much more attractive way to utilize these agriculturally beneficial organisms. The selection of mutants, which continuously release ammonia, may also help to overcome some of the difficulties involved in this phycotechnology. Strain improvement through genetic engineering/manipulation and tissue culture (for establishing artificial plant-cyanobacterial symbiosis) of a wide array of luxuriantly available beneficial N_2-fixing cyanobacteria of the tropical wet fields is an area of considerable scope when we think of advanced cyanobacterial types as the biofertilizer components. Strain improvement is needed not only with respect to rapid growth, constitutive nitrogen fixation and pesticide tolerance under photoautotrophic, photoheterotrophic and chemoautotrophic conditions, but also for the ability to fix dinitrogen equally well under aerobic, micro-aerobic and anaerobic conditions under a wide range of oxygen tension found in rice fields. The non-heterocystous N_2-fixing cyanobacteria need to be critically screened for their ability whether to be included as a cyanobacterial biofertilizer component. An analysis of the processes like nitrification and denitrification as a result of cyanobacterial biofertilization of agricultural fields can be helpful in having a real picture of the actual contribution of available cyanobacterial N to the crop beyond all shadows of an under/over estimate. Further, concurrent to N_2 fixation, the cyanobacteria synthesize and excrete a variety of organic/growth promoting substances, a detailed systematic characterization of which is still needed to explore their possible role in breaking the physiological yield barrier in rice on application of cyanobacterial biofertilizer at recommended inorganic N level.

Despite significant progress in the fundamental and applied research on *Azolla-Anabaena* symbiotic N_2-fixing complex, e.g., molecular biology of symbiosis, interchange of endosymbiotic N_2-fixing partner, sexual hybridization and mutagenesis for beneficial agronomic traits, and application of this mutualistic association as an efficient biofertilizer with rice, the recognition processes between *Azolla* and *Anabaena* need to be established. Probably, the inability to maintain a pure and confirmed *ex-planta* culture of *Anabaena azollae* has been a major limitation in this regard which needs to be resolved for a clear understanding of the

Comparative response of two rice genotypes to *Azolla*

Both the rice varieties responded to *Azolla* maximally at 60 kg N ha^{-1} level. Although, HUR-36 rice variety is a better yielder than Saket-4, yet the per cent increase in grain yield of Saket-4 rice variety through *Azolla* application was higher than that of HUR-36 by:

(i) 12.26% (with WT *Azolla*), 12.69% (with mutant *Azolla*) at 0 N level;

(ii) 12.59% (with WT *Azolla*), 13.09% (with mutant *Azolla*) at 40 N level;

(iii) 14.01% (with WT *Azolla*), 14.14% (with mutant *Azolla*) at 60 N level;

(iv) 3.46% (with WT *Azolla*), 7.57% (with mutant *Azolla*) at 80 N level.

symbiotic process as well as a routine genetic improvement at the microsymbiotic level. Apart from this, there are still persisting constraints in relation to eco-physiological research on this form of cyanobacterial symbiosis in rice soil-water ecosystem. This often causes an inconsistency in the response of rice crops to *Azolla* inoculation. A number of management problems are also associated with the use of *Azolla* in rice culture and the successful management scheme will vary in accordance with the availability of opportunity cost labour, water, available P, propagation space and rice propagation schedules. A systematic approach needs to be opted at smaller levels to manageably implement *Azolla-Anabaena* biofertilizer technology. *Azolla* germplasm containing various species and worldwide collections are now available at the International Rice Research Institute, Manila, Philippines. On priority basis these need to be tested at different locations for their possible inherent ability to resist the various biotic and abiotic stress factors, such as high temperatures in tropics (which also indirectly influence greater damage of *Azolla* fronds owing to vigorous insect-pest infestation with rise in heat), desiccation, salinity and low phosphorus level for sustenance of their biofertilizer properties in one and/or the other area(s). This is particularly important because mutants, resistant to some of these stress factors, may have a competition problem of adaptability on soil inoculation. Further, numerous reports are available now on *Azolla* spore technology for its potentially more economic, easy and direct mass propagation in transplanted rice fields, but raising spore-mediated full-fledged *Azolla* sporophytes from application point of view is still awaited. More concerted efforts are, thus needed to make the very concept of *Azolla* spore technology practically possible for significantly reducing the cost of mass vegetative *Azolla* production and transportation. This would also help an easy clonal mutagenesis and interspecific crosses, using potentially germinating spores as the starting material, for developing new *Azolla* with specifically designed traits.

Through a long-term trial, a screening of the area-specific rice genotypes, responsive to the free-living or symbiotic cyanobacterial biofertilizer is essential to have good results, as more of the inorganic N can be saved in those cases through cyanobacterial biofertilization. Quality control procedures need to be developed and specified. It has to be kept in mind that effect of these microbial biofertilizer inoculants on crop growth and yield is seldom dramatically visible in one or two trials. But, of course, what is not visible even after the very first year of biofertilization is the truth that the soil is conserved, i.e., for achieving the optimum grain yield, less of the in-built soil N will be lost through cyanobacterial biofertilization. The farmers, therefore, need to be educated in this regard through close interactions with extension personnel at user's level, followed by efforts towards creating a consumer demand for such economic, self-renewable and soil-healthy biofertilizer through field-level extension workers and mass media. Our ultimate aim is to supply our farmers with inexpensive indigenous biofertilizers and lessen their burden on expensive chemical fertilizers. Thus, any practice aimed at increasing biological nitrogen fixation through cyanobacterial biofertilization must be acceptable to a large spectrum of farmers. The present status is conservation by supplementation and the future hope is total substitution of chemical fertilizers by biofertilizers, probably still a long way to go!

ACKNOWLEDGEMENTS

A portion of unpublished work incorporated in this article was financially supported by (i) University Grants Commission, Govt. of India in the form of a Career Award Research Project sanctioned to the author, and (ii) Indian Council of Agricultural Research, New Delhi in the form of an All

India Coordinated Research Project on Biological Nitrogen Fixation sanctioned to the Department of Genetics and Plant Breeding, Institute of Agricultural Sciences, Banaras Hindu University, India.

REFERENCES

Anand, N. 1998. Blue-green algae (cyanobacteria) as biofertilizers - Retrospects and prospects. In: **Microbes : For Health, Wealth and Sustainable Environment** (eds.) Ajit Varma, Malhotra Publishing House, New Delhi (India), pp. 65-71.

Bottomley, P.J., J.F.Grillo, C.van Baalen and F.R. Tabita. 1979. Synthesis of nitrogenase and heterocysts by *Anabaena* sp. CA in presence of high levels of ammonia. **J.Bacteriol., 140:** 938-943.

Brotonogero, S., M.Sudjadi, S.Partharjono, H.Sukiman, T. Prihatini and V. Hendrik. 1982. Some experiments on the use of *Azolla* for rice production in Indonesia. In : **Biological Nitrogen Fixation Technology for Tropical Agriculture** (eds.) P.H. Graham and S.C. Harris, Centro International de Agricultura Tropical Cali, Columbia, pp. 567-573.

Espinosa-Abarca, S.A., S.P. Myozga and M.M.Ortego. 1985. A study on the growth of *Azolla filiculoides* in culture media and in rice field soil of Morelos State of Mexico. **Review of Latinoam-Microbiologie, 27 :** 61-67.

Ferrera-Cerrato, R. and A.M. Romero. 1982. Propagation of an *Azolla* sp. and its application as green manure for corn in Mexico. In: **Biological Nitrogen Fixation Technology for Tropical Agriculture** (eds.) P.H. Graham and S.C. Harris, Centro International de Agricultura Tropical, Cali Columbia, pp. 561-564.

Goyal, S.K. 1989. Stress compatability in Cyanobacteria. **Phykos, 28:** 267-273.

Kannaiyan, S. 1994. .Sporulation and spore germination in symbiotic nitrogen-fixing water fern *Azolla.* In: **Biology and Application of Nitrogen-fixing Organisms: Problems and Prospects** (eds.) A.B. Prasad and A. Vaishampayan Scientific Publishers, Jodhpur (India), pp. 71-94.

Kundu, D.K. and J.K. Ladha. 1995. .Enhancing soil nitrogen use and biological nitrogen fixation in wetland rice. **Experimental Agriculture, 31 :** 261-277.

Lee, C.C. and C.J. Lin. 1981. The possibility of using *Azolla* as a source of nitrogen for rice in Taiwan. In : **Soils and Fertilizers in Taiwan**, pp. 35-41.

Lumpkin, T.A. 1982. Chinese technology for the cultivation of *Azolla*. In : **Biological Nitrogen Fixation Technology for Tropical Agriculture** (eds.) P.H. Graham and S.C. Harris, Centro International de Agricultura Tropical, Cali, Columbia, pp. 117-135.

Matsuguchi, T. 1979. Factors affecting heterotrophic nitrogen fixation in submerged rice soils. In: **Nitrogen and Rice,** International Rice Research Institute, Los Banos, Laguna, Philippines, pp. 207-222.

Meeting, B. 1988. Micro-algae in agriculture. In: **Micro-algal Biotechnology** (eds.) Borowitzka, M.A. and Borowitzka, L.J., Cambridge University Press, Cambridge, U.K., pp. 288-304.

Mikheeva, L.E., S.D.Pyramchuk, I.N. Gogotov and S.V. Shestakov. 1990. *Anabaena azollae* cyanobacterial mutants with derepressed nitrogenase synthesis. **Mikrobiologie, 59:** 35-39.

Mishra, A.K., A. Vaishampayan and V.P. Singh. 1991. A herbicide-resistant cyanobacterial mutant derepressed for nitrogen fixation under ammonia-mediated condition. In: **Proceedings of Golden Jubilee Symposium of the Indian Society of Genetics and Plant Breeding (Genetic Research and Education : Current Trends and the Next Fifty Years)**, Indian Agricultural Research Institute, New Delhi (India), pp. 98-100.

Moore, A.W. 1969. *Azolla* : Biology and agronomic significance. **Botanical Review, 35 :** 17-34.

Prasad, A.B. and A. Vaishampayan. 1984. Genetic recombination in *Nostoc muscorum* strains. In: **Perspectives in Cytology** *and Genetics* (eds.) G. Manna and U. Sinha, Typographers Rashtravani Press, New Delhi (India) pp. 179-184.

Prasad, A.B. and A. Vaishampayan. 1994a. **Biology and Application of Nitrogen-fixing Organisms : Problems and Prospects** (edited Book), Scientific Publishers, Jodhpur (India), pp. 1-332 (ISBN: 81-72-33-084-7).

Prasad, A.B., A.Vaishampayan, M.L. Vishwakarma and D.K.Shrivastava. 1991. Development of fungicide-resistance, coupled with N_2-fixing ability during NH_4^+ mediated growth, as a mutational event, in a cyanobacterium *Nostoc muscorum*. **Biomedical Letters (Cambridge), 46:** 245-251.

Rai, A.N. 1990. Cyanobacteria in symbiosis. In: **Handbook of Symbiotic Cyanobacteria** (Ed. A.N. Rai), CRC Press, In., Boca Raton, Florida (USA), pp. 1-7.

Reddy, P.M. and P.A. Roger. 1988 Dynamics of algal populations and acetylene reducing activity in five rice soils inoculated with blue-green algae. **Biol. Fertil. Soils., 6:** 14-21.

Reynaud, P.A. 1982. The use of *Azolla* in West Africa. In : **Biological Nitrogen Fixation Technology for Tropical Agriculture** (eds.) P.H. Graham and S.C. Harris), Centro International de Agricultura, Cali, Columbia, pp. 565-566.

Singh, A.L., P.K. Singh and P.L. Singh.1988. Effect of different herbicides on the *Azolla* and blue-green algal biofertilization of rice. **J Agric. Sci., 111:** 451-458.

Singh, H.N. and K.C.Sonie. 1977. Isolation and characterization of chlorate-resistant mutants of the blue-green alga *Nostoc muscorum*. **Mutation Research, 43:** 205-212.

Singh, H.N., A. Vaishampayan and R.K.Singh. 1978a. Evidence for the Involvement of a genetic determinant controlling functional specificity of group VI-B elements in N_2 and NO_3^- metabolism in a blue-green alga *Nostoc muscorum*. **Biochemical and Biophysical Research Communications (Lond.), 81:** 67-74.

Singh, H.N., A.Vaishampayan and K.C.Sonie. 1978b. Mutation from Mo-dependent growth to W-dependent growth and further evidence for a genetic determinant common to nitrogenase and nitrate reductase in a blue-green alga *Nostoc muscorum*. **Mutation Research (Elsevier), 50:** 427-432.

Singh, P.K. 1975. Fertilizer tolerance of blue-green algae and their effect on heterocyst differentiation. **Phykos, 14:** 81-88.

Singh, P.K. 1979. Use of *Azolla* in rice production in India. In: **Nitrogen and Rice,** International Rice Research Institute, Los Banos, Laguna, Philippines, pp. 407-418.

Stewart, W.D.P. and P. Rowell. 1975. Effects of L-methionine-DL-sulfoxamine on the assimilation of newly fixed NH_3, acetylene reduction and heterocyst production in *Anabaena cylindrical.* **Biochem. Biophys. Res.Commun., 65:** 846-856.

Tuan, D.T. and T.Q. Thuyet. 1979. Use of *Azolla* in rice production in Vietnam. In : **Nitrogen and Rice,** International Rice Research Institute, Manila, Philippines, pp. 395-405.

Vaishampayan, A. 1982b. A methylamine-resistant mutant of the blue-green alga, *Nostoc muscorum*: possible involvement of glutamine synthetase in methylamine metabolism. **New Phytologist (Cambridge), 91:** 607-613.

Vaishampayan, A. 1994b. Recent advances in the molecular biology of *Azolla-Anabaena* symbiotic nitrogen-fixing complex and its use in agriculture. In: **Microbes and Environment** (eds.) A.B. Prasad and R.S. Bilgrami, Narendra Publishing House, New Delhi (India), pp. 121-143.

Vaishampayan, A. 1996. Mineral requirements of the free-living and symbiotic cyanobacteria. In: **Advancements in Micronutrient Research** (eds.) A. Hemantaranjan. Scientific Publishers, Jodhpur (India), pp. 103-126.

Vaishampayan, A. 1998. Physiological responses of genetically improved nitrogen-fixing cyanobacteria to agro-chemicalization in relation to paddy culture : Prospects as a source material for engineering herbicide sensitivity and resistance in plants. In: **Adv. Plant Physiology** (eds.) A. Hemantaranjan, Scientific Publishers, Jodhpur (India), pp. 191-220.

Vaishampayan, A. and A.K.Awasthi. 1996. Micropropagation of *Azolla-Anabaena* symbiotic nitrogen-fixing mutant somaclones for an economic rice culture. **Presentation at the First International Symposium on Microbial Exploitation,** Centre for Advanced Studies in Botany, Banaras Hindu University, Varanasi (India), March 13-16, pp. 9-10.

Vaishampayan, A., Dey, Tuhina and A.K.Awasthi. 1998b. Advances in molecular biology, agronomics and somaclonal mutagenesis of *Azolla-Anabaena* symbiotic N_2-fixing complex. In: **Microbes: For Health, Wealth and Sustainable Environment** (eds.) Ajit Varma, Malhotra Publishing House, New Delhi (India), pp. 81-114.

Vaishampayan, A., T.Dey, R.P. Sinha and D.P. Haeder. 1998c. Successful rice cultivation with genetically manipulated thermotolerant *Azolla* as a bio-N fertilizer. **Acta Hydrobiologica (Cracow, Poland), 40:** 207-213.

Vaishampayan, A., R.P.Sinha. and D.P. Haeder. 1998d. Use of genetically improved nitrogen-fixing cyanobacteria in rice paddy-fields: prospects as a source material for engineering herbicide sensitivity and resistance in plants. **Botanica Acta (George Thieme Verlag, Stuttgart, Germany and New York, U.S.A.), 111:** 176-190.

Vaishampayan, A. and K.K. Banerjee. 1995. Genetic approaches to accomplish reduced phosphate-dependence of *Azolla-Anabaena* symbiotic nitrogen-fixing complex in wet agriculture. In: **Algal Ecology : An Overview** (eds.) A.N.

Kargupta and E.N. Siddique, International Book Publishers and Distributors, Dehradun (India), pp. 161-193.

Vaishampayan, A., Dey, Tuhina and A.K. Awasthi. 1996. Genetic improvement of nitrogen-fixing cyanobacteria in response to modern rice agriculture. In: **Cytology, Genetics and Molecular Biology of Algae** (eds.) B.R. Chaudhary and S.B. Agrawal, SPB Academic Publishing bv, Amsterdam (The Netherlands), pp. 395-421.

Vaishampayan, A., Y.R.Reddy, B.D. Singh and R.M.Singh. 1992. Reduced phosphorus requirement of a mutant *Azolla-Anabaena* symbiotic N_2-fixing complex. **J.Exp. Botany (Oxford), 43:** 851-856.

Vaishampayan, A. and A.B.Prasad. 1982. Blitox-resistant mutants of the N_2-fixing blue-green algae *Nostoc linckia* and *Nostoc muscorum*. **Environmental and Experimental Botany (Pergamon, USA), 22:** 427-435.

Vaishampayan, A. and A.B. Prasad. 1984. Inter-strain transfer of a pesticide-resistance marker during N_2 and NO_3^- metabolism in various strains of a N_2-fixing cyanobacterium *Nostoc muscorum*. **Molecular and General Genetics (Springer-Verlag), 193:** 195-197.

Vaishampayan, A. and H.N. Singh. 1981a. Mutation in the blue-green alga *Nostoc muscorum*. I. Methylamine-resistant mutants. **Biochemie und Physiologie der Pflanzen (Germany), 176:** 625-630.

Vaishampayan, A. and H.N. Singh. 1981b. Mutation in the blue-green alga *Nostoc muscorum*. II. L-methionine-DL-sulfoxamine-resistant mutants. **Biochemie und Physiologie der Pflanzen (Germany), 176:** 631-637.

Vaishampayan, A., R.P.Singh, Dey, Tuhina and B.K. Prasad. 2000a. Use of improved phototrophic and heterotrophic N_2 fixers in rice, wheat and barley cultivation. In: **Proceedings (Extended Summary) of International Conference on Managing Natural Resources, Indian Agricultural Research Institute** (eds.) Singh, G.B., Indian Council of Agricultural Research, New Delhi (India), February 14-18, **2:** 662-664.

Vaishampayan, A., R.P.Sinha, A.K. Gupta and D.P. Haeder. 2000b. A cyanobacterial mutant resistant against a bleaching herbicide. **J. Basic Microbiol., (Germany), 4:** 1-10.

Vaishampayan, A., R.P.Sinha, A.K. Gupta and D.P.Haeder. 2000c. A cyanobacterial recombination study, involving an efficient N_2-fixing non-heterocystous partner. **Microbiol. Res., (Germany), 155:** 1-5.

Venkataraman, G.S. 1981. Blue-green algae : A possible remedy to nitrogen scarcity. **Curr. Sci., 50:** 253-256.

Venkataraman, G.S. 1993. Blue-green algae (cyanobacteria). In: **Biological Nitrogen Fixation** (eds.) S.N. Tata, A.M. Wadhwani and M.S. Mehdi, Indian Council of Agricultural Research, New Delhi, India, pp. 45-76.

Watanabe, I., K.K.Lee, B.V.Alimagno, M.Sato, D.C. Del Rosario and M.R. De Gozman. 1977. Biological N_2 fixation in paddy field studies by *in situ* acetylene reduction assays. **Inte. Rice Res. Inst. Paper Series, 3:** 1-16.

Whitton, B.A. 2000. Soils and rice-fields. In: **The Ecology of Cyanobacteria** (eds.) B.A. Whitton and M. Potts, Kluwer Academic Publishers, The Netherlands, pp. 233-255.

CHAPTER - 3

CERTAIN RECENT DEVELOPMENTS IN USING CYANOBACTERIAL BIOFERTILIZERS

V. Gayathri, V.S.R. Murugesan, G. Selvi and N. Anand

INTRODUCTION

Cyanobacteria are diazotrophic prokaryotic microorganisms with oxygen evolving photosynthetic systems. The free-living cyanobacteria could be used directly as biofertilizer (Kannaiyan *et al.*, 1980) or indirectly by the release of fixed nitrogen in the form of ammonia (Guerrero *et al.*, 1982). Field trials using a combination of cyanobacteria have shown that about 30% of chemical fertilizer input could be conserved by using cyanobacteria as biofertilizer (Kannaiyan, 1985). The uses of heterocystous cyanobacteria in biofertilizer programmes have been well established by Kannaiyan (1993). Recently, Kaushik (2000) reported that the application of cyanobacterial biofertilizer in rice field could give 10-15% increased crop yield and also conserve 20-30 Kg N/ha of chemical fertilizers.

Generally, cyanobacteria exhibit considerable tolerance to salt or osmotic stress. Salt tolerant cyanobacteria include both nitrogen fixing as well as non-nitrogen fixing forms, but the ability to resist salt stress seems to be more common in the unicellular cyanobacteria (Thomas and Apte, 1984). In salt sensitive strains, salinity inhibits nitrogen fixation, while in salt tolerant strains, initially higher rates of nitrogen fixation was observed followed by loss of nitrogen fixation at much higher salinity range (Apte *et al.*, 1987). The ability to maintain low intracellular Na^+ levels seems to decide the tolerance limit of cyanobacteria to salt stress. The reductions in Na^+ influx contribute to salt tolerance by minimizing the energy requirement for subsequent Na^+ efflux. The inhibition of Na^+ influx either by alkaline pH or high external K^+ substantially enhances the salt tolerance of both salt tolerant strains and salt sensitive strains of cyanobacteria. Cyanobacterial nitrogen fixation is also proportionately enhanced (Apte *et al.*, 1987). Earlier, it has been observed that the presence of combined nitrogen in the external medium reduces Na^{+} influx and thereby helps osmoregulation, but represses nitrogenase biosynthesis (Reddy *et al.*, 1989). Nitrogen fixation by cyanobacteria is known to occur under a wide pH range i.e. from acidic to alkaline in alkalotolerant forms (Anand and Revathy, 1992; Singh *et al.*, 1995).

The addition of heavy dose of pesticides in the form of insecticides, herbicides, fungicides and algicides affect the growth of blue-green algae in the rice field ecosystem (Roger and Kulasooriya, 1980). Earlier, Anand and Veerappan (1980) have shown that the pesticides (Dimecron 100 and Ekalux 25) and fungicides (Bliotox and Dithane) were toxic to *Anabaena vaginicola, Cylindrospermum muscicola* and *Nostoc entophytum* at higher concentrations. Anand and Subramanian (1989) have reported that the pesticides affect various physiological processes like nitrogenase activity, photosynthesis, carbon fixation, protein synthesis, enzymes of assimilatory nitrate reduction and enzymes of ammonia assimilation. The relative tolerance of ten species of heterocystous cyanobacteria belonging to *Anabaena, Nostoc, Calothrix, Scytonema* and *Westiellopsis* to 3 pesticides namely Sevin, Rogor and Hildan have been studied by Das and Adhikary (1996).

Cyanobacterial biofertilizer programmes have continuously been subjected to review and modification, since the ultimate aim has been efficiently use the system for sustained agriculture. Some of the important findings in the laboratory are being applied in the field for improved crop yield. Certain aspects initiated in the last few years of the previous century and the future promise is given in detail.

NITROGEN FIXATION BY NON-HETEROCYSTOUS CYANOBACTERIA

Nitrogen fixation by cyanobacteria that have the specialized cells called "heterocysts" was unequivocally proved by the work of Fogg and his co-workers (Fogg *et al.*, 1973; Stewart, 1980). The finding that unicellular and colonial organism like *Gloeothecae* (Mullineaux *et al.*, 1981; Gallon and Chaplin, 1988) and non-heterocystous filamentous forms also fix nitrogen aerobically (Stal and Krumbein, 1985; Gallon, 1989; Anand, 1997) incited much attention on these forms. Besides the acetylene reduction assay identification of the nitrogenase protein by western blot analyses has been carried out in these forms.

The western-blot analysis of nitrogenase regulation in *Anabaena* sp shows that oxygen represses nitrogenase synthesis and inactivates and degrades the existing nitrogenase proteins (Apte, 1994). The modification of Fe protein of nitrogenase using western-blot technique has been reported in *Trichodesmium thiebautii* (Capone *et al.*, 1990; Ohki *et al.*, 1992; Zehr *et al.*, 1993). Du and Gallon (1993) studied the modification of Fe protein of nitrogenase of *Gloeothecae* sp grown under alternating light and dark. Recently, it has been observed that the cultures of *Oscillatoria foreaui* and *Lyngbya spiralis* incubated in light showed no Fe protein band, whereas the same cultures when they enter the dark phase, the nitrogenase protein was activated (Murugesan, 1998). This was confirmed by western-blot analysis of the Fe protein of nitrogenase (Plate 1 a&b). Maximum nitrogenase activity in *O. foreaui* was recorded in dark phase of alternating light/dark cycle and resembles other non-heterocystous nitrogen fixers (Gallon, 1989; Anand, 1998). The use of non-heterocystous forms as biofertilizer appears to be a prospective step in sustained agriculture. Experiments conducted in pots and in field under different treatments of chemical fertilizer-biofertilizer combination have shown that the two varieties of rice, ADT 43 and ADT 36 (short duration crops) performed well when combination of farmyard manure and the associating non-heterocystous cyanobacteria *Oscillatoria foreaui* or *Oscillatoria tenuis* was used. Earlier, it has been shown that the combined inoculation of algae and *Azolla* performed better than individual inoculation (Kannaiyan *et al.*, 1982).

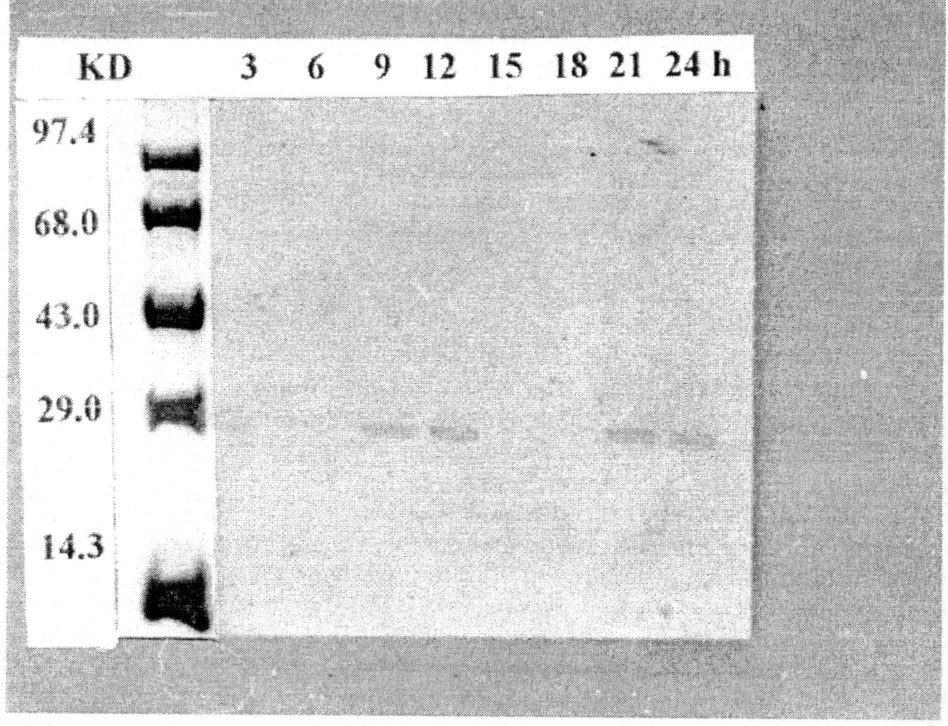

Plate 1a. Immuno-western blot analysis of Fe proteins in *O. foreaui* under alternating 6h L/ 6h D cycles. The Fe protein bands are clearly seen only when the cultures are incubated in the dark (9-12 h and 21-24h). Marker in SDS-PAGE is given for reference.

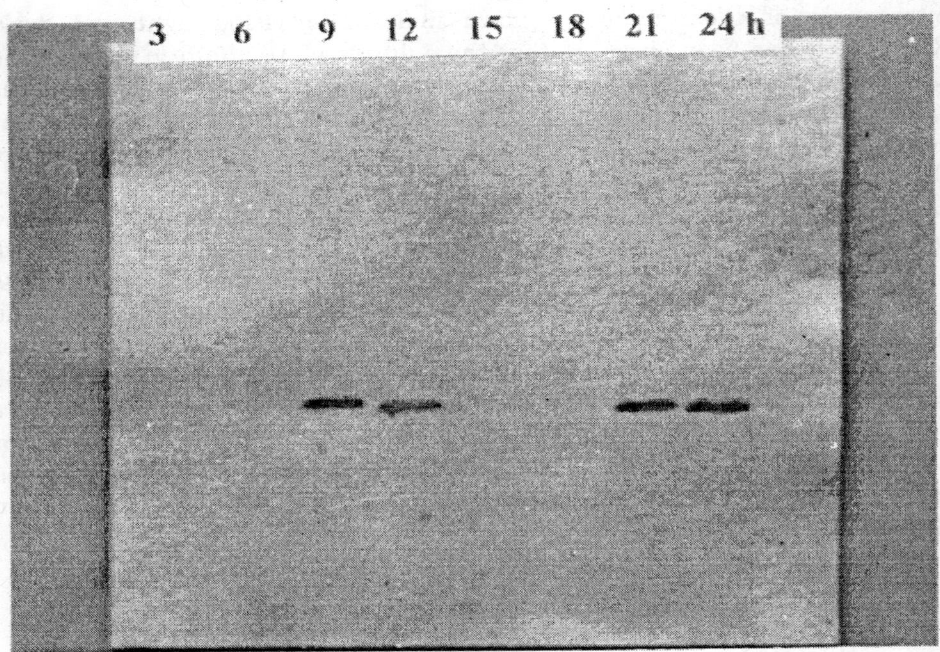

Plate 1b. Immuno-western blot analysis of Fe proteins in *L. spiralis* under alternating 6h L/ 6h D cycles. The Fe protein bands are clearly seen only when the cultures are incubated in the dark (9-12 h and 21-24 h)

Ammonia releasing strains and immobilization

Strains that release the fixed ammonia have been identified (Subramanian and Shanmugasundaram, 1986; Spiller *et al.*, 1986; Anand and Parameswaran, 1992; Goyal and Goyal, 1998). All the earlier reports on ammonia release was mainly by heterocystous filamentous cyanobacteria. Recently, the release of ammonia by a non-heterocystous filamentous cyanobacterium, *Lyngbya spiralis* was reported from our laboratory (Anand and Gayathri, 1999). Samal and Kannaiyan (1992) have shown the release of ammonia by the algal symbiont *Anabaena azollae* under immobilized condition. The photoproduction of ammonia by the cyanobacteria immobilized in polyurethane foam and polyvinyl foam have been observed by Kannaiyan *et al.* (1992,1994). Uma and Kannaiyan (1994) have found stimulation of ammonia excretion when treated with the insecticide carbofuran. Balachandar and Kannaiyan (1998) observed that the ammonia release was higher when the cyanobacteria were immobilized in solid matrices. It was observed that the growth of the cyanobacterium, *Calothrix elenkinii* was higher in alginate bead immobilization than polyurethane foam immobilization. But the release of ammonia was higher on the 28th day in PU-foam immobilized *C. elenkinii* (Fig. 1).

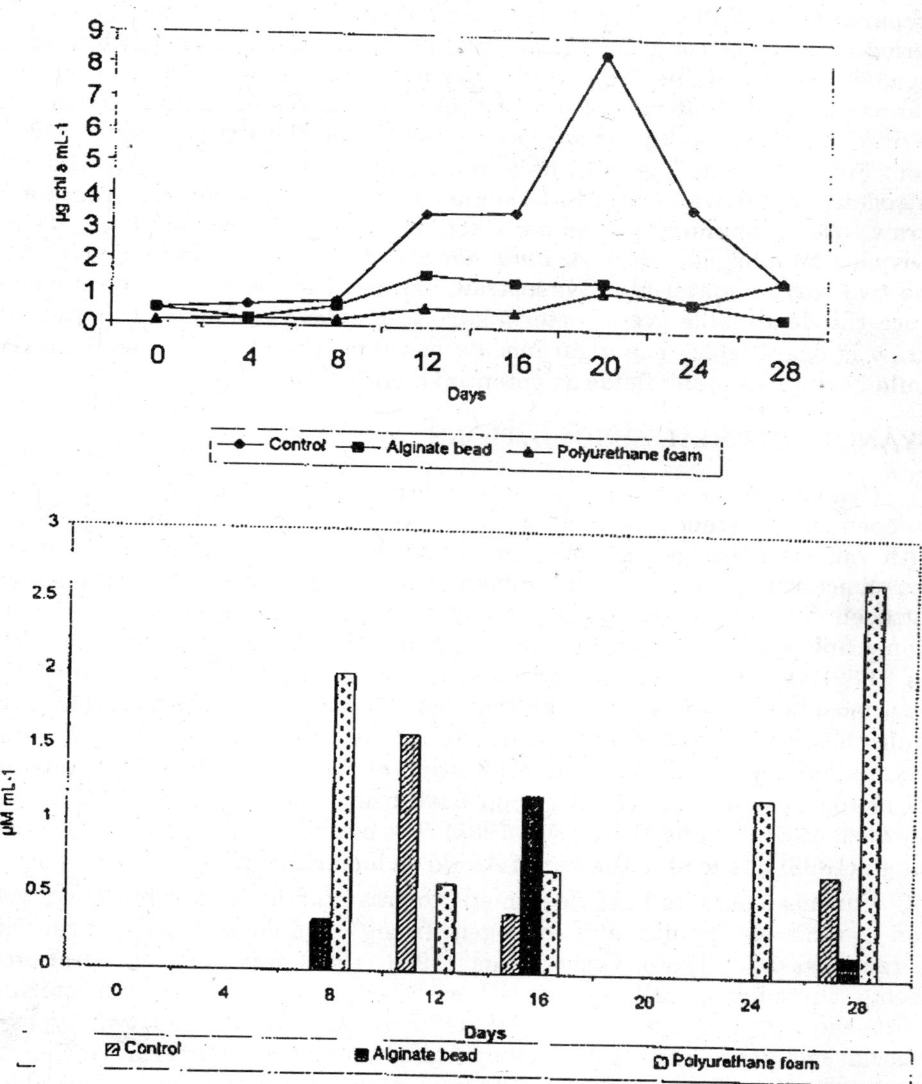

Fig. 1. Growth and ammonia release of *Calothrix elenkinii* under immobilization.

The Carrier material

An important aspect of an effective biofertilizer programmes has been the method of distribution of blue-green algae developed by mass multiplication. Initially, soil served as the carrier material. But due to its liability to get contaminated, the farmers did not prefer the use of soil (Kaushik, 1998). So, various other carrier materials such as straw and Fuller's earth (multani mitti) were selected. Unlike soil based inoculum, straw can be supplied as ready to use quality material. The algal flakes prepared using Fuller's earth as carrier material could be stored for a longer period at ambient conditions (Kaushik, 1996). Immobilizations of blue-green algae in polyurethane foam and alginate beads have been suggested by Kannaiyan (1992). It was observed that the cultures of *Anabaena ambigua* revived from the paddy straw could grow better than the cultures revived from Fuller's earth. The total protein content, nitrogenase activity and total nitrogen content was found to be higher in the cultures revived from paddy straw. The maximum nitrogenase activity was observed on the 16th day (Divya, unpublished). Both *A. ambigua* and *O. foreaui* showed revival from the two carrier materials. Cyanostraw has been found to be a good carrier since the degradable cyanobacteria survive on the straw and ultimately the straw is degradable. The algal inocula based on the above carrier materials could be applied to the fields as potential biofertilizer.

CYANOBACTERIAL ASSOCIATIONS

Cyanobacteria are almost free-living and are ideally suited to an independent existence. Nevertheless, a few species form specific associations with various plant groups (Fogg *et al.*, 1973; Peters, 1977). Sometimes, the cyanobacteria occur as endosymbionts that are able to fix atmospheric nitrogen (Meeks *et al.*, 1978). One of the symbiotic association on which pioneering work has been done by Kannaiyan and his group (Kannaiyan *et al.*, 1982) was on *Azolla – Anabaena* symbiotic association. *Azolla* is known to fix atmospheric nitrogen through the algal symbiont *Anabaena azollae* which multiplies rapidly and supplies nitrogen to the rice crop after decomposition (Kannaiyan, 1985). Earlier, mineralization of *Azolla* in submerged soils and its nitrogen availability to rice crop have been studied in detail (Mian and Stewart, 1984; Ito and Watanabe, 1985). The contribution of Goyal (1987) and Singh (1998) has lead to the use of *Azolla* as biofertilizer.

Various attempts have been made to create artificial symbiosis between non-leguminous plants and nitrogen fixing microorganisms (Gusev and Korzhenevskaya, 1990). Gusev *et al.* (1982, 1984) reported regeneration of shoots in tobacco callus cultures associated with the cyanobacterium, *Anabaena variabilis*. Gantar *et al.* (1991a) have recognized two types of associations (loose and tight) by the heterocystous cyanobacteria, *Nostoc* and *Anabaena* with wheat seedlings. The non-symbiotic colonization of rice plants by cyanobacteria is a new area of research. The association of cyanobacteria with the aerial roots (pneumatophores) of mangrove plant, *Avicennia* sp contribute to the nitrogen status of mangrove ecosystem. Recently, Toledo *et al.* (1995) have shown that *Avicennia* seedlings colonized artificially with cyanobacteria promote the growth of the seedlings. Attempts have been made in our laboratory to create association of cyanobacteria with the root system of rice plant, *Oryza sativa* L. var. ADT 36. The scanning electron microscopic observation of *Calothrix elenkinii* with the root system of the rice plant shows intertwining of the cyanobacterium with the root hairs. The filaments remain tightly attached to the root surface (Plate 2a). SEM observations on the cross-section of the root of *O. sativa* L. var. ADT 36 associated with *C. elenkinii* reveals aseriate stage of the cyanobacterium within the circular band of filaments (Plate 2b). Spherical bodies were found within each cell of the root occupying the entire cell lumen.

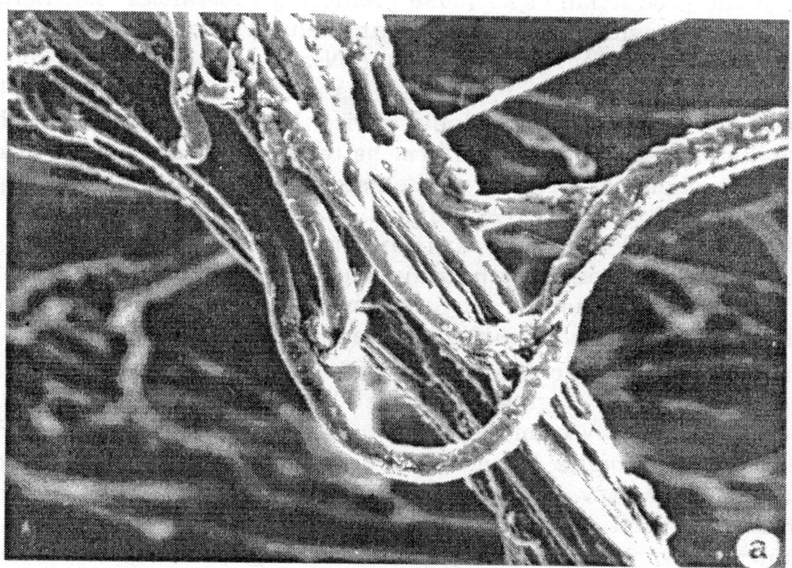

Plate 2a. Scanning Electron Micrograph showing the association of *Calothrix elenkinii* with the root system of *Oryza sativa* var. ADT 36 (x 1000)

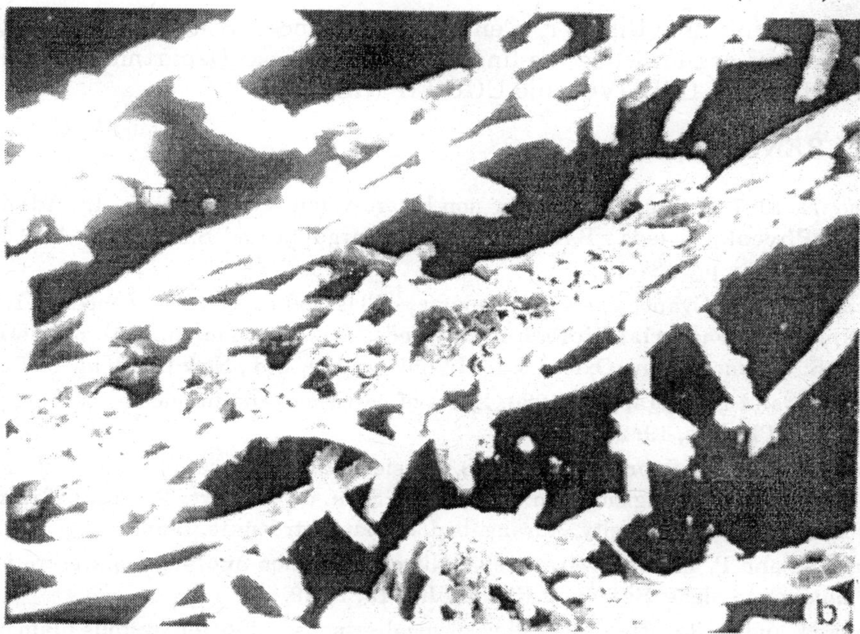

Plate 2a. Scanning Electron Micrograph of T.S. of root of *O. sativa* var. ADT 36 showing aseriate stage of the cyanobacterium, *C. elenkinii* (x 900)

Earlier, Gantar *et al.* (1991b, 1993, 1995a,b) have carried out detailed studies on colonization of *Triticum vulgare* L. by cyanobacteria. SEM observations showed cyanobacterial cells inside the cortical cells. They also suggested the production of hormone promoting substances that help in the life cycle of the organism to associate with the root system at different stages. The colonizing strains of *Phormidium* and *Nostoc* have been found to enhance the nitrogenase activity and nitrogen assimilation in rice plants, thereby increasing the grain and straw yield of rice. Recently, it has been observed that non-heterocystous forms perform better than heterocystous forms and certain other chemical fertilizers. The association of the rice seedlings with non-heterocystous forms, *Oscillatoria foreaui* and *O. tenuis* before transplantation and supply of farmyard manure during the growing stage could increase the shoot length, grain and straw yield of rice compared to control (without addition of any fertilizer), heterocystous forms (*Nostoc calcicola* and *Calothrix elenkinii*) and other chemical fertilizers.

The use of cyanobacterial extracellular products also appears to hold promise as promoters of growth of vegetables, fruits and medicinal plants. The use of culture filtrates of *Oscillatoria foreaui* was found to perform better than *Anabaena ambigua* in the growth of the medicinal plants, *Solanum nigrum* and *Withania somnifera*. Cyanobacterial biofertilizer technology developed and used in a judicious way appears to hold the key for a successful management of the crop fields and sustainable agriculture.

ACKNOWLEDGEMENTS

We thank the Director, Centre for Advanced Studies in Botany for facilities and acknowledge the financial support from Department of Science and Technology, CSIR (VG) and UGC-FIP (GS).

REFERENCES

Anand, N. 1997. Nitrogen fixation by non-heterocystous cyanobacteria. In: **Advances in Phycology.** (eds.) B.N. Verma, A.N. Kargupta and S.K. Goyal. APC Publ., New Delhi. pp. 187-197.

Anand, N. 1998. Cyanobacterial Taxonomy – Classical concepts and modern trends. In: **Cyanobacterial Biotechnology**. (eds.) G. Subramaniam, B.D. Kaushik and G.S. Venkataraman. Oxford & IBH Publ. Co. Pvt. Ltd., New Delhi. pp. 337-340.

Anand, N. and B. Veerappan. 1980. Effect of pesticides and fungicides on blue-green algae. **Phykos, 19:** 210-212.

Anand, N. and T.D. Subramanian. 1989. Physiological responses of *Nostoc calcicola* to pesticides and fungicides. In: **Abst. Natl. Sem. on Biofert. Technol. Transfer,** Marathwada University, Aurangabad, Maharashtra state, India.

Anand, N. and P. Parameswaran 1992. Nitrogen fixation and ammonia excretion by blue-green algae. **Seaweed Res. Utiln., 15:** 15-19.

Anand, N. and G. Revathy. 1992. Physiological responses of heterocystous filamentous blue-green algae *Nostoc calcicola* Breb. ex. Born. et. Flah to varying pH. In:

Proc. Natl. Symp. Nitrogen fixation. (ed.) B.D. Kaushik, IARI, New Delhi. pp. 151-161.

Anand, N. and V. Gayathri. 1999. Ammonia release by a non-heterocystous filamentous cyanobacterium *Lyngbya spiralis* Geitler (A-6). **Phykos, 38:** 27-30.

Apte, S.K. 1994. Cyanobacterial nitrogen fixation: Molecular genetic aspects. **Proc. Ind. Natl. Sci. Acad., B59:** 367-386.

Apte, S.K., B.R. Reddy and J. Thomas. 1987. Relationship between sodium influx and salt tolerance of nitrogen fixing cyanobacteria. **Appl. Environ. Microbiol., 53:** 1934-1939.

Balachandar, D. and S. Kannaiyan 1998. Effects of immobilization on growth and ammonia excretion by nitrogen fixing cyanobacteria. **Phykos, 37:** 69-76.

Capone, D.G., J.M. O'Neil, J. Zehr, and E.J. Carpenter. 1990. Basis for dual variation in nitrogen activity in the marine planktonic cyanobacterium *Trichodesmium thiebautii*. **Appl. Environ. Microbiol., 56:** 3532-3536.

Das, M.K. and S.P. Adhikary. 1996. Toxicity of three pesticides to several rice field cyanobacteria. **Trop. Agric., 73:** 155-157.

Divya, R. 1999. **Choice of carrier material for cyanobacterial biofertilizers.** M.Sc. Dissertation submitted to the University of Madras (unpublished).

Du, C. and J.R. Gallon. 1993. Modification of the Fe protein of the nitrogenase of *Gloeothece* (Nageli) sp. ATCC 27152 during growth under alternating light and darkness. **New Phytol., 125:** 121-129.

Fogg , G.E., W.D.P. Stewart, P. Fay, and A.E. Walsby. 1973. **The blue-green algae.** Academic press, London, England.

Gallon, J.R.1989. The physiology and biochemistry of nitrogen fixation by non-heterocystous cyanobacteria. **Phykos, 28:** 18-46.

Gallon, J.R and A.E. Chaplin. 1988. Recent studies on nitrogen fixation by non-heterocystous cyanobacteria. In: **Nitrogen fixation: Hundred Years After.** (ed.) H. Bothe, F.J. de Bruijn and W.E. Newton. Stuttgart, New York: Gustav Fisher. pp 183-188.

Gantar, M., N.W. Kerby, P. Rowell and Z.Obreht.1991a. Colonization of wheat (*Triticum vulgare* L.) by nitrogen fixing cyanobacteria: I. A survey of soil cyanobacterial isolates forming associations with roots. **New Phytol., 118:** 477-483.

Gantar, M., N.W. Kerby, and P. Rowell. 1991b. Colonization of wheat (*Triticum vulgare* L.) by nitrogen fixing cyanobacteria: II. An ultrastructural study. **New Phytol., 118:** 485-492.

Gantar, M., N.W. Kerby and P. Rowell. 1993. Colonization of wheat (*Triticum vulgare* L.) by nitrogen fixing cyanobacteria: III. The role of a hormogonia promoting factor. **New Phytol., 124:** 505-513.

Gantar, M., P. Rowell, N.W. Kerby and I.W. Sutherland. 1995a. Role of extracellular polysaccharides in the colonization of wheat (*Triticum vulgare* L.) roots by nitrogen fixing cyanobacteria. **Biol. Fertil. Soils., 19:** 41-48.

Gantar, M., N.W. Kerby, P. Rowell, Z. Obreht and C. Scrimgeour. 1995b. Colonization of wheat (*Triticum vulgare* L.) by nitrogen fixing cyanobacteria: IV. Dark nitrogenase activity and effects of Cyanobacteria on natural ^{15}N abundance in the plants. **New Phytol., 129:** 337-343.

Goyal, S.K. 1987. Algal biofertilizers- An appraisal. **J.Maharashtra Agri. Univ., 12:** 11-19.

Goyal, S.K. and S.K. Goyal. 1998. Biotechnological potential of algae. In: **Advances in Phycology.** (eds.) B.N. Verma, A.N. Kargupta and S.K. Goyal. Pub. APC Publications, New Delhi. pp. 1-22.

Guerrero, M.G., J.L. Ramos and M. Losada. 1982. Photosynthetic production of ammonia. **Experentia, 38:** 53-58.

Gusev, M.V. and T.G. Korzhenevskaya. 1990. Artificial associations. In : **CRC Handbook of Symbiotic Cyanobacteria**. (eds.) A.N. Rai, CRC Press, Inc. Boca Raton. pp. 173-230.

Gusev, M.V., R.G. Butenko, M.N. Agafodorova, T.G. Korzhenevskaya and O.I. Baulina. 1982. Morphogenesis in mixed callus culture of Tobacco and cyanobacterium. (In Russ) **Dokl Akad. Nauk. USSR. 272:** 1007-1010.

Gusev, M.V., R.G. Butenko and T.G. Korzhenevskaya. 1984. **Cyanobacteria in association with cultivated cells of higher plants**. Soviet Sci. Rev. Sect. D: Physicochem. Biol. Revs. (eds) V.P. Skulachev, Harwood Academic Publs. New York. 4: 1-40.

Ito O. and I. Watanabe. 1985. The relationship between combined nitrogen uptakes and nitrogen fixation in *Azolla-Anabaena* symbiosis. **New Phytol., 95:** 647-654.

Kannaiyan, S. 1985. Studies on the algal application for low land rice crop. In: Tamil Nadu Agric. Univ. Bull., Coimbatore, Tamil Nadu. pp.16.

Kannaiyan, S. 1992. ***Azolla* Biofertilizer Technology for Rice.** In : **Tech. Bull.** Tamil Nadu Agric. Univ. Coimbatore, Tamil Nadu, India. pp. 56.

Kannaiyan, S. 1993. Algal biofertilizer technology for rice crop in India. **Phycol. Rev., 2:** 73-82.

Kannaiyan, S., K. Govindarajan and H.D. Lewin. 1980. Studies on the survival of blue-green algae under dried condition. In: **Highlights of Research, Paddy Experiment station,** Tirur, Tamil Nadu, India. pp.54-59.

Kannaiyan, S., M. Thangaraju and G. Oblisami. 1982. Influence of *Azolla* biofertilizer application as green manure and dual cropping on rice. **Natl. Sci. Acad. Newslett., 5:** 31-37.

Kannaiyan, S., B. Sopkin, K.K. Rao and D.O. Hall. 1992. Ammonia excretion by the algal symbiont *Anabaena azollae* (AS-DS). In: **Biological Nitrogen Fixation and Biogas Technology**. (eds.) S. Kannaiyan, S. Ramasamy, K. Ilamurugu and K. Kumar, TNAU, Tamil Nadu, India. pp. 12-15.

Kannaiyan, S., K.K. Rao and D.O. Hall. 1994. Immobilization of *Anabaena azollae* from *Azolla filiculoides* in polyvinyl foam for ammonia production in a photobioreactor system. **World J. Microbiol. Biotechnol., 10:** 55-58.

Kaushik, B.D. 1996. Consolidated Report: Mission-mode project on Technology: Development and Demonstration of Algal Biofertilizers. IARI, New Delhi. India.

Kaushik, B.D. 1998. Blue-green Algal Biofertilizers for rice - A technology. In: **Modern methods of Plant analysis.** (eds.) K. Peack and M.R. Tracy, Springer-verlag, Berlin. Vol.I, pp. 469-502.

Kaushik, B.D. 2000. Role of cyanobacteria in sustainable agriculture. In: **Microbiotech 2000 (Abst.).** 41st Annual conference, Birla Institute of Scientific Research, Jaipur. pp.15.

Meeks, J.C., C. P. Wolk, H. Lockau, N.Schilling, P.W. Shaffer and W.S. Chien 1978. Pathways of assimilation of (^{12}N) N_2 and $^{13}NH_4^+$ by Cyanobacteria with and without heterocysts. **J. Bacteriol., 134:** 125-130.

Mian, M.H. and W.D.P. Stewart. 1984. A study on the availability of biologically fixed atmospheric dinitrogen fixation by the *Azolla-Anabaena* complex to flooded rice crop. In: **Practical application of *Azolla* for rice production.** (eds.) W.S. Silver and T.C. Schroeder, Martinus Nijhoff Publ., Dordrecht. pp. 168-178.

Mullineaux, P.M., J.R. Gallon and A.E. Chaplin. 1981. Nitrogen fixation in cultures of the cyanobacterium *Gloeocapsa (Gloeothece)* sp. 1430/3 incubated in the dark. **J. Gen. Microbiol., 124:** 141-146.

Murugesan, V.S.R. 1998. **Nitrogen fixation by non-heterocystous cyanobacteria (blue-green algae)** Ph.D., thesis, University of Madras, Chennai, India.

Ohki, K., J.P. Zehr and Y. Fujita. 1992. Regulation of nitrogenase activity in relation to light-dark regime in the filamentous non-heterocystous cyanobacterium *Trichodesmium* sp. NIBB 1067. **J. Gen. Microbiol., 138:** 2679-2685.

Peters, G.A. 1977. The *Azolla-Anabaena azollae* symbiosis. In: **Genetic Engineering for nitrogen fixation.** (eds.) A. Hollaender, Plenum press, New York. pp. 231-258.

Reddy, B.R., S.K. Apte and J. Thomas. 1989. Enhancement of cyanobacterial salt tolerance by combined nitrogen. **Plant Physiol., 89:** 204-210.

Roger, P.A. and S.A. Kulasooriya. 1980. **Blue-green algae and rice**. International Rice Research Institute. Los Banos, Laguna, Philippines. pp. 112.

Samal, L.C. and S. Kannaiyan. 1992. Ammonia excretion by *Anabaena azollae* immobilized in alginate and its effect on the growth of rice seedlings. **Inte. Rice Res. Newslett., 17:** 21-22.

Singh, P.K. 1998. Water fern *Azolla*. A proven biofertilizer for wetland rice. In: **Soil plant microbe interaction in relation to integrated nutrient management**. IARI, New Delhi. pp. 64-71.

Singh, B.V., S.K. Goyal and Tasneem Fatma. 1995. pH tolerance in certain cyanobacteria. **Phykos, 34:** 97-103.

Spiller, H., H.C. Lattore, M.E. Hassan and K.T. Shanmugam. 1986. Isolation and characterization of nitrogenase depressed mutants of cyanobacterium *Anabaena variabilis*. **J. Bacteriol., 165:** 412-419.

Stal, L.J. and W.E. Krumbein. 1985. Nitrogenase activity in the non-heterocystous cyanobacterium *Oscillatoria* sp. grown under alternating light dark cycles. **Arch. Microbiol., 143:** 67-71.

Stewart, W.D.P. 1980. Some aspects of structure and functioning of N_2 fixing Cyanobacteria. **Ann. Rev. Microbiol., 34:** 497-536.

Subramanian, G. and S. Shanmugasundram. 1986. Uninduced ammonia release by a nitrogen fixing cyanobacterium *Anabaena*. **FEMS Microbiol. Lett., 37:** 151-154.

Thomas, J. and S.K. Apte. 1984. Sodium requirement and metabolism in nitrogen-fixing cyanobacteria. **J. Bio.Sci., 6:** 771-794.

Toledo, G., Y. Bashan and A. Soeldner. 1995. *In vitro* colonization and increase in nitrogen fixation of seedling roots of black mangrove inoculated by a filamentous cyanobacteria. **Can. J. Microbiol., 41:** 1012-1020.

Uma, D. and S. Kannaiyan. 1994. Effect of the insecticide carbofuran on ammonia production by cyanobacteria under immobilized state in PU foam. Paper presented in the **National Seminar on *Azolla* and Algal Biofertilizer for rice.** March 1-3, 1994, Tamil Nadu Agric. Univ. Coimbatore, Tamil Nadu, India.

Zehr, J.P., M. Wyman, U. Miller, L. Duguay and D.G. Capone. 1993. Modification of the Fe protein of nitrogenase in natural populations of *Trichodesmium thiebautii*. **Appl. Environ. Microbiol., 59:** 669-676.

<u>CHAPTER - 4</u>

CYANOBACTERIAL BIOFERTILIZER TECHNOLOGY

B.D. Kaushik

INTRODUCTION

Traditionally agriculture was dependent on natural inputs like farm yard manure, green manure or on microbial activity. However, with the introduction of inorganic fertilizer as well as fertilizer responsive varieties of various crops, organic inputs continue to decline. It resulted in decreased organic matter (humus) in soil which affected fertility of soil severely. For example, the organic carbon content of Punjab soils was 0.5% in 1980 whereas, today it is only 0.2%. Hence serious attention was paid by scientists to exploit micoorganisms which are directly responsible to replenish the soil with certain nutrients like fixation of atmospheric nitrogen to ammonia, solubilization of phosphorus as well as addition of organic carbon. With the introduction of high yielding fertilizer responsive varieties of rice, demand for nitrogenous chemical fertilizers has increased several-fold.

Nitrogen is an essential nutrient required for plant growth. It is present as diatomic elemental nitrogen and constitute more than 79 per cent of total gases. On global basis out of 180 million tonnes of nitrogen added annually on the earth's surface, the industrial fertilizer nitrogen accounts for only 1/3, while 2/3 comes from biological reduction of atmospheric nitrogen. The role of the diazotrophic reduction in sustainability of soil fertility is therefore, more important. The dinitrogen reduction so far is confined to prokaryotic cellular forms in nature. Among these prokaryotes, the photosynthetic diazotrophs - cyanobacteria, more commonly known as blue-green algae, are free living as well as symbionts with lichens, ferns and cycads and their contribution in total nitrogen fixation by biological means probably comes next to the legume-*Rhizobium* symbiosis.

NITROGEN FIXING CYANOBACTERIA

As early as 1939 (De, 1939) first paper appeared that the fertility of Indian rice field is maintained due to the presence of blue-green algae (BGA) now known as cyanobacteria. Since then large amount of work has been done to exploit this renewable biological resource as nitrogen input in rice cultivation. The aerobic filamentous heterocystous as well as colonial forms represent the common flora of arable lands and are considered to be the important in nitrogen economy of rice cultivation. The most common filamentous heterocystous species include *Anabaena, Anabaenopsis, Aulosira, Calothrix, Cylindrospermum, Fischerella, Hapalosiphon, Mastigocladus, Nostoc, Scytonema, Scytonematopsis, Stigonema, Tolypothrix, Westiella* and *Westiellopsis* (Stewart *et al.*, 1979) and the unicellular or colonial are the species of *Aphanothece, Chroococcidiopsis, Cyanothece, Dermocarpa, Gloeothece, Myxosarcina, Synechoococcus, Synechocystis* and *Xenococcus* (Ripkka *et al.*, 1979; Prasanna and Kaushik, 1994). The non-heterocystous filamentous forms capable of nitrogen fixation are species of *Oscillatoria, Trichodesmium, Microcoleus, Pseudanabaena, Plectonema* and *Lyngbya* (Wyatt and Silvey, 1969).

NITROGEN FIXATION POTENTIAL

The paddy field ecosystem provides an environment favourable for the growth of BGA with respect to their requirement for light, water, temperature, humidity and nutrient availability. This may account for the higher abundance of BGA in paddy soils than in cultivated soils (Watanabe and Yamamoto, 1971) as reported under widely different climatic conditions of India, Japan, and Ukraine. Tropical conditions ensure increased incidence of cyanobacteria in the rice fields soils and high humidity and temperature and shade provided by the crop canopy favor the luxuriant growth of cyanobacteria (Roger and Reynaud, 1979).

Nutrients fixed by algae are released either through exudation or through microbial decomposition after the cells die (Roger and Reynaud, 1979). Wilson (1980) recovered from rice crop 39% of the nitrogen from ^{15}N-labelled *Aulosira* spp. spread on the soil and 51% from the algae incorporated into the soil. Quantitative estimates from different countries are Philippines 18-33 kg ha^{-1}, Senegal 0-30 kg ha^{-1}, Mali 50-80 kg ha^{-1} and Japan 11-23 kg ha^{-1}. Yoshida *et al.* (1973) reported 40-80 Kg N ha^{-1} through algal application. Henriksson (1971) showed annual fixation rate of 15-51 Kg N ha^{-1} yr^{-1} in an agricultural field where *Nostoc* was abundant, and 4.44 Kg N ha^{-1} yr^{-1} in a lakeside meadow containing *Nostoc, Anabaena, Cylindrospermum* and *Calothrix*.

Cyanobacterial N_2 fixation and crop response

From the results of field experiments done earlier, it appears that average algal inoculation, when effective, causes a relative increase in grain yield of paddy about 14% over the treatments and 16% over the control, corresponding to about 450 kg grain per hectare per crop. No significant difference was observed between average yields in fields receiving nitrogen fertilizer and fields where none was applied.

ALGALIZATION IN THE PRESENCE OF FERTILIZERS OTHER THAN NITROGEN

Phosphorus, lime and sometimes molybdenum application has been demonstrated to frequently have a beneficial effect on the establishment and growth of nitrogen fixing algal flora. In field experiments, algalization in combination with lime, phosphorus and molybdenum application was more efficient than algalization alone (Relwani, 1965; Relwani and Subrahmanyan, 1963; Sankaram *et al.*, 1966).

ALGALIZATION IN THE PRESENCE OF NITROGEN FERTILIZERS

There are numerous reports on the beneficial effect of algalization in the presence of fertilizer-N. Under field conditions, supplementation of urea at 60 kg ha^{-1} with algal inoculum resulted in a grain yield comparable to the obtained with 120 kg N as urea (Singh *et al.*, 1972). In a large number of Indian field trials, recommended high levels of nitrogen fertilizers complemented with algal inoculation resulted in yield significantly higher than that without algalization. Yield increases ranged from 2.2 to 28.9% and averaged 7.2% (Venkataraman, 1979). In recently concluded field trials in Haryana showed 10-15% increase in rice yield in presence of fertilizer nitrogen upto 150 kg ha^{-1} (Kaushik, 1998). From the recorded data, it appears that response to algalization was positive at any level of nitrogen; however, this response was generally lower at high levels. As BGA are generally inoculated 1 week after fertilizer N application, a decrease in mineral nitrogen concentration in floodwater can be expected. Venkataraman (1979) reported that nitrogenase activity was not depressed in soil-rice-algae system in the presence of less than 40 ppm ammonium-N. However, in recent studies it was observed that urea nitrogen upto 30 ppm can be applied without any loss of ammonia liberated through cyanobacterial nitrogen fixation.

Application of cyanobacteria along with inorganic fertilizers always show a positive gain yield over fertilizer application alone. Such gains are higher at lower fertilizer doses and marginal at optimum doses of even upto 100-120 kg N ha^{-1}. Based on 118 field application of cyanobacteria in Tamilnadu, Venkataraman (1992) reported an average increase in rice yield by 7.29% when 100 kg N ha^{-1} was applied along with the algae as compared to yields obtained with 100 kg N ha^{-1} fertilizer alone. Recently algalization trials were conducted with straw based cyanobacterial inoculum (1kg ha^{-1}) in Haryana state a net saving of 25-30 kg N ha^{-1} per cropping season as well as 10-15% increased yield of paddy was obtained.

PHOSPHORUS AVAILABILITY

A portion of the applied phosphorus is fixed in the soil and is immediately unavailable to the crop. However, due to the activities of some of the microorganisms, the fixed phosphorus is released slowly and becomes available to the crop. Solubilisation of mussorrie rock phosphate - a source of P_2O_5 and a raw material for fertilizer industry, by nitrogen fixing BGA *Tolypothrix tenuis, Scytonema cincinnatum* and *Hapalosiphon fontinalis* was observed *in vitro* studies (Roychoudhury and Kaushik, 1989).

Growth promoting substances

In addition to N-contributions, presoaking of rice seeds in BGA culture or extracts has decreased losses from sulphate reducing processes and this has been attributed to the enhancement of germination and faster seedling growth due to algal exudates (Jacq and Roger, 1977). Water soluble products from *Calothrix, Anabaena* and *Stratonostoc* has rhizogenous effect and stimulated plant organs. The growth promoting substances produced by the algae have been well identified and are vitamins (Vitamin B_{12}, folic acid, nicotinic acid and pantothenic acid), auxins (indole acetic acid and 3-methyl indole), amino acids and sugars (Misra and Kaushik, 1989a, b). Additionally, cyanobacteria excrete complex organic carbon compounds polysaccharides, these compounds bind the soil particles and improve soil aggregation, hence soil structure and an improved infiltration rate, soil permeability and water holding capacity of soil.

Production of cyanobacterial biomass as a source of biofertilizer

The idea of utilizing algal biofertilizer to supplement the nitrogen requirement of rice crop ,was mooted as early as 1939. However, the systematic approach was undertaken at IARI and an algal biofertilizer technology was developed (Venkataraman, 1972). The technology was named as "Rural oriented mass multiplication technology" and the application in field was often referred as algalization. In this process cyanobacteria were allowed to grow with soil as carrier and then utilized as inoculum in rice fields. This was adopted by farmers all over India because of its simple technology and economic benefits. Depending upon the requirement of algal biofertilizer, the production technology could be scaledup to-troughs, pits, field plots/cemented tanks/polythene layered tanks. However its wider adaptations was not possible (Kaushik, 1994) because of the following possible reasons.

- Fluctuations in quality and quantity of biomass due to nature of soil used and variations in environmental parameters like light, rainfall, temperature and wind velocity.
- Aerial and soil based contamination.
- Therefore in order to obviate these problems, research was directed towards:

FORMULATION OF MEDIUM AND OPTIMIZATION OF ENVIRONMENTAL PARAMETERS

In the traditional method of biomass production, soil served as both the nutrient source and inoculum carrier. In order to reduce problems related to the use of soil, like poor quality of inoculum with higher level of contamination a growth media was formulated with the objectives that it is cheaper and supports the growth of BGA equal or better than the known growth media. The medium was based on the commercial fertilizers available. The growth kinetics in the final formulated medium was found to be significantly superior to the control (basal medium). The cost of 1000L FM was US$ 0.28 against US$ 1.07 for the BG_{11} medium (BM). Hence, the formulated medium (FM) was much cheaper than the routinely used growth medium for BGA. Utilizing this medium, the important environmental parameters - pH, light, and temperature were optimized. The BGA strains were grown in media buffered with pH ranging from 7.0 to 10.0. A pH of 7.3 ± 0.2 was found to be the most suitable. Although the cultures were able to tolerate a temperature of 40°C, sustained biomass production and nitrogen fixation was observed at 30 ± 1°C.

STRAINS USED FOR BIOFERTILIZER

Anabaena variabilis, *Nostoc muscorum*, *Aulosira fertilissima* and *Tolypothrix tenuis*.

STRATEGIES TO REDUCE THE CONTAMINATION IN FINAL INOCULUM

The soil based, open air technology resulted in contamination of final product through air and soil thus a need was felt to develop closed chambers/ environment to avoid air borne contamination. To overcome this, multiplication units (ponds) were constructed in a polyhouse/greenhouse. Therefore, BGA was grown in polythene lined shallow pits containing formulated medium in a polyhouse/ greenhouse. The height of liquid medium was kept at 5 cm. A thick mat of BGA was observed to form with in 5 days of incubation and was ready to mix with a suitable carrier.

INOCULUM CARRIER

A large variety of carriers including thermocoal, riverbed sand, rice straw, wheat straw and saw dust were tested but wheat straw performed best. The individual strains were grown in the formulated medium in polyhouse/greenhouse independently in a multiplication unit and after 4-5 days mixed with wheat straw (400 g/pond of $1.5m^2$). BGA immobilized in wheat straw is sundried and can be packed in polythene packets for storage. The straw based BGA biofertilizer was found to contain more than 10^5 to 10^6 propagules per gram of carrier, therefore 1 kg BGA biofertilizer/ha (400 g/acre) was enough as against 10 kg/ha soil based BGA biofertilizer.

PACKING AND STORAGE OF CYANOBACTERIAL BIOFERTILIZER

The individual strains of cyanobacteria mixed in carrier, sundried and mixed all the four strains in equal proportion are packed at 400 g per pack and sealed. Such material could be stored for a long time (more than 2 years) in a dry state at normal room temperature in shade without any loss in quality of the inoculum. Shelf life is about 2 years.

FIELD APPLICATION

- BGA biofertilizer produced in a formulated medium in indoor culture facility contains 100,000 to 1,000,000 units (propagules) per gram carrier hence 1 kg/ha (400g/acre) straw based BGA inoculum is enough.
- Broadcast the dry inoculum over the standing water in the rice field at a rate of 400 g/acre. Application should be done immediately after transplantation is completed. Application of excess BGA inoculum is not harmful and will accelerate the fast growth of BGA in the fields.
- Ensure that there is enough water atleast 1 week after cyanobacterial biofertilizer application.
- In case, problem of broadcasting 400 g material in a acre mix with dry soil from the field or ash or saw dust and then use for even distribution.
- Recommended pest control measures and other management practices do not normally interfere with the establishment and activity of the BGA in the field.
- BGA biofertilizer should be used in every rice crop as a kind of security to the crop yield as well as to give a stop to the deterioration of soil physico-chemical properties.
- Use of BGA fertilizers is mainly recommended for rice. However, it has given good results with crop which require large amount of water like sugarcane.

REFERENCES

De, P.K. 1939. The role of blue green algae in nitrogen fixation in rice fields. **Proc Roy Soc, London. 127B:** 129-139.

Henriksson, E. 1971. In: **Biological nitrogen fixation in natural and agricultural habitats** (eds.) TA Lie and EG Mulder. p. 415, Plant Soil, Special Volume.

Jacq, V and Roger, P.A. 1977. Decrease of losses due to sulphate reducing processes in the spermosphere of rice by presoaking seeds in a culture of blue green algae (in French) *Cahiers* **ORSTOM Ser Biol., 12:**101-108.

Kaushik, BD. 1994. Blue green algae and sustainable agriculture. In: **Natural Resource Management for Sustainable Agriculture and Environment**, pp. 403-416, (ed.) DL Deb, Angkor Publs (P) Ltd. New Delhi.

Kaushik, B.D. 1998. Use of cyanobacterial biofertilizer in rice cultivation: A Technology Improvement. In: **Cyanobacterial Biotechnology** (eds.) G Subramanian, B.D. Kaushik, G.S Venkataraman, pp. 211-222, Science Publs. Inc. USA.

Misra, S., B.D. Kaushik. 1989 a. Growth promoting substances of cyanobacteria. I. Vitamins and their influence on rice plant. **Proc Indian Natn Sci Acad. B55:** 295-300.

Misra, S., B.D. Kaushik. 1989 b. Idems. II. Detection of amino acids, sugars and auxins. **Ibid., B55:** 499-504.

Prasanna, R., B.D. Kaushik. 1994. Physiological and molecular genetic aspects of nitrogen fixation in non-heterocystous cyanobacteria. **Ind. J. Exptl. Biol., 32:**248-251.

Relwani, L.L. 1965. Response of paddy varieties to blue green algae and method of propagation. **Curr Sci., 34:**188-189.

Relwani, L.L, R. Subrahmanyan. 1963. Role of blue green algae, chemical nutrients and partial soil sterilization on paddy yield. **Curr Sci., 32:**441-443.

Rippka, R., J. Deruelles, J.B. Waterbury, M. Herdmann, R.Y. Stanier 1979. Generic assignment, strain histories and properties of pure culture of cyanobacteria. **J Gen Microbiol., 111:**1-61.

Roger, P.A, Reynaud, P.A. 1979. **Ecology of blue green algae in paddy fields.** p.289-309 In: Nitrogen and Rice International Rice Res. Institute. Los Banos, Philippines.

Roychoudhury, P, Kaushik B.D. 1989. Solubilization of Mussorie rock phosphate by cyanobacteria. **Curr Sci., 58:**569-570.

Sankaram, A, M.J. Mudholkar, M.N.Sahay. 1966. Algae as aid in rice production. **Indian Farming, 16:**37-38.

Srinivasan, S. 1979. Algae multiplication fertilizer practices. **Int Rice Res Newl., 4:**3.

Stewart, W.D.P., Rowell, P, Ladha, J.K, M.J.A. Sampio. 1979. Blue-green algae (cyanobacteria) - some aspects related to their role as source of fixed nitrogen in paddy soils. In: **Nitrogen and Rice,** International Rice Research Institute, Los Banos, Philippines, p. 263.

Venkataraman, G.S. 1972. **Algal biofertilizer and rice cultivation.** Today and Tommorow`s publ., New Delhi.

Venkataraman, G.S. 1979. Algal inoculation of rice fields. **Nitrogen and Rice,** Los Banos, International Rice Res. Institute. Philippines, 311-321.

Venkataraman .G.S. 1992 In: **Algal Biofertilizer Technology for Rice**-Sankar Printer Madurai, 9.

Watanabe. A .1973. On the inoculation of paddy, fields in the Pacific area with nitrogen fixing blue green algae. **Soil Biol Biochem ., 5.** 161-162.

Watanabe. A., and Y. Yamamoto 1971. Algal nitrogen fixation in the tropics. **Plant. Soil Specl vol: 403-413.**

Wilson. J.T. 1980. In: **Blue green algae and Rice.** (eds.) P.A Roger and S.A Kulasooriya. International Rice Research Institute, Los Banos, Philippines.

Wyatt, J.T., J.K.G. Silvey. 1969. Nitrogen fixation by *Gloeocapsa*. **Science, 165**: 908-909.

Yoshida. T., R.A, Roncal and E.M. Bantista 1973. In: **Proc. 2nd Asian Soil Conf.** Indonesia.

CHAPTER - 5

GROWTH RESPONSE AND NITROGEN FIXATION OF RICE FIELD CYANOBACTERIA TO PESTICIDES

J.K. Sahu and S.P. Adhikary

ABSTRACT

Relative tolerance of eight selected species of heterocystous cyanobacteria belonging to four different genera, e.g. *Anabaena, Nostoc, Calothrix* and *Aulosira* to various concentrations of two commonly used insecticides *viz.* Balwan (monocrotophos, 36 SL) and Thiokill (endosulfan, 35 EC) and the herbicide Hiltachlor (50 EC) was studied. These cyanobacterial species differed widely with respect to their tolerance to different pesticides. Variation in the tolerance level exists among the species of the same genus and also within genera to a specific pesticide. Of the eight cyanobacteria tested, *Nostoc* sp. UU 29130 and *Aulosira* sp. UU 25118 were comparatively more tolerant to all the pesticides; *Nostoc* sp. UU 24100 and *Calothrix* sp. UU 24112 were less tolerant. Pesticide induced changes in the chlorophyll-*a* pigment content and rate of photosynthetic oxygen evolution followed a similar trend as that of their respective growth rate. Nitrogenase activity was stimulated after exposure of the organisms to recommended dose of Balwan, though there was a depression in the activity within 24 hours. ARA of the insecticide sensitive species was not affected adversely even in presence of their EC 50 dose though in the tolerant species nitrogenase activity was decreased to about 50 per cent at the same dose of the insecticide. Similar comparatively higher nitrogen fixing capacity of certain organisms sensitive to Thiokill and Hiltachlor than the tolerant species in the pesticide supplemented media was also observed. The results suggested that pesticide induced growth inhibition of a specific organism did not always correlate with the decrease of nitrogen fixing capability of rice field cyanobacteria.

INTRODUCTION

In modern agriculture, application of pesticides and herbicides has become inevitable and these agrochemicals are widely used throughout the world on a large scale to combat pests and diseases of rice. These agrochemicals also damage a wide variety of beneficial microorganisms due to their persistence in the environment. In rice field ecosystem, pesticides and herbicides applied to control selective pests or weeds might be affecting adversely on the non-target diazotrophic cyanobacteria. Reduction in the number of cyanobacteria in pesticide applied fields have been observed earlier (Subramanian, 1988). This shows that the insecticides and herbicides used in routine application in rice fields have important ecological effects in addition to those usually intended.

A large number of commercial grade insecticides and pesticides have been used in studies for understanding their influence on various cyanobacteria in laboratory culture in the last two decades. These studies include growth responses and tolerance levels and changes in the physiological characteristics of cyanobacteria when exposed to insecticides and herbicides. In most of these studies, single species toxicity tests have been made and little emphasis have been paid to the diversity response among cyanobacterial taxa to various pesticides (Rath and Adhikary, 1994). It has been suggested that a variety of taxa should be used when making ecotoxicological impact assessment (Hamilton *et al.*, 1988). Further, though in several reports toxicity of pesticides to diazotrophic cyanobacteria are published (Ahmad and Venkataraman, 1973; Singh, 1978; Da Silva *et al.*, 1975; Kolte and Goyal, 1992; Das and Adhikary, 1996; Anand and Subramanian, 1997; Kaur *et al.*, 1997) seldom toxicity studies have been made taking into account of cyanobacterial species isolated from rice field habitats.

In the paddy cultivation in Orissa state the organophosphate, Balwan (monocrotophos, 36 SL) and organochlorine, Thiokill (endosulfan, 35 EC) insecticides and the herbicide, Hiltachlor (50 EC) are widely used. Nothing is known regarding the tolerance of diazotrophic cyanobacteria of the rice fields to the commercial formulation of these chemicals and toxic effects of these chemicals on them. Recently one hundred and twenty species of heterocystous cyanobacteria belonging to the genera *Anabaena, Calothrix, Cylindrospermum, Nostoc, Aulosira, Scytonema, Tolypothrix, Rivularia, Hapalosiphon* and *Westiellopsis* have been isolated from the rice fields of different agroclimatic regions of Orissa state and their growth characteristics, nitrogen fixing ability together with strain history has been documented (Nayak *et al.*, 1996). In order to find out the insecticides and herbicide tolerance capacity of several of these cyanobacteria, eight species belonging to four different genera were selected and their growth response, photosynthetic oxygen evolution, respiratory oxygen uptake and acetylene reduction activity to different concentrations of Balwan (monocrotophos, 36 SL), Thiokill (endosulfan, 35 EC) and Hiltachlor (50EC) was studied.

MATERIALS AND METHODS

Eight species of heterocystous cyanobacteria belonging to the genera *Anabaena, Nostoc, Calothrix* and *Aulosira* isolated from rice field soils of Orissa state were used as the experimental material. Selection of these eight species, two species from each genus, was based on their relatively better growth rate, wider occurrence in different agroclimatic zones and nitrogenase activity .Two commercial grade insecticides, e.g. Balwan (monocrotophos, 36 SL) (Rallis India Ltd., Bombay) and Thiokill (endosulfan, 35 EC) (United phosphorus Ltd., Bombay) and the herbicide, Hiltachlor (50 EC) (Hindustan Insecticides Ltd., New Delhi) were used in the investigation. Fresh stock solutions of these pesticides were prepared in double distilled water and added to the culture medium to obtain the desired concentration. pH of all the media was adjusted to 7.8 prior to sterilization. Experiments were conducted in 15 x 150 mm Borosil test tubes containing 10 ml of nitrogen free BG 11medium (Rippka *et al.* 1979) and by inoculating equal amounts of homogenized culture suspension (absorbance of the inoculum of each organism from their exponential growth phase at 760 nm was 0.4 always). The medium contained various concentrations of Balwan: 1, 5, 10, 50, 100, 200, 300 and 400 ppm, Thiokill: 1, 2.5, 5, 10, 25, 50 and 75 ppm or Hiltachlor: 10, 50, 100, 200, 300, 400 and 500 ppm. Cultures were incubated at 25 ± 1°C under 7.5 W/m^2 light intensity up to 10 days. Growth was recorded as absorbance of the homogenized culture suspension at 760 nm in a Systronics 105 spectrophotometer. The EC 50 concentration of the insecticides and herbicide for each organism was inversely estimated from the regression lines of percentage inhibition of growth on log dose of the agrochemicals. Chlorophyll-*a* pigment of the cyanobacterial cells was extracted with 80% (v/v) acetone. Absorbance of the acetone extract was recorded at 660 nm in a Hitachi U-2000 UV visible spectrophotometer and the amount of chlorophyll-*a* was determined using extinction coefficient of Mackinney (1941). For measurement of the photosynthetic rate of O_2 evolution and respiratory O_2 uptake, healthy cyanobacterial cell suspension was taken in the periglass vessel of the Hitachi DW 1 oxygen electrode and exposed to various concentrations of the insecticides and herbicide. Changes in the rate of respiration and photosynthesis was measured soon after their exposure to the agrochemicals. Acetylene reduction activity of the pesticide treated cultures was measured after 24 hours and 10 days of incubation using a Aimil Nucon Gas Chromatograph with FID detector and Porapak-R column. Triplicates were set up for each experiment. The mean value ± S.D. of triplicate determinations has been given in the text.

RESULTS

Growth response of eight different species of heterocystous cyanobacteria to different concentrations of Balwan (monocrotophos, 36 SL), Thiokill (endosulfan, 35 EC) and the herbicide, Hiltachlor (50 EC) is given in Tables 1 to 3. Among the eight species tested against commercial grade monocrotophos (36 SL), both the species of *Anabaena, Nostoc* sp. UU29130, *Calothrix* sp. UU

29139 and *Aulosira* sp. UU 25118 tolerated up to 300 ppm, whereas the other three species survived in presence of up to 200 ppm of the insecticide. The lethal concentration of the insecticide to the respective species of cyanobacteria was 400 and 300 ppm in sequence. Lower concentration of the insecticide (5 ppm) stimulated the growth of *Calothrix* sp. UU 29139 by 6 per cent over its control value. Growth of *Aulosira* sp. UU 25118 did not decrease in presence of up to 50 ppm of Balwan in the medium; its growth was increased by 36.6 per cent over the control when the medium contained 10 ppm of the insecticide. The organochlorine insecticide, Thiokill (endosulfan, 35 EC) was relatively more toxic than the organophosphate, Balwan. *Nostoc* sp. UU 29130 tolerated up to 50 ppm of Thiokill followed by *Aulosira* sp. UU 25118 (Table 2). Both the species of *Anabaena, Calothrix* sp. UU 29139 and *Aulosira* sp. UU 29147 tolerated up to 10 ppm of commercial grade endosulfan (35 EC), while the same dose was lethal for *Nostoc* sp. UU 24100 and *Calothrix* sp. UU 24112. The commercial grade herbicide Hiltachlor (50 EC) inhibited the growth of *Anabaena* sp. UU 2493, *Aulosira* sp. UU 29147 and both the *calothrix* species when the medium contained 200 ppm of the chemical. *Anabaena* sp. UU 147 tolerated upto 200 ppm Hiltachlor, *Nostoc* sp. UU 24100 up to 300 ppm, *Nostoc* sp. 29130 and *Aulosira* sp. UU 25118 up to 400 ppm of the herbicide. Further increase in the dose of agrochemical was lethal. Lower concentration of the Hiltachlor up to 100 ppm did not decrease growth of *Aulosira* UU 25118 over the control rather at lower concentration of 10 ppm increased the growth of the organism by 30 per cent than the control. Presence up to 200 ppm of the herbicide did not show any adverse effect on its growth, however, further increase of the concentration in the medium showed adverse effect and 500 ppm of the chemical was lethal for all the organisms (Table 3).

Table 1. Growth response (absorbance of the culture at 760 nm) of several species of nitrogen fixing cyanobacteria of rice fields to the organophosphate insecticide, Balwan (monocrotophos, 36 SL). Cultures were incubated at 25 ± 1°C under continuous light 7.5 W/m^2 up to 10 days.

Organism	Concentration of Balwan (Monocrotophos, 36 SL), ppm							
	Control	5	10	50	100	200	300	400
Anabaena sp. UU 2493	0.44 ± 0.06	0.35 ± 0.04	0.26 ± 0.05	0.2 ± 0.02	0.16 ± 0.02	0.09 ± 0.01	0.07 ± 0	L
Anabaena sp. UU 147	0.58 ± 0.04	0.56 ± 0.04	0.43 ± 0.05	0.31 ± 0.03	0.24 ± 0.04	0.2 ± 0.02	0.15 ± 0.02	L
Nostoc sp. UU 24100	0.71 ± 0.08	0.57 ± 0.06	0.52 ± 0.04	0.4 ± 0.03	0.26 ± 0.03	0.17 ± 0.01	L	
Nostoc sp. UU 29130	0.52 ± 0.03	0.46 ± 0.02	0.42 ± 0.03	0.32 ± 0.03	0.25 ± 0.02	0.17 ± 0.02	0.11 ± 0.01	L

Calothrix sp. UU 24112	0.34 ± 0.05	0.29 ± 0.04	0.26 ± 0.03	0.25 ± 0.03	0.22 ± 0.02	0.11 ± 0.02	L	
Calothrix sp. UU 29139	0.49 ± 0.03	0.52 ± 0.04	0.4 ± 0.02	0.24 ± 0.02	0.22 ± 0.02	0.17 ± 0.01	0.11 ± 0.01	L
Aulosira sp. UU 25118	0.3 ± 0.03	0.35 ± 0.03	0.41 ± 0.05	0.31 ± 0.03	0.27 ± 0.02	0.21 ± 0.02	0.13 ± 0.02	L
Aulosira sp. UU 29147	0.61 ± 0.05	0.55 ± 0.03	0.46 ± 0.03	0.31 ± 0.03	0.22 ± 0.02	0.11 ± 0	L	

Absorbance of culture of each species soon after inoculation into the experimental vessels at 760 nm = 0.04; Values represent mean of three replicates ± S.D., L = lethal concentration (filamentous structure lost and the organism was bleached).

Table 2. Growth response (absorbance of the culture at 760 nm) of several species of nitrogen fixing cyanobacteria of rice fields to the organochlorine insecticide, Thiokill (endosulfan, 35 EC). Cultures were incubated at 25 ± 1°C under continuous light 7.5 W/m^2 up to 10 days.

Organism	Concentration of Thiokill (endosulfan, 35 EC), ppm							
	Control	1	2.5	5	10	25	50	75
Anabaena sp. UU 2493	0.44 ± 0.06	0.3 ± 0.02	0.25 ± 0.04	0.08 ± 0.01	0.04 ± 0	L		
Anabaena sp. UU 147	0.58 ± 0.04	0.43 ± 0.03	0.12 ± 0.01	0.05 ± 0	0.04 ± 0.05	L		
Nostoc sp. UU 24100	0.71 ± 0.08	0.75 ± 0.07	0.5 ± 0.03	0.35 ± 0.04	L			
Nostoc sp. UU 29130	0.52 ± 0.03	0.48 ± 0.04	0.46 ± 0.02	0.39 ± 0.04	0.38 ± 0.03	0.35 ± 0.03	0.21 ± 0.03	L
Calothrix sp. UU 24112	0.34 ± 0.05	0.33 ± 0.04	0.22 ± 0.03	0.06 ± 0	L			
Calothrix sp. UU 29139	0.49 ± 0.03	0.37 ± 0.03	0.16 ± 0.02	0.14 ± 0.01	0.08 ± 0	L		
Aulosira sp. UU 25118	0.3 ± 0.03	0.28 ± 0.03	0.26 ± 0.02	0.23 ± 0.02	0.21 ± 0.01	0.18 ± 0.01	L	
Aulosira sp. UU 29147	0.61 ± 0.05	0.49 ± 0.04	0.4 ± 0.05	0.27 ± 0.03	0.12 ± 0	L		

Absorbance of culture of each species soon after inoculation into the experimental vessels at 760 nm = 0.04; Values represent mean of three replicates ± S.D., L = lethal concentration (filamentous structure lost and the organism was bleached).

Table 3. Growth response (absorbance of the culture at 760 nm) of several species of nitrogen fixing cyanobacteria of rice fields to the herbicide, Hiltachlor (50 EC). Cultures were incubated at 25 ± 1°C under continuous light 7.5 W/m² up to 10 days.

Organism	Concentration of Hiltachlor (50 EC), ppm							
	Control	10	50	100	200	300	400	500
Anabaena sp. UU 2493	0.44 ± 0.06	0.43 ± 0.03	0.23 ± 0.01	0.11 ± 0	L			
Anabaena sp. UU 147	0.58 ± 0.04	0.45 ± 0.03	0.4 ± 0.01	0.37 ± 0.02	0.09 ± 0	L		
Nostoc sp. UU 24100	0.71 ± 0.08	0.55 ± 0.02	0.5 ± 0.01	0.44 ± 0.04	0.33 ± 0.03	0.12 ± 0.01	L	
Nostoc sp. UU 29130	0.52 ± 0.03	0.5 ± 0.005	0.48 ± 0.04	0.45 ± 0	0.32 ± 0.03	0.24 ± 0.05	0.07 ± 0.01	L
Calothrix sp. UU 24112	0.34 ± 0.05	0.25 ± 0	0.14 ± 0.01	0.11 ± 0.05	L			
Calothrix sp. UU 29139	0.49 ± 0.03	0.2 ± 0.005	0.15 ± 0.01	0.08 ± 0	L			
Aulosira sp. UU 25118	0.3 ± 0.03	0.39 ± 0.01	0.35 ± 0.05	0.31 ± 0.03	0.28 ± 0	0.27 ± 0.02	0.12 ± 0	L
Aulosira sp. UU 29147	0.61 ± 0.05	0.22 ± 0.02	0.18 ± 0.01	0.08 ± 0.05	L			

Absorbance of culture of each species soon after inoculation into the experimental vessels at 760 nm = 0.04; Values represent mean of three replicates ± S.D., L = lethal concentration (filamentous structure lost and the organism was bleached).

EC 50 and lethal concentrations of these three pesticides to eight different species of cyanobacteria was calculated and given in Figs. 1 to 3. Lethal dose of Balwan though varied from 300 to 400 ppm for almost all the species, EC 50 dose of the same chemical showed a wide range of variation between 30 to 300 ppm. *Anabaena* sp. UU 2493 was the most sensitive organism to Balwan, whose 50 per cent cells killed in presence of about 30 ppm of the insecticide. EC 50 dose of *Calothrix* sp. UU 29139 and *Aulosira* sp. UU 29147 was about 50 ppm, for *Anabaena* sp. UU 147 and *Nostoc* sp. UU 24100 about 70 ppm, for *Nostoc* sp. UU 29130 about 90 ppm, for *Calothrix* sp. UU 24112 nearly 150 ppm and the same dose for *Aulosira* sp. UU 25118 was 250 ppm of the agrochemical. Recommended dose of the insecticide under rice field condition was 4 ppm which did not affect adversely to any of the organisms under our experimental condition. A wide range of variation was observed with regard to EC 50 and lethal dose of Thiokill (endosulfan, 35 EC) for different cyanobacteria. In *Anabaena* sp. UU 147, *Calothrix* sp. UU 29139, *Anabaena* sp. UU 2493 and *Calothrix* sp. UU 24112, EC 50 dose of the organochlorine insecticide was about 1.7, 1.9, 2.8 and 3.2 ppm respectively, which was much less than its recommended dose, of 4 ppm. Even at its recommended dose, about 50 per cent cells of *Aulosira* sp. UU 29147 and *Nostoc* sp. UU 24100 were killed in the culture. However EC 50 dose of *Aulosira* sp. UU 25118 and *Nostoc* sp. 29130 was much higher than the other organisms. EC 50 dose of Thiokill for these organisms was 31 and 38 ppm respectively where other organisms could not survive in culture. Similar wide range of variation in the tolerance of different cyanobacteria to the herbicide, Hiltachlor (50 EC) was also observed. Fifty per cent inhibition in growth of *Calothrix* sp. UU 29139 and *Aulosira* sp. UU 29147 was observed in presence of the recommended dose of 10 ppm Hiltachlor in the medium. EC 50 dose of *Anabaena* sp. UU 2493 and *Calothrix* sp. UU 24112 was about 50 ppm of the herbicide. EC 50 dose for *Anabaena* sp. UU 147 was 88 ppm. Both the species of *Nostoc* and *Aulosira* sp. UU 25118 tolerated higher concentration of the herbicide. EC 50 dose of the chemical for *Nostoc* sp. UU 24100 was about 200 ppm and for *Nostoc* sp. UU 29130 was nearly 300 ppm, at which concentration most of the organisms except *Aulosira* sp. UU 25118 were killed in the culture. EC 50 dose of the latter species was about 400 ppm and lethal dose of the chemical for the same organism was 500 ppm.

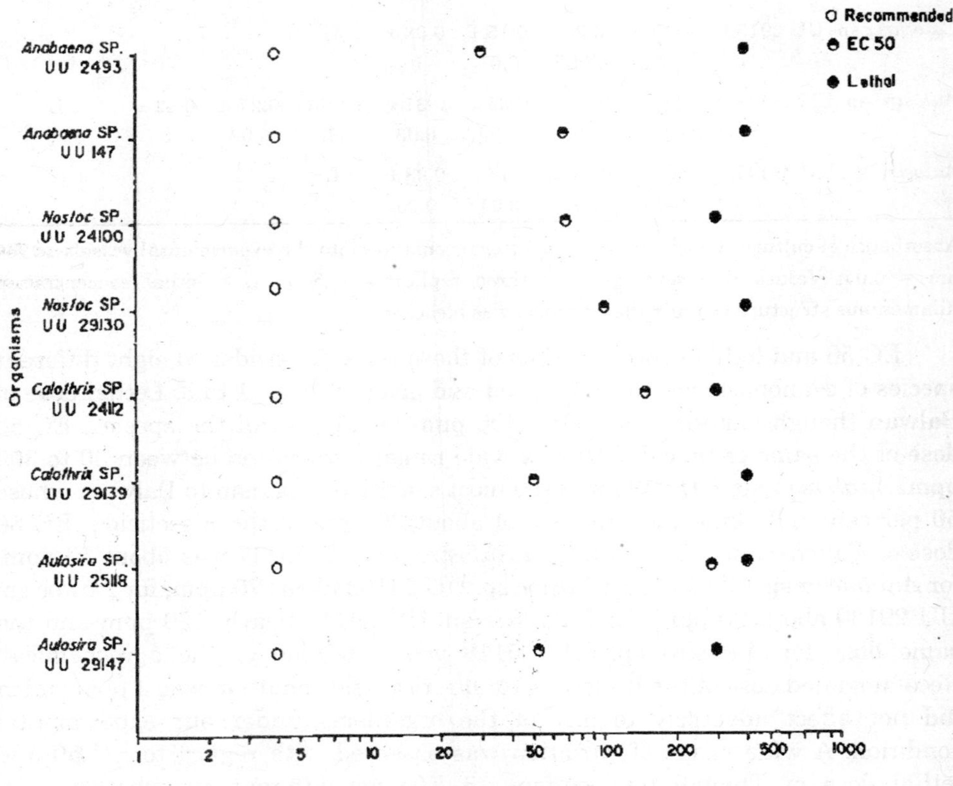

Fig. 1. Recommended, EC 50 and lethal dose (ppm) of Monocrotophos (36 SL) on the growth of eight different species of cyanobacteria.

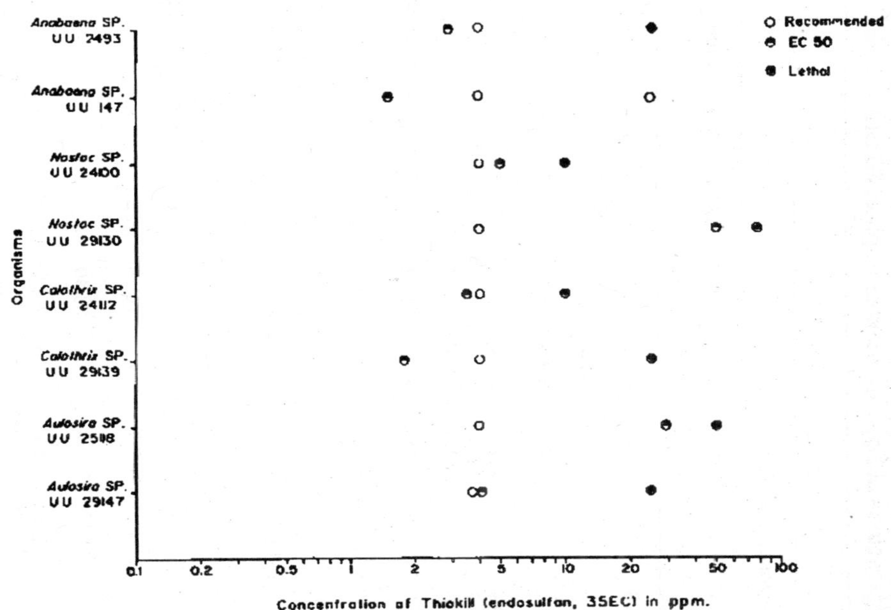

Fig. 2. Recommended, EC 50 and lethal dose (ppm) of Endosulphan (35 EC) on the growth of eight different species of cyanobacteria.

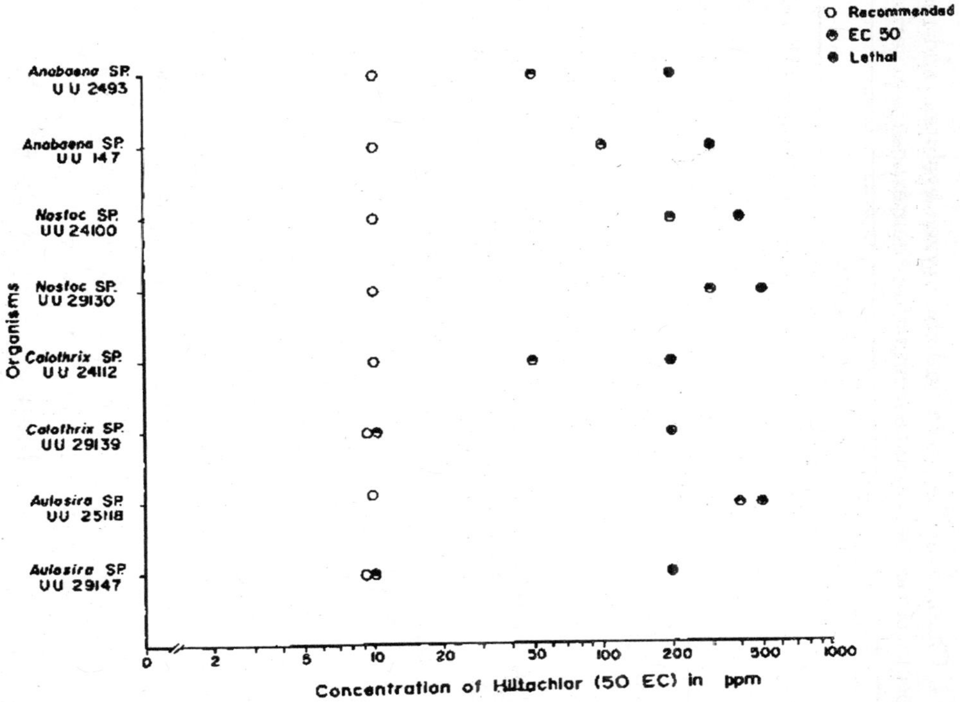

Fig. 3. Recommended, EC 50 and lethal dose (ppm) of Hiltachlor (50 EC) on the growth of eight different species of cyanobacteria.

Table 4. Changes in the photosynthetic oxygen evolution and respiratory oxygen uptake (ì mol. O_2. mg chl-a^{-1}.h^{-1}) of several species of rice field cyanobacteria after their exposure to various concentrations of Balwan (monocrotophos, 36 SL)

Organism	Photosynthesis								Respiration							
	Concentration of Balwan (monocrotophos, 36 SL), ppm															
	Control	5	10	50	100	200	300	400	Control	5	10	50	100	200	300	400
Anabaena sp.	270.76		219.74	160.88	164.80		153.03	0	50.76		76.71	80.98	84.59		88.83	0
UU 2493	± 2.1		± 1.4	± 1.1	± 2.7		± 2.3		± 4.1		± 2.3	± 3.3	± 4.6		± 3.6	
			(-18.8)	(- 40.5)	(-39.1)		(-43.4)				(+51.1)	(+59.5)	(+66.6)		(+75)	
Anabaena sp.	141.71		474.43	345.03	262.65		118.91	0	80.98		242.94	80.98	70.15		25.90	0
UU 147	± 1.3		± 5.6	± 3.2	± 2.6		± 2.7		± 6.7		± 1.8	± 4.3	± 3.7		± 1.1	
			(+ 234.7)	(+143.4)	(+85.3)		(-16)				(+200)	(+ 0)	(-13.3)		(-68.01)	
Nostoc sp.	212.13	175.76	127.27		163.64	193.94	0		141.42	311.14	254.58		182.27	226.29	0	
UU 24100	± 4.1	± 3.1	± 2.1		± 1.8	± 2.7			± 5.3	± 3.1	± 2.2		± 6.5	± 2.2		
		(-17.1)	(- 40)		(-22.8)	(-8.5)				(+120.01)	(+80.01)		(+28.88)	(+60.01)		
Nostoc sp.	279.22	421.24	492.19		478.89	461.15	251.1	0	101.53	156.91	161.52		152.29	124.60	60.90	0
UU 29130	± 4.6	± 5.1	± 5.3		± 3.4	± 2.3	± 2.1		± 7.6	± 8.2	± 3.6		± 2.6	± 1.3	± 1.2	
		(+50.8)	(+76.2)		(+ 71.5)	(+65.1)	(-10)			(+54.5)	(+59.08)		(+49.9)	(+22.7)	(-40.01)	
Calothrix sp.	144.1	144.1	182.76		115.98	70.29	0		72.05	72.20	74.93		63.40	86.46	0	
UU 24112	± 1.5	± 3.7	± 2.3		± 3.7	± 2.1			± 4.3	± 1.8	± 2.1		± 2.7	± 1.5		
		(± 0)	(+26.8)		(-19.5)	(-51.2)				(+0.20)	(+3.9)		(-12)	(+20)		

Calothrix sp.	170.41	143.13	136.32	143.14	95.41		17.00	0	92.95	56.80	77.45	149.75	191.50		18.58	0
UU 29139	± 1.7	± 4.2	± 1.4	± 1.1	± 3.1		± 1.1		± 3.7	± 3.5	± 2.1	± 1.2	± 1.3		± 1.3	
		(- 16)	(-20)	(-16)	(- 44)		(-90)			(-38.8)	(-16.6)	(+ 61.1)	(+106.02)		(- 80.01)	
Aulosira sp.	317.64	283.30	283.30	279.00	276.11		274.71	0	158.82	254.11	301.76	314.17	333.53		349.41	0
UU 25118	± 2.8	± 1.7	± 2.2	± 2.1	± 1.7		± 2.7		± 8.5	± 2.3	± 2.8	± 3.1	± 3.5		± 2.5	
		(- 10.8)	(-10.8)	(-12.1)	(-13)		(-13.5)			(+59.9)	(+90)	(+ 97.8)	(+110)		(+120)	
Aulosira sp.	298.18	217.59	177.29	89.12	53.68	94.06	0		139.15	139.23	164.91	139.15	50.10	16.60	0	
UU 29147	± 2.1	± 1.8	± 1.3	± 1.1	± 2.1	± 1.3			± 2.8	± 1.3	± 1.2	± 1.3	± 1.4	± 1.2		
		(-27)	(-40.5)	(-70.1)	(-81.9)	(-94)				(+0.05)	(+18.5)	(± 0)	(-63.9)	(-88.07)		

Figures in parenthesis represent per cent increase or decrease of photosynthetic oxygen evolution and respiratory oxygen uptake by the experimental organisms after exposure to various concentration of the insecticide over their respective control value.

Table 5. Changes in the photosynthetic oxygen evolution and respiratory oxygen uptake (ì mol. O_2. mg chl-a^{-1}.h^{-1}) of several species of rice field cyanobacteria after their exposure to various concentrations of Thiokill (endosulfan, 35 EC)

Organism	Photosynthesis								Respiration							
	Concentration of Thiokill (endosulfan, 35 EC), ppm															
	Control	1	2.5	5	10	25	50	75	Control	1	2.5	5	10	25	50	75
Anabaena sp.	270.76	179.75	188.35	196.20	219.74	0			50.76	49.95	67.67	84.59	131.13	0		
UU 2493	± 2.1	± 2.9	± 1.8	± 2.8	± 3.1				± 4.1	± 2.5	± 2.3	± 3.3	± 1.7			
		(- 33.6)	(- 30.4)	(- 27.5)	(- 18.8)					(- 1.5)	(+ 33.3)	(+ 66.6)	(+ 158.3)			
Anabaena sp.	141.71	92.41	43.12	31.18	17.00	0			80.98	80.98	101.22	69.64	38.87	0		
UU 147	± 1.3	± 2.7	± 3.7	± 1.2	± 1.1				± 6.7	± 3.3	± 4.1	± 1.9	± 1.3			
		(- 34.7)	(- 69.5)	(- 77.9)	(- 88)					(± 0)	(+ 24.9)	(- 14)	(- 52)			
Nostoc sp.	212.13	193.94	193.94	296.98	0				141.42	226.29	198.00	17.96	0			
UU 24100	± 4.1	± 4.5	± 1.5	± 3.3					± 5.3	± 6.1	± 3.5	± 1.1				
		(- 8.5)	(- 8.5)	(+ 39.9)						(+ 60.01)	(+ 40)	(- 87.3)				
Nostoc sp.	279.22	319.25	366.15	557.71	292.65		58.61 ±	0	101.53	101.66	96.91	60.01	70.00		14.21	0
UU 29130	± 4.6	± 3.5	± 2.9	± 3.6	± 1.6		1.3		± 7.6	± 5.3	± 2.1	± 4.2	± 4.1		± 1.4	
		(+ 14.3)	(+ 31.1)	(+ 99.7)	(+ 4.8)		(- 79)			(- 0.12)	(- 4.5)	(- 40.8)	(- 31.05)		(- 86)	
Calothrix sp.	144.1	182.76	200.33	158.16	0				72.05	115.28	132.57	129.69	0			
UU 24112	± 1.5	± 1.3	± 2.9	± 2.7					± 4.3	± 5.3	± 2.7	± 3.7				
		(+ 26.8)	(+ 39.0)	(+ 9.7)						(+ 60)	(+ 83.9)	(+ 80)				
Calothrix sp.	170.41	156.77	184.03	127.79	74.99	0			92.95	201.52	242.70	171.00	193.29	0		

UU 29139	± 1.7	± 1.9	± 2.1	± 1.4	± 1.1		± 3.7	± 2.1	± 2.7	± 1.7	± 2.1	
		(- 8)	(+ 7.9)	(- 25)	(- 55.9)			(+ 116.8)	(+ 161.1)	(+ 83.9)	(+ 107.9)	
Aulosira sp.	317.64	154.52	154.53	60.09	44.45	0	158.82	158.82	190.59	254.11	336.70	0
UU 25118	± 2.8	± 1.6	± 1.3	± 2.1	± 1.2		± 8.5	± 3.7	± 1.8	± 2.2	± 2.1	
		(- 51.3)	(- 51.3)	(- 81)	(- 86)			(± 0)	(+ 20)	(+ 59.9)	(+ 112)	
Aulosira sp.	298.18	249.82	177.29	125.20	71.57	0	139.15	159.76	175.22	121.00	69.52	0
UU 29147	± 2.1	± 5.3	± 1.9	± 1.4	± 1.3		± 2.8	± 2.5	± 4.5	± 1.1	± 1.2	
		(- 16.2)	(- 40.5)	(- 58)	(- 75.9)			(+ 14.8)	(+ 25.9)	(- 13.04)	(- 50.03)	

Figures in parenthesis represent per cent increase or decrease of photosynthetic oxygen evolution and respiratory oxygen uptake by the experimental organisms after exposure to various concentration of the insecticide over their respective control value.

Table 6. Changes in the photosynthetic oxygen evolution and respiratory oxygen uptake (ì mol. O_2. mg chl-a^{-1}.h^{-1}) of several species of rice field cyanobacteria after their exposure to various concentrations of Hiltachlor (50 EC).

Organism	Photosynthesis								Respiration							
	Concentration of Hiltachlor (50 EC), ppm															
	Control	10	50	100	200	300	400	500	Control	10	50	100	200	300	400	500
Anabaena sp.	270.76	219.99	203.00	0					50.76	84.61	89.50	0				
UU 2493	± 2.1	± 2.9	± 1.7						± 4.1	± 2.1	± 2.5					
		(- 18.7)	(- 25)							(+ 66.6)	(+ 76.3)					
Anabaena sp.	141.71	182.2	53.85	0					80.98	114.99	36.44	0				
UU 147	± 1.3	± 3.1	± 1.2						± 6.7	± 5.2	± 1.3					
		(+ 28.5)	(- 61.9)							(+ 41.9)	(- 55)					
Nostoc sp.	212.13	141.42	141.42	141.42	141.42	0			141.42	74.33	107.47	123.00	141.42	0		
UU 24100	± 4.1	± 1.5	± 1.7	± 1.2	± 1.4				± 5.3	± 4.6	± 1.4	± 3.7	± 7.3			
		(- 33.3)	(- 33.3)	(- 33.3)	(- 33.3)					(- 47.44)	(- 24)	(- 13.02)	(± 0)			
Nostoc sp.	279.22	380.76	330.87	309.93	290.38	279.22	203.07	0	101.53	177.69	254.85	244.69	233.52	126.92	126.92	0
UU 29130	± 4.6	± 4.7	± 3.3	± 3.6	± 2.8	± 3.3	± 3.1		± 7.6	± 1.2	± 2.3	± 2.1	± 1.8	± 1.3	± 2.7	
		(+ 36.3)	(+ 18.4)	(+ 10.9)	(+ 3.9)	(± 0)	(- 27)			(+ 75.01)	(+ 151)	(+ 141)	(+ 130)	(+ 25)	(+ 25)	
Calothrix sp.	144.1	144.1	144.1	72.05	0				72.05	100.87	100.87	72.05	0			
UU 24112	± 1.5	± 5.2	± 1.2	± 1.4					± 4.3	± 3.3	± 2.6	± 5.1				
		(± 0)	(± 0)	(- 50)						(+ 40)	(+ 40)	(± 0)				

Calothrix sp.	170.41	170.41	123.94	0					92.95	123.94	123.94	0				
UU 29139	± 1.7	± 2.6	± 5.7						± 3.7	± 1.3	± 1.2					
		(± 0)	(- 27.2)							(+ 33.3)	(+ 33.3)					
Aulosira sp.	317.64	176.46	173.12	173.12	170.01	170.01	158.82	0	158.82	88.23	96.88	100.06	103.22	105.88	104.33	0
UU 25118	± 2.8	± 2.9	± 1.5	± 1.8	± 2.2	± 3.1	± 4.1		± 8.5	± 2.3	± 1.2	± 1.4	± 2.1	± 3.7	± 2.7	
		(- 44.4)	(- 45.4)	(- 45.4)	(- 46.4)	(- 46.4)	(- 50)			(- 44.4)	(- 39)	(- 36.9)	(- 35)	(- 33.3)	(- 34.3)	
Aulosira sp.	298.18	218.66	135.63	99.39	0				139.15	139.15	111.31	99.39	0			
UU 29147	± 2.1	± 3.7	± 2.3	± 3.5					± 2.8	± 4.3	± 2.6	± 2.2				
		(- 26.6)	(- 54.5)	(- 66.6)						(± 0)	(- 20)	(- 28.5)				

Figures in parenthesis represent per cent increase or decrease of photosynthetic oxygen evolution and respiratory oxygen uptake by the experimental organisms after exposure to various concentration of the herbicide over their respective control value.

Table 7. Nitrogenase activity (n mol. C_2H_4/ìg Chl-*a*/h) of several species of nitrogen fixing cyanobacteria of rice fields in presence of the organophosphate insecticide, Balwan (monocrotophos, 36 SL). Cultures were incubated at 25 ± 1°C under 7.5 W/m^2 light intensity up to 24 hours or 10 days.

Organism	24 hours						10 days					
	Concentration of Balwan (monocrotophos, 36 SL)											
	Control	Recommended dose (4 ppm)	EC 50 dose		Sublethal dose		Control	Recommended dose (4 ppm)	EC 50 dose		Sublethal dose	
Anabaena sp. UU 2493	3.47 ± 0.1	3.58 ± 0.3 (+ 3.1)	4.2 ± 0.3 (+ 21)	[30]	2.34 ± 0.02 (- 32.5)	[300]	0.41 ± 0.03	0.24 ± 0.01 (- 41.5)	0.23 ± 0.02 (- 43.9)	[30]	0.04 ± 0.005 (- 90.2)	[300]
Anabaena sp. UU 147	0.18 ± 0.01	0.2 ± 0.02 (+ 16.6)	0.17 ± 0.01 (- 5.5)	[62]	0	[300]	2.16 ± 0.02	0.61 ± 0.05 (- 71.7)	0.4 ± 0.02 (- 81.4)	[62]	0.07 ± 0 (- 96.7)	[300]
Nostoc sp. UU 24100	0.54 ± 0.03	0.51 ± 0.05 (- 5.5)	0.46 ± 0.02 (- 14.8)	[70]	0.27 ± 0.01 (- 50)	[200]	1.72 ± 0.1	1.86 ± 0.01 (+ 8.1)	1.72 ± 0.1 (± 0)	[70]	0.59 ± 0.04 (- 65.6)	[200]
Nostoc sp. UU 29130	0.98 ± 0.005	0.75 ± 0.05 (- 23.4)	0.59 ± 0.03 (- 39.7)	[100]	0	[300]	2.96 ± 0.1	4.24 ± 0.02 (+ 43.2)	1.74 ± 0.1 (- 41.2)	[100]	0.45 ± 0.04 (- 84.7)	[300]
Calothrix sp. UU 24112	2.96 ± 0.01	1.17 ± 0.1 (- 60.4)	1.08 ± 0.01 (- 63.5)	[150]	0.72 ± 0.05 (- 75.6)	[200]	5.14 ± 0.3	5.35 ± 0.3 (+ 4)	4.78 ± 0.2 (- 7)	[150]	1.65 ± 0.05 (- 67.8)	[200]
Calothrix sp. UU 29139	0.12 ± 0.01	0.12 ± 0.01 (± 0)	0.12 ± 0.05 (± 0)	[49]	0.01 ± 0.001 (- 91.6)	[300]	12.45 ± 0.7	9.37 ± 0.5 (- 24.7)	5.61 ± 0.3 (- 54.9)	[49]	0.63 ± 0.06 (- 94.9)	[300]
Aulosira sp. UU 25118	0.53 ± 0.05	0.1 ± 0.005 (- 81.1)	0.01 ± 0.001 (- 98.1)	[290]	0.01 ± 0.001 (- 98.1)	[350]	2.66 ± 0.3	3.69 ± 0.1 (+ 38.7)	1.35 ± 0.1 (- 49.2)	[290]	1.35 ± 0.02 (- 49.2)	[350]
Aulosira sp. UU 29147	2.12 ± 0.02	2.67 ± 0.2 (+ 25.9)	2.65 ± 0.01 (+ 25)	[55]	1.49 ± 0.01 (- 29.7)	[200]	0.78 ± 0.05	0.69 ± 0.02 (- 11.5)	0.63 ± 0.05 (- 19.2)	[55]	0.12 ± 0.01 (- 84.6)	[200]

Figures in parenthesis represent per cent increase or decrease of ARA by the experimental organisms after exposure to various concentration of the insecticide over their respective control values. Figures in brackets under EC 50 and sublethal dose column represent the EC 50 and sublethal concentrations of the insecticide for respective organism

Table 8. Nitrogenase activity (n mol. C_2H_4/ìg Chl-*a*/h) of several species of nitrogen fixing cyanobacteria of rice fields in presence of the organochlorine insecticide, Thiokill (endosulfan, 35 EC). Cultures were incubated at 25 ± 1°C under 7.5 W/m^2 light intensity up to 24 hours or 10 days.

Organism	24 hours						10 days					
	Concentration of Thiokill (endosulfan, 35 EC)											
	Control	Recommen-ded dose (4 ppm)	EC 50 dose		Sublethal dose		Control	Recommen-ded dose (4 ppm)	EC 50 dose		Sublethal dose	
Anabaena sp. UU 2493	3.47 ± 0.1	2.22 ± 0.2 (- 36)	3.33 ± 0.2 (- 4)	[2.8]	0.007 ± 0.005 (- 99.7)	[10]	0.41 ± 0.03	0.35 ± 0.03 (- 14.6)	0.4 ± 0.02 (- 2.4)	[2.8]	0.18 ± 0.01 (- 56)	[10]
Anabaena sp. UU 147	0.18 ± 0.01	0.15 ± 0.01 (- 16.6)	0.27 ± 0.01 (+ 50)	[1.6]	0	[10]	2.16 ± 0.02	0	1.1 ± 0.1 (- 49)	[1.6]	0	[10]
Nostoc sp. UU 24100	0.54 ± 0.03	0.65 ± 0.03 (+ 20.3)	0.49 ± 0.04 (- 9.2)	[5]	0.04 ± 0.005 (- 92.5)	[9]	1.72 ± 0.1	1.29 ± 0.1 (- 25)	0.97 ± 0.05 (- 43.6)	[5]	0.15 ± 0.01 (- 91.2)	[9]
Nostoc sp. UU 29130	0.98 ± 0.005	0.69 ± 0.05 (- 29.5)	0.29 ± 0.01 (- 70.4)	[41]	0.15 ± 0.01 (- 84.6)	[60]	2.96 ± 0.1	3.08 ± 0.2 (+ 4)	1.18 ± 0.1 (- 60.1)	[41]	0.61 ± 0.05 (- 79.3)	[60]
Calothrix sp. UU 24112	2.96 ± 0.01	2.05 ± 0.02 (- 30.7)	3.09 ± 0.02 (+ 4.3)	[3.3]	1.39 ± 0.1 (- 53)	[5]	5.14 ± 0.3	2.85 ± 0.2 (- 44.5)	4.47 ± 0.3 (- 13)	[3.3]	2.11 ± 0.1 (- 58.9)	[5]
Calothrix sp. UU 29139	0.12 ± 0.01	0.11 ± 0.01 (- 8.3)	0.12 ± 0.005 (± 0)	[1.8]	0.004 ± 0.001 (- 96.6)	[10]	12.45 ± 0.7	2.38 ± 0.1 (- 80.8)	4.88 ± 0.3 (- 60.8)	[1.8]	1.44 ± 0.1 (- 88.4)	[10]
Aulosira sp. UU 25118	0.53 ± 0.05	0.55 ± 0.03 (+ 3.7)	0.11 ± 0.05 (- 79.2)	[26]	0.04 ± 0 (- 92.4)	[40]	2.66 ± 0.3	3.27 ± 0.2 (+ 22.9)	2.99 ± 0.01 (+ 12.4)	[29]	0.85 ± 0.04 (- 68)	[40]
Aulosira sp. UU 29147	2.12 ± 0.02	2.51 ± 0.1 (+ 18.3)	1.23 ± 0.01 (- 41.9)	[4.1]	0.05 ± 0.002 (- 97.6)	[10]	0.78 ± 0.05	0.69 ± 0.04 (- 11.5)	0.58 ± 0.05 (- 25.6)	[4.1]	0.32 ± 0.3 (- 58.9)	[10]

Figures in parenthesis represent percent increase or decrease of ARA by the experimental organisms after exposure to various concentration of the insecticide over their respective control values. Figures in brackets under EC 50 and sublethal dose column represent the EC 50 and sublethal concentrations of the insecticide for respective organism.

Table 9: Nitrogenase activity (n mol. C_2H_4/ìg Chl-*a*/h) of several species of nitrogen fixing cyanobacteria of rice fields in presence of the herbicide, Hiltachlor (50 EC). Cultures were incubated at 25 ± 1°C under 7.5 W/m² light intensity up to 24 hours or 10 days.

Organism	24 hours						10 days					
	Concentration of Hiltachlor (50 EC)											
	Control	Recommended dose (4 ppm)	EC 50 dose		Sublethal dose		Control	Recommended dose (4 ppm)	EC 50 dose		Sublethal dose	
Anabaena sp. UU 2493	3.47 ± 0.1	0	0	[50.6]	0	[100]	0.41 ± 0.03	0.38 ± 0.02 (- 7.3)	0.33 ± 0.02 (- 19.5)	[50.6]	0	[100]
Anabaena sp. UU 147	0.18 ± 0.01	0.17 ± 0.01 (- 5.5)	0	[10]	0	[200]	2.16 ± 0.02	3.53 ± 0.03 (+ 63.4)	1.37 ± 0.01 (- 36.6)	[10]	0	[200]
Nostoc sp. UU 24100	0.54 ± 0.03	0.88 ± 0.05 (- 62.7)	0	[30]	0	[300]	1.72 ± 0.1	1.22 ± 0.01 (- 29.1)	0.24 ± 0.02 (- 86.04)	[30]	0	[300]
Nostoc sp. UU 29130	0.98 ± 0.005	1.08 ± 0.08 (+ 10.2)	0.72 ± 0.05 (- 26.5)	[120]	0.36 ± 0.04 (- 63.2)	[400]	2.96 ± 0.1	2.36 ± 0.02 (- 20.3)	1.18 ± 0.08 (- 60.1)	[120]	0.88 ± 0.05 (- 70.2)	[400]
Calothrix sp. UU 24112	2.96 ± 0.01	1.48 ± 0.06 (- 50)	1.18 ± 0.1 (- 60.1)	[10]	0.59 ± 0.05 (- 80.1)	[100]	5.14 ± 0.3	4.11 ± 0.03 (- 20)	1.61 ± 0.1 (- 68.6)	[10]	0.75 ± 0.04 (- 85.4)	[100]
Calothrix sp. UU 29139	0.12 ± 0.01	0.05 ± 0.005 (- 58.3)	0.04 ± 0.001 (- 66.6)	[10]	0	[100]	12.45 ± 0.7	14.52 ± 0.08 (+ 16.6)	12.45 ± 0.09 (± 0)	[10]	0	[100]
Aulosira sp. UU 25118	0.53 ± 0.05	0.26 ± 0.03 (- 51)	0	[400]	0	[400]	2.66 ± 0.3	2.66 ± 0.01 (± 0)	1.1 ± 0.1 (- 58.6)	[400]	0	[400]
Aulosira sp. UU 29147	2.12 ± 0.02	0	0	[10]	0	[100]	0.78 ± 0.05	1.04 ± 0.01 (+ 33.3)	0.26 ± 0.01 (- 66.6)	[100]	0	[100]

Figures in parenthesis represent percent increase or decrease of ARA by the experimental organisms after exposure to various concentration of the herbicide over their respective control values. Figures in brackets under EC 50 and sublethal dose column represent the EC 50 and sublethal concentrations of the herbicide for respective organism.

Changes in the rate of photosynthetic oxygen evolution and respiratory oxygen uptake of the experimental organisms soon after their exposure to various concentrations of the insecticides and herbicide is shown in Tables 4 to 6. Result showed that exposure of most of the organisms except. *Nostoc* sp. UU 24100, *Calothrix* sp. UU 24112 and *Aulosira* sp. UU 29147 to 400 ppm of Balwan completely suppressed their photosynthetic oxygen evolution capabilities. The rate of photosynthesis of the rest three species was ceased when were exposed to 300 ppm of Balwan (monocrotophos, 36 SL). Respiratory oxygen uptake was also ceased at similar concentration of the insecticide which inhibited the photosynthesis of the respective organisms, however, respiratory oxygen uptake was higher in *Anabaena* sp. UU 2493, *Nostoc* sp. UU 24100, *Calothrix* sp. UU 24112 and *Aulosira* sp. UU 25118 over the control value when the organism was exposed to the dose prior to the lethal dose of the insecticide. When *Nostoc* sp. UU 24100 and *Calothrix* sp. UU 24112 were exposed to even 10 ppm of Thiokill their photosynthesis was completely ceased. Except that of *Nostoc* sp. UU 29130 which continued to photosynthesize even in presence of 50 ppm of Thiokill, photosynthesis of all other species were ceased soon after their exposure to 25 ppm of the insecticide. Similar was the result of the inhibition of the insecticide on the rate of respiratory oxygen uptake of the respective species in presence of this organochlorine insecticide. Respiratory oxygen uptake increased several fold over the control in *Anabaena* sp. UU 2493, *Calothrix* sp. UU 29139 and *Aulosira* sp. UU 25118 when exposed to the concentrations prior to their lethal dose. Rate of photosynthetic oxygen evolution and respiratory oxygen uptake in presence of several concentrations of the herbicide, Hiltachlor is given in Table 6. *Nostoc* sp. UU 29130 and *Aulosira* sp. UU 25118 continued to photosynthesize and respire even when exposed to up to 400 ppm of Hiltachlor; lethal dose for both the species was 500 ppm of the herbicide. Lethal dose of the agrochemical to cease the photosynthesis and respiration for both the species of *Anabaena* and *Calothrix* sp. UU 29139 was 100 ppm which was several fold less than the former organisms. *Nostoc* sp. UU 24100 continue to photosynthesize and respire in presence of 200 ppm of Hiltachlor whereas the same concentration of the chemical ceased the metabolic rate of *Calothrix* sp. UU 24112 and *Aulosira* sp. UU 29147. In this case also respiratory oxygen uptake of *Anabaena* sp. UU 2493, *Nostoc* sp. UU 29130 and both the species of *Calothrix* was much higher than the control value in presence of the herbicide prior to the lethal doses.

Nitrogenase activity (acetylene reduction activity) of the experimental organisms after 24 hours or 10 days of exposure to recommended, EC 50 and sublethal dose of the insecticides Balwan and Thiokill and the herbicide, Hiltachlor has been given in Tables 7 to 9. Nitrogenase activity of *Nostoc* sp. UU 29130 and *Aulosira* sp. UU 25118 was stimulated up to 43 and 38 per cent respectively in presence of recommended dose of Balwan in the culture and incubated up to 10 days in the light. But in presence of EC 50 dose of the same insecticide, about 50 per cent inhibition of the ARA was observed under similar experimental conditions. Acetylene reduction activity of *Nostoc* sp. UU 24100 and *Calothrix* sp. UU 24112 was not adversely affected in presence of the recommended dose of the insecticide. Even in presence of EC 50 dose of the chemical, no sharp depression in their acetylene reduction activity was observed. Almost a similar trend that acetylene reduction activity was less effected in the insecticide sensitive *Nostoc* sp. UU 24100 and *Calothrix* sp. UU 24112 than the tolerant species, *Nostoc* sp. UU 29130 and *Aulosira* sp. UU 25118 up to their respective EC 50 dose was observed even within 24 hours of exposure of the organism to the commercial grade monocrotophos (36 SL). In *Nostoc* sp. UU 24100, which was sensitive to the insecticide, an inhibition of 5 and 14 per cent at the recommended and EC 50 dose, while in the resistant species, *Aulosira* sp. UU 25118, inhibition of ARA up to 81 and 98 per cent at the recommended and EC 50 dose of the same insecticide respectively was observed. Almost similar effect of Thiokill on nitrogenase activity of the experimental organisms was noticed. Though ARA of *Calothrix* sp. UU 24112 was decreased over control in presence of recommended dose of Thiokill, the agrochemical at its EC 50 dose did not cause any drastic inhibition in the ARA as that observed in case of *Nostoc* sp. UU 29130; in the latter species 60 per cent inhibition of acetylene reduction activity was observed in presence of EC 50 dose of the insecticide in 10 days. However, ARA of *Aulosira* sp. UU 25118, *Aulosira* sp. UU 29147 and *Nostoc* sp. UU 24100 was stimulated up to 3, 18 and 20 per cent in presence of recommended dose of Thiokill within 24 hours of their exposure, though in presence of EC 50 dose of the chemical, in the former two species, the adverse effect was more pronounced. In those species, e.g. *Anabaena* sp. UU 2493, *Anabaena* sp. UU 147 and *Calothrix* sp. UU 29139 where the recommended dose of the insecticide was higher than their respective EC 50 dose, their nitrogenase activity was decreased considerably even in presence of recommended dose of the insecticide within few hours or after 10 days of incubation. In almost all the organisms, except both the species of *Calothrix* and *Nostoc* sp. UU 29130, nitrogenase activity was completely stopped within 24 hours exposure to EC 50 dose of the herbicide Hiltachlor. Under similar experimental conditions even recommended dose of the chemical inhibited the ARA of *Anabaena* sp. UU 2493 and *Aulosira* sp. UU 29147. In both the *Calothrix* species, ARA was inhibited by over 50 per cent within 24 hours of their exposure to recommended dose of Hiltachlor, but the same concentration of the herbicide stimulated the ARA of *Nostoc* species at recommended dose followed by a decrease in the activity in presence of higher concentration of the chemical within 24 hours of incubation in light. However, with prolonged incubation in the presence of different concentrations of the herbicide up to 10 days, adverse effect of the herbicide on the ARA of almost all organisms was considerably removed. On the 10th day recommended dose of Hiltachlor did not show any adverse affect on the ARA of *Aulosira* sp. UU 25118 and increased the ARA of *Calothrix* sp. UU 29139, *Aulosira* sp. UU 29147 and *Anabaena* sp. UU 147 by 16.33 and 64 per cent respectively over the control. ARA of all the organisms except that of *Nostoc* sp. 29130 and *Calothrix* sp. UU 24112 was completely ceased in presence of sublethal dose of the herbicide.

Per cent inhibition of nitrogenase activity in presence of EC 50 dose of the insecticides and herbicide depicted in Figs. 4 and 5 showed that ARA of *Anabaena* sp. UU 147, *Calothrix* sp. UU 24112 and *Calothrix* sp. 29139 was not at all adversely affected in the presence of EC 50 dose of Thiokill within 24 hours. Similarly ARA of *Anabaena* sp. UU 2493, *Calothrix* sp. UU 29139 and *Aulosira* sp. UU 29147 continued to show nitrogenase activity on par with control in the presence of EC 50 dose of Balwan. However ARA was completely ceased in four out of eight species within 24 hours of their exposure to EC 50 dose of Hiltachlor. On further incubation up to 10 days adverse effect of Hiltachlor on the ARA was reduced; nitrogenase activity of *Calothrix* sp. UU 29139 and *Aulosira* sp. UU 29147 were on par with control in the presence of EC 50 dose of Hiltachlor on the 10th day. ARA of *Aulosira* sp. UU 25118 and *Anabaena* sp. UU 2493 was not decreased in presence of EC 50 dose of Balwan. Similar dose of Thiokill also did not decrease ARA of *Nostoc* sp. UU 24100 on the 10th day. However ARA of *Aulosira* sp. UU 25118, *Calothrix* sp. UU 24112 was decreased by 100, 88 and 69 per cent in presence of EC 50 dose of Hiltachlor and of *Anabaena* sp. UU 147 by 82 per cent in presence of EC 50 dose of Thiokill in 10 days.

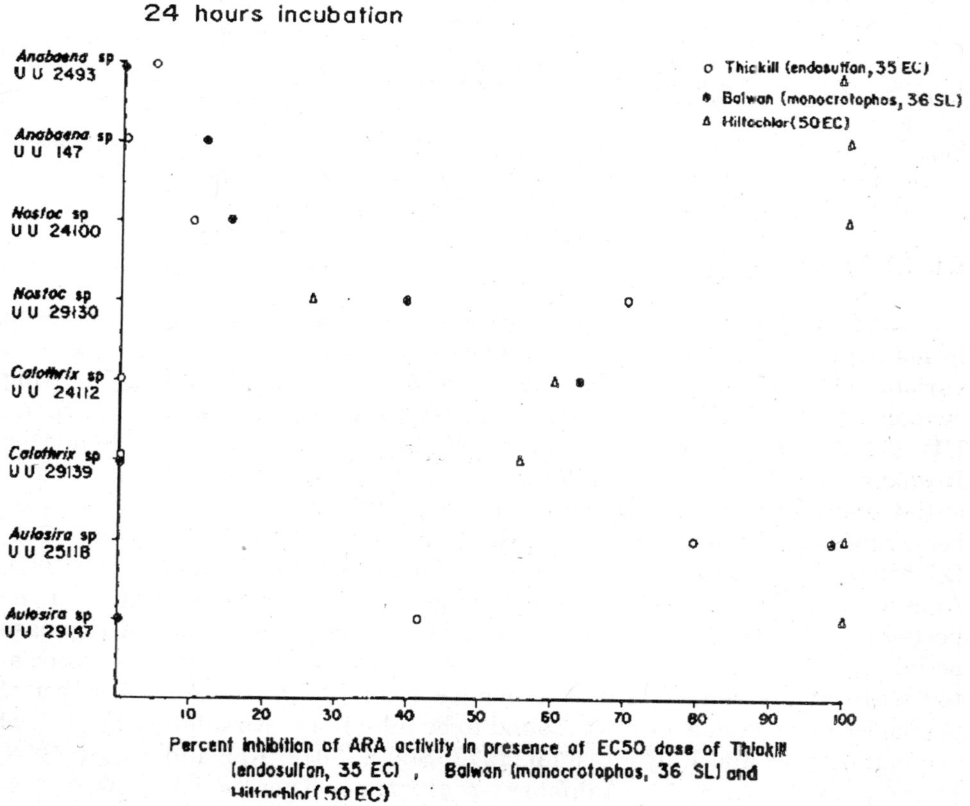

Fig. 4. Percent inhibition of nitrogenase activity (ARA) of eight different species of cyanobacteria in presence of EC50 dose of Endosulfan (35 EC), Monocrotophos (35 SL) and Hiltachlor (50 EC) after 24 hours of incubation.

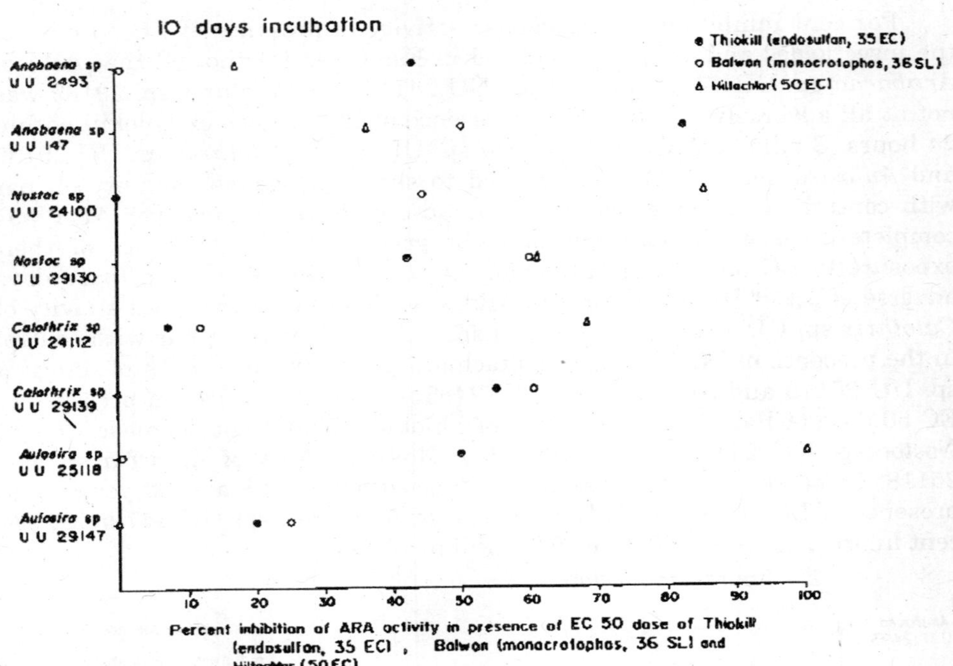

Fig. 5. Percent inhibition of nitrogenase activity (ARA) of eight different species of cyanobacteria in presence of EC50 dose of Endosulfan (35 EC), Monocrotophos (35 SL) and Hiltachlor (50 EC) after 10 days of incubation

DISCUSSION

Results of the present investigation revealed that the cyanobacterial forms occurring in the rice field soils showed different resistance capacity to various doses of the insecticides and herbicide. Out of eight nitrogen fixing cyanobacteria used in the experiment, *Nostoc* sp. UU 29130 and *Aulosira* sp. UU 25118 were comparatively more tolerant species to the agrochemicals; *Nostoc* sp. UU 24100 and *Calothrix* sp. UU 24112 were less tolerant. Differential effect of various pesticides on the growth of cyanobacteria has also been reported earlier by several authors (Hamdi *et al.* 1970; Ibrahim, 1972; Da Silva *et al.*, 1975; Singh and Vaishampayan, 1978 and Goyal, 1987). Anand and Subramanian (1997) had shown that higher concentration of the pesticides affected the various physiological processes like nitrogenase activity, photosynthesis, carbon fixation, protein synthesis, enzyme associated with nitrate reduction and ammonia assimilation. In general, effect of pesticides on cyanobacteria was found to be inhibitory, stimulatory or neutral (Venkataraman and Rajyalakshmi, 1972; Singh, 1973; Kar and Singh, 1978; Begum and Islam, 1993; Adhikary *et al.*, 1984; Thomas and Shanmugasundaram, 1986; Das and Adhikary, 1996). The pesticides are known to interfere with the synthesis of chlorophyll pigment by inhibiting the formation of porphyrin and biosynthetic steps in the carotenoid synthesis pathway which ultimately leads to chlorosis (Moreland, 1980). There are reports that the lethal action of pesticides at higher concentration is due to the imbalance in the function of the thylakoid by making polyhedral body incapable of performing their functions (Mehta and Hawxby, 1979). These results showed that the agrochemicals adversely affected chlorophyll-*a* synthesis which in turn reduced the rate of photosynthesis, cell division and finally the rate of growth of the population. Based on the result of the growth response of several species of cyanobacteria to commercial grade Furadon (carbofuran, 3% G), it was proposed that those species which possess a well defined sheath around their trichome could tolerate higher concentrations of the insecticide (Rath and Adhikary, 1994; Sahu *et al.*, 1992). But the results of the present investigation on the influence of monocrotophos, endosulfan and Hiltachlor on the growth and chlorophyll-*a* content of eight different cyanobacteria showed that though the *Calothrix* and *Aulosira* species possesses a well defined external envelope layer, it seldom protects their cells from the adverse effect of the insecticides and herbicide more effectively than that of the unsheathed forms like *Anabaena* and *Nostoc*. Further, toxicity of a particular agrochemical also varied between two different species under the same genus suggesting that pesticide tolerance by a specific organism is genetically determined and is not affected due to the presence of external envelope layer always.

All these results on the growth response of cyanobacteria to agrochemicals and influence of various concentrations of the pesticides on the rate of photosynthetic oxygen evolution, respiratory oxygen uptake and nitrogenase activity showed that there exists a great deal of variation among the species of the same genus and between genera with regard to their tolerance and metabolic activity to a particular insecticide and herbicide. Further, the pesticide tolerant species were never always accompanied by protection to their nitrogen-fixing capacity with increase in the pesticide dose, rather at times acetylene reduction activity of pesticide sensitive species was less affected than the tolerant species in presence of higher concentrations of the pesticides. These results indicate that growth and nitrogen fixing capacity of rice field cyanobacteria under the influence of various doses of insecticides and herbicide did not go parallely. Thus if a specific pesticide applied to control a target organism though showed adverse effect on the growth of a non-target species, might not decrease its nitrogen fixing ability.

The recommended level of the insecticide monocrotophos (36 SL) and endosulfan (35 EC) to control rice pests is 1 l/ha and that of the herbicide Hiltachlor is 2.5 l/ha. At these levels, the concentration of Balwan (monocrotophos, 36SL) and Thiokill (endosulfan, 35 EC) in the rice field water of different regions will be 4 ppm and of Hiltachlor (50 EC) will be 10 ppm. Results of the present investigation showed that lethal concentrations of the insecticides and herbicide for the experimental organisms were much higher than those arising from their recommended rates in the field. Thus, application of these commercial grade insecticides and herbicide at the recommended levels may not affect adversely the growth of these beneficial soil organisms under test. However, an increase in the concentration of these agrochemicals above the recommended levels or their excessive and indiscriminate use may adversely affect cyanobacterial growth and decrease their nitrogenase activity thereby adversely affecting the overall nitrogen economy of rice field soils.

ACKNOWLEDGEMENTS

We thank the Head of the Department of Botany, Utkal University for providing laboratory facilities. Financial assistance for this work from the Department of Biotechnology, Govt. of India is gratefully acknowledged.

REFERENCES

Adhikary, S.P., P. Das and H. Pattnaik.1984. Effect of the carbamate insecticide Sevin on *Anabaena* sp. and *Westiellopsis prolifica*. **Acta Microb. Hungarica. 31:**335-338.

Ahmad, M.H. and G.S. Venkataraman. 1973. Tolerance of *Aulosira fertilissima* to pesticides. **Curr. Sci., 42:**108-114.

Anand, N. and T. D. Subramanian. 1997. Effect of certain pesticides on the physiology of *Nostoc calcicola*. **Phykos, 36:**15-20.

Begum, Z.N.T. and M.M.S. Islam. 1993. Hypertrophic and polluted freshwater Ecosystems: Ecological Bases for Water Resource Management. In: **Proc. Int. Symp. Limnol.,** Department of Botany, University of Dhaka, Dhaka, Bangladesh. pp. 103-109.

Das, M.K. and S. P. Adhikary. 1996. Toxicity of three pesticides to several rice-field cyanobacteria. **Trop. Agric. (Trinidad) 73:**155-157.

Da Silva, E.J., L. E. Henriksson and E. Henriksson. 1975. Effects of pesticides on blue-green algae and nitrogen fixation. **Arch. Environ. Contam. Toxicol., 3:**193-204.

Edwin, S. 1977. **Action of methyl parathion on algal metabolism**. M. Phil. dissert. Madurai Kamraj Univ., India.

Goyal, S.K. 1987. Alga biofertilizer: An appraisal. **J. Maharashtra Agric. Univ., 12:**11-19.

Hamdi, Y.A., A.S. El-Nawawy and M. S. Tewfik. 1970. Effect of herbicides on growth and nitrogen-fixation of alga, *Tolypothrix tenuis*. **Acta Microbiol. Pol. B., 2:**53-55.

Hamilton, P.B., G. S. Jackson, N.K. Kaushik, K. R. Solomon and G. L. Stephenson. 1988. The impact of applications of atrazine on the planktonic communities of *in situ* enclosures. **Aquat. Toxicol., 13:**123-140.

Ibrahim, A.N. 1972. Effect of certain herbicides on growth of nitrogen-fixing algae and rice plants. **Symp. Biol. Hung., 11:**445-448.

Kar, S. and P.K Singh. 1978. Toxicity of carbofuran to blue-green alga *Nostoc muscorum*. **Bull. Environ. Contam. Toxicol., 22:**707-714.

Kaur, M., S. Dahuja and A. S.Ahluwalia 1997. Response of diazotrophic cyanobacteria to Butachlor. **Phykos, 36:**93-101.

Kolte, S.O. and S. K. Goyal 1992. On the effect of herbicides on growth and nitrogen-fixation by cyanobacteria. **Acta Bot. Ind., 20:**225-229.

Mackinney, G. 1941. Absorption of light by chlorophyll solutions. **J. Biol. Chem., 140:**315-322.

Mehta, R.S. and K. W. Hawxby. 1979. Effect of Simazina on the blue-green alga *Anacystis nidulans*. **Bull. Environ. Contam. Toxicol., 23 :** 319-326.

Moreland, D.E. 1980. Mechanism of action of herbicides. **Ann. Rev. Plant Physiol., 31:**597-638.

Rath, B. and S.P. Adhikary. 1994. Relative tolerance of several nitrogen-fixing cyanobacteria to commercial grade Furadon (carbofuran, 3%). **Ind. J. Expt. Biol., 32:**213-215.

Rippka, R., J. Deruelles, J.B Waterbury, M. Herdman, R. Y. Stainer. 1979. Generic assignments, strain histories and properties of pure culture of cyanobacteria. **J. Gen. Microbiol., 111:**1-61.

Sahu, J., M. K. Das, and S. P. Adhikary. 1992. Reaction of blue-green algae of rice field soils to pesticide application. **Trop. Agric., (Trinidad) 69 :** 362-364.

Singh, H.N. and A. Vaishampayan. 1978. Biological effect of rice field herbicide Machete on various strains of the nitrogen-fixing blue-green alga *Nostoc muscorum*. **Environ. Expt. Bot., 18:**87-94.

Singh, P.K. 1973. Effect of pesticides on blue-green algae. **Arch. Mikrobiol., 89:**317Ä320.

Singh, P.K. 1978. Nitrogen economy of rice soils in relation to nitrogen fixation by blue-green algae and *Azolla*. In: **National Symposium on increasing rice yield in Kharif,** Central Rice Research Institute, Cuttack, India.

Subramanian, T.D. 1988. Blue-green algae of paddy fields: **Ecological studies on natural populations and physiological responses of certain isolates.** Ph.D. Thesis, University of Madras.

Thomas, P.S. and S. Shanmugasundaram 1986. Interaction of agrochemicals with the cyanobacterium *Anacystis nidulans* and patterns of growth. **Microbiol Lett., 33:**115-120.

Venkataraman, G.S. and B. Rajyalakshmi. 1972. Relative tolerance of nitrogen-fixing blue-green algae to pesticides. **Ind. J. Agric. Sci., 42:**119-121.

CHAPTER - 6

GROWTH, NITROGEN FIXATION AND EXTRACELLULAR AMINO ACIDS OF CYANOBACTERIA FROM RICE FIELDS AT DIFFERENT TEMPERATURES

H. Nayak and S.P. Adhikary

ABSTRACT

Growth, ARA and liberation of extracellular amino acids of three different species of cyanobacteria viz *Aulosira* sp. UU25118, *Nostoc* sp. UU24100 and *Nostoc carneum* UU29130 during their incubation period up to 90 days at 25°C and 32°C was investigated. Variation in the temperature regime, though did not influence much on biomass increase of the experimental organisms, showed a differential impact on their chlorophyll-*a* pigment content and N_2 fixation rates. In all the three species ARA was maximum within 10 days followed by decline up to the last day of incubation. The rate of N_2 fixation as well as the chlorophyll-*a* content of the cells was greatly hampered at high temperature in *Anabaena variabilis* UU147 and the *vice versa* in *Nostoc* sp. UU24100. Marked increase in the quantity of extracellular amino acids was observed in *Anabaena variabilis* UU147 and *Nostoc* sp. UU24100 in the late stationary phase. *Aulosira* sp. UU25118 was a slow growing form and liberated amino acids over a prolonged period without being absorbed by the cells. The quantity of extracellular amino acids of *Aulosira* sp. UU25118 was higher than that obtained in the 90th day of the culture of former two species. These criteria were taken into consideration while selecting strains for using as biofertilizer in rice cultivation.

INTRODUCTION

In the rice fields, higher water temperature usually showed a significant correlation with biomass production and N-yield of cyanobacteria, and low temperature showed a negative correlation (Roger and Reynaud, 1979; Bisoyi and Singh, 1988). Cyanobacteria when fluorish in the fields are known to secrete a wide variety of nitrogenous substances (Roger and Kulasooriya, 1980; Venkataraman, 1981), polysaccharides (Marathe, 1972; Bertocchi *et al.*, 1990), hormones and many other plant growth promoting substances (Venkataraman and Neelakanthan, 1967; Singh and Trehan, 1973; Jacq and Roger, 1977), which are taken up by rice plants. Further, such secretions are also absorbed during the exponential growth of these and other microorganisms there by regulate the microbial webs of the ecosystems (Adhikary, 1998). Nothing much is known about the extracellular releases by rice field cyanobacteria during their different growth phases which might be influencing their growth and nitrogen fixation over prolonged periods in the rice fields.

In the present work, growth response and nitrogenase activity of three cyanobacterial species isolated from different agroclimatic zones of Orissa state to different temperatures for a prolonged period of incubation was studied.

MATERIALS AND METHODS

For studying the effect of two different temperatures (25±1°C, 32±2°C) under various growth periods (up to late stationary phase) of cyanobacteria, three selected species e.g. *Anabaena variabilis* UU147, *Nostoc* sp. UU24100 and *Aulosira* sp. UU25118 were grown in 50ml BG 11-N medium (Rippka *et al.*, 1979) in 100ml capacity conical flasks. Since the organisms were to be grown for up to 90 days, the cells in the inoculum was kept low (absorbance of the culture soon after inoculation at 760 nm was 0.04). The cultures were harvested after 10, 20, 30, 40, 60 and 90 days of incubation and their growth, chlorophyll-*a* content, nitrogenase activity and extracellular amino acids was determined.

The cultures were incubated at 25±1°C under 7.5 $W.m^{-2}$ irradiance in the culture room for definite periods as per the design of the experiments. Triplicates were set up for each set of experiment and mean value of three readings ± S.D. has been presented in the text. Growth was determined by absorbance of the homogenised culture suspension at 760 nm. The cells from culture vessels were collected by centrifugation and the pellet was suspended in 80% (v/v) acetone for chlorophyll-*a* extraction. Chlorophyll-*a* content of the cultures was estimated following Mackinney (1941). Nitrogen fixing capacity was determined by incubating the cultures with acetylene under light for definite period before measuring the ARA in a AIMIL-NUCON Gas Chromatograph using FID detector. Nitrogenase activity of the organisms was expressed as n mol C_2H_4 $culture^{-1}$ h^{-1}. Quantity of extracellular amino acids was estimated as per the method of Spies (1957).

RESULTS AND DISCUSSION

Growth, nitrogenase activity, liberation of extracellular amino acids and absorption spectra of the whole cell suspensions of three selected species of cyanobacteria, *viz.*, *Anabaena variabilis* UU147, *Nostoc* sp. UU24100 and *Aulosira* sp. UU25118 at two different temperatures i.e. 25°C and 32°C up to a period of 90 days under continuous light was studied. The objective was to screen those species capable of growing and fixing N_2 at different temperature ranges and thriving for prolong periods. Such species are expected to colonize and dominate other species in nature and thus can be screened for their use as biofertilizer.

All these species of cyanobacteria grew well in nitrogen free BG 11 medium and showed different behaviour in their biomass increase in culture at two different temperatures. Growth of *Anabaena variabilis* UU147 and *Nostoc* sp. UU24100 was slowed down after 20 days of incubation and did not change much for the next 40 days. Though similar trend in their growth was marked at both the temperatures, lower temperature of 25°C was favourable for their growth. However, after 60 days of incubation, the biomass increase at low temperature was decreased but at higher temperature the growth was shot-up. Microscopic observation of the filaments showed abundance of hormogonia formation in the culture at 32°C which were over two months old. But in *Aulosira* sp. UU25118, higher temperature of 32°C favoured the growth of the organism though the value was almost similar to the biomass increase that obtained at 25°C. Further, the growth of the organism increased gradually up to 3 months almost without any change in the rate during the growth period (Fig. 1). The growth pattern of the organisms was also reflected in the chlorophyll synthesis of their respective cells during the exponential period. Chlorophyll-*a* content of *Aulosira* sp. UU25118 increased gradually up to 90 days and the higher amount of chlorophyll-*a* per cell was obtained in the cultures grown at low temperatures (Fig. 2). Chlorophyll-*a* content of *Nostoc* sp. UU24100 was higher within 10 days of incubation at both the temperatures followed by a decrease, and the same decreased there after up to 40 days during which the organism sporulated and remained as such for about a month. The cyanobacterium neither revived its growth nor its chlorophyll-*a* content upon further incubation at 25°C, rather the cells become bleached and disintegrated (Fig. 2). However, the cells grew after 60 days of incubation at 32°C corresponding to the germination of spores, production of hormogonia and synthesis of chlorophyll-*a* pigment. In *Anabaena variabilis* UU147 chlorophyll-*a* content was increased up to 30 days followed by a decline during the next 10 days corresponding to the sporulating phase of the organism. Like *Nostoc* sp. UU24100, in this organism also chlorophyll-*a* content increased during the next 20 days corresponding to formation of fresh filament due to germination of spores. After 60 days of culture, the organism again sporulated and remained as such for the next 30 days which led to decrease in chlorophyll-*a* content of cells in the culture. In this organism low temperature was favourable for its growth and sporulation enabling the organism to thrive for prolonged periods since at higher temperature of 32°C the germination percentage of spores was negligible (Fig.2).

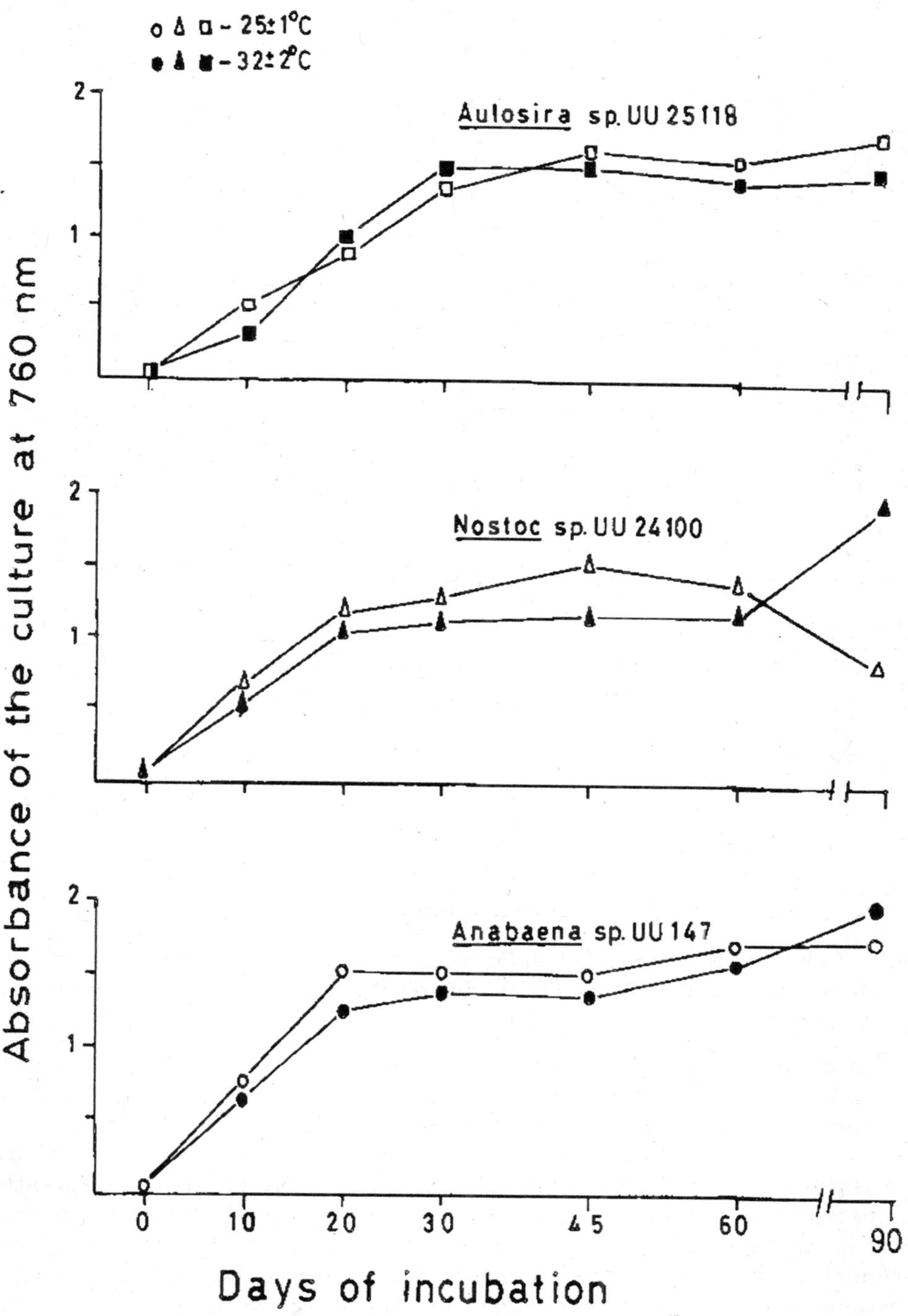

Fig. 1. Growth (absorbance of the culture suspension at 760 nm) of *Aulosira* sp. UU25118, *Nostoc* sp. UU24100 and *Anabaena variabilis* UU147 at different temperatures during prolonged incubation periods (up to 3 months).

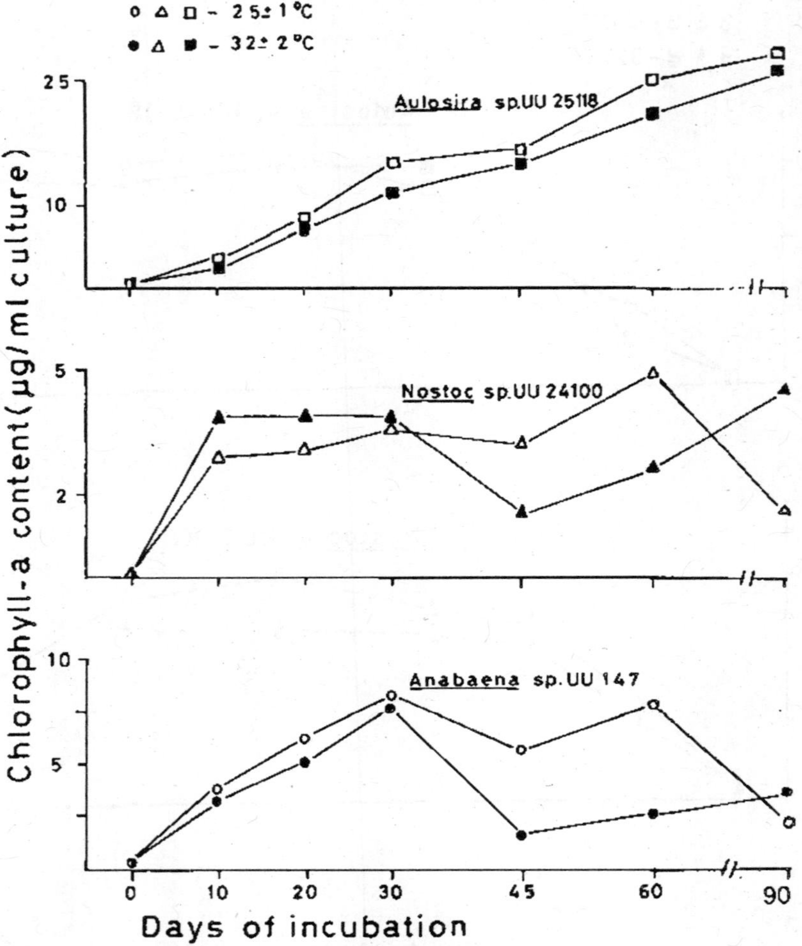

Fig. 2. Chlorophyll- *a* content of *Aulosira* sp. UU25118, *Nostoc* sp. UU24100 and *Anabaena variabilis* UU147 at different temperatures during prolonged incubation periods (up to 3 months).

In all the three species nitrogenase activity was maximum within 10 days followed by a decline up to the last day of incubation; the rate of decrease of nitrogenase activity upon increased days of incubation was slowed down especially at low temperatures in *Anabaena variabilis* UU147 and *Aulosira* sp. UU25118 (Fig. 3). The rate of N_2 fixation was greatly hampered at higher temperature in *Anabaena variabilis* UU147 and the *vice versa* in *Nostoc* sp. UU24100. However, when the spores of these organisms germinated and formed new filaments, the corresponding increase in the nitrogenase activity was negligible. In *Anabaena variabilis* UU147 the extracellular amino acids content increased gradually up to 30 days and then remained constant up to 60 days followed by significant increase in the next 30 days of incubation. In this organism extracellular amino acid content was always higher during the entire growth period (Fig. 4). In *Nostoc* sp. UU24100 the extracellular amino acids content was gradually higher in the culture up to 20 and 30 days at low and high temperatures respectively (Fig.4). The extracellular substances were then utilized by the cells during their sporulating phase up to 60 days and then liberated corresponding to the reappearance of growth at high temperature and the death of the organism at 25°C. But in *Aulosira* sp. UU25118 the quantity of extracellular amino acids was less in the culture up to 40 days of the culture. Upon further incubation, the extracellular substances increased steadily up to the last day of incubation. On the 90th day of culture, the quantity of extracellular amino acids in *Aulosira* sp. UU25118 was higher than that of the cultures of *Nostoc* sp. UU24100 and *Anabaena variabilis* UU147 respectively (Fig.4). Like the other two species, in *Aulosira* sp. UU25118 the extracellular amino acids was higher when the cultures were incubated at high temperature. *Aulosira* sp. UU25118 was a slow growing form and liberated the amino acids over a prolonged period into the culture without being absorbed by the cells during its growth up to 90 days.

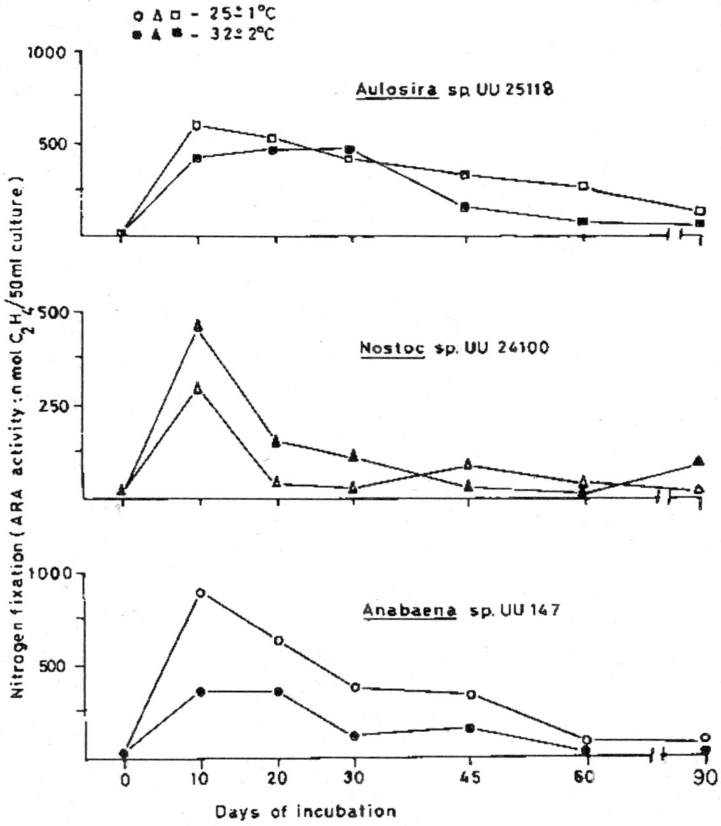

Fig. 3. Nitrogenase activity (n mol. C_2H_4/50 ml culture/h) of *Aulosira* sp. UU25118, *Nostoc* sp. UU24100 and *Anabaena variabilis* UU147 at different temperatures during prolonged incubation periods (up to 3 months).

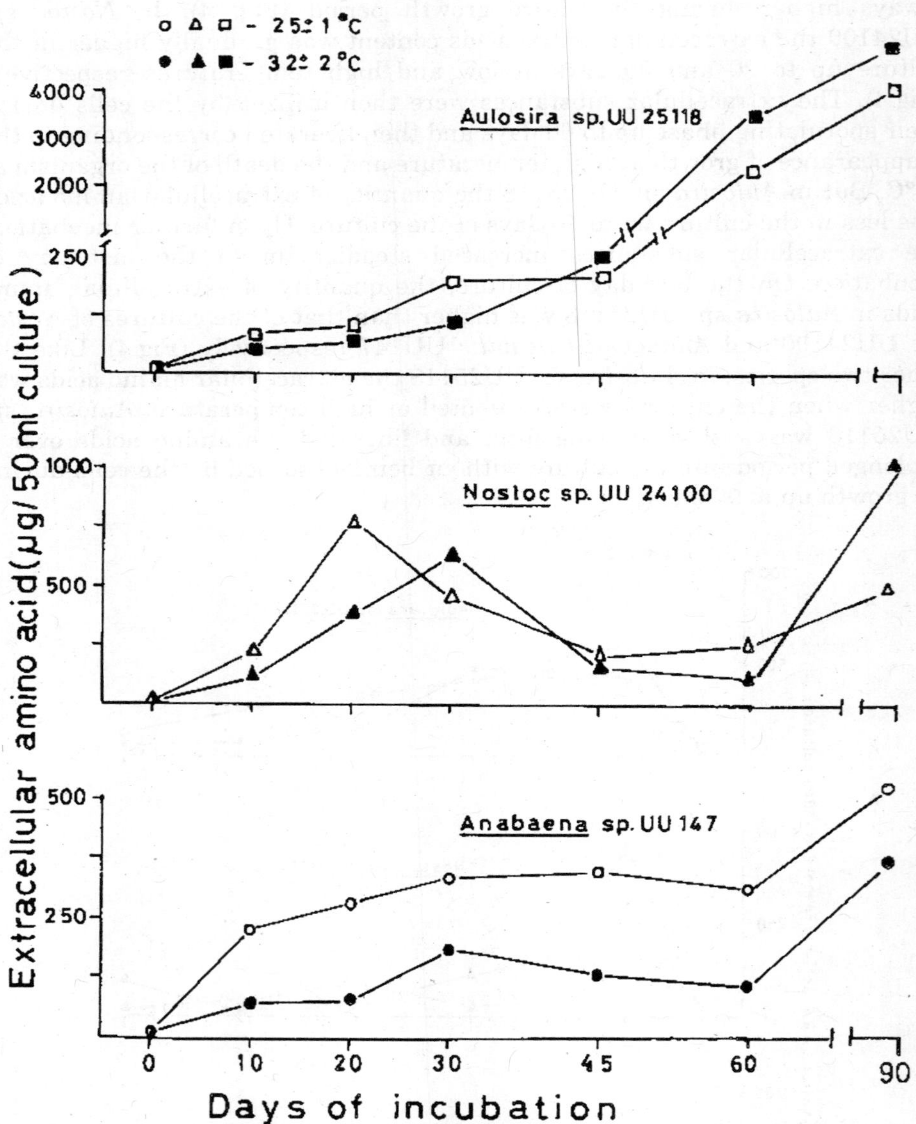

Fig. 4. Effect of different temperatures on the liberation of quantity of extracellular amino acids by *Aulosira* sp. UU25118, *Nostoc* sp. UU24100 and *Anabaena variabilis* UU147 during prolonged incubation periods (up to 3 months)

Several reports have shown that the optimal temperature for cyanobacterial growth is about 30-35°C (Roger and Reynaud, 1982). In tropical rice fields the temperature often remains within this range. In submerged soils daily variations in the temperature are moderated by the buffering effect of flood water. Thus, temperature is rarely a limiting factor for growth of cyanobacteria in the rice fields. This is also because the range of temperatures permitting the growth of cyanobacteria is larger than that required by rice. However, temperature variations influence both cyanobacterial biomass composition and productivity. In the present experiments though the organisms showed similar growth pattern at two different temperatures, this variation in temperature showed a marked effect on the chlorophyll-*a* content, N_2 fixing capacity, regeneration of the spores after attaining the stationary phase and also in the liberation of extracellular amino acids to the outer environment. Variations among the cyanobacterial taxa to two different temperatures with respect to their nitrogenase activity was observed; though *Anabaena variabilis* UU147 could grow and fix N_2 well at 25°C, *Nostoc* sp. UU24100 isolated from the rice fields of the same area showed a reverse response and also the same was the case with *Aulosira* sp. UU25118. These results together with the field observations agree with the finding of earlier workers that low temperature usually favours eukaryotic algae over that of cyanobacteria and high temperatures favour cyanobacterial growth as well as productivity of phytoplankton in the natural environments.

The growth behaviour and nitrogenase activity of these organisms over a prolonged period of incubation up to 90 days showed considerable variations among themselves. Though *Aulosira* sp. could grow steadily in the culture up to 3 months, simultaneously fixing N_2 for about 30 days followed by a slow decline, the other two species showed maximum growth and nitrogenase activity within few days of incubation followed by a decline; the decrease correlated well with the sporulating condition of the cells. The higher extracellular amino acid content of *Anabaena variabilis* UU147 as well as of *Nostoc* sp. UU24100 after 60 days may be due to cell lysis leading to higher amount of amino acid production. Microscopic examination of the cells also showed that they were disorganised and bleached, but in *Aulosira* sp. UU25118 the higher amount of extracellular amino acid production even up to 90 days of incubation was from the healthy cells. These results showed that environmental conditions especially the temperature and the nutrient level of the surrounding medium influences significantly on the survival, growth, and nitrogen fixation of cyanobacteria. *Aulosira* sp. UU25118 seems to tolerate well to such conditions and thus can be employed as an inoculant along with other species for biofertilization of rice fields.

ACKNOWLEDGEMENTS

We thank the Head of the Department of Botany, Utkal University for providing laboratory facilities. One of us (SPA) thank the Department of Biotechnology, Government of India for financial assistance to carry out this work.

REFERENCES

Adhikary, S.P. 1998. Polysaccharides from mucilaginous envelope layers of cyanobacteria and their ecological significance. **J. Sci. Ind. Res., 57:** 454-466.

Bertocchi, C., L. Navarini, A.Cesaro and M. Anastasio. 1990. Polysaccharides from cyanobacteria. **Carbohydrate Polym., 12:** 127-153.

Bisoyi, R.N. and P.K. Singh. 1988b. Effect of seasonal changes on cyanobacterial production and nitrogen yield. **Microb. Ecol., 16:** 149-154.

Jacq, V. and P.A. Roger. 1977. Decrease of losses due to sulphate reducing processes in the spermosphere of rice by pre-soaking seeds in a culture of blue-green algae. **Cab ORSTOM Ser Biol., 12:** 101-108.

Mackinney, G. 1941 Absorption of light by chlorophyll solutions. **J. Biol. Chem., 140:** 315-322.

Marathe, K.V. 1972. Role of some BGA in soil aggregation. In: **Taxonomy and Biology of BGA**. (ed.) Desikachary, T.V., Madras Univ. Press, Madras. pp 328-331.

Rippka, R., J. Deruelles, J.B. Waterbury, M. Herdman and R.Y. Stanier 1979. Generic assignments, strain histories and properties of pure cultures of cyanobacteria. **J. Gen. Microbiol., 111:** 1-61.

Roger, P.A. and S.A. Kulasooriya 1980. **Blue-green algae and rice.** International Rice Research Institute. Los Banos, Laguna, Phillippines, p.112.

Roger, P.A. and P.A. Reynaud 1979. Ecology of blue-green algae in paddy fields. In: **Nitrogen and Rice**, International Rice Research Institute, Los Banos, Phillipines. pp. 289-309.

Roger, P.A. and P.A. Reynaud 1982. Free-living blue-green algae in tropical soils. In: **Microbiology of tropical soils and plant productivity**. (eds.) Dommergues, Y.R. and Diem, H.G., Martinus Nijhoff/Dr. W. Junk, The Hague, Netherlands. pp. 147-168.

Singh, V.P. and T. Trehan 1973. Effect of extracellular products of *Aulosira fertilissima* on the growth of rice seedlings. **Plant Soil, 38:** 457-464.

Spies, J.R. 1957. Colorimetric procedures for amino acids. In: **Methods in Enzymology-3**. (eds.) Colowick S.P. and Kaplan, N.O., Academic Press, New York. pp. 467-471.

Venkataraman, G.S. 1981. Blue-green algae for rice production - a manual for its promotion. **FAO Soils Bulletin. 46**, Rome. p.102.

Venkataraman, G.S. and S. Neelakanthan 1967. Effect of cellular constituents of nitrogen-fixing blue-green alga *Cylindrospermum* on root growth of rice plants. **J. Gen. Appl. Microbiol., 13:** 53-62.

CHAPTER - 7

EFFECT OF PHYTO-EXTRACTS HARBOURING INSECTICIDAL PROPERTY ON GROWTH AND NITROGEN FIXATION OF CYANOBACTERIA

M.N. Jha, A.N. Prasad and S.K. Mishra

ABSTRACT

The effect of aqueous extract of fresh and dry leaf of neem (*Azadirachta indica*), Bel (*Aegale marmilose*), Moringa (*Moringa obifera*) and Gurich (*Tinospora cordifolia*) on growth and nitrogen fixation of cyanobactrial strains (*Nostoc hatei, Anabaena spherica, Cylindrospermum majus, Aulosira fertilissima, Westiellopsis prolifica*) was investigated. Cyanobacterial nitrogen fixation and growth were estimated by acetylene reduction activity and chlorophyll measurement respectively. Different strains respond differently to these phyto-extracts. *Anabaena spherica* was observed to be the most sensitive strains while *Westiellopsis prolifica* was recorded most resistant one. Among different phyto-extracts, moringa had least effect while gurich had maximum toxicity. The effect was recorded to be more toxic than fresh leaf extract. At 0.1% or 1000 ppm, the phyto-extract proved to be stimulatory for all the strains in both types of preparation. These extracts were also tested for molluscicidal properties because one of the major hurdle in the establishment of cyanobacteria in paddy fields of Bihar is the prevalence of grazers particularly snail 5.0 per cent mortality of snail was recorded in the laboratory from the preparation of neem, bel and gurich. However moringa extract was observed to be stimulatory for snalis growth.

INTRODUCTION

Nitrogen is an integral part of agriculture. Indian soil system is negative in nitrogen balance. Indian farming depends on chemical nitrogenous fertilizer. But rising fertilizer prices and their limited availability and the low purchasing power of marginal and small farmers have demonstrated the importance of biofertilizer especially cyanobacteria as a source of nutrients in paddy ecosystem. This is true for most of the developing world but even more so in heavily populated country such as India.

The inherent fertility of paddy field is attributed due to diazotrophic cyanobacterial activity. Introduction of efficient strains resulted in the increase in grain yield and improvement in soil health (Venkataraman, 1979; Roger and Kulasooriya, 1980; Jha *et al.,* 1999; Kannaiyan, 2000). The establishment of introduced strains is the basic requirement for getting the desired level. Grazers population did not allow to establish cyanobacteria in many cases. Suppression of Ostracoda and Molluscan by carbofuran, parathene or neem seeds tripled cyanobacterial biomass and increased nitrogen fixation by ten folds (Grant *et al.*, 1983). Besides neem, other plant products like bel, gurich are also identified as a potential candidate for biopesticide. Moringa was also identified to have allelopathic potential. Thus, cyanobactera has to interact with these biopesticide and so we studied the effect of phyto-extracts from said plant on cyanobacterial growth and nitrogen fixation in the laboratory.

MATERIALS AND METHOD

Five cyanobacterial strains, isolated from paddy field soils of Pusa, were selected for the present investigation. The isolates were *Nostoc hatei, Anabaena spherica Aulosira fertilissima, Cylindrospermum majus* and *Westiellopsis prolifica.* These cyanobacterial isolates were grown in triplicate in 100 ml conical flasks containing 50 ml of nitrogen free BG11 medium amended with various concentrations of aqueous extracts of neem, bel, moringa and gurich.

Fresh, healthy and young leaves of *Azadirachta indica* (neem), *Aeglemarmelos* sp. (bel), *Morringa obifera* (Drumstick) and *Tenospora cordifolia* (gurich) were collected and divided into two parts. One part is sun dried. Both fresh and dried leaves were used for preparing aqueous extract. All these were ground separately with sterile distilled water at room temperature, filtered through whatman filter paper aseptically and the resultant clear extract was added to BG11 medium to obtain the desired concentration. One gram leaf with one ml water was considered as 100 per cent concentration. The flasks were inoculated with 5 ml of homogenised 12 days old growth of each cyanobacteria and incubated for 28 days at 2000 lux light intensity with 16 / 8 h light / dark cycle at 28± 2°C. Initial observation on growth was taken at 0 day. From each replication, two 10 ml portions were taken for chlorophyll estimation (McKinney, 1941) and ARA measurement (Jha, 2001). The results reported are mean of three observations.

In order to assess the extent of infestation properties of these leaves by snails and experiment was conducted in the laboratory. Adult snails collected from the field were starved for 24 hours. A batch of 5 snails were released in the different feeding chamber with the above mentioned leaves. A controlled chamber was also maintained in similar conditions to know the loss in weight of leaves by other means. The sails were also introduced at different concentration of phytoextracts to examine its mortality.

Table 1. Effect of neem (*Azadirachta indica*), leaf extract on growth (Chl μg / ml) and ARA (nmol $C_2H_4hr^{-1}$ μg Chl^{-1}) of diazotrophic cyanobacteria

A. Fresh leaf

Treatment	Growth					ARA				
	Nostoc hatei	*Aulosira fertilissima*	*Cylindro spermum majus*	*Anabaena spherica*	*Westiellopsis prolifica*	*Nostoc hatei*	*Aulosira fertilissima*	*Cylindro spermum majus*	*Anabaena spherica*	*Westiellopsis prolifica*
Control	4.04	3.3	3.5	4.91	3.16	3.28	2.88	3.42	6.6	2.84
0.1%	7.83	5.28	6.40	10.36	5.81	5.26	4.12	6.3	14.16	4.6
0.5%	9.85	7.68	7.36	3.25	7.46	7.54	5.94	8.4	6.4	5.9
1.0%	2.41	7.34	2.47		8.83	3.20	8.43	4.2		6.3
2.0%		1.67			1.64		2.52			3.2
4.0%					1.01					3.4

B. Dry leaf

Treatment	Growth					ARA				
	Nostoc hatei	*Aulosira fertilissima*	*Cylindro spermum majus*	*Anabaena spherica*	*Westiellopsis prolifica*	*Nostoc hatei*	*Aulosira fertilissima*	*Cylindro spermum majus*	*Anabaena spherica*	*Westiellopsis prolifica*
Control	3.89	3.6	4.1	4.9	3.0	4.2	3.65	3.9	6.1	2.8
0.1%	7.3	5.8	7.6	8.4	6.3	6.4	4.11	5.8	10.05	4.9
0.5%	7.1	6.9	5.2	2.1	7.8	7.3	5.4	7.6	7.5	5.6
1.0%	1.4	2.1	1.3		4.2	3.6	4.3	4.0		6.4
2.0%					1.7					3.1

Table 2. Effect of bel (*Aegale marmilose*), leaf extract on growth (Chl μg / ml) and ARA (nmol $C_2H_4hr^{-1}$ μg Chl^{-1}) of diazotrophic cyanobacteria

A. Fresh leaf

Treatment	Growth					ARA				
	Nostoc hatei	*Aulosira fertilissima*	*Cylindro spermum majus*	*Anabaena spherica*	*Westiellopsis prolifica*	*Nostoc hatei*	*Aulosira fertilissima*	*Cylindro spermum majus*	*Anabaena spherica*	*Westiellopsis prolifica*
Control	3.2	3.5	3.8	4.5	2.6	2.9	2.3	4.8	5.3	2.7
0.1%	6.3	4.9	5.1	6.9	3.7	5.4	3.6	5.3	8.9	4.5
0.5%	7.4	5.6	6.3	5.6	4.5	6.3	4.3	6.4	9.4	5.3
1.0%	4.3	6.4	4.0	1.3	5.4	7.1	6.4	7.3	3.6	6.2
2.0%	0.64	4.1	1.1		3.2	1.2	3.9	2.8		5.4
4.0%		1.2			2.2		2.4			3.8

B. Dry leaf

Treatment	Growth					ARA				
	Nostoc hatei	*Aulosira fertilissima*	*Cylindro spermum majus*	*Anabaena spherica*	*Westiellopsis prolifica*	*Nostoc hatei*	*Aulosira fertilissima*	*Cylindro spermum majus*	*Anabaena spherica*	*Westiellopsis prolifica*
Control	3.3	3.1	4.2	5.06	3.8	2.3	2.2	3.2	7.0	2.7
0.1%	4.7	5.3	5.3	6.9	4.9	3.6	4.8	4.9	10.9	3.3
0.5%	7.9	6.4	4.0	2.0	5.6	6.1	5.3	6.3	6.1	4.0
1.0%	2.4	3.1	1.8		3.9	3.2	6.3	2.9		5.2
2.0%		2.1			2.4		4.3			4.7

Table 3. Effect of gurich (*Tenospora cordifolia*), leaf extract on growth (Chl µg / ml) and ARA (nmol $C_2 H_4 hr^{-1}$ µg Chl^{-1}) of diazotrophic cyanobacteria

A. Fresh leaf

Treatment	Growth					ARA				
	Nostoc hatei	*Aulosira fertilissima*	*Cylindro spermum majus*	*Anabaena spherica*	*Westiellopsis prolifica*	*Nostoc hatei*	*Aulosira fertilissima*	*Cylindro spermum majus*	*Anabaena spherica*	*Westiellopsis prolifica*
Control	4.0	3.7	3.9	5.0	3.0	3.7	3.5	3.9	5.3	2.6
0.1	5.9	5.2	5.4	6.2	4.4	6.6	5.6	6.7	9.6	4.4
0.5	2.7	2.8	2.0	2.0	3.7	4.2	4.6	4.4	3.7	5.5
1.0					1.0					3.0

B. Dry leaf

Treatment	Growth					ARA				
	Nostoc hatei	*Aulosira fertilissima*	*Cylindro spermum majus*	*Anabaena spherica*	*Westiellopsis prolifica*	*Nostoc hatei*	*Aulosira fertilissima*	*Cylindro spermum majus*	*Anabaena spherica*	*Westiellopsis prolifica*
Control	4.3	3.0	3.5	4.8	2.8	4.0	2.8	3.7	5.3	3.2
0.1	3.0	4.8	2.9	3.0	4.9	7.1	3.9	5.5	6.1	5.8
0.5	1.3	1.2	2.0	1.0	2.16	3.2	3.6	4.6	4.2	4.9

Table 4. Effect of moringa (*Moringa obifera*), leaf extract on growth (Chl μg / ml) and ARA (nmol $C_2H_4hr^{-1}$ μg Chl^{-1}) of diazotrophic cyanobacteria

A. Fresh leaf

Treatment	Growth					ARA				
	Nostoc hatei	*Aulosira fertilissima*	*Cylindro spermum majus*	*Anabaena spherica*	*Westiellopsis prolifica*	*Nostoc hatei*	*Aulosira fertilissima*	*Cylindro spermum majus*	*Anabaena spherica*	*Westiellopsis prolifica*
Control	3.6	3.4	4.0	4.5	3.2	3.2	2.9	3.8	5.4	2.8
0.1%	4.3	4.8	5.1	5.9	3.9	4.8	3.6	5.3	6.8	3.4
0.5%	5.6	5.2	5.8	6.8	4.3	5.3	4.3	5.9	7.7	4.9
1.0%	5.9	5.9	6.4	7.3	5.6	5.8	5.2	6.8	8.8	6.1
2.0%	4.8	6.1	5.3	3.9	5.9	6.1	5.9	7.3	5.8	6.8
4.0%	3.3	4.2	3.9	2.8	4.0	4.1	3.8	5.1	4.3	4.5
8.0%	1.8	2.6	2.1		3.1	2.0	2.5	3.2		4.2

B. Dry leaf

Treatment	Growth					ARA				
	Nostoc hatei	*Aulosira fertilissima*	*Cylindro spermum majus*	*Anabaena spherica*	*Westiellopsis prolifica*	*Nostoc hatei*	*Aulosira fertilissinma*	*Cylindro spermum majus*	*Anabaena spherica*	*Westiellopsis prolifica*
Control	3.7	3.5	4.2	4.6	3.0	3.7	3.4	4.2	5.2	3.6
0.1%	4.4	4.3	5.0	6.9	4.8	4.1	4.5	5.1	8.4	4.2
0.5%	6.9	6.6	6.0	5.3	6.2	7.7	5.3	5.4	10.3	5.2
1.0%	4.0	5.3	2.3	1.7	6.6	5.4	6.5	4.3	3.2	6.0
2.0%	0.7	3.7	1.1		2.1	1.6	4.5	2.8		4.0

RESULTS AND DISCUSSION

The growth of test organisms were augmented at 0.1 per cent of fresh leaf neem extract (Table 1). Relatively higher concentration (>4 per cent) proved lethal for *Westiellopsis prolifica,* whereas extensive fragmentation followed by death was noticed in *Anabaena spherica* even at 1 per cent. The lethal dose for *Nostoc hatei, Cylindrospermum majus* and *Aulosira fertilissima* were 2 per cent and 4 per cent respectively. Nitrogenase activity also followed the similar trend but even at higher level the activity was either high or at least at par with the level of control. This might be due to the higher nitrogen requirement in stress condition. Dry leaf extract was observed to be more lethal than fresh leaf extract. Drying of leaves results loss of water and in such situation, the expected value of alkaloids will be more than fresh leaf. This might be the reason for failure of the tested organism to grow beyond one per cent level. However, the exception was *Westiellopsis prolifica* which showed its growth upto 2 per cent. Trend of nitrogenase activity was similar to that of fresh leaf extract.

Increase in the growth of test organisms were recorded upto 0.5 per cent of bel leaf extract (Table 2). Algicidal dose vary for different strains. For *Westiellopsis prolifica* and *Aulosira fertilissima,* the level was beyond 4 per cent while for *Anabaena spherica* it was beyond one per cent. Nitrogenase activity increased over control with an increase in the concentration of extract upto one per cent. The organism were more sensitive to dry leaf preparation.

Gurich leaf extract was observed to have more pronounced effect on cyanobacterial growth and ARA (Table 3). The algicidal value was recorded to be one per cent and stimulatory value upto 0.1 per cent. However, even at higher level nitrogenase activity was observed to be enhanced.

Among all phytoextracts, moringa extract had effect on cyanobacterial nitrogenase activity and growth (Table 4). No reduction in the growth of *Westiellopsis prolifica* was recorded even upto 8 per cent. *Anabaena spherica* which was recorded to be most sensitive strains, can tolerate upto 2 per cent. Nitrogenase activity was also observed to be increased upto 2 per cent level for all strains and thereafter varies with respect to strain variation. However, in dry leaf none of the test organism grew beyond 2 per cent.

The results of the study of food preferences showed that among all four plant leaves moringa was most preferred food for the snails (Tables 5 & 6). No infestation was recorded in neem, bel and gurich leaf. The extracts of neem, gurich and bel were found to be moderate effective in controlling the snails population.

The results in this paper clearly demonstrated that (i) phyto-extracts showed both stimulatory and inhibitory effects on cyanobacterial growth and nitrogen fixation, depending on species and extract concentration (ii) At the lowest concentration i.e. 0.1 per cent stimulation was recorded for all the species in all phytoextracts (iii) Gurich extract was most lethal and moringa extract was least lethal to the test organism (iv) Growth was more sensitive than cyanobacterial nitrogen fixation (v) Dry leaf extract had more effect than fresh leaf (vi) *Westiellopsis prolifica* was most resistant strain while *Anabaena spherica* was most sensitive (vii) Neem, bel and gurich leaf was not preferred by snails and its aqueous extract lead to have moderate control on snails population (viii) Recommended level of these products were stimulatory for cyanobacterial growth and ARA.

Some of these plant products have been shown to be a good biopesticide and organic manure (Devkumar and Goswami, 1992), reduces loss of nitrogen fertilizers by inhibiting population of nitrifiers (Chonker and Mishra, 1978) and increases the population of cyanobacteria in flooded soil. The neem seed was also observed to control cyanobacterial grazers like ostracod, tubificids and molluscean which resulted manyfold increase in growth and nitrogen fixation by cyanobacteria under natural condition (Grant *et al.*, 1983). Mishra and Adhikary (1997) also observed an enhancement in growth and nitrogen fixation of *Anabaena variabilis* upto 0.5% level of the aqueous extract of neem leaves and neem cake. Normally, these leaves or its products are introduced in the paddy soil at the time of transplanting and thus alkaloids may leach out and stimulated cyanobacterial growth directly or indirectly by controlling grazers. The recommended level of some of these products are 100-120 kg / ha or 60 ppm, the level observed to be stimulatory for cyanobacterial growth and ARA. In the future, these leaves might play an important role if it is used as carriers for cyanobacterial biofertilizers. However, the identification of active substances and their behaviour in the soil should be determined.

Table 5. Extent infestation by snails

Hort plant	Scientific name	Extent of infestation
Drumstick	*Moringa obifera*	XX
Neem	*Azadirachta indica*	Nil
Bel	*Aegale marmilose*	Nil
Gurich	*Tinospora cordifolia*	Nil

XX: Moderate infestation

Table 6. Effectiveness of phytoproducts against snail

Hort plant	Scientific name	Extent of infestation
Drumstick	*Moringa obifera*	XX
Neem	*Azadirachta indica*	Nil
Bel	*Aegale marmilose*	Nil
Gurich	*Tinospora cordifolia*	Nil

XX: 50%

REFERENCES

Chonkar, P. K. and K. C. Mishra. 1978. Possible utilization of neem cake for inhibiting nitrification in soil. **J. Ind. Soc. Soil Science, 26:** 90-92.

Devkumar, C. and B. K. Goswami. 1992. Neematicidal principles from neem – isolation and bioassay of some meliacins. **J. Pesticide Residue, 4:** 79-84.

Grant, I. F., A. C. Tiroi, T. Aziz and I. Watanabe, I. 1983. Regulation of invertebrate grazers as a means to enhance biomass and nitrogen fixation of cyanophyceae in wet land rice fields. **Soil Sci. Soc. of American Journal, 47 (4) :** 669-675.

Jha, M. N., S. Sriram and G. S. Venkataraman. 1999. Cyanobacteria a potential binitrogen source in integrated nutrient management systems for rice crops. In **Agromicrobes** (eds.) M. N. Jha, S. Sriram, G. S. Venkataraman, S. G. Sharma, Today & Tomorrow Publishers, New Delhi, pp.191-203.

Kannaiyan, S. 2000. Biological fertilizers for sustainable rice (*Oryza sativa*) production. **Advances in Agricultural Research in India, 8:** 67-107.

MacKinney, G. 1941. Absorption of light by chlorophyll solution. **J. Biol. Chemi., 140:** 315-322.

Mishra, A. and S. P. Adhikary. 1997. Effect of different neem preparations on growth, protein content, nitrogen fixation and extracellular amino acids of *Anabaena variabilis* Uu 147. **Phykos, 36 (1 & 2) :** 73 –80.

Roger, P. A. and Kulasooriya, S. A. 1980. **Blue green Algae and Rice Pub.** International Rice Research Institute, Manila, Philippines, pp. 1-112.

Venkataraman, G. S. 1979. Algal inoculation of rice fields. In : **Nitrogen and Rice,** International Rice Research Institute, Los Banos, Philippines, pp. 311-321.

CHAPTER – 8

EFFECT OF PHYSICO-CHEMICAL PARAMETERS ON AKINETE DIFFERENTIATION AND GERMINATION AND ITS APPLICATION IN BIOMASS PRODUCTION IN *Anabaena* sp.

M.K. Shivaprakash, Vidya Kulkarni and Binu Koshy

ABSTRACT

The present study was undertaken with an intention of looking into the possibilities of inducing akinetes in *Anabaena* sp. and studying the factors which favour the germination of akinetes namely nutrient concentrations, pH levels, wave length and intensity of light and making use of akinete induction and regermination as a means in the production of pure culture inoculum of cyanobacteria. The detailed methodology employed and the interesting finding are discussed.

INTRODUCTION

Cyanobacteria are a group of photosynthetic prokaryotes many of which are capable of fixing atmospheric nitrogen (Subbarao *et al.*, 1993). It is a well known fact that blue-green algae could easily contribute up to 25-30 kg of nitrogen per hectare per cropping season (Venkataraman, 1981). In addition, by their photosynthetic activity they also add considerable quanity of organic matter, which helps in improving soil properties. In heterocystous cyano-bacteria, which are the main source of blue-green algal biofertilizers, in the latter part of their life cycle, a resistant spore called `akinetes' is formed which after a spell of dormancy, germinate and give rise to new filaments. However, under certain cultural conditions majority or even all the vegetative cells are known to be temporarily transformed into akinetes which are known to germinate immediately without any dormancy (Nichols and Adams, 1982; Sutherland *et al.*, 1985; Agarwal and Sharma, 1994), where individual spores give rise to separate filaments after germination, resulting in many fold increase in cyanobacterial population. In the present study, an attempt has been made to evaluate the effect of physico-chemical parameters on akinete differentiation and germination and its application in biomass production in *Anabaena* sp.

MATERIAL AND METHODS

Cyanobacterial culture used for the study

The cyanobacterial culture used for the study *viz., Anabaena* sp., was obtained from culture collection center NCCUBGA, IARI, New Delhi

Effect of wavelength and intensity of light on akinete induction

To 100 ml of Fogg's medium one ml of homogenized cyanobacterial culture was inoculated. The flasks were covered with cellophane coloured sheets in triplicates with different shades namely blue, green, yellow and red. The light spectral range of these colours was measured by spectrophotometer. In another experiment, flasks were incubated in light chamber at different distance from the source of illumination such that each set of three flasks received different light intensity of 2000 lux, 1500 lux, 1000 lux and 500 lux. The spore count was taken in each treatment on 30th day and expressed as per cent akinete induction.

$$\text{Akinete induction frequency} = \frac{\text{Total number of akinetes}}{\text{Total number of vegetative cells}} \times 100$$

Effect of pH and nutrient concentration on akinete differentiation

Fogg's liquid medium was used for the study. In every experiment, one of the components of the Fogg's medium was used as the variable factor from complete absence to many fold higher concentration than the recommended level. The type of treatment used were as follows.

1. pH of the medium – 4.5, 5.5, 6.5, 7.5, 8.5, 9.5
2. KH_2PO_4 – 0, 0.01, 0.1, 0.2, 0.3, 0.4, 0.5 g/l
3. $MgSO_4.7H_2O$ – 0, 0.01, 0.2, 0.3, 0.4, 0.5 g/l
4. $CaCl_2$ – 0, 0.01, 0.1, 0.2, 0.3, 0.4 g/l
5. KNO_3 – 0, 0.25, 0.50, 0.75, 1.00, 1.50 g/l

One hundred ml volume of the medium with different treatments were inoculated with homogenized cyanobacterial culture at 1ml/flask in triplicates and were incubated in illumination chamber at 1500 lux light intensity for 30 days period. The algal suspensions were observed under microscope to decide the number of vegetative cells transformed into akinetes and per cent akinete frequency was calculated using the above mentioned formula

Isolation of akinetes

The culture medium containing spores were subjected to low speed centrifugation and the pellet formed was resuspended in sterile distilled water. It was subjected to mild sonication to disrupt the cell debris and few remaining vegetative cells. The akinetes were harvested by low speed centrifugation and stored in sterile distilled water at room temperature for germination studies.

Effect of wavelength and intensity of light on akinete germinaton

To 100 ml of Fogg's medium one ml of spore suspension was inoculated. The flasks were covered with cellophane coloured sheets in triplicates with different colour namely blue, green, yellow and red and incubated in light chamber at 1500 lux light intensity. The culture suspension was observed under the microscope to decide the time taken for initiation of spore germination and on 15th day percent spore germination was calculated. Similarly another set of culture flasks was maintained in triplicates at 2000 lux, 1500 lux, 1000 lux and 500 lux. Time taken for initiation of spore germination was expressed in days and spore germination was calculated in percentage.

$$\text{Percent akinete germination} = \frac{\text{Number of germinated akinetes}}{\text{Total number of akinetes}} \times 100$$

Effect of pH and nutrient concentration on akinete germination

To 100 ml of medium receiving different treatment as mentioned earlier, 1 ml of akinete suspension was inoculated in triplicate and the flasks were incubated in illumination chamber at 1500 lux light intensity. The culture suspension were observed periodically to determine the number of days taken for the initiation of spore germination which was indicated by change in colour of spores from yellowish green to blue green and division of spore protoplast into four germ cells. Germination was completed by the emergence of germling through ruptured spore cell wall. The percent spore germination was calculated using the above-mentioned formula.

Role of akinete induction and regermination on biomass production in cyanobacteria

In order to know the effect of repetition of sporulation and germination cycle on the built up of biomass of cyanobacteria, two experiments were conducted. In one series, to a set of six flasks containing 100 ml Fogg's medium, one ml of homogenized cyanobacterial cultures was added. Once in 15 days the culture media was carefully siphoned out and replaced with fresh medium. On 30th day first batch of three flasks were taken out, the algal contents were filtered which was then dried up at 60°C for six hours and the dry weight was noted. On 60th day the dry weight of the algae in the remaining three flask were determined. In another series, a batch of six flasks containing 100 ml of sporulation medium was inoculated with one ml of homogenized cyanobacterial culture. The flasks were incubated at 500 lux light intensity for 15 days. On 15th day, the medium in the flask was siphoned out and replaced with spore germination medium. The light intensity was increased to 2000 lux and incubated for next 15 days. On 30th day, a batch of three flasks were taken out and the algal dry matter was estimated. The remaining three flasks were subjected to another cycle of sporulation and germination for next 30 days and dry matter was determined on 60th day.

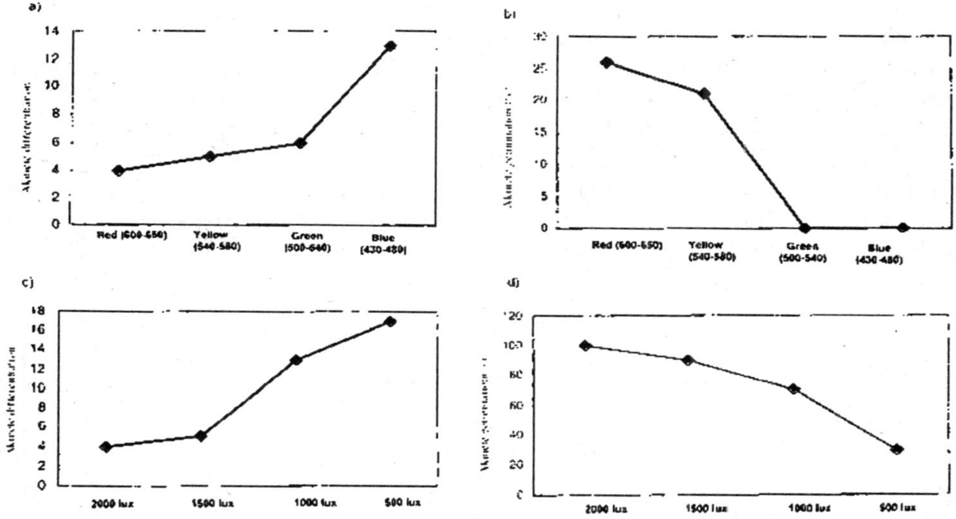

Fig. 1. Effect of wave length and intensity of light on akinete differentiation and germination in *Anabaena* sp.

RESULTS AND DISCUSSIONS

Influence of quality and intensity of light on akinete induction and germination

In the present study, it was observed that among the light wavelengths, blue light encouraged akinete induction whereas yellow and red light did not (Fig. 1a). Blue light being non photosynthetic wavelength, might play a role in switching on the spore formation process. As it has been shown by many workers that akinetes are not active photosynthetically and hence they contain less of photosynthetic and accessory pigments (Sutherland *et al.*, 1979 and 1985). Out of the four light wave length used, red light favoured germination followed by yellow (Fig. 1b), where as there was no germination in blue and green wavelengths. Red and yellow being photosynthetically active wavelengths, might be involved in the activation of biosynthesis of crucial enzymes and other molecules required for photosynthesis. Akinete formation was negatively correlated to light intensity. At lower light intensity, sporulation was maximum and at higher intensity sporulation was minimum (Fig. 1c). Sutherland *et al.* (1979) have shown in his studies that at higher light intensity akinete production occurred at high culture density. Germination percentage had direct correlation with light intensity (Fig. 1d). The lowest germination percentage was observed at 500 lux light intensity and highest percent germination was observed at 2000 lux. The works of Yamamoto (1976) and Agarwal and Sharma (1974) substantiates the present findings, in that spore germination was dependent on light intensity.

Culture media condition on akinete differentiation and germination

Among the pH levels studied, alkaline range of pH (Fig 2a) from 7.5 to 8.5 was found to favour akinete induction in *Anabaena* sp which has been observed by the previous workers (Nichols and Adams, 1982; Agarwal and Sharma, 1994). Alkaline range of pH encouraged the germination process and acidic range inhibited the same (Fig 2b). Since cyanobacteria prefer alkaline pH for their growth, cyanobacterial spores also might require the same conditions for the germination process. Similar observation were also made earlier by Rai and Pandey, 1981.

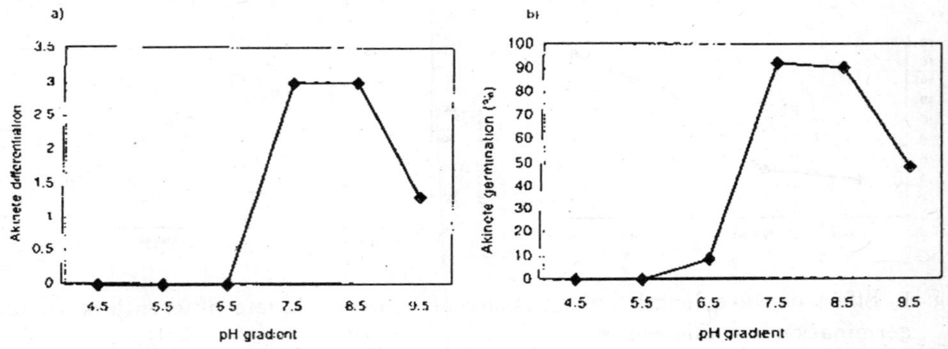

Fig. 2. Effect of pH gradient on akinete differentiation and germination in *Anabaena* sp.

Potassium nitrate increased the sporulation percentage at higher concentrations (Fig 3a) and recorded maximum sporulation at 1g/l level. There are numerous research evidences available for the role of nitrate in akinete induction process (Sutherland *et al.*, 1979, Sharma and Khattar, 1993). The results from Fig 3b showed that nitrate is required for germination at 0.25 g/l concentration at which maximum germination percentage was recorded. Since organisms require vital proteins during germination stage which are yet to start nitrogenase activity, exogenous supply of little nitrogen helps them to cope up with the situation. The work of Rai and Pandey (1981) also showed the same results. They reported that addition of nitrate to the medium induced early and maximum germination in *Anabaena vaganicola*.

Among the factors studied, phosphate deficiency was found to favour akinete formation (Fig. 4a). The treatments without phosphate as well as with 0.01g/l favoured akinete induction. The previous works of Wolk (1965), Agarwal and Sharma (1994) substantiates the results, in that a deficiency of phosphate stimulates akinete formation. Pandey *et al.* (1991) reported that an increase in phosphate concentration delayed spore initiation process and also decreased the frequency. The results from Fig. 4b showed that higher phosphate concentration encouraged quick and maximum germination. Since phosphate is involved in energy metabolism, it is most essential for young seedlings to emerge. Rai and Pandey (1981) noted that omission of phosphate resulted in a loss in germination of akinetes in *Anabaena vaginicola.*

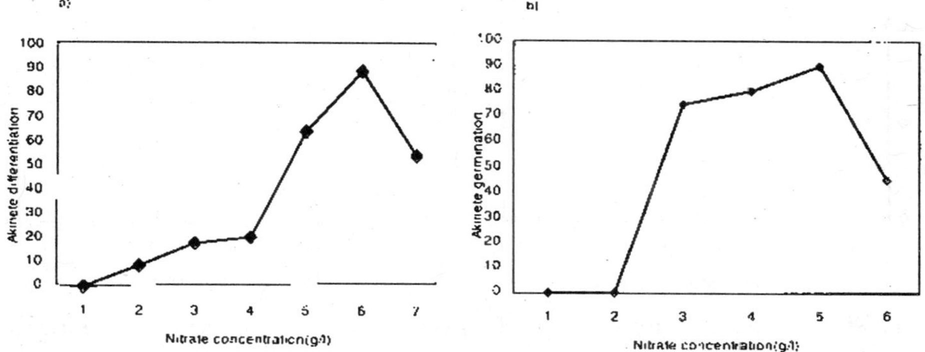

Fig. 3. Effect of Nitrate (KNO_3) concentration on akinete differentiation and germination in *Anabaena* sp.

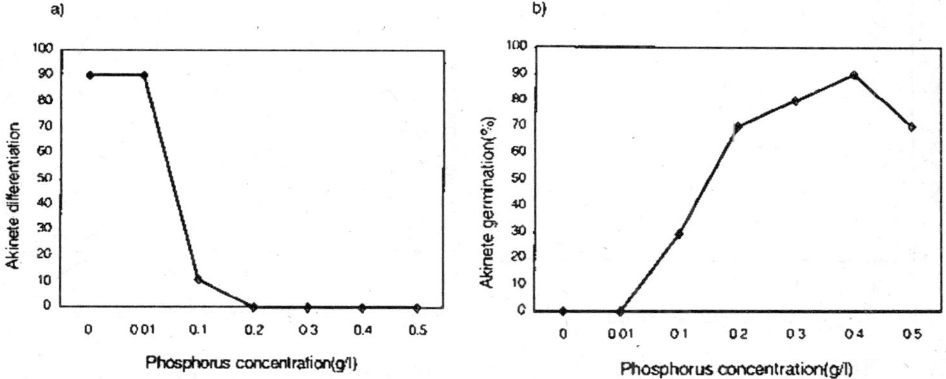

Fig. 4. Effect of Phosphorus (KH_2PO_4) concentration on akinete differentiation and germination in *Anabaena* sp.

Among the other two nutrient elements included in Fogg's medium, magnesium and calcium were found to stimulate the akinete induction at slightly higher concentration (Fig. 5a and Fig. 6a), than recommended in the medium. Calcium at 0.2 g/l level and magnesium at 0.4 g/l favoured akinete formation. In a similar study, Sinclair and Whitton (1977) have shown that a deficiency of calcium and magnesium decreased the frequency of akinetes. The data have indicated that both magnesium (Fig. 5b) and calcium (Fig. 6b) were required for germination. Magnesium sulphate at 0.4 g/l concentration encouraged germination. Its absence or very low concentration stopped germination process. Calcium chloride at 0.2 g/l concentration favoured germination process and at zero concentration there was no germination. Similar observations were made by Agrawal and Sharma (1994). Both calcium and magnesium being catalytic agents in many processes, may play a role in spore formation and spore germination.

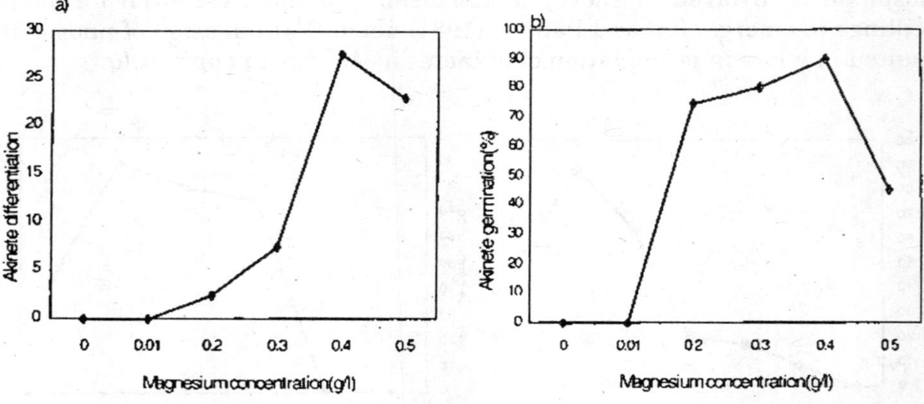

Fig. 5. Effect of Magnesium ($MgSO_4.7H_2O$) concentration on akinete differentiation and germination in *Anabaena* sp.

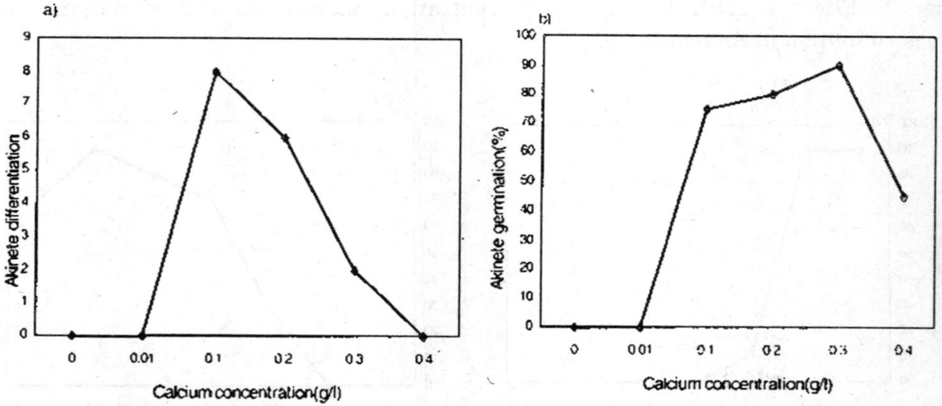

Fig. 6. Effect of Calcium ($CaCl_2.2H_2O$) concentration on akinete differentiation and germination in *Anabaena* sp.

Akinete induction, germination and cyanobacterial built up

The increase in biomass through a regular cycle of akinete induction and germination added to a fast built up of biomass compared to continuous vegetative growth (Fig. 7). Increase in biomass through sporulation and germination cycle was more than 50 times as compared to around 20 times increase in continuous vegetative growth. The reason behind this is, in case of sporulation followed by germination, every individual cell of the filament acts as a propagule, which will further give rise to new filaments. This increase takes place in logarithmic proportion.

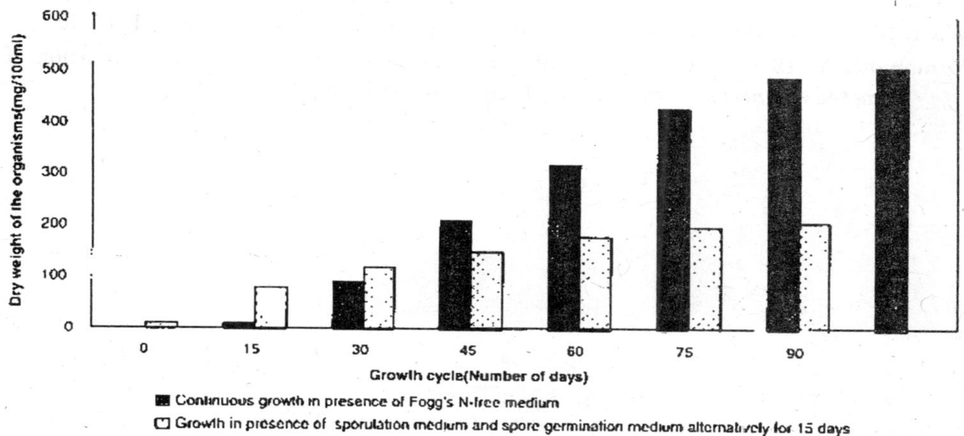

Fig. 7. Effect of akinete induction and regermination on biomass in *Anabaena* sp.

REFERENCES

Agarwal, S.C. and U.K. Sharma. 1994. Sporulation and spore germination in *Westiellopsis prolifica* in various culture condition. **Phykos, 33** (1-2) : 32-38.

Nichols, J.M. and D.C. Adams. 1982. Akinetes. In: **The biology of cyanobacteria** (eds.) Carr. N.G. and B.A. Whitton, Blackwell scientific publications, Oxford, pp. 384-412.

Pandey, K.D., S. Sarkar and A. K. Kashyap. 1991. Role of inorganic phosphate and alkaline phosphatase in sporulation of *Anabaena doliolum*. **Biochemie. Und. Physiologie. Der. Pelanzen., 187:** 439-445.

Rai, A.K. and B. P. Pandey. 1981. Influence of environmental stress on the germination of *Anabaena vaganicola* akinetes. **Ann. Bot., 48:** 361-370.

Sarma, T.A. and J.I.S. Khattar. 1993. Akinete differentiation in phototrophic, photoheterotrophic and chemoheterotrophic conditions in *Anabaena torulosa*. **Folia. Microbiologica., 38:** 335-340.

Sinclair, C. and B. A. Whitton. 1977. Influence of nutrient deficiency on hair formation in Rivuloriaceae. **Br. Phyol. J., 13**: 335-340.

Subbarao, N. S., G.S. Venkataraman and S. Kannaiyan. 1993. **Biological Nitrogen Fixation.** IARI publication.

Surendra Singh and A. K. Kashyap. 1998. Metabolic characteristics of akinetes of the cyanobacterium *Fischerella musicola*, **New. Phytol., 110** : 97-100.

Sutherland, J.N., W.D.P. Stewart and N. Herdman. 1979. Akinetes of the cyanobacterium *Nostoc pcc* 7524. Macromolecular composition, structure and control of differentiation. **J. Gen. Microbiol., 115** : 273-287 .

Sutherland, J.N., J. Herton, W.D.P. Stewart and N. Herdman. 1985. Akinetes of the cyanobacterium *Nostoc pcc* 7524. Macromolecular and biochemical changes during synchronous germination. **J. Gen. Microbiol., 131** : 2855-2863.

Venkataraman, G.S. and S. Shanmugasundaram 1992. **Algal biofertilizer technology for rice.** DBT centre of BGA biofertilizers, Madurai Kamraj University, Madurai.

Wolk, C.P. 1965, Control of sporulation in blue- green algae. **Dev. Biol., 12:** 15-35.

Yamamoto, Y. 1976. Effect of some physical and chemical factors on germination of akinetes of *Anabaena cylindrica*. **J. Gen. Microbiol., 22:** 311-323.

CHAPTER - 9

CYANOBACTERIAL BIOFERTILIZERS FOR PROBLEM SOILS

G. Gopalaswamy

INTRODUCTION

The agronomic potential of cyanobacterial N_2 fixation in rice fields was first recognized in India during 1939 by De, who attributed the natural fertility of tropical rice fields to N_2 fixing blue green algae. Diazotrophic cyanobacteria have been found to play a vital role in sustaining the cultivation of wet land rice without added nitrogen fertilizer due to their inherent ability to add fixed nitrogen to such habitats under aerobic photosynthetic conditions (Singh, 1961). Many cyanobacteria fix nitrogen under aerobic conditions in specialized cells called heterocysts, which comprise 5-10% of cells in a filament (Fleming and Haselkorn, 1973). Besides fixing atmospheric nitrogen, cyanobacteria play a major role in reducing soil erosion, increasing the soil organic matter content and in producing certain substances which enhance the growth of higher plants (Venkataraman *et al.*, 1974). In recent years, the practice of utilizing cyanobacteria as an efficient source of biofertilizer for rice has been advocated and adopted in India (Venkataraman, 1981, Kannaiyan, 1990). The rice field ecosystem provides an environment favourable for the growth of cyanobacteria with respect to their requirements for light, water, high temperature and nutrient availability. However, cyanobacteria in rice fields are subjected to various stresses like salinity, acidity *etc.* which reduce the population and nitrogen fixation. Many agricultural ecosystems are increasingly becoming salt affected, which ultimately has an adverse effect on crop production. Salt affected soils collectively called as 'usar' soils comprising saline, saline sodic and sodic soils cover more than 7 million ha in India (Bhumbla and Abrol, 1978). High pH, low organic matter content and high calcium carbonate in soil strongly limit the availability of micronutrients to crops grown in sodic soil. Singh (1961) was the first to suggest the feasibility of reclaiming such salt affected soils with nitrogen fixing blue green algae (cyanobacteria). The salt tolerant cyanobacteria immobilize Na^+ in an organic form and thereby play a major role in saline soil condition. The performance of cyanobacteria in normal and saline soils are reviewed by Kannaiyan (1995). However, very little information is available on the performance of cyanobacteria in acid soils. In Tamil Nadu, acidic soils are prevalent in Kanyakumari district, parts of Pudukkottai district, Pattukkottai and Peravurani taluks. Selection of cyanobacteria cultures for problem soils is worth investigating.

CYANOBACTERIA FOR SALINE SOILS

Cyanobacteria react to salt stress either through osmotic adjustment or by secretion of excess polysaccharide (Saxena and Kaushik, 1991). Salt tolerance in cyanobacteria is mediated by an osmo regulatory mechanism which involves the accumulation of carbohydrates, amino acids and quaternary ammonium compounds (Reed and Stewart, 1988). Salt adaptation strategy of cyanobacteria includes two main processes *viz.*,(i) Maintenance of low internal contents of inorganic ions by active export mechanism and (ii) Synthesis and accumulation of osmoprotective compounds corresponding to the osmotic potential of the surrounding medium. The first process is achieved by Na^+ / H^+ antiporters that are coupled to cytochrome oxidase (Molitor *et al.*, 1986).

Several compounds like sucrose, trehalose, glucosyl glycerol and glycine acting in the second process were identified in salt loaded cyanobacteria (Reed and Stewart, 1988). The physiological basis of salt tolerance may be due to extrusion of Na^+ and maintenance of low intracellular Na^+, accumulation of K^+ or ionic balance between K^+ and Na^+ (Apte and Thomas, 1986).

There is an increasing evidence that respiration plays a primary role in the adaptation of cyanobacteria to higher concentration of salt (Wastyn *et al.*, 1987). In *Anacystis nidulans* increasing salt concentration led to decreased photosynthesis and respiration (Vonshak *et al.*, 1988). When cyanobacterial cells were exposed to salt stress, a short term increase in the cellular sodium concentration was noticed which was due to a transient increase in the permeability of plasma membrane during the first few seconds of exposure to high salt and that led to the immediate inhibition of photosynthetic and respiratory systems (Ehrenfeld and Cousin, 1984).

With increased sodium chloride concentration, the filaments of *Anabaena doliolum*, a salt sensitive fresh water cyanobacterium was shorter with less number of heterocysts which was an indirect evidence of reduced nitrogen fixation, decreased protein and chlorophyll-a contents (Rai and Abraham, 1993). *Synechocystis* sp. PCC 6803 when grown in different salt concentrations, exhibited optimum growth with maximum photosynthesis per cell volume, chlorophyll and phycocyanin at 34.2 mM Nacl (Schubert *et al.* 1993).

Anand *et al.* (1994) have reported that *Chroococcus minor* and *Oscillatoria salina* tolerated upto 3% salinity while, *Nostoc piscinale, Tolypothrix tenuis* and *Gloeocapsa polydesmatica* failed to grow beyond 1.5% salinity. The cells of *Anabaena* strain L-31 when exposed to 350 mM sucrose, 10 mM KNO_3, osmotic stress induced proteins (OSPS) were stimulated which resulted in significant osmoprotection (Iyer *et al.*, 1994). Several saline tolerant cyanobacterial cultures were isolated from saline soils. They were characterized biochemically. Based on the performance, Anabaena, Nostoc and Westiellopsis were selected for saline soil inoculations (Amsaveni and Kannaiyan, 1995). Inoculation of composite culture of saline tolerant cyanobacteria with nitrogen and phosphorus sources increased the yield of CO 43 rice significantly (Tables 1 and 2).

Table 1. Effect of inoculation of salt tolerant cyanobacteria and fertilizer nitrogen on grain and straw yield of CO 43 rice

Treatment	Grain yield (g / pot)	% increase over control	Straw yield	% increase over control
Control	17.51	-	49.10	-
Cyanobacteria	19.64	12.16	51.59	5.07
Prilled urea (60 kg Nha^{-1})	19.88	13.53	51.73	5.36
PU (60 kg Nha^{-1}) + Cyanobacteria	21.13	20.67	54.97	11.95
Urea super granule (60 kg N ha^{-1})	20.90	19.36	53.46	8.87
USG (60 kg N ha^{-1}) + Cyanobacteria	22.95	31.03	50.28	14.62

Table 2. Effect of inoculation of salt tolerant cyanobacteria and phosphorus on grain and straw yield of CO43 rice

Treatment	Grain yield (g / pot)	% increase over control	Straw yield	% increase over control
Control	17.93	-	48.95	-
Cyanobacteria	20.15	12.38	51.84	14.15
SSP (50kg P / ha)	20.57	14.72	54.64	9.94
SSP (50kg P / ha) + cyanobacteria	24.62	37.31	64.93	15.50

CYANOBACTERIA FOR ACID SOILS

In India, out of a total of 157 million hectares of cultivable area, 49 million hectares of land are acidic in nature (Das, 1996). Generally they are deficient in available nutrients which are important for the life of all organisms. Only iron and aluminum are available in excess and are toxic to many microorganisms. Moore (1963) reported the occurrence of many cyanophyceae in acid soils. Many cyanobacterial cultures were isolated from acid soils of Kerala having a pH of 3.8 (Aiyer, 1963; Madhusoodanan and

Dominic, 1999). The survival of cyanobacteria under acid soil rice field condition was studied by Jurgensen and Davey (1968). Later Sardeshpandey and Goyal (1981 a) observed that the native isolates survived better under acid soils than the introduced cultures. Singh *et al.* (1997) isolated many acid tolerant strains of cyanobacteria from acid soils of Nagaland.

Recently, several efficient cyanobacterial cultures were isolated from the acid soils of Ambasamudram, Tamil Nadu. (Table 3). The predominant genera were *Anabaena* (50%), *Westiellopsis* (17.5%), *Nostoc* (15%), *Oscillatoria* (5%) and other algae (12.5%). Interestingly *Westiellopsis* was found from pH of 4.9-6.2, indicating its adaptability. This study showed the presence of cyanobacterial cultures in acidic soils and is in agreement to the findings of Aiyer (1965) and Madhusoodanan and Dominic (1999) who were successful in isolating the cyanobacterial cultures from acidic soils.

Acid tolerant cyanobacterial cultures produced higher growth and biomass than non acid soil isolate *Anabaena* - NS-TGK-A1. Among the acid tolerant cultures, *Westiellopsis* - AT-TGK-W1, showed a significant increase in growth and biomass production compared to non acid soil *Anabaena* - NS-TGK-A1 (Table 4). At acidic pH (4.0 and 5.0) *Anabaena* - NS-TGK-A1 turned to brown colour, while it retained the blue green colour at pH 7.0. On the other hand *Westiellopsis* –

Table 3. Acid tolerant cyanobacterial cultures isolated from acid soils

Sample No.	Soil pH range	Cyanobacterial isolates
A1-A10	5.02 - 6.02	*Anabaena*
		Oscillatoria
		Microcystis
		Pseudoanabaena
		Westiellopsis
		Nostoc
		Arthrospira
		Synechococcus
		Calothrix
B1-B10	4.91-6.19	*Anabaena*
		Nostoc
		Gleotheca
		Oscillatoria
		Westiellopsis
		Phormidium
		Aphanocapsa
C1-C10	5.14 - 7.10	*Nostoc*
		Oscillatoria

Anabaena
Westiellopsis
Microcystis
Gleocapsa
Chroococcus

Table 4. Growth and biomass production by acid tolerant cyanobacteria at low pH

Cyanobacterial cultures	pH	Growth (OD at 750 nm)	Biomass (mg ml^{-1})
Anabaena – NS-TGK-A1	4	0.405	34.18
	5	0.500	47.64
	6	1.197	53.71
Anabaena-AT-TGK-A1	4	0.767	38.78
	5	1.086	49.25
	6	1.280	60.93
Nostoc-AT-TGK-N1	4	1.059	45.50
	5	1.227	56.89
	6	1.314	64.58
Oscillatoria –AT-TGK-O1	4	1.332	65.17
	5	1.377	66.14
	6	1.401	66.80
Westiellopsis-AT-TGK-W1	4	1.159	64.58
	5	1.166	68.10
	6	1.626	67.84
Westiellopsis-AT-TGK-W2	4	0.761	65.93
	5	1.206	69.22
	6	1.465	70.96

	SEd	CD	SEd	CD
PH	0.02	0.04	0.99	1.97
Cultures	0.02	0.05	1.40	2.78
pH x Cultures	0.03	0.06	2.43	4.82

AT-TGK-A1 retained the green colour even at pH 4.0. Reduction in pH to the acidic range led to the sudden reduction in growth and chlorophyll - a content with alteration in branching habit of *Westiellopsis prolifica* (Biswal *et al.,* 1975). However, due to adaptability of cyanobacterial cultures to acidic soils having a pH of 4.9 to 6.2, no such reduction was observed in *Westiellopsis* cultures isolated from acidic rice soils. This stressed the importance of location specific strains for desirable response to acidity, which have high adaptability to native acidic soils (Sardeshpandey and Goyal,1981b).

Cyanobacteria harbour light harvesting pigments like, chlorophyll-a, phycocyanin, allophycocyanin, phycoerythrin and other phycobiliproteins (Smith and Roger, 1988). All acid tolerant cyanobacterial cultures recorded higher chlorophyll-a and phycobilin pigments. Among them, *Westiellopsis* - AT-TGK-A1 recorded the highest chlorophyll-a and phycobilin pigments at all acidic pH ranges (Table 5). This is in agreement with earlier findings where maximum chlorophyll-a was recorded in *Anabaena* ARM-220 at acidic pH of 5.0 (Subramanian and Shanmugasundaram, 1987).

Table 5. Photosynthetic pigments production by acid tolerant cyanobacteria at low pH

Cultures	pH	Chlorophyll-a*	Phycocyanin*	Allophycocyanin*	Phycoerythrin*
Anabaena – NS-TGK-A1	4	32.35	11.6	7.67	7.08
	5	37.78	8.10	6.54	5.56
	6	41.32	7.98	8.19	5.11
Anabaena- AT-TGK-A1	4	52.63	9.28	6.54	6.47
	5	40.81	9.08	9.17	7.48
	6	33.33	8.58	10.21	4.88
Nostoc-AT-TGK-N1	4	44.4	8.08	8.29	5.20
	5	40.81	7.87	7.56	5.19
	6	33.33	7.59	8.16	4.90
Oscillatoria – AT-TGK-O1	4	22.22	7.70	8.25	5.82
	5	20.40	8.87	9.59	5.79
	6	37.80	10.0	9.67	6.57
Westiellopsis- AT-TGK-W1	4	46.15	6.69	7.61	4.32
	5	43.47	8.04	7.90	4.60
	6	42.80	10.48	10.72	5.00
Westiellopsis- AT-TGK-W2	4	46.87	6.64	7.34	4.28
	5	43.47	7.78	7.54	4.52
	6	42.88	9.08	9.56	5.07

	SEd	CD	SEd	CD	SEd	CD	SEd	CD
pH	0.81	1.66	0.18	0.36	0.18	0.38	0.14	0.28
Cultures	1.61	2.35	0.25	0.52	0.26	0.53	0.20	0.04
pH x Cultures	2.01	4.08	0.44	0.90	0.46	0.93	0.36	0.70

* Values in mg g^{-1}

Inoculation of acid tolerant a composite (cyanobacterial) culture in plastic trays with acidic soils had very good growth. Quick growth in trays was obtained in three weeks after the inoculation by adding water in the trays continually. In general, the acid tolerant cyanobacterial cultures applied individually and in composite recorded significant increase in plant height compared to uninoculated control at all observation periods. Between the individual and combined inoculation, application of composite cultures at the rate of 1 g fresh wt per 0.06 m^2 (16.6 kg dry wt ha^{-1}) significantly increased the plant height which was 39% over uninoculated control on the 40th day. Among the individual applications, *Westiellopsis* AT-TGK-W1 culture increased plant height significantly. However the effect was very much pronounced with the composite inoculum. The acid tolerant cultures grew well and covered the whole surface of the soil in 30 days after inoculation. The higher accumulation of the chlorophyll and nitrogen content in rice seedlings might possibly be attributed to the vigorous growth of cyanobacteria, which facilitated the uptake of nutrients, by rice seedlings. The results clearly revealed that the inoculation of acid tolerant cyanobacteria would increase the growth of rice seedlings by virtue of their continuous photo production of ammonia in the flood water system.

Inoculation of acid tolerant cyanobacteria to rice crop in acid soil caused desirable quantum jump in the growth and yield of ASD 20 rice (Tables 6 and 7). The yield increase might be due to secretion of vitamins, plant hormones and increased uptake of nutrients by rice. However, Nair *et al.* (1993) and Watanabe (1993) reported negative yield increase due to algalization in acid soils. This might be due to the use of cultures, which were not native to acid soils. This underlines the importance of using locally adapted strains for a desired response. Prasad *et al* (1978) noticed luxuriant growth and active N_2 fixation by some algal form at low pH, this points towards the possibility of isolating acid tolerant forms which can be used in cyanobacterial rice fields with low pH.

Table 6. Effect of inoculation of composite acid tolerant cyano-bacterial cultures on plant height and number of tillers per hill of ASD 20 rice.

Treatments	Plant height (cm)				No. of tillers $hill^{-1}$			
	30th day	60th day	90th day	At harvest	30th day	60th day	90th day	At harvest
Composite culture	10.60	29.80	49.70	53.70	2.28	3.71	4.02	4.12
Uninoculated	9.80	25.63	43.77	45.60	1.42	2.71	3.13	3.28
		SEd		CD		SEd		CD
Days		0.16		0.43		0.31		0.43
Treatment		0.11		0.23		0.22		0.23
Days x Treatment		0.23		0.46		0.44		0.46

Table 7. Effect of inoculation of composite culture of acid tolerant cyanobacteria on the nitrogen status and grain yield of ASD 20 rice.

Treatment	Nitrogen status			
	Nitrogen (%)	Nitrogen uptake (Kg ha^{-1})	Expected yield (g m^{2-1})	% increase over control
Uninoculated	0.52	3.631	270.24	-
Composite culture	0.63	4.725	478.80	43.5
SEd	0.002	0.463	0.1089	-
CD	0.001	0.334	0.2308	-

MASS MULTIPLICATION UNDER FIELD CONDITIONS

Mass production of dried soil based algal flakes under rice field condition was developed by Kannaiyan (1979). The field selected for mass multiplication of cyanobacteria is prepared well and levelled uniformly. The field is divided into small plots size of 20m x 2m (40 sqm), by providing suitable bunds and irrigation channels. Water is maintained to a height of 10 cm. Super phosphate at 2 kg is added to the plot and inoculated with 5 kg of starter culture of cyanobacteria. Carbofuran (Furadan) at 200 g is also added to prevent the activities of soil insects. Water is let in at periodical intervals, so that the height of water level is always maintained at 10cm. The growth of algae is rapid and in about 10 days, a thick mat is formed which floats on the surface of the water. After 15 days of inoculation, the plots are allowed to dry up in the sun and the algal flakes collected and stored. From each harvest, 30 - 40 kg of dry algal flakes are obtained per plot.

MODE OF NITROGEN AVAILABILITY FOR RICE

It is now well established that N_2 fixation by cyanobacteria plays a vital role in the build up and maintenance of soil fertility but it is equally important to understand how much, when and in what ways the fixed N is made available to rice plants. Nitrogen fixed by the cyanobacteria are released either through exudation or through microbial decomposition after cell death. Laboratory experiments have frequently shown that cyanobacteria liberate large portions of their assimilated nitrogenous substances. Release of nutrients through microbial decomposition after the death of the cyanobacteria appears to be the principal means by which N is made available to the rice crop. Some cyanobacteria decomposed in 2-3 days, while others withstand microbial digestion for more than 4 weeks. A strain of *Bacillus subtilis* was found to decompose several N_2 fixing cyanobacteria very rapidly, about 40% of cyanobacterial nitrogen being converted to ammonia within 10 days. Grazers through their digestive tracts also make cyanobacterial nitrogen available to rice.

NITROGEN AND PHOSPHORUS CONSERVATION BY CYANOBACTERIA

Certain N_2 fixing as well as non N_2 fixing cyanobacteria were known to immobilize nitrogen during the decomposition of organic materials. By the process of temporary immobilization of nitrogen by cyanobacteria conserve the nitrogen in their cell system and prevent nitrogen loss in rice field. The N_2 fixing *Oscillatoria sp* plays significant role in conserving nitrogen in rice soil. N_2 fixing cyanobacteria also conserve the phosphorus in soil. The organic acids excreted by them solubilise the unavailable phosphorus and utilise for its growth and N_2 fixation. Besides phosphorus is released into the soil solution during decomposition. Cyanobacteria store the phosphorus in the form of polyphosphate granules in their cell system and it releases into the soil during decomposition.

CONCLUSION

Cyanobacteria are ideal biological systems under tropical situation for better nitrogen fixation and conservation in rice soil. They are renewable, self-supporting biological agents. The development of cyanobacteria for problem soils has several advantages like high adaptability and biomass production. Rice cropping system provides favourable aquatic environment for the growth, multiplication and nitrogen fixation by utilizing the solar energy. Thus, these potential group of nitrogen fixing cyanobacteria may be better exploited for harnessing atmospheric nitrogen so as to increase the productivity of rice in problem soils.

ACKNOWLEDGEMENT

The authors are thankful to the Indian Council of Agricultural Research, New Delhi for the financial assistance under National Agricultural Technology Project-Team of Excellence on Biofertilizers for Rice Based Cropping System.

REFERENCES

Aiyer, R. S. 1965.Comparative algalogical studies in rice fields of Kerala State. **Agri. Res. J. Kerala 3**: 100-104.

Amsaveni, P. and S. Kannaiyan. 1995. Influence of nitrogen and phosphorus sources on the growth, nitrogen fixing activity and ammonia excretion of salt tolerant cyanobacteria. In: **National Seminar on *Azolla* and Algal Biofertilizers for rice,** (ed). S. Kannaiyan, Tamil Nadu Agri, Univ., Coimbatore, Tamil Nadu India, pp.10.

Anand, N., R.S.S.Hopper and G.Jagatheswari. 1994. Response of certain blue-green algae (cyanobacteria) to salinity. In : **Recent Advances in Phycology** (eds.) A.K.Kashyap and H.D.Kumar, Rastogi Publications, Varanasi, India. pp.125-129.

Apte, S.K. and J.Thomas. 1986. Membrane electrogenesis and sodium transport in filamentous nitrogen fixing cyanobacteria. **Eur. J. Biochem., 154**:395-401.

Biswal, U.G., S. N., Padhy, and H. Pattnaik. 1975. Growth and branching frequency of *Westiellopsis prolifera* Janet at different pH levels. **Phykos, 14** : 113 - 115.

Bhumbla, D. R. and I. P. Abrol. 1978. Saline and sodic soils. In: **Soils and Rice**, Inte. Rice Res. Inst., Los Banos, Philippiness, pp.718-738.

Das, D.K.1996. Soil acidity. In: **Introductory soil science**. Kalyani Publishers. Ludhiana, India, P. 498.

De, P.K. 1939. The role of blue green algae in nitrogen fixation in rice fields, **Proc. Res. Soc. Lond., 127**: 121-139.

Ehrenfeld, J. and J.L.Cousin. 1984. Ionic regulation of the unicellular green alga, *Dunaliella teritolecta* : response of hypertonic shock. **J. Membre. Biol., 77**:45-55

Fleming, H. and R. Haselkorn.1973. Differentiation in *Nostoc muscorum* – Nitrogenase is synthesized in heterocysts. **Proc. Natl. Acad. Sci.** USA **70**: 2727-2731.

Iyer, V., T. Fernandes and S. K. Apte. 1994. A role for osmotic stress induced proteins in the osmotolerance of nitrogen fixing cyanobacterium, *Anabaena* sp. strain L-31. **J. Bacteriol., 176(18)**:5868-5870.

Jurgensen, M. F. and C. B. Davey. 1968. Nitrogen fixing blue green algae in acid forest and nursery soils. **Can. J. Microbial., 14**: 1179-1183.

Kannaiyan, S. 1979. Blue-green algal biofertilizer for rice crop. **Farmers and Parliament, 14**(11): 19-25.

Kannaiyan, S. 1990. **Biotechnology of biofertilizer for rice crop**. Tamil Nadu Agric. Univ., Coimbatore, Tamil Nadu, India. p.212.

Kannaiyan, S. 1995. Biofertilizers for rice-status and scope. In: **Rice Management Biotechnology** (ed). S. Kannaiyan, Assoc. Pub. Co., New Delhi, pp. 237-265.

Madhusoodanan, P. V. and T.K. Dominic. 1999. Cyanobacteria from extreme acidic environments. **Curr. Sci., 77**:1021-1023.

Molitor, V., W.W. A. Erber and G. A. Peschek. 1986. Increased levels of cytochrome-c-oxidase and sodium antiporter in the plasma membrane of *Anacystis nidulans* after growth in sodium enriched media. **FEBS Lett., 204**:151-255.

Moore, A.W. 1963. Occurrence of non-symbiotic nitrogen fixing microorganisms in nitrogen soils. **Plant and Soil, 19** : 385-395.

Nair, S.K., R.S.Shenana, V.K.Girija and K.S.Meena Kumari. 1993. Sensitivity of blue green algae to soil reaction - a factor affecting its efficient use as biofertilizer **J. Tropical Agric., 31**: 116 - 118.

Prasad, E.N. and R.K.Mehrotra and Y. Singh. 1978. On pH tolerance of some soil blue green algae. **Acta Botanica Indica, 6**:130-138.

Rai, A. K. and G. Abraham. 1993. Salinity tolerance and growth analysis of the cyanobacterium, *Anabaena dolionum*. **Bull. Environ. Cont. and Toxicol., 51**(5):724-731.

Reed, R.H. and W.D.P. Stewart. 1988. The responses of cyanobacteria to salt stress. In : **Biochemistry of the Algae and Cyanobacteria**. (eds.) L. J. and J. R. Gallon, Clarendon Press, Oxford, pp.217-237.

Sardeshpandey, J.S. and S.K. Goyal. 1981a. Distributional Pattern of blue green algae in rice field soils of Konkan region of Maharastra State. **Phykos, 20**: 102-106.

Sardeshpandey, J.S. and S.K. Goyal. 1981b. Effect of pH on growth and nitrogen fixation by blue green algae. **Phykos 20**:107-113.

Saxena, S. and B.D.Kaushik. 1991. Polysaccharides (biopolymers) from halotolerant cyanobacteria. **Indian J. Exptl. Biol., 30**:433-434.

Schubert, H.S. Fulda and M. Hagemann. 1993. Effects of adaptation of different salt concentration on photosynthesis and pigmentation of the cyanobacterium, *Synechocystis* sp. PCC 6803. **J. Plant Physiol., 142**:291-295.

Singh, N.I., H. Dorycanta, G.A. Devi, N.S. Singh and M. S. Singh. 1997. Blue green algae from rice field soils of Nagaland. **Phykos, 36(182)**.: 115-120.

Singh, R.N. 1961. The role of blue-green algae in nitrogen economy of Indian Agriculture, ICAR publication, New Delhi, India.

Smith, A.J. and L.J. Rogers. 1988. Tetrapyrrole biosynthesis of the C_5 pathyway. In: **Biochemistry of the algal and cyanobacteria** (eds.) Clarendon Press, Oxford. pp. 69-96.

Subramanian, G. and S. Shanmugasundaram. 1987. The influence of nitrate, pH and light duration on the growth of *Anabaena*. **Phykos, 26**: 159-164.

Venkataraman, G.S. 1981. Blue green algae: a possible remedy to nitrogen scarcity. *Curr. Sci.*, **50:** 253-256.

Venkataraman, G.S., S.K. Goyal, B.D. Kaushik and P. Roychoudhury. 1974. Algal: Forum and Function. Today and Tomorrow printers and publishers, New Delhi, pp.562.

Vonshak, A., R.Guy and M.Guy. 1988. The response of the filamentous cyanobacterium, *Spirulina platensis* to salt stress. **Arch. Microbiol., 150**: 417-420.

Wastyn, M., A.Achatz, M.Trnka and G.A.Peschek. 1987. Immunological and special characterisation of partly purified cytochrome oxidase from the cyanobacterium, *Synechocystis* 6714. **Biochem, Biophys. Res. Commun., 149**:102-111.

Watanabe, A. 1973. On the inoculation of paddy fields in the Pacific area with nitrogen fixing blue green algae. **Soil Biol. Biochem.,** 5:161-162.

CHAPTER - 10

ASSESSMENT OF GROWTH AND BIOMASS PRODUCTION BY CYANOBACTERIA ISOLATED FROM DIFFERENT SOIL ENVIRONMENTS

G. Selvakumar, G. Gopalaswamy and S. Kannaiyan

ABSTRACT

Cyanobacteria are unique photosynthetic diazotrophs that have contributed to the fertility of rice fields for centuries. The mass multiplication of cyanobacteria for field application has gained momentum and is seen as a viable option to cut down fertilizer costs. The soil based cyanobacterial inoculum is one of the primary modes of algalization of the rice fields. In such a scenario the actual biomass production by the cyanobacteria needs to be estimated. Thirteen elite strains of cyanobacteria, isolated from diverse environments *viz.*, acid soils, saline soils, herbicide applied soils and normal soils were used to study the biomass production in soil (pH 7.6) under *in vitro* conditions for a period of 30 days. They were incubated under sunlight and a uniform water level was maintained throughout the experimental period. At end of the experimental period the cyanobacterial biomass was estimated and the growth parameters *viz.,* doubling time and relative growth rate were worked out. Among the isolates, *Westiellopsis* – PA-SK-ST_3 isolated from the saline soils of Tamil Nadu recorded the maximum biomass (21.22g m^{2-1}). The relative growth rate of the culture *Westiellopsis*–PA-SK-ST_3 was found to be highest (0.10 g g^{-1} day^{-1}), which also registered the lowest doubling time (6.75 days) when soil was used as a medium. The results clearly indicated that the saline tolerant isolate, *Westiellopsis*-PA-SK-ST_3, performed better than the other isolates in terms of growth rate and biomass production under normal soil conditions.

INTRODUCTION

The existence of an agronomic potential for cyanobacteria was at first recognized by De (1939) who attributed the fertility of tropical rice fields to biological nitrogen fixation by these organisms. The extent to which the cyanobacteria contribute to the nitrogen requirement of the crop is determined mostly by the cyanobacterial standing biomass, rate of nitrogen fixation, turnover of the fixed nitrogen and the extent to which the fixed nitrogen becomes available to the plant (Roger, 1991). An attempt was made to determine the biomass production by the cyanobacterial cultures isolated from various ecosystems in order to evaluate their agronomic potential.

MATERIALS AND METHODS

Wetland soil (pH 7.6) was filled in tubs of 9 cm ht and 29 cm dia @ 2 kg per tub. The tubs were copiously watered, sprinkled with super phosphate and incubated in the sunlight for a period of 21 days. All the surfacing natural population was removed, and it was made sure that the soil was fairly free of any residual natural population. The tubs were then inoculated with the cyanobacterial cultures *viz.*, *Westiellopsis* – AT –TGK $4A_2$, *Westiellopsis* - AT – TGK – $4A_7$, *Westiellopsis* – AT –TGK - 5 B_2 (isolated from acid soils); *Westiellopsis* - HT - SGK -1, *Westiellopsis* – HT – SGK – 2 (isolated from herbicide applied soils), *Westiellopsis*–PA – SK- ST_3 (isolated from saline soils) *Westiellopsis* – NSG 22 (normal soil isolate), *Westiellopsis prolifica* IARI 906, *Westiellopsis* –ARM 48, *Anabaena variabilis* PCC 7120 @ 0.5g fresh wt per tub. A composite inoculum comprising all the cultures was also maintained. The tubs were copiously watered and incubated in sunlight for a period of 30 days. On the 30th day, the algal biomass was carefully skimmed off and washed repeatedly with water by sedimentation technique to get rid of all the adhering soil particles. The weight of the harvested biomass was then recorded and the growth parameters *viz.*, doubling time and the RGR were worked out using the formula,

DT = t/r

t = Experimental period;

$$r = \frac{\log (W_1/W_0)}{0.301}$$

W_1 = final weight of the cyanobacterial biomass

W_0 = weight of the inoculum

RGR = 0.693 / DT

RESULTS AND DISCUSSION

Among the various cultures *Westiellopsis* - PA –SK - ST_3 isolated from the saline soils of Tamil Nadu recorded the maximum biomass of 21.22 $gm^{2\text{-}1}$ (Table 1). The cultures also registered the least doubling time of 6.75 days and the maximum RGR of 0.10 g day^{-1}. This was closely followed by the composite inoculum which registered a biomass of 20.06 gm^{2-1} with a doubling time of 9.83 days and a RGR of 0.07 g day^{-1}.

Table 1. Biomass production and growth rates of the cyanobacterial isolates from different soil environments

	Inoculum wt (g)	Final wt (g)	Doubling time (days)	RGR (g day^{-1})	Biomass Production g m^{2-1}
Westiellopsis-AT-TGK-4A7	0.5	9.50	7.07	0.09	18.36
Westiellopsis-AT-TGK-5B2	0.5	5.28	8.82	0.07	9.59
Westiellopsis-AT-TGK-4A2	0.5	7.81	7.57	0.09	14.91
Westiellopsis-HT-SGK-1	0.5	5.30	8.82	0.07	9.79
Westiellopsis- HT-SGK-2	0.5	4.40	9.58	0.07	8.97
Westiellopsis-ST-PA-SK-3	0.5	10.90	6.75	0.10	21.22
Westiellopsis prolifica (IARI 906)	0.5	3.50	10.71	0.06	7.14
Westiellopsis ARM 48	0.5	1.57	18.12	0.03	3.20
Westiellopsis NSG 22	0.5	4.60	9.37	0.07	8.36
Anabaena PCC7120	0.5	9.70	7.02	0.09	18.77
Composite inoculum	0.5	7.25	9.83	0.07	20.06

Determination of the biomass production by a standing mass of inoculated cyanobacteria becomes essential in the context of N contribution by the cyanobacteria, since most of the fixed nitrogen by the cyanobacteria becomes available only after the death and decomposition of the cyanobacteria. Values reported for the standing cyanobacterial biomass in rice fields have ranged from a few kg ha^{-1} to 0.5 t ha^{-1} dry wt (Roger *et al.,* 1988). Assuming a maximum biomass of 0.5 t dry wt. per ha and using average ash and N values obtained, the potential average contribution of a bloom of cyanobacteria could be estimated at 15 kg N ha^{-1} (Roger and Watanabe, 1986). In the present investigation the maximum biomass recorded was 2.12 t ha^{-1} on a fresh wt basis which is in agreement with previously recorded values. The commonly followed practice of using a composite inoculum for algalization produced 2 t ha^{-1}, which may correspond to 10-15 kg N ha^{-1}. While a visible bloom of cyanobacteria may contribute less than 10 kg N ha^{-1}. A dense bloom of cyanobacteria may correspond to 10-20 kg N ha^{-1}. Such biomass measurements gain significance since they provide a fairly accurate estimate of the contribution by the inoculated cyanobacteria. Hence the potential of the cyanobacterial cultures to contribute to the long term fertility of the rice fields is amply proved.

REFERENCES

De, P.K. 1939. The role of Blue Green Algae in nitrogen fixation in rice fields. Proc. Roy. Soc. London, 127 : 121-139.

Roger, P.A. 1991. Reconsidering the utilization of blue green algae in wet land rice cultivation. In: **Biological Nitrogen Fixation Associated With Rice Production.** (ed.) S.K. Dutta and Charles Sloger. Oxford and IBH Publishing Company, New Delhi.

Roger, P. A. and I. Watanabe 1986. Technologies for utilizing biological nitrogen fixation on wetland rice potentialities current usage and limiting factors. In: **Nitrogen Economy and Flooded Rice Soils** (eds.) S.K. Dutta and W.H. Partic, Jr., Martinus Nijhoff Pbul. Dordrecht, The Netherlands, pp. 38-47.

Roger, P.A., S.S. Ardales, P.M. Reddy and I. Watanabe 1988. The abundance of heterocystous blue green algae in rice soils and inocula used for application in rice fields. **Biol. Feril. Soils.** 5(2): 98-105.

CHAPTER - 11

CHARACTERISATION OF POTENTIAL CYANOBACTERIAL STRAINS FOR BIOFERTILIZER AND THEIR EFFECT ON PADDY

Chhaya D. Saboji and G.S. Geeta

ABSTRACT

Paddy and sugarcane are the major crops of North Karnataka where flooded conditions persist. In order to develop the cyanobacterial biofertilizer for paddy, the soil samples from districts of North Karnataka were surveyed and analysed for the presence of Blue green algae under N-free conditions. The cyanobacteria belonging to the genera *Anabaena, Calothrix, Cylindrospermum, Haphalosiphon, Nostoc* and *Westiellopsis* were isolated and identified from all the samples. The morphological and biochemical characterization of these strains were done. Their efficiency in nitrogen fixation was tested initially by estimating total nitrogen which ranged between 0.1-0.4%. The performance of these strains along with standard cultures were tested in pot experiment using paddy (variety Amrut). The yield and yield parameters were non-significant as compared to the regular fertilizer dose indicating the high efficiency of the isolated strains.

INTRODUCTION

In Karnataka, paddy is one of the major crops in Dharwad, Belgaum, Sirsi, Kumta, Gangavathi, Raichur etc., covering 60% of the total areas. Some of the places for eg. Gangavathi are highly alkaline. Its already established that BGA can substitute 25 Kg N ha^{-1} beside providing growth promoting substances and oxidised conditions for mobilising nutrients. A large variety of cyanobacteria have been reported to colonize the rice field soils. Though the technology has been well established (Venkatraman,1981, Tiwari, *et al.*,1991,

Kannaiyan, 1993, Goyal, *et al.*, 1997, Singh, 1992, Kaushik and Radha Prasanna, 1998) it is essential to isolate local strains of N_2 fixing BGA from various sources and screen for their efficiency in N_2 fixing ability and plant growth promoting substances.

Hence, the present study was conducted to isolate efficient BGA strains and screen for various properties. Their efficiency was tested based on their performance under pot culture conditions.

MATERIALS AND METHODS

Fresh soil samples were collected from paddy and sugarcane growing fields from Dharwad, Belgaum, Sirsi, Kumta, Gangavathi, Raichur. The N_2 fixing BGA were isolated using nitrogen-free BG-11 medium and made axenic by repeated streaking. The isolates were identified based on morphological and physiological characters following Desikachary (1959) manual. The total nitrogen in the biomass was estimated for screening efficient strains (Bremner, 1960). The isolates were mass cultivated and tested for their efficiency in pot experiment using Amrut variety of paddy in terms of yield and yield parmeters. These isolates were compared with the standard strains obtained from IARI, New Delhi and TNAU, Coimbatore.

RESULTS AND DISCUSSIONS

All the soil samples showed presence of BGA which were filamentous forms dominant in rice fields. Tiwari, (2000) stated that filamentous forms were dominant mostly in rice fields. One of the soil sample (Madikere) showed a composite of 4 cyanobacterial strains coexisting in soil without antgonism. Totally 18 strains were isolated belonging to the genera *Anabaena, Nostoc, Calothrix, Westiellopsis, Scytonema and Cylindrospermum.* They were identified based on their morphology. The isolates were also characterized for their chlorophyll and phycocyanin pigments which varied among strains (data not shown). Table 1 shows that N content of the BGA biomass ranged from 0.11 to 0.34 indicating their efficiency to use as biofertiliser. Out of 18 strains, 10 efficient isolates were tested under pot culture experiment. From the pot experiment (Table 2), it is observed that there was significant difference at 1% in yield and yield components of paddy. The cultures showed similar response to that of recommended dose of chemical fertilizer indicating their high efficiency as N-fixers. Among the local isolated strains, *Calothrix, Anabaena* and *Westiellopsis* showed maximum yield. From the study it is evident that the isolated strains can be further tested under field conditions for using as biofertilisers for rice.

Table 1. Screening of BGA strains for N_2 fixation

Sl. No.	Cultures	% N	Sl. No.	Cultures	% N
1	*Anabaena*	0.16	10	*Calothrix*	0.32
2	*Anabaena*	0.13	11	*Cylindrospermum*	0.11
3	*Anabaena*	0.12	12	*Haphalosiphon*	0.16

4	*Anabaena*	0.11	13	*Haphalosiphon*	0.12
5	*Calothrix*	0.34	14	*Nostoc*	0.16
6	*Calothrix*	0.18	15	*Scytonema*	0.11
7	*Calothrix*	0.17	16	*Westiellopsis*	0.22
8	*Calothrix*	0.11	17	*Westiellopsis*	0.15
9	*Calothrix*	0.16	18	*Westiellopsis*	0.18

Table 2. Effect of isolated BGA strains on yield of paddy

Sl. No.	Cultures	Grain yield (kg/ha)	Straw yield (kg/ha)
1	*Anabaena (B)*	1383.2	1485
2	*Anabaena (G)*	1088.9	1100
3	*Calothrix (S)*	1683.5	1287
4	*Calothrix (M)*	1340.7	1323
5	*Calothrix (T)*	1638.0	1474
6	*Cylindrospermum*	1037.4	1338
7	*Scytonema*	1377.1	1100
8	*Westiellopsis (B)*	1098.0	1283
9	*Westiellopsis (M)*	1062.0	1573
10	*Westiellopsis (S)*	1613.7	1287
11	*Westiellopsis (IARI)*	1056.0	1030
12	*Anabaena variabilis (TNAU strain)*	1092.0	1430
	SEm±	0.24	0.41
	CD at 1%	0.93	1.62

REFERENCES

Bremner, J.M. 1960. Determination of nitrogen in soil by the kjeldhal method. **Journal of Agricultural Sciences,55**:11-13.

Desikachary, T.V. 1959. **Cyanophyta.** Indian Council of Agriculture Research, New Delhi.

Goyal, S.K., B.V. Singh, Veena Nagapal and T.S. Marwaha 1997. An improved method for production and algal biofertiliser. **Indian J. of Agri. Sci.,67**: 314.

Kannaiyan, S. 1993. **Biofertilisers for rice.** Tamilnadu Agricultural University, Coimbatore.

Kaushik, B.D. and Radha Prasanna 1998. Blue green algal biofertliser for rice -A technology. In: **Mass production of blue-green algal biofertiliser,** (eds.) P. K. Singh and Sunil Pabbi IARI, New Delhi, pp. 28.

Singh, P.K. 1992. Biofertilizers for flooded rice ecosystem. In: **Fertilizers, Organic Manures, Recycleable Wastes and Biofertilizers,** (ed.) H.C.S. Tabdon New Delhi, pp: 113-131.

Tiwari, D. N., A. Kumar and A.K. Mishra. 1991. Use of cyanobacterial diazotrophic technology in rice Agriculture. **Applied Biochemistry and Biotechnology,29** :387-396.

Tiwari, D. N. 2000. Current areas of research in cyanobacteria with special reference to their use biofertilizers. In: **Biofertilizers - Blue green algae and *Azolla*.,** (chief ed.) P.K. Singh, pp. 55.

Venkataraman, G.S. 1981. Blue- green algae for rice production. **FAO Soils Bulletin, 16:** 33-42.

CHAPTER - 12

POTENTIAL OF COIR WASTE BASED BGA FERTILIZER FOR PADDY CULTIVATION

P. Malliga

ABSTRACT

The present investigation has attempted to introduce coir waste as an inexpensive carrier material for the BGA biofertilizer for rice. So far *Azolla* and BGA are being used as biofertilizers where as, the present study is an attempt for the first time to use *A. azollae* with coir waste as a carrier which revealed good results. *A. azollae* when inoculated with coir dust got penetrated and immobilized in the matrix and the growth pattern and ARA showed an interesting results. As a results of lignin degradation and the probable increase in the concentration of phenolics, almost all the cells were converted into spores. The sporulation resulted in increased inoculum potential.

INTRODUCTION

The world population is increasing at an alarming rate and is expected to increase from 6.5 billion at present to 7.5 billion by 2025. Most of the population live in the rural areas in the developing countries when poverty, food, security and nutritional deficiencies are the major problem. Low crop productivity, limited use of inputs such as fertilizer and pesticides or over doses due to biotic and abiotic stress factors are major constraint to increase production and productivity of crops (Sharma, *et.al.*, 2001).

At the same time, one of the major problem in today's world is the pollution and contamination of soil. Frequent use of chemical fertilizer has caused tremendous nuisance to the environment. A remedy to this is the biofertilizer, an environmentally eco-friendly fertilizer, which is now being used in most countries. Biofertilizers are microorganism that enrich the nutrients quality of soil without affecting the environment of soil in any way. Among the biofertilizer, blue green algae are considered to be the simplest living autotrophs, capable of building up food material from inorganic matter and fixing nitrogen from atmospheric gas.

Cyanobacterial Biofertilizer

Singh (1998) has reported that BGA is an important biofertilizer which fixes nitrogen to the extent of 20-30 kg N/ha. Besides nitrogen, the algae excrete vitamin B_{12}, auxins and ascorbic acid which may contribute to the growth of rice plants. Thus, by using BGA as a source at the rate of 30 kg N / ha, an average increase of 4 kg / ha in paddy yield could be obtained which results in additional income of Rs. 1630/ha ie. Rs. 1400 in terms of increase in yield and Rs.230/ha saving of N fertilizer. The use of BGA depends upon the selection of suitable strain, correct method of production and its application in fields. The results of a large number of experiments and field demonstrations on farmers fields are giving strong signal for more and more use of BGA as a potential biofertilizer without deteriorating the ecosystem.

The cyanobacterial biomass when incorporated in the soil conserves organic matter, nitrogen, phosphorus and moisture and improves the chemical and physical structure of the soil. This characteristic of several strains of cyanobacteria may be very much useful in reclaiming saline and alkaline soil. This aspect has been well substantiated and documented under different agroclimatic conditions (Kaushik, 1983 and 1993, Kaushik and Ummat 1992 and Boussiba, 1999).

The role of nitrogen fixing cyanobacteria in the maintenance of the fertility of rice fields has been well substantiated and documented all over the world. In India alone, the beneficial effect of cyanobacteria on the yield of many rice varieties has been demonstrated in a number of localities. (Saga and Mandal, 1980; Santra 1993; and Sri Lavanya Priya, 1997). In many areas, 10 to 20% increase in the grain yield has been observed as a result of cyanobacterial application in the absence of any added chemical nitrogen fertilizer (Santra, 1993). Rice field ecosystem provides favourable environment for the growth of cyanobacteria with respect to their requirements of light, water, temperature and nutrient availability (Kannaiyan, 1999). Kaushik (1996) has reported that a large number of field trials were conducted in different agro climatic regions of our country to assess the effect of algalisation on rice yield.

The combination of nitrogen fixing and non-nitrogen fixing algae give more effect in terms of growth and yield of paddy. The non-nitrogen fixing algae have also independently played an important role in the growth and yield of paddy. The combined effect of nitrogen fixing and non-nitrogen fixing cyanobacteria have an overall beneficial effects on soil enrichment. The non-nitrogen fixing cyanobacteria also enriched the phosphorus and potassium content in the soil. This may be due to release of growth promoting substance from cyanobacteria (Selvarani, 1983).

Application of heterocystous and non-heterocystous cyanobacterial biofertilizers to the rice field clearly showed increase in number of grains and weight of the grains and also enhanced the NPK content of the soil (Selvarani, 1983). Immobilized as well as free-living cyanobacteria application was found to be distinctly advantageous over control, as it enhanced significantly the various parameters of growth of rice plants such as shoot and root length, fresh and dry weight of the plants, chlorophyll and protein content over a period of 30 days (Sophia Rajini, 1995). From the above investigations, it is proved that cyanobacteria serve as natural renewable biological sources as biofertilizers for rice.

Carrier Material

A large number of carrier materials have been developed as media for bacterial inoculants. These are peat, straw, husk, peanut, hulks, corn cobs, lignite, charcoal, polymer compound, rice husk, vermiculite, soil, press mud etc. Indeed, the quality of the carrier is the most vital factor which will ascertain the quality of the inoculant with regard to microbial load and also the shelf-life. However, the selection of suitable carrier is the difficult task, because desirable success has not achieved especially in India (Vyas, *etal.*, 1998). Conventionally, cyanobacterial biofertilizers were used as soil based inocula (Subba Rao, 1982) resulting in considerable loss of viability. So the soil based BGA biofertilizer was not successful in certain trails due to the viability loss caused by the hydrophilic nature of soil that removes moisture beyond the minimum requirement for retaining the viability in cells. Watanabe (1959) devised the lyophilization method and also culturing on the moist surface of porous volcanic gravels. Venkataraman (1961) evaluated the washed, boiled in distilled water and dried quartz sand as a seeding material for BGA. The sand was mixed with equal amount of concentrated algal material and sun dried. It was found that *Nostoc* retained its viability for two years in a dry state, whereas the greens, *Chlorella, Selenastrum* and *Scenedesmus* failed completely to grow even after the treatment. These results indicated the possibility of supplying BGA-biofertilizer in a sun dried state with a suitable carrier.

From the above findings, not only selection of BGA but selection of carrier material is also a very impotant process for storage and field application of biofertilizer for improving yield.

An ideal carrier material should possesses the following features. Highly absorptive and easy to process, non-toxic to microbes, easy to sterilize, freely available and inexpensive, increase in the shelf life of the inoculum, provide good adhesion to the seed, have a good buffering capacity, easy germination during field application.

Hence a new inexpensive carrier material namely coir waste was selected, because, *Cocos nucifera* Linn. is the widely cultivated palm in the tropics and India is the principal producer and exporter of coir products. The by-products of coir industry are pith and coir dust. Coir dust though in the form of "Mulch" is used to improve soil tilth, also is considered as a major waste, as it is accumulated enormously making its disposal difficult due to the slow biodegradation in the natural environment. Coconut shells contain more lignin content (39.8%) when compared to other plants. So, this coir waste was preferred because it can increase the shelf life, viability and inoculum potential of the BGA biofertilizer. (Malliga *et al.*, 1996). Hence, the present study was undertaken and the following experiments were conducted.

MATERIALS AND METHODS

Organism used and Growth Condition

Anabaena azollae ML2 an isolate from leaf cavities of *Azolla pinnata* R. Br. (Malliga, 1993) was used. The culture was maintained in nitrogen free Bothe's medium (1968) under white fluorescent light of 1500 lux of 25±2° 14/10 LD cycle regime.

Measurement of Growth (Mackinney, 1941)

Growth of cyanobacterium alone and with coir waste was measured for 30 days in terms of chlorophyll *a*.

Estimation of Protein

The method of Lowry *et.al.* (1951)was used. Bovine serum albumin served as standard.

Nitrogenase assay (Stewart *et al.*,1967)

Fresh cyanobaterial cultures were added to 13 ml glass vials with Bothe's nitrogen free medium an aerobic and microaerophilic conditions were provided. In microaerophilic conditions, Argon gas was bubbled for 1 min. period and vials were sealed with rubber septa and secured with aluminium cap. The vials were incubated under white fluorescent light of 1,500 lux for 24 hrs. For every 3 hrs., the vials were taken from the incubator and tested for acetylene reduction activity.

Nitrogenase activity of *A. azollae* with coir waste

With and without coir waste based *A. azollae* cultures were taken and tested for ARA for 30 days period.

Spectral Analysis (Goldschmid, 1971)

A. azollae in the free living form and also with coir dust were grown for a period of 30 days. The clear supernatant was collected from the above cultures for every 5 days and measured phenol content at 280 nm by using Spectrophotometer Beckman DU-64 USA.

RESULTS AND DISCUSSION

Microscopic observations revealed that *Anabaena azollae* soon after inoculation with coir dust commenced adhering to the surface and immobilized inside the particles. After a period of 15 days, it slowly started sporulation and within 30 days almost all the vegetative cells of the cyanobacteria were converted into spores. Probably this could be due to the stress created by the release of phenolics during degradation (Malliga *et. al.*, 1996).

In the case of cyanobacterium without addition of coir, the maximum growth was observed on 12th day and then entered the senile phase. The growth of coir based cyanobacteria was significantly more and had no extended log phase till 30th day (Fig. 1). This could be due to the consequence of penetration, immobilization and degradation of cyanobacteria of the coir particles. Finally all the vegetative cells were converted into spores. The reason may be due to the degradation of lignin and utilization of phenolic content which was present in coir waste. The previous investigations also reported that degradation of phenol and lignin is possible by cyanobacterial strains. (Shashi Reha 1994 and Bharati *et. al.*, 1992).

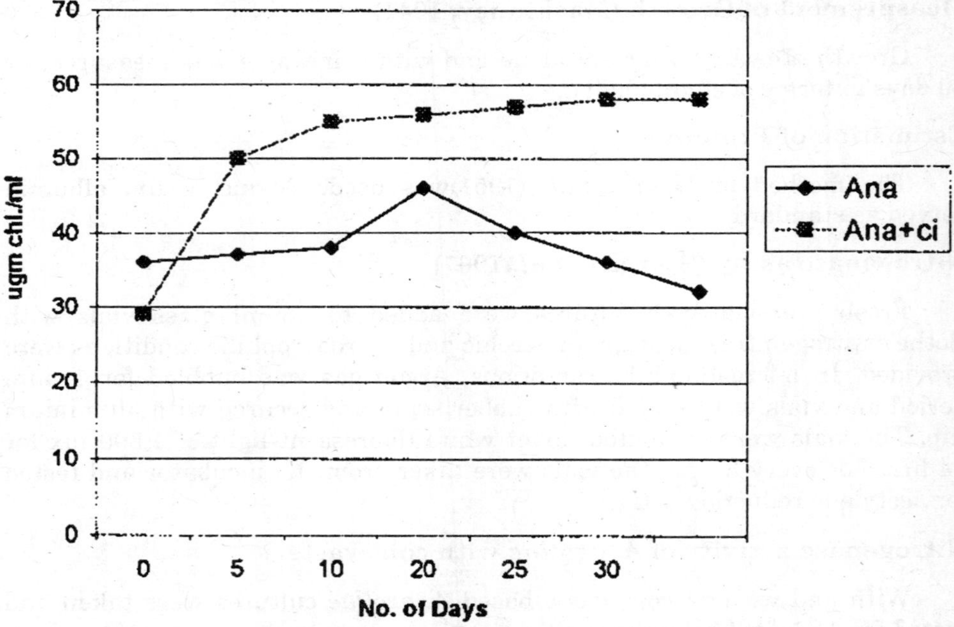

Fig. 1. Chlorophyll content in *A. azollae* with coir waste

Nitrogenase activity of cyanobacteria was at peak during initial days and gradually decreased, whereas in cyanobacteria with coir dust, low activity was measured at the beginning when compared to only cyanobacteria. There was a drastic reduction till 15th day and there was no activity after a period of 20 days (Fig. 2). This clearly indicated the result of sporulation of the organism.

A clear diurnal pattern of nitrogenase activity in terms of acetylene reduction was found. Under aerobic conditions, in light the maximum activity was found to at 9th hr. and there was a sudden decrease thereafter in dark (Fig. 3). The shift in maxima can be explained by the fact that the transition from light to dark. Most probably, the decrease in activity can be suggested by assuming limitation in energy and electrons. Nevertheless, the high activity observed in the light showed that nitrogenase provided energy and electrons by photosynthesis. This phenomenon has been noted previously in various experiments. Nitrogenase activity decreased when cultures were transferred from light to dark (Pei-Chung Chen 1986, and Huber, 1986) which indicated that diurnal changes in nitrogen fixation were observed maximum occurring early in the morning. Activity of nitrogenase decreased after changing to anaerobic condition and suddenly declained for rest of dark incubations (Lucas and Heyer, 1987). In case of microaerophilic condition, initially there was a high nitrogenase activity which gradually decreased during 3rd & 6th hr. and marginally increased at the 9th hr. (Fig. 4). However, nitrogenase activity of *Anabaena azollae* in aerobic and light conditions was more when compared to microaerophilic condition. So, heterocystous cyanobacteria are considered to be the best adopted for nitrogen fixation under aerobic and oxygenic photo production conditions.

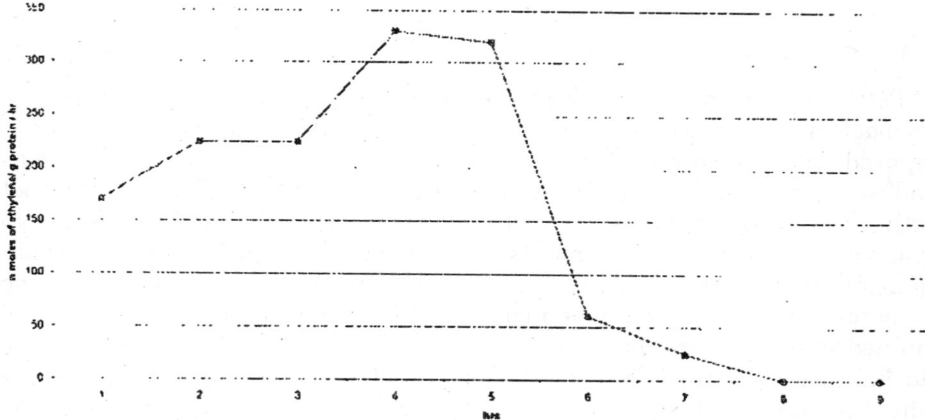

Fig. 2. Nitrogenase activity in *Anabaena azollae*

Direct monitoring of phenol degradation was carried out by observing the changes in absorption in the UV region due to the release of soluble phenolic compounds during lignin depolymerization (Goldschmid, 1987). The degradation process by cyanobacteria significantly increased the phenol content on the 30th day at 280 nm when compared to control. This could be explained only due to the lignin deploymerising ability of *Anabaena azollae.* Kelly (1988) has demonstrated that the extracellular ligninase are capable of degrading a variety of chemical compounds. Detectable polyphenol oxidase and laccase activities were recorded only from the 10th day and their maximal activity was recorded on the 40th day. The ligninase activity of the organism with coir waste measured in terms of ethylene released was 146.8n mol ml^{-1} in contrast to conditions of KTBA alone, organism alone and lignin alone where the release was 30-80, 24.92 and 12.41 nmol ml^{-1}, respectively (Malliga *et al.*, 1996).

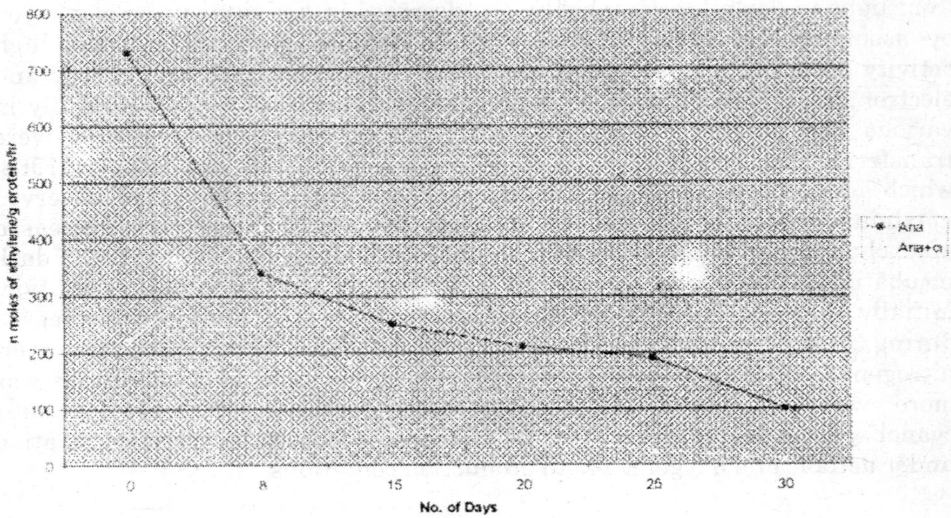

Fig. 3. Nitrogenase activity in *A. azollae*

Pot culture experiments were performed to test coir as a carrier for cyanobacteria in rice cultivation and it revealed appreciable results, compared to uninoculated control and the cyanobacterial application was found to slightly increase the various parameters of colour of the leaves, root length, shoot length, number of roots and culm circumference (Sri Lavanya Priya, 1997). The field experiments also showed interesting results. Application of recommended doses of chemical fertilizer enhanced all the parameters and grain yield when compared to control plants. In the case of combined effect of 50% chemical fertilizer and coir based biofertilizer more yield has been obtained than in the case of chemical fertilizer. The reason might be due to the combination of degradation of organic waste by cyanobacteria and secretion of variety of substances such as growth hormones, vitamins and polysaccharides by cyanobacteria and this could be utilized by paddy plants. Though the yield obtained in the case of only coir based biofertilizer is lesser than in the combined effect case, it is moderately more than the chemical fertilizer (Krishnaveni, 1999). The production of growth promoting substance by cyanobacteria and their direct growth promoting effect on rice crop has been well elucidated by Kannaiyan *et al.* (1997). But, whether these growth promoting substances were hormones, vitamins or amino acids or any other related compounds and their mode of action is not very clear (Kannaiyan *et al.*, 1997). In his report inoculation of polyurethane - immobilized cyanobacteria and mixture of *Anabaena* species with sugar cane waste increased the straw and grain yield of the rice plants in pot culture experiment and also in the field trails Samal and Kannaiyan (1992) had also noted ammonia excretion by *A. azollae* immobilized in alginate and its positive effect on the growth of rice seedlings. Latorre *et al.* (1986) observed that the growth of rice plants in the laboratory was improved by the inoculation of *A. variabilis*. Further Kannaiyan *et al.*, (1997) showed that the combined use of fertilizer nitrogen and inoculation of immobilized cyanobacteria gave the highest grain yield.

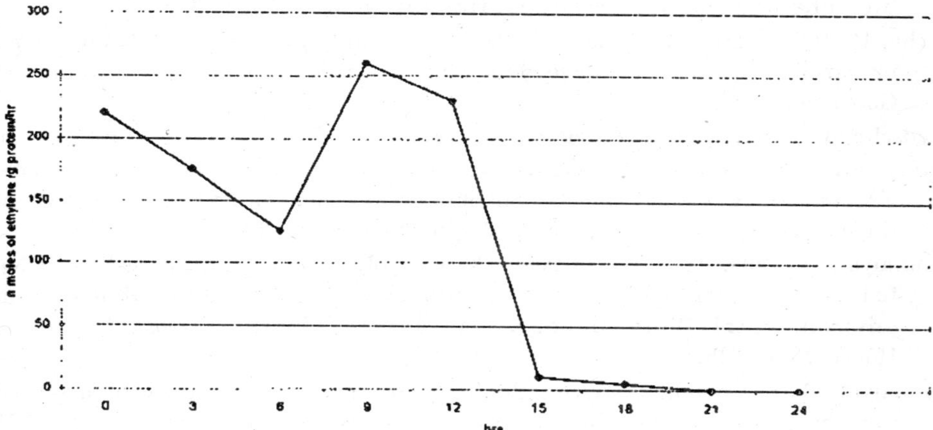

Fig. 4. Diurnal pattern of microaerophlic nitrogenase activity in *A. azollae*

Use of coir waste based BGA was found to be distinctly advantageous over the use of cyanobacteria alone and this study has revealed that if coir waste is used as a carrier material there are several merits *viz.*,

1. Inoculum gets immobilized.
2. Enhanced growth during initial days.
3. Sporulation started at the end of the degradation process (Malliga *et. al.*, 1996).
4. Sporulation results in maximum viability and increased inoculum potential.
5. Akinete germination started when transferred to new medium (Udayabala, 1998).
6. Solid waste can be easily disposed
7. Coir waste by virtue of its high organic content and water holding capacity proved to be an ideal carrier material when compared to soil.

From the above findings, it can be concluded that the application of selected strain with suitable carrier based BGA fertilizer can increase the crop yield in a considerable amount.

ACKNOWLEDGEMENT

The author is grateful to the Department of Biotechnology, Government of India, New Delhi for funding the facility.

REFERENCE

Bharati, S.D., A.S. Salanki, T.C. Taranath and M.V.R.N. Acharyulu. 1992. Role of cyanobacterial in the removal of lignin from paper mill waste waters. **Bull. Environ. Contam. Toxicol.**, 49: 738-42.

Bossiba, S. 1999. Cyanobacterial for agricultural applications. In: **International workshop and training course on micro algal biology and biotechnology.** Mosonmagyarover, Hungary. June 13-26. pp-6.

Bothe, H. 1968. Ferrodoxin and phytoflavin in photosynthetischen reakionen einer preparation aus der blanalgne *Anacystis nidulance.* Dissertation University of Gottingen.

Goldschmid, O.1948. **J. Ann. Chegoc., 75:** 3780.

Huber, A.L. 1986. Nitrogen fixation by *Nodularia spumigena* mertens (Cyanobacteriaceae) I. Field studies and the contribution of blooms to the Nitrogen budget of the peel Harvy estuary western. **Hydrologica 131:** 193-203

Kannaiyan, S. 1999. Ammonia excretion by the polyurethane foam immobilized N_2-fixing cyanobacterial for Rice. In: **Cyanobacterial and algal metabolism and environmental Biotechnology.** (ed.) Tasneem Fatma, Narosa Publishing House. New Delhi. pp. 20-27.

Kannaiyan, S., S.J. Aruna, S. Merina Prem kumari and D.O. Hall. 1997. Immobilized cyanobacteria as a biofertilizer for rice crops. **J. Appl. Phycology. 7:**1-8.

Kaushik, B.D. 1983. Amelioration of salt affected sol. In: **Proceedings of the All India Applied Phycology Congress,** Kanpur, India. 60-66.

Kaushik, B.D. 1993. Cyanobacterial research- An IARI Pursuit. In: **Proceedings of the National Seminar On Cyanobacterial Research - Indian Scene,** Jan 19-21, 1993. (ed.) G. Subramanian, Bharathidasan University, Tiruchirapalli pp. 32-49.

Kaushik, B.D. and J. Ummat. 1992. Reclamation of salt affected soils with blue green algae (cyanobacterial): A Technology development. In: **Proc. Natl. Seminar Biofertilizer Technology Transfer.** (eds.) by L.V. Gangawne, Marathwada University, Aurangabad.

Kaushik, B.D. 1996. Use of cyanobacterial biofertilizer in rice cultivation: A technology improvement. In: **Cyanobacterial Biotechnology. Proceedings of the International Symposium.** September 18-21. (eds.) G. Subramanian, B. D. Kaushik, and G.S. Venkataraman Oxford & IBH publishing Co PVT LTD. New Delhi. pp. 211-222India.157.

Kelly R.I. 1988. Lignolytic activity in *Phaenorochaete chryssporium* measured as ethylene production from –Keto, -methyl butyric acid. In: **Biomass (Lignin, pectin, chitin).** (eds.) W.A. Wood and S.T.Kellogg. **Methods Enzymol., 61:**79-82 Academic press New York.

Krishnaveni, M.1999. **Identification of the compounds responsible for the induction of sporulation and germination in *Anabaena azollae* ML2 during the degradation of coir waste & Field study of *A. azollae* and *Phormidium valderianum* BDU 20041 as coir waste based biofertilizer.** M.Sc. thesis. NFMC, Dept. of Microbiology, Bharathidasan University, Tiruchirapalli 24.

Latorre, C., J.H. Lee, H. Spiller and K.T. Shanmugam 1986. Ammonium ion excreting cyanobacterial mutant as a source of nitrogen for growth of rice: A feasibility study. **Biotechnol. Lett., 8:** 507-512.

Lowry, T., N.J. Rosenbrough, A.C. Farr and R.J. Randall 1951. Protein measurements with the Folin-Phenol reagent. **J.Biol.Chem., 193 :** 265-275.

Lucas J.S. and H. Heyer 1987. Dark anaerobic nitrogen fixation (acetylene reduction) in the cyanobacterium *Oscillatoria* sp. **FEMS Microbiol. Ecol., 45:** 227-232 Published by Elsevier.

Mackinney, G. 1941. Absorption of light by chlorophyll solution. **J. Biol. Chem., 140:** 315-322.

Malliga, P. 1993. **Microbial association in the leaf cavities of *Azolla pinnata*.** R.Br. Ph. D. Thesis, Bharathidasan University Tiruchirapalli.

Malliga, P., L. Uma and G. Subramanian. 1996. Lignolytic activity of the cyanobacterium *Anabaena azollae* ML2 and the value of coir waste as a carrier for BGA biofertilizer. **Microbios, 86:**175-183.

Pei-chung Chen, 1986. Alteration of Nitrogenase activity of *Anabaena* CH1 and CH2 during Light-dark transition. Reprinted from the Chinese. **J. Microbiol. Immunology, 19(3):** P 36-41.

Saha, K. C., and Mandal. 1980. A green house study on the effect of inoculation of N_2 fixing blue green algae. in an alluvial soil treated with P and Mo on the yield of rice and changes in the nitrogen content of soil. **Plant Soil, 57:**23-30.

Samal, K.C. and S. Kannaiyan. 1992. Ammonia excretion by *Anabaena azollae* immobilized in alginate and its effect on the growth of rice seedlings. **Inte. Rice Res. News Letters** Rice Res. Inst., P.O. Box 933, Manila 1099, Philippines, **17:**21-22.

Santra, S.C. 1993. Importance of blue green algae. In: **Biology of rice-fields blue green algae**. (eds.) S.C. Santra. Daya publishing house. Delhi-110035.

Selvarani, V. 1983. Studies on the influence on nitrogen fixing and non-nitrogen fixing blue-green algae on the soil, growth and yield of paddy (*Oryza sativa* -IR50). M.Sc. thesis. Madurai Kamaraj University, Madurai.

Shashi Reha, S. 1994. Utilization of phenol by marine cyanobacterium "*Phormidium valderianum*" BDU 30501. M.Sc. thesis. Bharathidasan University, Tiruchirapalli.

Sharma, H. C., K.K. Sharma, N. Seetharaman and Rodomiro Ortiz, 2001. Genetic transformation of crop plants: Risks and opportunities for the rural poor. **Curr. Sci., 80(12):** 1495-1508.

Singh, A.N. 1998. Use of Blue Green Algae for nitrogen fixation in paddy fields. **Indian J. Agric. Chem., 31(1):** 47-57.

Sophia Rajini, V. 1995. **Studies on immobilization of cyanobacteria**. Ph.D. thesis, Department of Microbiology, Bharathidasan University, Tiruchirapalli.

Sri Lavanya Priya, S 1997. **Efficacy of coir waste based *Anabaena azollae* ML2 as a biofertilizer for rice.** M.Sc. thesis, Department of Microbiology, Bharathidasan University, Tiruchirapalli.

Stewart, W.D.P., G.P. Fitzerald and R.A. Buries 1967. *In situ* studies on nitrogen fixation using acetylene reduction technique. **Proc. Natl. Acad. Sci. USA. 58:** 2071-2078.

Subba Rao, N.S. 1982. **Biofertilizers in agriculture**. Oxford and IBH publishing Co.

Udayabala, M.1998. **Induction of sporulation and spore germination of *A. azollae* with coir waste.** M.Sc thesis, Department of Microbiology, Bharathidasan University, Tiruchirapalli.

Venkataraman, G.S. 1961. A method of preserving blue green algae for seeding purposes. **J. Gen. Appl. Microbiol., 7:96.**

Vyas, S.C., S. Vyas and H. A. Modi. 1998. In: **Biofertilizer and organic farming.** Educational Publishers & distributors. Akta Prakasan, Nadiad-1.

Watanabe, A. 1959. Distribution of Nitrogen fixing blue-green algae in various areas of South and West Asia. **J. Gen. Appl. Microbiol., Tokyo, 5:21-29.**

CHAPTER - 13

A NEW APPROACH FOR INOCULATING CYANOBACTERIA FOR RICE CULTIVATED IN SODIC SOILS

P. Pandiyarajan and P. Nagarajan

ABSTRACT

Cyanobacteria can withstand salt stress by its physiological ability for osmotic adjustment and its capability to scavenge sodium ions from alkaline soils. These potentials render cyanobacteria to be used as biological reclamation as well as biofertilizer, in salt affected soils. Experiments were conducted to assess the extent of these dual benefits accrued by inoculating cyanobacteria for rice in sodic soils. Salt tolerant cyanobacterial cultures were identified and inoculated in the main field @ 10 $kgha^{-1}$ 20 days before transplanting and 7 days after transplanting along with the recommended dose of super PO_4 and potash. Nitrogen was applied @ 75 and 100 per cent of the recommended dose in three splits. Rice variety TRY 1 was used as a test crop. The results revealed that grain and straw yield (6000 kgs and 7814 kg ha^{-1} respectively) significantly increased over control as well as 100 per cent N alone. The soil pH significantly declined from 8.9 to 8.06 when cyanobacteria was inoculated before and after transplanting.

INTRODUCTION

Alkalinity, salinity and sodicity of soils are major problems for successful cultivation of any crop. This phenomenon is a major deterrent to agricultural productivity in India. Many research programmes are being carried out to reclaim these soils through various ways. Reclamation of these soils through biological systems is one sort of ecofriendly and cost effective approach.

Generally cyanobacteria exhibit considerable tolerance to salt stress and this may be due to the extrusion of sodium ions and maintenance of low intercellular sodium (Apte and Thomas, 1986). Singh (1961) suggested that cyanobacteria have the ability to remove sodium ions from the alkaline soils and thus reduces alkalinity. Subashini and Kaushik (1984) used the blue green algae *viz., Calothrix, Tolypothrix, Scytonema* and *Haplosiphon* for the reclamation of saline soils. Kannaiyan *et al.,* (1992) suggested that when algal biomass is incorporated in the soil, it converts the sodium clay into calcium type. Pandiyarajan and Kannaiyan (1996) reported that the inoculation of BGA alongwith gypsum increased the yield of paddy grown in sodic soils. Hence the present investigation was made with a view to use cyanobacteria both as biological reclamation agent and biofertilizer.

MATERIALS AND METHODS

Native cyanobacterial cultures growing in sodic soils were isolated and their tolerance to varying levels of NaCl was assessed. The efficient cultures were purified and characterized. The cultures were then tested for their chlorophyll content (Bennet and Bogored, 1971), protein content (Lowry *et al.,* 1951) and ARA activity (Turner and Gibson, 1980). These cultures were then grown in cement tanks as a composite culture in a soil having a pH of 8.9 and EC 2.91 dSm^{-1}.The composite cultures thus developed was used for the field study.

A field experiment was conducted to assess the ability of the salt tolerant composite culture of cyanobacteria as a reclamation agent.

The following treatments were imposed, in a randomized block design with three replications.

T1 75 % N alone
T2 100% N alone
T3 75% N + BGA (AT)*
T4 75% N + BGA (BT)**
T5 100% N + BGA (AT)
T6 100% N + BGA (BT)
T7 75% N + BGA (AT/BT)
T8 100% N + BGA (AT/BT)

*(AT - 7 days after transplanting) **(BT - 20 days before transplanting).

The salt tolerant composite culture of cyanobacteria was inoculated @ 10 kg ha^{-1} 20 days before transplanting (BT) and was allowed to multiply in the field.

At the time of planting the algal mat was trampled and rice seedlings (Variety TRY 1) were transplanted. Another dose of BGA (10 kg ha^{-1}) was inoculated 7 days after transplanting (AT). Nitrogen was applied at 75 and 100 percent of the nitrogen requirement as per the soil test recommendation, in three split doses. Phosphorous and potash were applied basally as per soil test recommendation. The plant height, tiller number, grain and straw yield were recorded and analysed. The soil pH after harvest was noted.

RESULTS AND DISCUSSION

The native effective cyanobacterial genera characterized were *Anabaena* (ACT-2), *Westiellopsis* (ACT -5), *Calothrix* (ACT- 6), *Tolypothrix* (ACT- 8), *Oscillatoria* (ACT- 10), *Nostoc* (ACT 12) (Table 1). Kannaiyan *et al.,* (1992) listed 31 most common saline tolerant cyanobacterial genera in which the above mentioned genera were found to be present.

Table 1. *In vitro* studies on the effective isolates of cyanobacteria

Isolate No	Genera	Chlorophyll content (μg / ml)	Protein content (μg / ml)	ARA activity (n. ml / hr/ gm fresh wt.)
ACT –2	*Anabaena*	3.0	12.1	9.4
ACT –5	*Westiellopsis*	2.9	9.8	12.6
ACT –6	*Calothrix*	2.8	10.6	16.1
ACT – 8	*Tolypothrix*	3.9	11.1	14.1
ACT – 10	*Oscillatoria*	3.8	16.8	12.1
ACT – 12	*Nostoc*	3.2	18.4	12.0

These cultures showed relatively high acetylene reduction activity and found to contain considerable amount of chlorophyll and protein (Table 1). This is in accordance with the results obtained by Apte and Thomas (1987) and Kannaiyan *et al.,* (1992). The cyanobacterial cultures ACT- 2, ACT- 5, ACT- 6, ACT- 8, ACT -10 and ACT- 12 were found to tolerate 300 mM of NaCl (Table 2).

Table 2. Effect of different concentration of NaCl on the growth of cyanobacteria

Isolate No	Control	100 mm	200 mm	300 mm	400 mm
ACT –2	++++	++++	++++	++++	+++
ACT –5	++++	+++	+++	++	++
ACT –6	++++	+++	+++	++	++
ACT – 8	+++	++++	+++	++	+
ACT – 10	++++	++++	++++	++++	+++
ACT – 12	++++	++++	+++	+++	++

++++ Very good +++ Good ++ Moderately tolerant + Poor

The results of the field experiment clearly revealed that saline tolerant cyanobacteria can be used as a biological reclamation agent in sodic soils. The inoculation of cyanobacteria @ 10 kg ha^{-1} before and after transplanting significantly increased the yield and reduced the soil pH from 8.9 to 8.15. This could be due to fact that cyanobcteria besides fixing nitrogen also add an appreciable quantity of bulk organic matter to the soil, which is likely to influence the physical and chemical properties of the soil. This was well evidenced by Kannaiyan *et al.,* (1992), wherein the inoculation of BGA significantly increased the organic carbon content from 0.313% to 0.488%. Successive algal inoculation resulted in the reduction of oxidizable matter, total sulphides and ferrous ions and the yield increase could be the result of improved soil aggregation, increased nitrogen availability and organic matter content as suggested by Kannaiyan (1990) (Table 3).

Table 3. Effect of inoculating cyanobacteria in rice grain in sodic soils.

Treatment	Soil pH (Initial 8.9)	Plant height (cm)	Tiller No.	Grain kg / ha	Straw kg / ha
75 % N	8.70	6.8	19.3	3370	5667
100 % N	8.64	70.6	21.2	3593	6185
75 % N + BGA (AT)	8.40	69.6	23.3	4007	6778
75 % N + BGA (BT)	8.32	72.0	24.6	4630	7296
100 % N + BGA (AT)	8.30	72.0	24.5	5074	7333
100% N + BGA (BT)	8.30	75.0	26.6	5259	7666
75% N + BGA (AT / BT)	8.21	76.6	29.9	5667	7666
100% N + BGA (AT/ BT)	8.15	77.0	32.0	6000	7814
CD	0.21	5.7	8.7	300.3	500.0

AT- After transplanting BT - Before transplanting

Blue green algae are renewable biofertilizer. They are also known to be associated with the formation of calcareous deposits. They can be very well exploited both as a biofertilizer and biological ameliorating agents in saline and alkali soils.

REFERENCES

Apte, S.K. and J. Thomas. 1982. In: **Proceedings of the symposium on biological nitrogen fixation.** Dept of Atomic energy. p.150.

Bennet, A and L. Bogored 1971. **Properties of subunits and aggregates of blue green algae biliproteins**. Health and Laboratory Science 3: 90-100.

Kannaiyan, S. 1990. **Blue green algal biofertilizer for rice crop.** TNAU, Coimbatore. pp. 212-215.

Kannaiyan, S.K. Kumar, and P. Pandiyarajan, 1992. The use of saline tolerant cyanobacteria for salt affected lands. **Proc. of the International Symposium on Strategies for utilizing salt affected lands.** 17-25 February, 1991. Bangkok. Thailand. pp. 394-404.

Lowry, O.H. N.J. Rosenbrough, A.L. Farr, and R.J. Randall. 1951. Protein measurement with Folin Phenol reagent. **J. Biol. Chem. 193**:265.

Pandiyarajan P and S. Kannaiyan, 1996. Effect of BGA inoculation with paddy straw and gypsum on rice in sodic soils. **Madras Agric. J. 83(8):** 478-547.

Singh, R.N. 1950. Reclamation of "Usar" lands in India through Blue green algae. **Nature. 165:** 325-326.

Subashini D and B.D. Kaushik. 1984. Amelioration of salt affected soils with blue green algae-1. Influence of algalization in the properties of saline and alkaline soils. **Phykos., 23:** 273-277.

Turner, G.L. and A.H. Gibson. 1980. In: **Methods of evaluating biological nitrogen fixation** (ed.) Bergerson, John Wiley and Sons NY. p. 111.

SECTION – II

AZOLLA BIOFERTILIZER

CHAPTER - 14

MINERALIZATION PATTERN OF NITROGEN FIXING GREEN MANURES *Sesbania rostrata* AND *Azolla microphylla* IN WETLAND RICE SOIL ECOSYSTEM AND THEIR EFFECT ON RICE

S. Sundaravarathan and S. Kannaiyan

ABSTRACT

The N demand of rice when supplemented with N supply from incorporated organic residues would increase N use efficiency of rice and reduce soil N losses. Under lowland rice soil ecosystem the mineralization rate primarily depends on the environmental factors, type of plant species and the growth state of crop incorporated in soil. Hence an attempt was made to study the mineralization pattern of *Sesbsnia rostrata* and *Azolla microphylla* with fertilizer N and neem cake and their effect on rice under submerged rice soil ecosystem.

Incorporation of N_2 fixing green manures *viz., S. rostrata* and *A. microphylla* with prilled urea recorded maximum ammonical N accumulation on 7th day after incorporation, while green manuring alone and with neem cake resulted in gradual accumulation of ammonical N. Maximum accumulation of ammonical N in flood water was noticed on 14th and 21st day after incorporation of *A. microphylla* and *S. rostrata* respectively. However, when combined with neem cake, incorporation of *A. microphylla* and *S. rostrata* recorded maximum ammonical N accumulation in flood water on 21st and 28th day respectively. Application of prilled urea alone showed maximum ammonical N accumulation on 7th day with drastic reduction in accumulation over subsequent days. The gradual accumulation of ammonical N in flood water upon application of neem cake along with green manure incorporation might be attributed to the nitrification inhibitory effect of neem cake.

Application of prilled urea at the time of incorporation of green manure *viz., S. rostrata* or *A. microphylla* alone, resulted in the maximum accumulation of nitrate N in flood water on 15th day after incorporation. However, green manure incorporation with prilled urea and neem cake resulted in gradual accumulation of nitrate N. *S. rostrata* incorporation has shown increased accumulation of nitrate N than incorporation of *A. microphylla* on 30th day after incorporation.

Application of prilled urea alone accumulated maximum nitrate N in flood water on 15th day with drastic decline on 30th day after application. Interestingly, application of neem cake along with green manure incorporation resulted in low accumulation of nitrate N in flood water on 15th day after incorporation than green manuring alone which implies the inhibitory effect of neem cake on nitrification process.

Green manure incorporation with prilled urea enhanced the population of nitrifiers than green manuring alone as application of prilled urea along with green manure incorporation may significantly reduce the C : N ratio and serve as an effective substrate for microbial decomposition. However, incorporation of green manure with or without fertilizer N but with neem cake significantly reduced the population of nitrifying bacteria. The nitrification retarding effect of neem cake is mainly attributed to its inhibitory effect on *Nitrosomonas* and *Nitrobacter* in soil.

The first phase of the decomposition of organic matter in rice soil is the rapid formation of organic acids followed by the formation of methane. Green manure incorporation registered increased accumulation of volatile fatty acids *viz.*, acetic, butyric and propionic acids in rice soil over application of prilled urea alone and control. Green manuring with prilled urea recorded maximim accumulation of volatile fatty acids in rice soil compared to green manuring alone. Among the green manures, incorporation of *A. microphylla* recorded higher accumulation of acetic acid while, *S. rostrata* incorporation resulted in relatively higher accumulation of butyric and propionic acids. Accumulation of volatile acids was higher on 15th day than on 30th day after incorporation. Among the volatile fatty acids, green manure incorporation recorded maximum accumulation of acetic followed by butyric and propionic acids. The amount of soil organics formed depends on the amount of organic matter added and its humification co-efficient.

Gaseous decomposition products of organic matter are important factors in biochemical and chemical changes in submerged soils and they directly or indirectly influence the availability of plant nutrients and the growth of rice. Incorporation of *S. rostrata* and *A. microphylla* resulted in increased evolution of CO_2 and CH_4 over application of fertilizer N alone. In general, green manure incorporation registered maximum evolution of CO_2 and CH_4 on 15th day compared to 30th day after incorporation. However, neem cake application along with green manure incorporation slightly reduced CO_2 and CH_4 evolution on 15th day while increasing slightly on 30th day after incorporation. The rate of CH_4 emission is a reflection of the complex interaction between the processes of plant primary production, soil organic carbon decomposition, methanogenesis and methanotrophic oxidation.

Addition of organic amendments to soil brought out changes in microbial population equilibrium due to change in the pH as well as changes in the C : N ratio. Incorporation of green manures significantly enhanced the microbial population compared to application of prilled urea alone. Incorporation of *A. microphylla* resulted in maximum population of bacteria, actinomycetes and fungi on 20th day while, *S. rostrata* incorporation showed maximum population on 30th day after incorporation. Application of neem cake along with green manure incorporation significantly reduced the microbial population on 10th day, while it increased on 20th day after incorporation. The initial reduction in microbial population might be attributed to the inhibitory effect of neem cake.

The extracellular enzymes produced by soil microorganisms are released into surrounding soil matrix which catalyse the depolymerization of high molecular weight compounds. Green manure incorporation with prilled urea recorded maximum soil urease and L-asparaginase activity on 15th day after incorporation, while incorporation of green manures without prilled urea showed maximum soil urease and L-asparaginase activity on 30th day. Application of neem cake along with green manure incorporation significantly reduced the soil urease and L- asparaginase activity on 15th day but the activity was increased significantly on 30th day after incorporation.

Green manure incorporation with prilled urea and neem cake recorded maximum plant height, tiller number, panicle numbers, grain and straw yield of rice followed by green manuring with prilled urea. Incorporation of *S. rostrata* significantly increased the plant height, tiller numbers, pancile numbers, grain and straw yield of rice over application of prilled urea alone and *A. microphylla* incorporation. In general, green manuring with neem cake significantly enhanced the growth and yield of rice compared to green manuring without neem cake. The biochemical components particularly alkaloids like azadirachtin azadirone, nimbin, nimbanidol etc. present in neem cake were likely to regulate the nitrogen use efficiency in rice soil eco-system and regulation of N reflects on yield and yield components of rice.

INTRODUCTION

The fast declining fertility status of cultivable soils, especially under high intensive cropping situations, brings into sharp focus the need to sustain soil productivity. Though inorganic nutrient sources bring significantly enhanced yield levels in the short term, the importance of organic sources in soil health maintenance on a long term basis has received greater attention and thrust in recent years.

Green manuring can be an important component of low input sustainable agriculture without sacrificing the level of productivity. Green manuring can be a suppliment to fertilizer application, partially substituting the fertilizer nutrients but not completely replacing them. A considerable reduction in the cost investment on fertilizers, which is ever increasing, could be achieved by green manuring. When compared to the different intensive cropping systems practiced, rice – based cropping systems are more amenable for the inclusion of green manures.

Unlike inorganic N fertilizers, the green manure biomass must decompose and undergo mineralization before its N becomes available to the rice plant. Synchronising N supply from incorporated organic residues with N demand of rice will increase N use efficiency of plants and reduce soil N losses (Becker *et al.,* 1994). Under waterlogged conditions, decomposition of *Azolla* and *Sesbania* to release ammonia and then made available as an inorganic ammonia fertilizer, has been the mode of nitrogen transfer to rice (Talley and Rains, 1980 and Kannaiyan, 1990). However, decomposition of organic residues (plant or animal origin) is essentially a microbiological process (Waksman, 1952). Under lowland rice soil ecosystem the decomposition rate primarily depends upon the environmental factors, type of plant species and the growth stage of the crop incorporated in soil (Roger and Watanabe, 1984) as well as soil physical, chemical and biological properties (Whitehead, 1963). Information on the pattern and amount of N release from green manure N and its recovery by rice is essential in developing management practices leading to the efficient use of green manure in lowland rice. Hence, an attempt was made to study the mineralization pattern of *Sesbania rostrata* and *Azolla microphylla* with fertilizer N and neem cake and their effect on rice in wetland rice soil ecosystem.

MATERIALS AND METHODS

An experiment was conducted in plastic tubs to study the mineralization pattern of N_2 fixing green manure crop, *Sesbania rostrata* and fern *Azolla microphylla*. Dried, powdered and sieved wetland rice soil was filled in plastic tubs at 6 kg per tub. The experiment was carried out in completely randomised design (CRD) with 3 replications. The treatments were (i) Neem cake (NC) alone, (ii) Prilled urea (PU) alone (iii) *S. rostrata* alone (iv) *A. microphylla* alone, (v) *S. rostrata* + NC, (vi) *S. rostrata* + PU, (vii) *S. rostrata* + NC + PU, (viii) *A. microphylla* + NC, (ix) *A. microphylla* + PU, (x) *A. microphylla* + NC + PU, (xi) Control. *S. rostrata* plants (45d old) were cut into small pieces of approximately 5 cm length for incorporation in to the soil. *Sesbania rostrata* and *Azolla microphylla* were incorporated @ 15 t ha^{-1} (45g tub^{-1}) respectively on wet weight basis. Neem cake and prilled urea were applied basally at 250 kg ha^{-1} (0.75g tub^{-1}) and 60 kg N ha^{-1} (0.4g urea tub^{-1}) respectively at the time of incorporation of *S. rostrata* and *A. microphylla.* The water level in plastic tubs were maintained at 2.5 cm above soil surface throughout the study period.

A field experiment was conducted with CO 43 rice in a randomized block design with three replications. Rice seedlings (25 day old) were transplanted with 15 x 10 cm spacing in one sq. m. plot. The treatments were i. Prilled Urea (PU) alone, ii) *Sesbania rostrata* alone, iii) *S.rostrata* + Neem cake (NC), iv) *S. rostrata* +PU, v) *S. rostrata* + NC + PU, vi) *Azolla microphylla* alone, vii) *A. microphylla* + NC, viii) *A. microphylla* + PU, ix) *A. microphylla* +

NC+PU and x) Control (Without N). Nitrogen (Prilled urea) was applied at 60 kg N ha^{-1} in three split doses at 30, 15 and 15 kg N ha^{-1} at basal, tillering and panicle initiation stage respectively. Neem cake was applied basally at 250 kg ha^{-1}. Phosphorus (Single Super Phosphate) and Potassium (Murate of Potash) were applied basally at 50 kg ha^{-1} P_2O_5 and K_2O respectively. *S. rostrata* and *A. microphylla* were incorporated at 15 t ha^{-1} a day before transplanting.

Estimation of NH_4^+ and NO_3^- N in flood water

The flood water samples (50ml) were collected from each treatment at periodic intervals (7-28d) after incorporation. The accumulation of NH_4^+ - N in flood water due to incorporation of N_2 fixing green manures was estimated as per the method developed by Solorzano (1969) and expressed as n moles ml^{-1}. Nitrate N was estimated at 15 and 30 d after incorporation by following the method developed by Sims and Jackson (1971). Chromotropic acid (CTA) stock solution (0.1%) was prepared by dissolving 1.84 g of CTA in 1.0 l of concentrated H_2SO_4. Concentrated HCl (10ml) was added to 100 ml of 0.1 % stock solution and the volume was adjusted to 1.0 l with concentrated H_2SO_4 to obtain 0.01 % CTA working solution. To 3.0 ml of flood water sample, 7.0 ml of 0.01% CTA working solution was added and thoroughly mixed. After cooling to room temperature, the colour intensity was determined spectro-photometrically at 430 nm. The NO_3^- N in flood water was determined using potassium nitrate as standard and expressed as ppm of NO_3^- N in flood water.

Enumeration of nitrifying bacteria

The nitrifying bacteria were enumerated by MPN method. Soil samples (1.0 g) were taken at 10, 20 and 30d after incorporation from each treatment and serially diluted in sterilized ammonium calcium carbonate broth (Skermann, 1967). Ten replications were kept for each dilution and were incubated at 28 ± 1°C for 3 weeks. The population of nitrifiers was arrived using MPN table as described by Cochran (1950).

Estimation of volatile fatty acids

Soil samples drawn at 15th and 30th d after incorporation of green manures were analysed for acetic, butyric and propionic acids as described by Paul and Beauchamp (1989) and expressed as n moles g^{-1} soil.

Estimation of carbondioxide and methane evolved

The CO_2 and CH_4 evolved from soil due to incorporation of N_2 fixing green manures were estimated at 15th and 30th day after incorporation by following the method described by Pramer and Schmidt (1956) and Bridgham and Richardson (1992) respectively.

Enumeration of microbial population

The population of total bacteria, fungi and actinomycetes of the soil samples drawn on 10th, 20th and 30th d after incorporation was enumerated by dilution plating using soil extract agar medium (Allen, 1950), Martin's rose Bengal streptomycin agar medium (Martin, 1950) and Kenknight's agar medium (Allen, 1950) respectively.

Estimation of soil enzyme activities

Soil samples drawn on 15th and 30th d after incorporation of green manures were analysed for the activity of urease and L-asparaginase by following the method of Tabatabai and Bremner (1972) and Frankenberger and Tabatabai (1991) respectively. The urease and L-asparaginase activities were expressed as µg of NH_3 released g^{-1} soil h^{-1}.

RESULTS AND DISCUSSION

Incorporation of *S. rostrata* and *A. microphylla* with prilled urea recorded maximum ammoniacal N accumulation on 7th d after incorporation, while green manuring alone and with neem cake resulted in gradual accumulation of ammoniacal N (Table 1). Maximum accumulation of ammoniacal N in flood water was noticed on 21st and 14th d after incorporation of *S. rostrata* and *A. microphylla* respectively. However, when combined with neem cake, incorporation of *S. rostrata* and *A. microphylla* recorded maximum ammoniacal N accumulation in flood water on 28th and 21st d respectively. Application of prilled urea alone showed maximum ammoniacal N accumulation on 7th d with drastic reduction in accumulation over subsequent days. Nagarajah *et al.* (1989) noticed faster and higher accumulation of soil ammonium during decomposition of *S. rostrata* than *A. microphylla*. The rate of mineralization of organic - N from *Azolla* proceeded at a much faster rate during the first two weeks after which the mineralization was more gradual (Watanabe,1982 and Subramani and Kannaiyan, 1987). In general, the release of ammoniacal N during mineralization of green manures increased sharply within 7 to 10 d of incorporation, maintained that level upto 20 d and then gradually declined (Buresh and De Datta, 1990). The gradual accumulation of ammoniacal N in flood water upon application of neem cake along with green manure incorporation might be attributed to the nitrification inhibitory effect of neem cake.

Application of prilled urea at the time of incorporation of green manures *viz., S. rostrata* and *A. microphylla* alone resulted in the maximum accumulation of nitrate N in flood water on 15th d after incorporation. However, green manure incorporation with prilled urea (PU) and neem cake (NC) resulted in gradual accumulation of nitrate N (Table 2). In general, green manuring has shown increased accumulation of nitrate N in flood water than control. *S. rostrata* incorporation has shown increased accumulation of nitrate N than incorporation of *A. microphylla* on 30th d after incorporation. Application of prilled urea alone accumulated maximum nitrate N in flood water on 15th d with drastic decline on 30th d after application. Interestingly, application of neem cake along with green manure incorporation resulted in low accumulation of nitrate N in flood water on 15thd after incorporation than green manuring alone which implies the inhibitory effect of neem cake on nitrification process.

Table 1. Effect of incorporation of *S. rostrata* and *A. microphylla* with fertilizer N and neem cake on the ammonical N content in flood water

Treatments	Ammonia (n moles ml^{-1})			
	7 d	14 d	21 d	28 d
Neem cake (NC) alone	90	132	203	174
Prilled Urea (PU) alone	955	684	343	227
S. rostrata alone	336	375	394	299
A. microphylla alone	369	408	354	275
S. rostrata + NC	287	345	372	398
S. rostrata + PU	1127	839	865	764
S. rostrata + NC + PU	1079	879	852	854
A. microphylla + NC	308	366	410	302
A. microphylla + PU	1237	1114	838	718
A. microphylla + NC + PU	1165	884	887	805
Control	137	186	155	140

	Treatment	Duration	Interaction
SE	2.78	1.67	5.55
CD	5.52	3.33	11.03

NC at 250 kg ha^{-1} PU at 60 kg N ha^{-1}

A. microphylla at 15 t ha^{-1} *S. rostrata* at 15 t ha^{-1}

Table 2. Effect of incorporation of *S. rostrata* and *A. microphylla* with fertilizer N and neem cake on the nitrate N content in flood water

Treatments	NO_3^- N (ppm)	
	15 d	30 d
Neem cake (NC) alone	0.13	0.09
Prilled Urea (PU) alone	0.39	0.29
S. rostrata alone	0.35	0.28
A. microphylla alone	0.36	0.25
S. rostrata + NC	0.31	0.25
S. rostrata + PU	0.51	0.40
S. rostrata + NC + PU	0.45	0.34
A. microphylla + NC	0.32	0.22
A. microphylla + PU	0.52	0.36
A. microphylla + NC + PU	0.46	0.31
Control	0.16	0.12
SE	0.010	0.013
CD	0.021	0.026

NC at 250 kg ha^{-1} PU at 60 kg N ha^{-1}

A. microphylla at 15 t ha^{-1} *S. rostrata* at 15 t ha^{-1}

Green manure incorporation with prilled urea enhanced the population of nitrifiers than green manuring alone as application of prilled urea along with green manure incorporation may significantly reduce the C : N ratio and serve as an effective substrate for microbial decomposition. However, incorporation of green manures with or without fertilizer N but with neem cake significantly reduced the population of nitrifying bacteria (Table 3). The nitrification retarding effect of neem cake is mainly attributed to its inhibitory effect on *Nitrosomonas* and *Nitrobacter* in soil (Watanabe *et al.*, 1981).

Decomposition of N_2 fixing green manures enhances microbial population, organic C and total N in soil, but N losses from flooded soils occur primarily through ammonia volatilization, nitrification, leaching and denitrification (Patrick, 1982). However, neem cake has been shown to slow down nitrification rates (Reddy and Prasad, 1975 and Sahrawat and Parmar, 1975) apparently through the action of the alkaloid azadirachtin and other organic compounds present in the neem on nitrifying bacteria (Mishra *et al.*, 1975) which are responsible for the conversion of NH_3 to NO_2, the first and rate limiting step of nitrification (Focht, 1979).

Table 3. Effect of incorporation of *S. rostrata* and *A. microphylla* with fertilizer N and neem cake on the population of nitrifying bacteria in submerged soil

Treatments	Nitrifying bacteria (x 10^3 g^{-1} oven dry soil)		
	10 d	20 d	30 d
Neem cake (NC) alone	4.5	7.8	8.6
Prilled Urea (PU) alone	32.5	24.8	12.4
S. rostrata alone	24.4	44.5	31.5
A. microphylla alone	29.7	37.2	20.7
S. rostrata + NC	11.4	22.6	17.4
S. rostrata + PU	41.0	56.3	38.2
S. rostrata + NC + PU	18.3	26.4	21.5
A. microphylla + NC	13.5	19.3	10.6
A. microphylla + PU	44.2	51.5	30.8
A. microphylla + NC + PU	21.5	24.1	16.0
Control	7.0	11.2	8.1
	Treatment	Duration	Interaction
SEd	0.47	0.25	0.82
CD	0.95	0.49	1.64

NC at 250 kg ha^{-1} PU at 60 kg N ha^{-1}

A. microphylla at 15 t ha^{-1} *S. rostrata* at 15 t ha^{-1}

The first phase of the decomposition of organic matter in rice soil is the rapid formation of organic acids followed by the formation of methane (Acharya, 1935). Green manure incorporation registered relatively higher quantities of volatile fatty acids *viz.*, acetic, butyric and propionic acids in rice soil compared to the application of prilled urea alone and control. Green manuring with prilled urea recorded maximum accumulation of volatile fatty acids on 15th d in rice soil than green manuring alone (Table 4). Among the green manures, incorporation of *A. microphylla* recorded higher accumulation of acetic acid while, *S. rostrata* incorporation resulted in relatively higher accumulation of butyric and propionic acids. Accumulation of volatile acids was higher on 15th d than on 30th d after incorporation. Among the volatile fatty acids, green manure incorporation recorded maximum accumulation of acetic acid followed by butyric and propionic acids.

Table 4. Effect of incorporation of *S. rostrata* and *A. microphylla* with fertilizer N & neem cake on the volatile fatty acids content in submerged soil

Treatments	Volatile fatty acids (n moles g^{-1} soil)					
	Acetic acid		Butyric acid		Propionic acid	
	15 d	30 d	15 d	30 d	15 d	30 d
Neem cake (NC) alone	43.2	11.5	19.7	5.6	14.0	3.5
Prilled Urea (PU) alone	77.9	21.0	26.4	12.5	20.7	9.8
S. rostrata alone	195.5	63.0	68.9	23.7	46.9	17.7
A. microphylla alone	237.1	75.1	62.3	18.3	42.2	13.2
S. rostrata + NC	191.3	68.8	66.5	24.5	45.1	19.1
S. rostrata + PU	248.2	72.4	87.7	29.8	59.3	25.3
S. rostrata + NC + PU	239.5	78.6	84.5	31.6	57.6	27.5
A. microphylla + NC	230.0	79.2	59.0	20.0	40.5	14.0
A. microphylla + PU	285.6	93.7	81.4	26.2	53.0	19.5
A.microphylla +NC+PU	273.9	96.3	79.0	27.5	50.8	21.0
Control	32.5	7.1	17.2	4.0	11.8	2.4
SEd	7.28	3.68	2.01	1.43	1.74	1.32
CD	14.83	7.51	4.10	2.91	3.54	2.69

NC at 250 kg ha^{-1} PU at 60 kg N ha^{-1}

A. microphylla at 15 t ha^{-1} *S. rostrata* at 15 t ha^{-1}

Formation of volatile and gaseous products in submerged soils is almost entirely due to facultative and obligate anaerobic bacteria (Tsutsuki, 1984). Organic acids produced during decomposition of organic materials under waterlogged and extremely reduced conditions may influence plant growth and nutrient uptake of rice (Takijima, 1963). The amount of soil organics formed depend on the amount of organic matter added and its humification coefficient (Chandrasekaran and Yoshida, 1973 and Qi-Xiau Wen, 1984).

Gaseous decomposition products of organic matter are important factors in biochemical and chemical changes in submerged soils and they directly or indirectly influence the availability of plant nutrients and the growth of rice. Most crop species respond positively to higher CO_2 concentration than to ambient CO_2 concentration (Witlewere and Robb, 1964). Rice soils release CO_2 into atmosphere ranging from 3.9 to 5.7 g m^{-2} day^{-1} under flooded condition and from 6.0 to 8.6 g m^{-2} day^{-1} under drained condition (Yoshida, 1981). Though atmospheric CO_2 is the general source of CO_2 in rice crop photosynthesis, direct use of soil CO_2 released into the atmosphere is the next most important source (Neue and Scharpenseel, 1984). The contribution of soil CO_2 to net or gross photosynthesis had been estimated for upland crops under varied weather conditions (Moss *et al.*, 1961 and Monteith *et al.*, 1964).

Fields planted with rice emitted more methane than unplanted fields (Adhya *et al.*, 1994). Mingkui Cao *et al.* (1996a) suggested that on an average, 12 per cent of the simple carbon substrate from decomposition of soil organic matter and rice plants was transformed into CH_4 and the emitted methane accounted for 6.5 per cent of the rice net primary production. The rate of CH_4 emission is a reflection of the complex interaction between the processes of plant primary production, soil organic carbon decomposition, methanogenesis and methanotrophic oxidation. (Mingkui Cao *et al.*, 1996b).

Incorporation of *S. rostrata* and *A. microphylla* resulted in increased evolution of CO_2 and CH_4 over application of fertilizer N alone. In general, green manure incorporation registered maximum evolution of CO_2 and CH_4 on 15th day compared on 30th day after incorporation. However, neem cake application along with green manure incorporation slightly reduced CO_2 and CH_4 evolution on 15th day while increased slightly on 30th day after incorporation (Table 5). The rate of CO_2 evolution from soils amended with *S. aculeata* increased initially, then reduced drastically and declined to more or less a constant value after 15 days of incubation (Sur *et. al.*, 1993).

Table 5. Effect of incorporation of *S. rostrata* and *A. microphylla* with fertilizer N and neem cake on carbondioxide evolution and methane emission in rice soil

Treatments	CO_2 evolved (mg day^{-1} 100g^{-1} soil)		CH_4 emitted (ppm 100g^{-1} soil)	
	15 d	30 d	15 d	30 d
Neem cake (NC) alone	15.0	13.8	18.0	12.1
Prilled Urea (PU) alone	19.3	13.5	44.5	26.3
S. rostrata alone	27.5	18.7	92.8	60.0
A. microphylla alone	29.2	17.5	107.1	45.8
S. rostrata + NC	26.4	20.5	87.3	68.2
S. rostrata + PU	33.6	26.5	130.5	81.2
S. rostrata + NC + PU	31.0	27.0	122.0	90.4
A. microphylla + NC	28.0	19.0	94.4	51.5
A. microphylla + PU	34.2	24.8	143.6	64.0
A. microphylla + NC + PU	32.3	25.2	133.3	72.5
Control	11.4	9.2	13.2	8.0
SEd	1.87	1.64	8.03	5.44
CD	3.81	3.35	16.37	11.08

NC at 250 kg ha^{-1} PU at 60 kg N ha^{-1}

A. microphylla at 15 t ha^{-1} *S. rostrata* at 15 t ha^{-1}

In general, acetic acid was known to be the substrate for CH_4 production (Takai, 1970). Incorporation of N_2 fixing green manures accumulated more acetic acid during decomposition and hence resulted in increased CH_4

emission (Watanabe, 1984). The amount of CH_4 released from plots amended with organic matter was higher than those applied with chemical fertilizers (Nugroho *et al.*, 1994). However, Lindau *et al.* (1993) found that some nitrification inhibitors and N sources can mitigate CH_4 emissions in rice fields. Also, addition of molybdate, a known inhibitor of sulfate reducing bacteria, retarded CH_4 emission in alluvial soil (Ramakrishnan *et al.*, 1995). The rate of release of CO_2 and CH_4 from soils incorporated with green manures is an index of the rate of decomposition of green manures employed. Variation in the evolution of CO_2 and CH_4 from soils incorporated with different green manures might be due to differences in the C:N ratio.

Addition of organic amendments to soil brought out changes in microbial population equilibrium due to changes in the pH as well as changes in the C:N ratio (Zentmeyer, 1963). The addition of organic matter stimulated the population of saprophytic fungi in soil, which played an important role in the decomposition of lignocellulosic compounds (Kannaiyan and Prasad, 1974). Incorporation of green manures significantly enhanced the microbial population than application of prilled urea alone and control. Incorporation of *A. microphylla* resulted in maximum population of bacteria, actinomycetes and fungi on 20th d while, *S. rostrata* incorporation showed maximum population on 30th d after incorporation (Table 6). Application of neem cake along with green manure incorporation significantly reduced the microbial population on 10th d, while it increased on 20th d after incorporation. The initial reduction in microbial population might be attributed to the inhibitory effect of neem cake. Shanthi (1985) found the deleterious effect of organic amendments like neem cake and neem leaf on bacterial population.

Table 6. Effect of incorporation of *S. rostrata* and *A. microphylla* with fertilizer N and neem cake on the bacteria, actinomycetes and fungal population in rice soil

Treatments	Bacteria ($x\ 10^7\ g^{-1}$ oven dry soil)			Actinomycetes ($x\ 10^5\ g^{-1}$ oven dry soil)			Fungi ($x\ 10^4\ g^{-1}$ oven dry soil)		
	10 d	20 d	30 d	10 d	20 d	30 d	10 d	20 d	30 d
Neem cake (NC) alone	11.2	23.1	21.5	4.5	9.2	7.9	2.8	9.0	8.0
Prilled Urea (PU) alone	19.4	25.5	24.0	7.1	10.8	9.7	8.3	11.5	10.1
S. rostrata alone	20.9	34.0	37.7	11.4	16.5	17.0	13.2	19.5	20.6
A. microphylla alone	27.5	38.2	32.3	12.6	18.3	15.5	14.6	22.4	18.1
S. rostrata + NC	17.2	39.4	43.9	7.2	20.5	23.3	8.7	25.0	27.7
S. rostrata + PU	36.8	52.1	48.2	16.4	27.2	25.6	18.5	32.7	31.2
S. rostrata + NC + PU	28.1	59.3	57.4	13.1	31.6	30.3	16.1	38.6	36.5
A. microphylla + NC	21.0	45.6	41.6	9.0	24.7	19.6	10.9	28.3	23.6
A. microphylla + PU	38.7	56.0	43.5	18.5	29.0	22.8	21.0	35.3	26.4
A. microphylla +NC + PU	29.3	61.5	52.0	14.0	33.2	28.5	17.6	41.2	33.0
Control	14.6	18.0	17.1	5.3	7.6	6.5	4.1	7.4	6.2

	T	D	I	T	D	I	T	D	I
SEd	0.16	0.09	0.28	0.12	0.06	0.20	0.95	0.49	1.65
CD	0.33	0.17	0.57	0.23	0.12	0.41	1.90	0.99	3.29

T - Treatment D - Duration I - Interaction
NC at 250 kg ha^{-1} PU at 60 kg N ha^{-1}
A. microphylla at 15 t ha^{-1} *S. rostrata* at 15 t ha^{-1}

The most accepted and valid indicator of sum total of soil biological activity is its enzyme activity. The extracellular enzymes produced by soil microorganisms are released into surrounding soil matrix which catalyse the depolymerization of high molecular weight compounds. L-asparaginase which plays an important role in N mineralization (Blagoveshchenskaya and Danchenko, 1974) was first detected in soils by Drobni'k (1956). Plants may contribute substantial amounts of L- asparaginase to soils which in turn release NH_4^+ to the inorganic N pool (Frankenberger and Tabatabai, 1991a). Incorporation of green manures significantly increased the urease and L-asparaginase activities in rice soil than control. Green manure incorporation with prilled urea recorded maximum soil urease and L-asparaginase activity on 15th d after incorporation, while incorporation of green manures without prilled urea showed maximum soil urease and L-asparaginase activity on 30^th^d (Table 7). Application of neem cake along with green manure incorporation significantly reduced the soil urease and L-asparaginase activity on 15th d but the activity was increased significantly on 30th d after incorporation. The L-asparaginase activity is positively correlated to soil organic carbon and total N content (Frankenberger and Tabatabai, 1991b).

Table 7. Effect of incorporation of *S. rostrata* and *A. microphylla* with fertilizer N and neem cake on the urease and L-asparaginase activities in rice soil

Treatments	Urease*		L-asparaginase*	
	15 d	30d	15 d	30 d
Neem cake (NC) alone	23.5	27.3	26.5	33.6
Prilled Urea (PU) alone	38.5	34.0	46.1	40.8
S. rostrata alone	37.6	43.0	45.0	51.6
A. microphylla alone	38.9	41.3	46.2	50.3
S. rostrata + NC	32.0	48.8	37.2	57.5
S. rostrata + PU	52.0	49.3	62.7	59.3
S. rostrata + NC + PU	43.9	60.1	50.5	71.0
A. microphylla + NC	33.5	47.6	38.7	55.7
A. microphylla + PU	54.2	48.6	64.2	58.0
A. microphylla + NC + PU	45.0	62.4	52.6	72.8
Control	24.7	23.2	29.7	27.3
SEd	1.49	1.91	1.89	2.06
CD	3.04	3.90	3.85	4.20

* - µg of NH_3 released g^{-1} soil h^{-1} NC at 250 kg ha^{-1} PU at 60 kg N ha^{-1}
A. microphylla at 15 t ha^{-1} *S. rostrata* at 15 t ha^{-1}

Addition of small amounts of urea to a silty clay loam soil induced urease activity by soil microorganisms (Paulson and Kurtz, 1969). The activity of soil urease was positively correlated to total organic matter and nitrogen content (Beri and Brar, 1978 and Reynolds *et al.,* 1985). The activity of urease enzyme was increased due to organic matter addition (Sahrawat, 1983) and green manure incorporation (Bolton *et al.,* 1985). Similar increase in the activity of soil enzymes *viz.,* urease and dehydrogenase due to incorporation of *S. rostrata* and *A. microphylla* has been reported (Kumar and Kannaiyan, 1992 and Kannaiyan and Kalidurai, 1995).

Green manure incorporation with prilled urea and neem cake recorded maximum plant height, tiller numbers, panicle numbers, grain and straw yield of rice followed by green manuring with prilled urea. Incorporation of *S. rostrata* significantly increased the plant height, tiller numbers, panicle numbers (Table 8), grain and straw yield (Table 9) of rice over application of prilled urea alone and *A. microphylla* incorporation. In general, green manuring with neem cake significantly enhanced the growth and yield of rice compared to green manuring without neem cake.

Table 8. Effect of incorporation of *S. rostrata* and *A. microphylla* with fertilizer N and neem cake on the growth and yield components of CO 43 rice

Treatments	Plant height (cm)		Tillers/hill	Panicles/hill
	40DAT	75DAT	40DAT	75DAT
Prilled Urea (PU) alone	51.2	72.5	7.8	6.0
S. rostrata alone	53.2	74.4	8.7	6.8
A. microphylla alone	50.8	72.2	7.9	6.1
S. rostrata + NC	57.5	78.8	10.1	8.2
S. rostrata + PU	67.1	88.1	13.0	11.3
S. rostrata + NC + PU	69.6	90.2	13.9	12.1
A. microphylla + NC	55.4	76.4	9.3	7.5
A. microphylla + PU	61.6	82.5	11.2	9.4
A. microphylla + NC + PU	66.2	87.0	12.7	11.1
Control	46.4	68.2	6.5	4.6
SE	0.95	1.00	0.31	0.30
CD	1.95	2.06	0.65	0.62

NC at 250 kg ha^{-1} PU at 60 kg N ha^{-1}

A. microphylla at 15 t ha^{-1} *S. rostrata* at 15 t ha^{-1}

Table 9. Effect of incorporation of *S. rostrata* and *A. microphylla* with fertilizer N and neem cake on the grain and straw yield of CO 43 rice

Treatments	Grain yield (t ha^{-1})	% increase over control	Straw yield (t ha^{-1})	% increase over control
Prilled Urea (PU) alone	5.6	18.6	10.60	10.0
S. rostrata alone	5.88	24.6	11.17	15.9
A. microphylla alone	5.65	19.7	10.64	10.4
S. rostrata + NC	6.36	34.7	12.32	27.8
S. rostrata + PU	7.41	57.0	14.58	51.2
S. rostrata + NC + PU	7.65	62.1	15.46	60.4
A. microphylla + NC	6.10	29.2	11.7	21.4
A. microphylla + PU	6.8	44.1	13.22	37.1
A. microphylla + NC + PU	7.32	55.1	14.32	48.5
Control	4.72	-	9.64	-
SEd	0.11		0.24	
CD	0.22		0.49	

NC at 250 kg ha^{-1} PU at 60 kg N ha^{-1}

A. microphylla at 15 t ha^{-1} *S. rostrata* at 15 t ha^{-1}

Green manuring with *S. rostrata* a day before transplanting of rice and application of fertilizer nitrogen at panicle initiation phase was found to be more efficient than spilt application of fertilizer nitrogen (Morris *et al.*, 1985). Dreyfus *et al.* (1985) reported that incorporation of *S. rostrata* at 20 t ha^{-1} increased the grain yield of rice by 53 per cent over fertilizer nitrogen at 60 Kg N ha^{-1}. Kalidurai and Kannaiyan (1989) have shown significant increase in rice crop response to application of *S. rostrata* at 20 t ha^{-1}. Incorporation of *S. rostrata* significantly increased the grain and straw yield of rice compared to *Azolla* and urea application alone (Ventura *et al.*, 1987; Kalidurai & Kannaiyan, 1989 & 1990). Application of *Sesbania* spp. enhanced rice crop response through its direct contribution of nitrogen and the response to inorganic nitrogen is also dependent on the *Sesbania* biomass incorporated in the rice field (Garrity and Flinn, 1988). Ventura & Watanabe (1993) reported that rice grain yield obtained due to green manuring with *Azolla* and *S. rostrata* was similar to or greater than that obtained in response to application of 60 kg N ha^{-1} as urea. Tiwari *et al.* (1995) noticed that N available to rice crop from 8.5 t fresh biomass of *S. rostrata* was equivalent to 80 kg urea-N.

Green manure incorporation significantly increased the N content of rice than control. Green manure incorporation with prilled urea and with or without neem cake recorded significantly higher N content in rice than green manuring with neem cake and green manure incorporation alone. Incorporation of green manures with prilled urea and neem cake recorded maximum N uptake of rice followed by green manuring with prilled urea, with neem cake and green manure incorporation alone. Incorporation of *S.*

rostrata registered higher N uptake of rice than *A. microphylla* incorporation and application of prilled urea alone (Table 10).

Table 10. Effect of incorporation of *S. rostrata* and *A. microphylla* with fertilizer N and neem cake on total N content and N uptake of CO43 rice

Treatments	Total N content (%)	N uptake (kg ha^{-1})	Percent increase over control
Prilled Urea (PU) alone	0.51	82.6	27.9
S. rostrata alone	0.52	88.7	37.3
A. microphylla alone	0.51	83.1	28.6
S. rostrata + NC	0.54	100.9	56.2
S. rostrata + PU	0.60	131.9	104.2
S. rostrata + NC + PU	0.60	138.7	114.7
A. microphylla + NC	0.54	96.1	48.8
A. microphylla + PU	0.59	118.1	82.8
A. microphylla + NC + PU	0.60	129.8	100.9
Control	0.45	64.6	-
SEd	0.009	1.77	
CD	0.019	3.65	

NC at 250 kg ha^{-1} PU at 60 kg N ha^{-1}

A. microphylla at 15 t ha^{-1} *S. rostrata* at 15 t ha^{-1}

Neem products are reported to have nitrification inhibiting properties (Reddy and Prasad, 1975). The nitrification retarding effect of neem cake is due to its inhibitory effect on nitrifying bacteria in soil (Khandelwal *et al.*, 1977). Blending of urea with neem cake inhibited nitrification by 70, 40 and 5 per cent at the end of 1, 2 and 14 days of incubation respectively (Reddy and Prasad, 1975). Usage of neem coated urea with *Azolla* application resulted in increased grain yield and NUE of rice (Mani *et al.*, 1982). Blending of urea with 15 per cent neem cake alone or with coal tar and kerosene increased the grain yield and NUE of rice (Lal *et al.*, 1982).

Mineralization of neem oil extract coated urea was slower than prilled urea (Shinde and Mangruee, 1983). Kannaiyan (1990b) observed that application of neem cake coated urea at 90 kg N ha^{-1} along with *A. pinnata* significantly increased the grain yield of rice. Application of neem cake with fertilizer N during incorporation of *S. rostrata* has recorded higher grain and straw yield of rice (Thangaraju and Kannaiyan, 1990) as incorporation of green manures with fertilizer N and neem cake resulted in gradual release of N continuously at critical stages of crop growth. The biochemical components particularly alkaloids like azadirachtin, azadirone, nimbin, nimbanidol etc., present in neem cake were likely to regulate the nitrogen use efficiency in rice soil ecosystem and such regulation of N reflects on yield and yield components of rice.

CONCLUSION

Incorporation of N_2 fixing green manures *viz., S. rostrata* and *A. microphylla* with or without fertilizer N and neem cake recorded. significantly higher ammoniacal N accumulation in flood water compared to control. Application of neem cake along with green manure incorporation resulted in significant reduction in the accumulation of nitrate N in flood water and the population of nitrifying bacteria in rice soil. The CO_2 and CH_4 evolution were more in green manure incorporated treatments than application of prilled urea alone. Accumulation of volatile fatty acids *viz.,* acetic, butyric and propionic acids in rice soil incorporated with green manures increased significantly compared to prilled urea application alone. Among the green manures, incorporation of *A.microphylla* and *S.rostrata* registered maximum accumulation of acetic; butyric and propionic acids respectively. Green manuring with prilled urea significantly increased the population of bacteria, actinomycetes and fungi in soil than green manuring without prilled urea. However, neem cake application along with green manure incorporation significantly reduced the soil microbial population on 10th day, which increased significantly thereafter. Incorporation of green manures significantly increased the soil urease and L-asparaginase activity compared to control. In general, neem cake application resulted in gradual accumulation of ammoniacal N in flood water; significantly inhibited the population of nitrifiers and the accumulation of nitrate N in flood water; regulated the accumulation of volatile fatty acids, total microbial population and their activity in rice soil.

Green manure incorporation significantly enhanced the growth, yield and yield components of rice than control. Incorporation of green manures with fertilizer N and neem cake recorded maximum plant height, tiller numbers, panicle numbers and grain and straw yield of rice. Of the green manures, *incorporation* of *S.rostrata* resulted in increased plant height, tiller numbers, panicle numbers and grain and straw yield of rice than *A. microphylla* incorporation. Maximum N content and N uptake of rice were recorded upon incorporation of green manures with fertilizer N and neem cake.

REFERENCES

Acharya, C.N. 1935. Studies on the anaerobic decomposition of plant materials. I. The anaerobic decomposition of rice straw (*Oryza sativa*), **Biochem, J.**, **29**: 528-541.

Adhya, T.K., A.K. Rath, P.K. Gupta, V.R. Rao. S.N. Das, K.M. Parida, D.C. Parashar and N. Sethunathan, 1994, Methane emission from flooded rice fields under irrigated conditions. **Biol Fertil. Soils, 18**:245-248.

Allen, O.N., 1953. **Experiments in Soil Bacteriology.** (II ed.) Burgees Publ. Co., Minneapolis, Minn., USA, p. 127.

Becker, M., J.K. Ladha and J.C.G. Ottow. 1994. Nitrogen losses and lowland rice yield as affected by residue nitrogen release. **Soil Sci. Soc, Am. J., 58(6)**: 1660-1665.

Beri, V. and S.S. Brar, 1978. Urease activity in subtropical, alkaline soils of India. **Soil Sci., 126(6):** 330-335.

Biagoveshchenskaya, Z.K. and N.A. Danchenko, 1974. Activity of soil enzymes after prolonged application of fertilizers to a corn monoculture and crops in rotation. **Soviet Soil Sci., 6**:569-575.

Bolton, H, Jr., L.F. Elliot, R.I. Papendick and D.F. Bezdicek, 1985. Soil microbial biomass and selected soil enzyme activities. Effect of fertilization and cropping practices. **Soil Biol. Biochem., 17**:297-302.

Bridgham, S.D. and C.J. Richardson, 1992. Mechanisms of controlling soil respiration (CO_2 and CH_4) In southern peatland. **Soil Biol. Biochem., 24**: 1089-1099.

Buresh, R.J. and S.K. De Datta, 1990. Nitrogen dynamics and management in rice legume cropping systems. **Adv. Agron., 45**: 1-59.

Chandrasekaran, S. and T. Yoshida, 1973. Effect of organic acid transformations in submerged soils on growth of the rice plant. **Soil Sci. Plant Nutr., 19**: 39-45.

Dreyfus, B.L., G. Rinaudo and Y.R. Dommergues, 1985. Observation on the use of *Sesbania rostrata* as green manure in paddy fields. **MIRCEN J., 1**:711-721.

Drobnik, J. 1956. Degradation of asparagine by the soil enzyme complex. **Cesk Mikrobiol., 1**:47.

Focht, D.D. 1979. Microbial kinetics of nitrogen losses in flooded soils. In : **Nitrogen and Rice**. Inte. Rice Res. Inst., Manila, Laguna, Philippines, pp. 119-134.

Frankenberger, W.T. Jr. and M.A. Tabatabai, 1991 a. Factors affecting L-asparaginase activity in soils. **Biol. Fertil. Soils, 11**:1-5.

Frankenberger, W.T. Jr. and M.A. Tabatabai, 1991b. L-Asparaginase activity of soils. **Biol. Fertil. Soils, 11**:6-12.

Garrity, D.P. and J.C. Flinn, 1988. Farm level management systems for green manure crops in Asian rice environments. In: **Sustainable Agriculture: Green manure in Rice Farming.** Inte. Rice Res. Inst., Manila, Philippines, p, 112-129.

Kalidurai, M and S. Kannaiyan, 1989. Effect of *Sesbania rostrata* on N uptake and yield of lowland rice. **J. Agron. Crop Sci., 163**: 284-288.

Kalidurai, M. and S. Kannaiyan, 1990. Effect of *Sesbania* and *Azolla* on rice grain and straw yields. **IRRN., 15(2)**:26.

Kannaiyan, S. 1990. **Biotechnology of Biofertilizer for Rice Crop.** Tamil Nadu Agrl. Univ., Coimbatore.p.212.

Kannaiyan, S. and M. Kalidurai, 1995. Nitrogen fixing potential of stem nodulating *Sesbania rostrata* and its effect on rice yield. In: **Rice Management Biotechnology,** (ed.) S. Kannaiyan. Assoc. Publ. Co., New Delhi.p. 173-195.

Kannaiyan, S. and N.N. Prasad, 1974. Effect of certain organic amendments on the population of *Fusarium oxysporum* f. melonis in soil. **Lab Dev. J. Sci. Technol., 44**:51-56.

Khandelwal, K.C., D.P. Singh and K.K. Kapoor, 1977. Mineralisation of urea coated with neem extract and response of wheat. **Indian J. Agric. Sci., 47**:267-270.

Kumar, K and S. Kannaiyan, 1992. Changes in the activity of soil enzymes during decomposition of N fixing green manures. In: **33rd Ann. Conf. of Assoc. of Microbiologists of India**. Goa Univ. Goa, (Abstr.) p.57.

Lal, P., G.L, Sharma, R.C. Gautam and P.S. Bisht, 1982. Increasing nitrogen use efficiency transplanted rice by blending urea with margosa seed cake powder, **IRRN., 7(6)**:18-19.

Lindau, C.W., P.K. Bollich, R.D. Dehaune, A.R. Mosler and K.R. Bronson, 1993. Methane mitigation Houisiana rice fields. **Biol. Fertil. Soils, 15**: 174-178.

Mani, S., P.P. Ramaswami and U.S. Sreeramulu, 1982. Study on nitrogen use efficiency with neem cake coated urea and *Azolla*. **Tamil Nadu Agric. Univ., Newslett., 12(2)**:3.

Martin, J.P. 1950. Use of acid, rose bengal and streptomycin in the plate method for estimating soil fungi. **Soil Sci., 69** : 215.

Mingkui Cao., K. Gregson, S. Marshall, J.B. Dent and O.W. Heal, 1996 a. Global methane emissions from rice paddies. **Chemosphere, 33(5):** 879-897.

Mingkui Cao., S. Marshall and K. Gregson, 1996b. Global carbon exchange and methane emissions from natural wetlands: Application of a process-based model. **J. Geophys. Res., 101**:14, 399-14, 414.

Mishra, M.M., S. Neelakantan, K.C. Khandelwal, S.K. Bhardwaj and S.R. Vyas 1975. Margosa (neem) seed cake as an inhibitor of nitrification. **Soil Biol. Biochem.,** 7 : 183-184.

Monteith, J.L., G. Szeic and K. Yabuki, 1964. Crop photosynthesis and flux of carbon dioxide below the canopy, **J. Appl. Ecol., 1**:321-337.

Morris, R.A., O.P. Meelu and R.E. Furoc, 1985. Green manuring for rice. Presented at the **UNDP sponsored BNF Project Advisory Group Meeting.** Inte. Rice Res. Inst., Manlla, Philippines.

Moss, D.N., R.B. Musgrave and E.R. Leinon, 1961. Photosynthesis under field conditions. III. Some effects of light, CO_2 temperature and soil moisture on photosynthesis, respiration and transpiration of corn. **Crop Sci., 1**:83-87.

Nagarajah, S., H.U. Neue and M.C.R. Alberto, 1989. Effect of Sesbania, *Azolla* and rice straw incorporation on kinetics of NH_4, K, Fe, Mn, Zn, and P in some flooded soils. **Plant soil, 116**:37-38.

Neue, H.U. and H.W. Scharpenseel, 1984. Gaseous products of the decomposition of organic matter in submerged soils. In: **Organic Matter and Rice,** Inte. Rice Res. Inst., Manila, Philippines, p. 311-328.

Nugroho, S.G., J. Lumbanraja, H. Suprapto, H. Sunyoto, W.S. Ardjasa, H. Haraguchi and M. Kimura, 1994. Effect of Intermittent irrigation on methane emission from an Indonesian paddy field. **Soil. Sci. Plant Nutr., 40(4)**:609-615.

Patrick, W.H.Jr. 1982. Nitrogen transformations in submerged soils. In : **Nitrogen in Agricultural soils. Agronomy Monograph No. 22,** ASA-CSSA-SSSA, Madison, U.S.A.

Paulson, K.N. and L.T. Kurtz, 1969. Locus of urease activity in soil. **Proc. Soil Sci. Soc. Am. J., 33**:897-901.

Pramer, D, and E.L. Schmidt, 1956. **Experimental Soil Microbiology.** Bulter Publ. Co., Minneapolis, USA, p. 107.

Qi-xiau Wen, 1984. Utilization of organic materials in rice production in China. In: **Organic Matter and Rice,** Inte. Rice Res. Inst., Manila, Philippines. p.419-433.

Ramakrishnan, B., S.N. Satpathy, P. Patnaik, T.K. Adhya, V.R. Rao and N. Sethunathan, 1995. Methane production in two Indian rice soils. **Geomicrobiol, J., 13**: 193-199.

Reddy, R.N.S. and R. Prasad, 1975. Studies on the mineralization of urea, coated urea and nitrification inhibitor treated urea in soil. **J. Soil Sci.,26**: 305-312.

Reynolds, C.M., D.C. Wolf and J.A. Armbruster, 1985. Factors related to urea hydrolysis in soil. **Soil Soi. Soc. Am. J., 49**: 104-108.

Roger, P.A. and I. Watanabe, 1984. Algae and Aquatic weeds as source of organic matter and plant nutrients for wetland rice. In: **Organic Matter and Rice,** Inte. Rice Res. Inst., Manila, Philippines, p.147-168.

Sahrawat, K.L. 1983. Nitrogen availability indices for submerged rice soils. **Adv. Agron., 36**:415-451.

Sahrawat, K.L. and B.S. Parmar. 1975. Alcohol extract of neem (*Azadirachta indica* L.) seed as nitrification inhibitor. **J. Indian Soc. Soil Sci., 23** : 131-134.

Shanthi, S.R. 1985. Influence of neem products on nitrogen use efficiency in lowland rice, M.Sc., (Ag.), thesis. TNAU, Coimbatore.

Shinde, B.N. and R.G. Mangruee, 1983. MIneralization of neem oil extract coated urea in upland rice. **Oryza, 30**:14-18.

Sims, J.R. and G.D. Jackson. 1971. Rapid analysis of soil nitrate with chromotrophic acid. **Soil Sci. Soc. Am. Proc., 35** : 603-606.

Skermann, V.B.A. 1967. In : **A Guide to Identification of the Genera of Bacteria,** (eds.) Williams and Wilkins, Baltimore.

Solorzano, L. 1969. Determination of ammonia in natural waters by the phenol hypochlorite method. **Limnol. Oceanogr., 14**:799-801.

Subramani, S and S. Kannaiyan. 1987. Effect of urea on decomposition of *Azolla.* **IRRN., 12(4)**:57.

Sur, H.S., R, Singh and Y.S. Deol, 1993. Decomposition kinetics of *Sesbania aculeata* ammended soil and its impact on integrated use with N fertilizers. **Ann. Agric. Res.,14(3)**:256-261.

Tabatabai, M.A. and J.M. Bremner, 1972. Assay of urease activity in soils. **Soil Biol, Biochem., 4**:479-487.

Takai, Y. 1970. The mechanism of methane fermentation in flooded paddy soil. **Soil Sci, Plant Nutr., 16**:238-244.

Takijima, Y., 1963. Studies on the behaviours of the growth inhibiting substances in paddy soils with special reference to the occurrence of root damage in the peaty paddy fields. **Bull. Natl. Inst. Agric. Sci., Japan Series B., 13**:117-252.

Talley, S.N. and D.W. Rains, 1980. *Azolla filliculoides* Lam. as a fallow season green manure for rice in temperate climate, **Agron. J., 72**:11-18.

Thangaraju, M. and S. Kannaiyan, 1990. Use of nitrification inhibitors with *Azolla* biofertilizer and fertilizer on nitrogen use efficiency in rice. In: **Biotechnology of Biofertilizers for Rice Crop** (eds.) S. Kannaiyan, Tamil Nadu, India, p. 131-140.

Tiwari, C., Y, Singh and D. Singh, 1995. Effect of *Sesbania* green manure and nitrogen fertilizer in rice (*Oryza sativa* L.) – Wheat (*Triticum aestivum*) cropping system. **Indian J. Agric. Sci., 659(10)**:708-711.

Tustsuki, K., 1984. Volatile products and low molecular weight phenolic products of the anaerobic decomposition of organic matter. In: **Organic Matter and Rice.** Inte. Rice Res. Inst., Manila, Philippines, p.329-343.

Ventura, W. and I. Watanabe, 1993. Green manure production of *Azolla microphylla* and *Sesbania rostrata* and their long term effects on rice yields and soil fertility. **Biol. Fertil. Soil., 15**:241-248.

Ventura, W., G. B. Mascarinia, R. E. Furoc and I. Watanabe, 1987. *Azolla* and *Sesbania* as biofertilizers for lowland rice. **Philipp. J. Crop Sci., 12**:61-69.

Waksman, S.A. 1952. **Soil Microbiology,** John Willey and Sons, Inc., New York.

Watanabe, I. 1982. *Azolla-Anabaena* symbiosis its physiology and use in tropical agriculture. In: **Microbiology of Tropical Soils and Plant Productivity,** (eds.) Y.R. Dommergues and H.G. Dlem Hague: Martinus Nijhoff/Dr. W. Junk Publ., The Netherlands, p. 169-185.

Watanabe, I. 1984. Anaerobic decomposition of organic matter in flooded rice soils. In: **Organic Matter and Rice.** Inte. Rice Res. Inst., Manila, Philippines. p. 237-258.

Watanabe, I., B.P.R. Subudhi and T. Aziz. 1981. Effect of neem cake on the population and nitrogen fixing activity of blue green algae in flooded soil. **Curr. Sci., 50(2)** : 937-949.

Whitehead, D.C. 1963. Some aspects of the influence of organic matter on soil fertility. **Soils and Fert., 26**: 217-223.

Witlewere, S.H. and W.H. Robb, 1964. CO_2 enrichment of green house atmospheres for food crop production. **Econ. Bot., 18**:34-56.

Yoshida, S. 1981. **Fundamentals of Rice Crop Science.** Inte. Rice Res. Inst., Manila, Philippines. p. 269.

Zentmeyer, A. 1963. Biological control of phytophthora root rot of avocado with alfalfa meal. **Phytopathology, 53**: 1383-1387.

CHAPTER - 15

ESTIMATION OF BIOLOGICAL NITROGEN FIXATION BY ^{15}N ISOTOPE TECHNIQUE

K. Arulmozhiselvan

INTRODUCTION

Wetland rice has been grown continuously for centuries with a reasonable yield level without much N fertilizer. It is also well recognized that two thirds or more of the N in a rice crop comes from the soil N, including soil biomass, even when inorganic N fertilizer is adequately used (Koyama and App, 1979). The major source of N replenishment in wetland soils is by the natural process known as biological N_2 fixation. Wetland rice field provides a unique set of conditions for N_2 fixation. Firstly, the aquatic plant habitat gives suitable sites for the activity of autotrophic nitrogen fixing blue green algae (BGA). Secondly, the anaerobic condition in submerged soil is suitable for heterotrophic N_2 fixing bacteria (App *et al.*, 1984).

Average rates of biological N fixation in flooded soils are around 50 kg N ha^{-1} yr^{-1}. These rates are much higher than rates in arable soils, that amount to about 12 kg N ha^{-1} yr^{-1}. Under these situations, for efficient N management, accurate measures of biological N_2 fixation are important prerequisite factors in various ecological systems (Hauck, 1973). Total N_2 fixed in lowland rice fields is calculated as 3.2 million t $year^{-1}$. Soyabean, which is grown in rice fields and uplands, accounts for 3.3 million t N $year^{-1}$. (Watenabe, 1986). Nevertheless, the research on N_2 fixation has yet to bring up more accurate values.

RICE FIELD ECOSYSTEM AND N_2 FIXING ORGANISMS

Flooding favours rice growth environments by bringing the soil pH near to neutral, making nutrients like P and Fe more available, depressing soil organic matter decomposition and thus maintain soil fertility. These conditions while promoting rice growth instantaneously support biological N_2 fixation. Reports available under long term fertility trials and N balance studies stand as evidences (Watanabe *et al.,* 1981) for N_2 fixation and state that under these conditions, the most important N_2 fixing cyanobacteria belonging to the genera *Anabaena*, *Nostoc* and *Rivularia* play the greatest role. Estimates on the average potential of N_2 fixation under green house conditions and actual assessment in field, usually show wide differences (Table 1). Hence application of appropriate technique is necessary to report exact values on biological N_2 fixation. Among the methods available, the stable isotope ^{15}N technique provides accurate estimate of the quantity of biologically fixed N. Labelled N is measured by either mass spectrometry or emission spectroscopy. The technique of assessing biological N_2 fixation can be either by direct or indirect methods (Danso, 1995). The direct method is known as enrichment technique, which involves the use of ^{15}N labelled N_2 gas. The indirect method is known as dilution technique in which labelled fertilizer or medium is used for the growth of the plant in the N_2 fixing environment.

Table 1. Potential and estimated amounts of biological N_2 fixation

Agents	Kg N ha^{-1} $crop^{-1}$	
	Theoretical	Measured value
Azolla (in crop of Azolla)	100 -190	30 -140
Blue green algae (in crop of rice)	42 - 150	80
Associative organisms	40	1.3 - 7.7

(Source: Watanabe, 1986)

NITROGEN TRACER TECHNIQUE

For nitrogen, although six isotopes are known, only those having mass numbers 14 and 15 are stable. The so-called long-lived of the four radioactive nuclides, ^{13}N isotope, which has been used in chemical and biological research has the half-life of only 10.05 minutes (Table 2). This necessitated using N compounds enriched with the naturally occurring stable isotope of ^{15}N, in most tracer investigations.

The methods available for detection and determination of stable isotopes require the use of mass spectrometer, which is an expensive instrument, demanding both continual care and considerable experience for its successful operation and maintenance. However, the stable isotope has several advantages over radioactive nuclides as tracer as it is not complicated by health hazards, dangers of radiation, damage to biological materials or limitation on duration of experiments.

A sample which is incubated in an atmosphere containing N_2 with 60% atom ^{15}N excess that would increase only 0.20% total N of the sample due to N_2 fixation will increase an ^{15}N atom% in the sample from 0.366 to 0.378%

and this increase (0.012 atom %) is readily detectable in mass spectrometer analysis.

Table 2. Isotopes of nitrogen and their half-life

Mass Number of N isotopes		Natural Abundance (Atom%)	Half-life
12	(Radioactive)	-	0.0125 seconds
13	(Radioactive)	-	10.05 minutes
14	(Stable)	99.634	-
15	(Stable)	0.366	-
16	(Radioactive)	-	7.36 seconds
17	(Radioactive)	-	4.14 seconds

(Source: Bremner, 1965)

EXPERIMENTAL METHODS OF ASSESSING BIOLOGICAL N_2 FIXATION

Laboratory and field methods involving ^{15}N technique offer direct and reliable means for measuring N_2 fixation in microbial biomass and plant tissues under symbiotic and non-symbiotic environments. It also permits to distinguish the proportions of N in cereal or legume components that those derived from soil, added fertilizer and atmosphere. The principle of the method is to compare the $^{15}N/^{14}N$ ratio of non-fixing and fixing components, and to relate the values obtained with any differences which may exist in isotopic composition of the primary sources of N available for plant growth (Liyanaga *et al.*, 1994, Sanginga *et al.*, 1995). The various methods adopted under direct and indirect measurements are detailed below.

The calculations are based on atom%^{15}N excess, which is the value indicating the surplus ^{15}N gained in the experimental environment above 0.366 atom% ^{15}N of the natural composition of Earth (Natural abundance). The amount of N_2 fixed is expressed as per cent N derived from atmosphere (%Ndfa)

DIRECT METHODS

Use of ^{15}N gas in growth chamber experiments

This direct enrichment technique involves the use of ^{15}N labelled N_2 gas. This enrichment is usually only feasible for short-term experiment under controlled conditions. Limitation for field applicability stems from the fact that plants need to be enclosed in sealed containers. In spite of that this method is most useful in exploratory studies of plants in ecosystems where not much is known about their N_2 fixing capabilities.

$$\% \text{ Ndfa} = \frac{\text{Atom\% }^{15}\text{N excess of plant tissue/microbial biomass}}{\text{Atom\% }^{15}\text{N excess of }^{15}\text{N enriched gas}} \times 100$$

Based on Natural Isotope Abundance

There are very slight isotopic discrimination effects during chemical or biological processes involving N compounds. Although the chemical properties of ^{14}N and ^{15}N are identical this discrimination by microbes occurs because of mass differences in the N cycle as isotopic shift, changes in diffusion or reaction rates, or changes in isotope distribution at equilibrium (La Rue and Patterson, 1981). These variations in ^{15}N abundance, which usually fall in the range of ± 10 σ ^{15}N units (± 0.0037 atom%^{15}N), originate from the isotope fractionation, and occur when nitrogenous compounds of the soil undergo chemical and biological transformations. Biological enzyme systems responsible for major soil N fluxes react more readily with ^{14}N containing molecules than with the equivalent ^{15}N species. Fractionation occurring as a result of such processes is expressed numerically as a fractionation factor (Table 3).

Table 3. Microbiological reactions resulting in fractionation of nitrogen isotopes

Process	Formal reaction	Organism	Fractionation factor (β)
N_2 fixation	$N_2 \rightarrow 2\ NH_4^+$	*Azotobacter*	1.0039
NH_4 assimilation	$NH_4^+ \rightarrow$ amine – N	*Azotobacter vinelandii*	1.0148
	$NH_4^+ \rightarrow$ amide – N	*Pichia terricola*	1.0018

(Source: Smith, 1983)

Under this circumstance, symbiotic fixation seems to favour very slight increase of ^{15}N. This can be determined by growing a plant on N free nutrient so that its entire N arises by fixation.

Similarly, in soils with active denitrification, there are chances for ^{15}N depleted N_2 gas production. This leads to nitrate remaining in soil slightly enriched. The degree of enrichment is measured as parts per thousand differences from the $^{15}N/\ ^{14}N$ ratio in a standard usually atmospheric N and assessed very accurately and expressed in parts per thousand (σ)

$$\sigma\ ^{15}N = \frac{(^{15}N/\ ^{14}N)\ \text{sample} - (^{15}N/\ ^{14}N)\ \text{atm}}{(^{15}N/\ ^{14}N)\ \text{atm}} \text{ X } 1000$$

The concentration of N deived from N_2 fixation is given by,

$$\%\text{Ndfa} = \left(1 - \frac{\sigma\ ^{15}\text{N in fixing system}}{\sigma\ ^{15}\text{N in non-fixing system}}\right) \text{ X } 100$$

A plant obtaining its entire N from soil (eg. Non-legume or non-nodulated control) will have a slightly enriched ^{15}N relative to the atmosphere. A plant obtaining N from symbiosis will have a lower ^{15}N composition. The ^{15}N enrichment is not uniform throughout the plant, although it is claimed that the seed content is representative of the whole legume plant.

The use of isotope dilution at natural abundance level has several advantages. Firstly, it is not necessary to label the soil with ^{15}N, which in turn eliminates the problems of uniform labeling and lack of homogeneity. This also eliminates the cost involved on labeling with costly ^{15}N compounds. Secondly, the most important is the fact that it permits %Ndfa in natural eco system. However, the constraints in quantifying this small variation, particularly in *in vivo* studies, limit its wide adoptability. Great care is necessary in sample preparation as well as contamination by ^{15}N-enriched material. This method must be considered as suitable one in semi quantitative procedures.

INDIRECT METHODS

Fixing and non-fixing systems

Recent improvements in ^{15}N tracer techniques permit quantification of N_2 fixation in legumes by labelling the soil or fertilizer N with ^{15}N. The technique of labelling the soil, instead directly labelling atmospheric N_2, invokes the principle of isotopic dilution technique, which has been very widely used to accurately quantify N_2 fixation in legumes under field conditions.

This indirect technique is suitable in various crop environments like growth chamber as well as actual field conditions. The method involves growing of plants in soil containing higher $^{15}N/^{14}N$ ratio than atmosphere. In N_2 fixing plants accumulation of N_2 by biological N_2 fixation results in a lower $^{15}N/^{14}N$ ratio than the non-fixing plants. Dilution technique involves artificial enrichment of culture medium or plant rooting medium with ^{15}N labelled fertilizer and subsequent study of dilution of ^{15}N in the biomass of the N_2 fixing organisms in a medium or plant tissue. This estimation is done both in fixing and non-fixing components and compared each other to assess N derived from atmosphere (%Ndfa).

The non-fixing system has two possible N sources for its assimilation and growth namely, the soil and fertilizer; while the fixing system has a third source, the atmosphere. Thus, plant N derived from the soil is slightly or highly enriched in ^{15}N equally in both systems, but will be diluted by $^{14}N_2$ from the atmosphere through N_2 fixation only in fixing system. Hence non-fixing system is essential to assess the contribution from soil (%Ndfs) and/or fertilizer (%Ndff) N to the fixing system.

Many non-fixing systems exist, but the need is that each must assimilate its N from the soil and/ or fertilizer N pools at similar rates (i.e. identical $^{15}N/^{14}N$ composition) but not necessarily in the identical quantity as the fixing system (Rennie *et al.*, 1978).

Several possibilities for establishing non-fixing systems are:

1. **Uninoculated:** An excellent control to study the effect of inoculation if bacterium added as an inoculant is not indigenous to the soil.
2. **Ineffective inoculation:** An excellent control if a bacterium exists that invades the plant but does not establish an effective N_2 fixing association.
3. **Non–nodulating isolines:** non–fixing clones of legume that function similar to non-nodulating legume isoline.
4. **Non-legume:** Non-legumes with similar N uptake patterns.
5. **High N:** Application of N from fertilizer at levels sufficient to inhibit N_2 fixation by the indigenous or added bacteria.

In order to estimate the extent of N_2 fixation the following function is used to compare the fixing and non-fixing systems:

$$\%Ndfa = \left(1 - \frac{\text{Atom\% } ^{15}\text{N excess in fixing system}}{\text{Atom \% } ^{15}\text{N excess in reference system}}\right) X\ 100$$

The fixing system would be cells in culture media, a legume, a N_2 fixing non-legume, which is effectively inoculated and labelled with ^{15}N. The non-fixing system would comprise, as categorized above, a plain media, a legume or non-legume, which is not inoculated and labelled with ^{15}N alone.

Merits and demerits of dilution technique

- Isotopic dilution assumes that the soil N pool is uniformly labelled with ^{15}N although this is virtually impossible.
- It also assumes that no significant isotopic discrimination occurs in any biological nitrogen cycle process.
- In case, N_2 fixation does occur in non-fixing system due to indigenous soil bacteria, isotopic dilution would underestimate, but never overestimate %Ndfa of fixing system.
- Further, the assumption is that the isotope tracer incorporated into soil N, is equally available to both crops. If the soil N is low, the non-fixing crop may grow lesser than the test crop and thus may not be a comparable control.

Ways to Increase Accuracy in Experiments

The majority of the N in soil is in the upper horizons and due to that the roots of the fixing system and non-fixing system may explore different soil volumes, if they have different rooting patterns (Witty, 1983). Since it is the isotope ratio that is critical and not the amount of soil N that is assimilated, the volume of soil explored by the roots of the fixing system and non-fixing system may be more crucial.

Further, for instance, when the ^{15}N abundance of available soil N decreases sharply with time, the differences in N uptake pattern results. These indicate the importance of selecting the reference crop with same rooting pattern. In order to avoid this problem it is necessary to make the change of ^{15}N abundance with time in soil to be slow, or more or less stabilized. For achieving this, various methods of application of labelled fertilizers to soil have been proposed (Witty and Ritz, 1984).

In addition, the accuracy depends very much on how accurate the soil $^{15}N/^{14}N$ ratio is assessed. Even after careful selection of reference plants is done, it may not work well and lead to errors. This is because in some non-nodulating isolines, there are possibilities for nodulation under certain circumstances. It is therefore necessary not to be taken for granted that non-nodulating isolines can be used as non-fixing reference plants without examining their roots for nodules. Hence sufficient isolines are to be grown to have multiple referencing.

Soil Availability Index Method by Tracer (A value technique)

In this method adequate labelled fertilizer N is added to the soil for the non-fixing control to promote growth (Danso, 1995). The assumption here is that, the available native (unlabelled N) soil N (A) is not altered through the addition of different rates of ^{15}N enriched fertilizer, and thus the A -value assessed by the reference plant is equal to A -value assessed by the fixing crop. The ^{15}N content of the plants is measured after harvest. For each crop and treatment, A-value is determined. The A-value is an indication of how much of a nutrient is available to a crop and is expressed as follows.

$$A = B\ (1\text{-}y)\ /\ y$$

Where B is the amount of fertilizer N added (kg ha^{-1}) and y is the proportion on N in plant derived from the added labelled fertilizer as given below.

$$y = \frac{\text{Atom\% } ^{15}\text{N excess in reference crop}}{\text{Atom \% } ^{15}\text{N excess in fertilizer}}$$

From this, the fraction of legume N from fixation is calculated as

$$\text{N fixed (kg ha}^{-1}) = \begin{matrix}\text{(A -for fixing plant)} - \\ \text{(A- for reference plant)}\end{matrix} \quad \text{X} \quad \frac{\text{N in crop (kg ha}^{-1})}{\text{N in fertilizer (kg ha}^{-1})}$$

By this it is possible to calculate for each crop and treatment the contribution made by the soil, fertilizer, and symbiotic organism.

Table 4. Estimates of N_2 Fixation by ^{15}N Techniques

Technique	Test Crop	Reference Crop	N_2 fixed
^{15}N dilution (Sanginga *et al.*,1989)	Gliricidia	*Cassia siamea* (Non N_2 fixing)	17.8 - 61.7 g $tree^{-1}$ (86-309 kg N at 5000 trees ha^{-1})
Natural Abundance (George *et al.*, 1993)	Soyabean	Soyabean (Non nodulating isoline)	71 % of plant N
^{15}N dilution (Toomsan *et al.*, 1995)	Groundnut	Groundnut (Non nodulating)	72 - 77% of plant N (150-200 kg N ha^{-1})

^{15}N dilution (Giller *et al.*, 1987)	Groundnut	Groundnut (non nodulating)	86 - 92% of plant N (100 -153 kg N ha^{-1})
Natural Abundance (Yoneyama *et al.*, 1991)	*Aeschynomene indica*	Non-nodulating soyabean and paddy weeds	93% (127 kg ha^{-1})
	Sesbania rostrata		78% (281 kg ha^{-1})
Natural Abundance (Singh, 1993)	*Saccharum spontaneum*	*Sclerotrachya fusae* (Non- fixing)	60% of plant N
	S. sinennse		54% of plant N
	S. barberi		35% of plant N

Laboratory Procedures for ^{15}N Estimation

To determine isotopic composition of stable elements it is necessary that the element under analysis be converted to a suitable gas. The gas should have a low molecular weight and be as simple as possible in structure from both the molecular and isotopic viewpoint. For ^{15}N isotope ratio analysis accordingly ^{15}N labelled nitrogen in the material is converted to N_2 gas (Buresh *et al.*, 1982). The analysis usually involves the following three steps.

1. Conversion of labelled N to ammonium.
2. Conversion of ammonium to N_2 by oxidation with alkaline sodium hypobromite in the complete absence of air.
3. Determination of isotopic composition of N_2 by mass spectrometer or optical emission spectrometer analysis.

Sample handling prior to laboratory analysis

For labelling media or soil, ^{15}N compounds that are commercially available, such as nitrate or sulphate of ammonium, calcium or potassium; ammonia, cynamide, urea, glycines and many other suitable compounds can be used, according to the experimental requirement. Sample collected must be meticulously processed. Generally speaking the sources of greatest error in ^{15}N research are not associated with analytical procedures, but are caused by poor sample preparation and sub sampling.

Careful consideration must be given to every facet of sample preparation sequence namely drying, grinding, weighing, sub sampling and storage because accurate analytical techniques can not correct errors made earlier in the experimental procedures. Further, the total soil or plant component in an experiment should be harvested, dried, weighed and ground, prior to sub sampling for analysis whenever possible to avoid sampling bias.

In particular, the total weight of the soil sampled and the weight of plant material harvested must be accurately measured at the termination of an experiment as these weights are used directly in the calculation of ^{15}N recovery. Extreme care must be taken during grinding and initial sub sampling. The grinding mill must be thoroughly cleaned between samples to avoid cross contamination. As much of the sample as possible should be recovered from the grinding mill. Coarse material remaining on the sieve should be crushed to pass through sieve and added to the rest of the sample rather than discarded.

After harvest, plant samples should be rapidly dried at 60°C in forced draft drier, weighed and then ground to pass completely through a 1mm sieve. Large samples should be reduced to approximately 70g by sub sampling and then stored in tightly sealed glass bottles.

Conversion of organic N to Ammonium

Kjeldahl digestion and distillation procedure is followed for conversion of organic N to ammonium. A sample containing about one mg of N is to be digested in a semi-micro kjeldahl flask with 10 ml of concentrated H_2SO_4 with 2g of a K_2SO_4 - $CuSO_4$ - selenium mixture. The digestion temperature of about 380°C for over four hours is employed. The digest is later steam distilled, and the distillate is preserved by acidification after titration, for ^{15}N assay (Table 5).

Conversion of Ammonium to Nitrogen Gas

The conversion of ammonium to N_2 for mass spectrometer analysis is accomplished by treatment of sample with alkaline hypobromite in the complete absence of air in the inlet system of the instrument.

Table 5. Steps in handling and analysis of samples containing ^{15}N

Steps	Soil and plant material
1	Sampling or harvest
2	Drying
3	Weighing
4	Grinding
5	Sub sampling for chemical analysis
6	Digestion to convert nitrogen to ammonium form
7	Moisture determination (soils only)
	(a) Steam distillation of ammonia
	(b) Titration with standard acid
	(c) Concentration of distillate
	(d) Transfer to a vial
	(e) Heating to dryness
	(f) Conversion of ammonium N to nitrogen gas
8	Isotope ratio analysis

Instrumentation Techniques for ^{15}N Estimation

The analysis of ^{15}N isotope of N is performed in Mass spectrometer and optical emission spectrometer. Each technique has unique advantages. In many instances either would perform the task equally well. Both instruments exploit a physical property of the N_2 molecule, to determine the relative amounts of each of the three possible species, $^{14}N_2$, $^{15}N^{14}N$ and $^{15}N_2$. In optical emission spectrometry N_2 molecules are separated on the basis of their vibration properties, while in mass spectrometry, charged ions are separated according to their mass to charge (m/e) ratio.

MASS SPECTROMETER

The development of the mass spectrometer followed the discovery of "positive rays" which seem to emerge from the perforated cathode of a partially evacuated discharge tube developed by Thompson and Aston during 1960s (Smith, 1983).

Mass spectrometer development, particularly over the last 4 decades has been extremely rapid. At present number of mass spectrometers are commercially available. The resolution and precision of an isotope ratio mass spectrometer to large extent determine how well it performs. The resolution, the ability of the instrument to separate the ion spectrum, is expressed in terms of Resolving Power, which is based on peak height and width. The resolution is affected by the collector aperture or slit width of a magnet deflector.

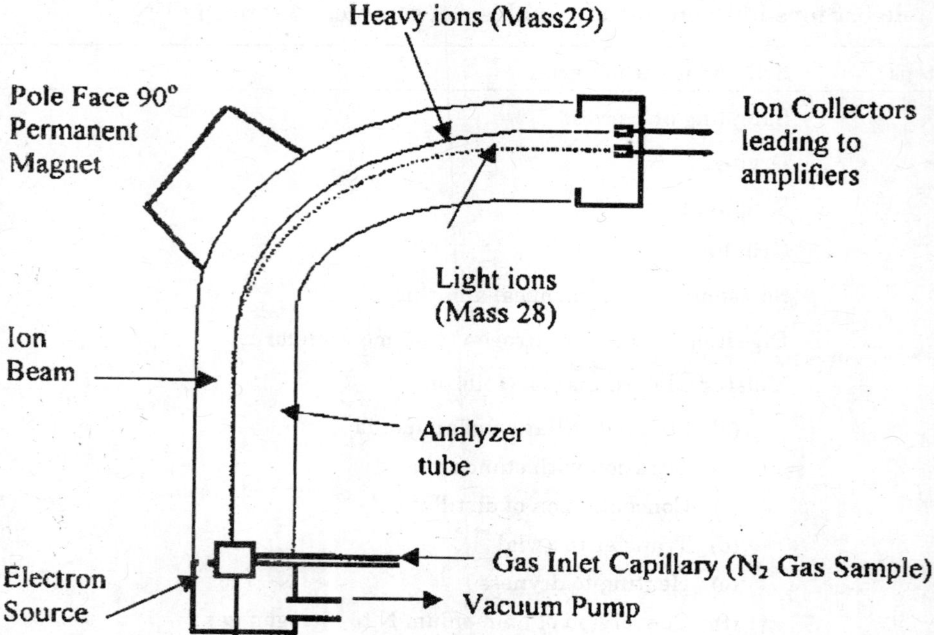

Fig. 1. Schematic diagram of 90° deflection type Mass Spectrometer equipped with double collector system

Following are the description of the simplest types (Fig.1) of mass spectrometers (Bremner, 1965). The whole of the source - analyzer – collector assembly is contained within an evacuated stainless steel enclosure, which is continuously pumped, so that during sample admission the pressure equilibrates between 10^{-4} to 10^{-7} m bars. Low pressure must be maintained in the mass spectrometer in order to have the mean free path of the ions (the distance an ion can move between collisions) is sufficiently large with respect to the geometry of the analyzer assembly. If this requirement is not met, the scattering due to ion-ion and ion-molecule collisions would prevent sharp beam focusing. Ions emerge under the influence of voltages from the source are further accelerated and collimated by a large electro static field established by half plates, and direct the ion beam towards the collector. A permanent magnet encloses the analyzer such that the homogenous magnetic field in perpendicular to ion beam plane deflects the ion beam 90° or 180° (as varying in different mass spectrometers) and focuses to the collector. The current discharged by the ion is amplified and referred in units in proportion to the magnitude of each atom or molecule ionized.

Optical Emission Spectrometer

Optical emission spectrometer analyses isotopic composition of samples based on their vibrational energy. The relationship is that the vibrational energy is proportional to the square root of the reduced mass. The reduced mass is a term indicative of the ratio between the product of masses of atoms and the sum of the masses of atoms in a molecule. In 1930s several bands of in the emission spectrum were isotopically analyzed to prove the existence of ^{15}N. The main components of an optical emission spectrometer (Fig. 2) used for the measurement of $^{14}N/^{15}N$ ratio are microwave generator, light dispersive system, high tension unit, photo multiplier tube amplifier and recorder.

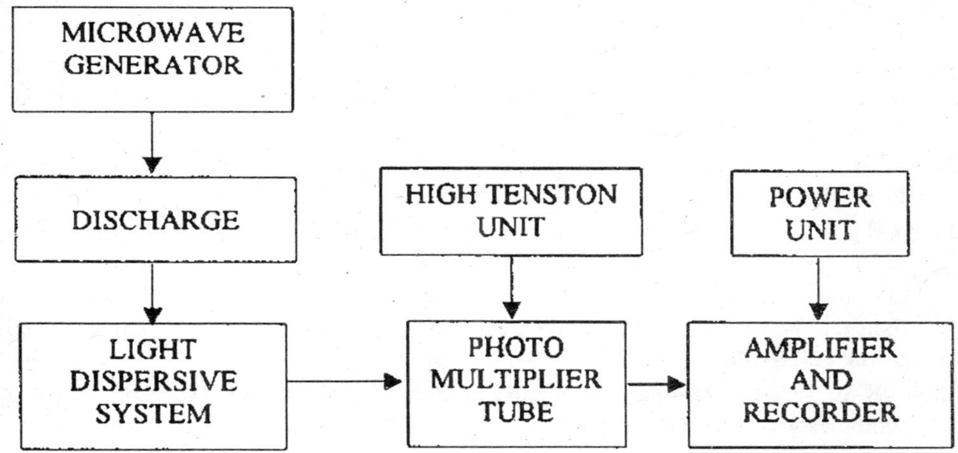

Fig. 2. Block Diagram of an Optical Emission Spectrometer used for ^{14}N / ^{15}N Ratio Analysis

Energy from a microwave generator (high frequency transmitter) excites the N_2 in the discharge tube to emit light. The light dispersive unit is usually a prism or a line grating. The spectral analysis of $^{14}N/^{15}N$ ratios in N_2 is based on an electronic transition, which gives rise to a group of bands in the ultra violet second positive N system region. A band of 298 nm (2977A) has been found to give the best results.

COMPARISON OF MASS SPECTROMETRY WITH OPTICAL EMISSION SPECTROMETRY

Mass spectrometry is indispensable for the analysis of extremely small variations in ^{15}N abundance that occur in nature, and optical emission spectrometry is uniquely applicable to isotopic analysis of nanomole quantities of N. These circumstances are largely due to the following two facts. (1) The accuracy of mass spectrometer is superior to that of an optical emission spectrometer in the analysis of $^{14}N/^{15}N$ ratios, and (2) The N_2 gas sample is consumed during analysis by mass spectrometry. It is conserved (in the discharge tube) during analysis by optical spectrometry and hence a non-destructive method.

There are a number of techniques being practiced for estimating N_2 fixation in legumes, cereals and other non-legumes. Among them, the techniques using ^{15}N are much suitable to derive information of plants from seed to seed. The results would be much accurate and complete to link the N_2 fixing behaviour of organisms associating with plants and the physiology of plants.

REFERENCES

App. A., T. Santiago, C. Daez, C. Manguito, W. Ventura, W. Titol, A. Do.I, Watanabe, S.K. De Datta and P. Roger, 1984. Estimation of the nitrogen balance for irrigated rice and the contribution of phototrophic nitrogen fixation. **Field Crop Res., 9:** 17-27.

Bremner, J.M., 1965. Isotopic-Ratio Analysis of Nitrogen in Nitrogen-15 Tracer investigations. In: **Methods of Soil analysis**. Ed. C.A. Black. 1965. Amer. Soc. Agron. Madison, Wisconsin, USA. Part I: pp 1257-1285.

Buresh. R.J., E.R. Austin and E.T. Craswell 1982. Analytical methods of ^{15}N research. **Fert. Res., 3:** 37-42.

Danso, S.K.A. 1995. Assessment of biological nitrogen fixation. **Fert. Res., 42:** 33-41.

George, T., P.W. Sinleton and C.V. Kessel, 1993. The use of nitrogen-15 natural abundance and nitrogen yield of non-nodulating isolines to estimate nitrogen fixation by soybeans (*Glycine max* L.) across three elevations. **Biol. Fertil. Soils. ,15:** 81-86.

Giller, K..E., P.T.C. Nambiar, B. Sirnivasa Rao. P.J. Dart and J.M. Day, 1987. A comparison of nitrogen fixation in genotypes of groundnut (*Arachis hypogaea* L.) using ^{15}N -isotope dilution. **Biol. Fertil. Soils., 5:** 23-25.

Hauck, R.D. 1973. Nitrogen tracers in nitrogen cycle studies- past use and future needs. **J. Environ. Quality 2:**317-327.

Koyama, T. and A.A. App, 1979. Nitrogen balance in flooded rice soils. In: **Nitrogen and Rice.** IRRI, Manila. pp 95-104.

La Rue, T.A. and T.G. Patterson, 1981. How much nitrogen do legumes fix? **Adv. Agron. 34:**15-38.

Liyanaga, M.S., S.K.A. Danso and H.P.S. Yajasundara, 1994. Biological nitrogen fixation in four *Giricidia sepium* genotypes. **Plant Soil, 161:** 267-274.

Rennie, R.J., D.A. Rennie, and M. Fried, 1978. **Concepts of ^{15}N Usage in Dinitrogen fixation.** International Atomic Energy Agency, Vienna. pp. 107-133.

Sanginga, N., K. Mulongoy and A. Ayanaba. 1989. Nitrogen fixation of field-inoculated *Leucaena leucocephala* (Lam.) de Wit, estimated by the ^{15}N and the difference methods, **Plant Soil, 117:** 269-274.

Sanginga, N., B. Vanlauwe and S.K.A. Danso, 1995. Management of biological N_2 fixation in alley cropping systems: Estimation and contribution to N balance. **Plant Soil, 174:** 119-141.

Singh, A.L. and P.K. Singh, 1987. Nitrogen fixation and balance studies in rice soil. **Biol. Fertil. Soils., 4 :** 15-19.

Singh, M., 1993. Estimation of Nitrogen Fixation in *Saccharum* spp. by ^{15}N Dilution Method. **J. Nuc. Tech. Agri. Biol., 1-5.**

Smith, K.A. 1983. **Soil Analysis.** Marcel Dekker Inc. Newyork and Basel. pp 555-404.

Toomsan B., J.F.Mc Donash, V. Limpi Nuntana and K.E. Giller, 1995. Nitrogen fixation by groundnut and Soybean and residual nitrogen benefits to rice in farmers fields in North East Thailand. **Plant Soil, 175:** 45-56.

Watanabe, I. and E.T. Craswell and A.A. App 1981. Nitrogen cycling in wetland rice fields in South-East Asia. In: **Nitrogen Cycling in South-East Asian Wet Monsoonal Eco-System.** (Ed.) R. Wetselaar. Aust. Acad. Sci. Canberra. pp 4-17.

Watanabe, I., 1986. Nitrogen fixation by non-legumes in tropical agriculture with special reference to wetland rice. **Plant Soil, 90:** 343-35.

Witty, J.F. 1983. Estimation of N_2 fixation in the field using ^{15}N – labelled fertilizer. Some problems and solutions. **Soil Biol.Biochem., 15:** 631-640.

Witty, J.F. and K. Ritz, 1984. Slow-release nitrogen-15 fertilizer formulations to measure nitrogen fixation by isotopic dilution. **Soil Biol. Biochem., 16:** 657-66.

Yoneyama, T., T. Uchiyama, H. Sasakawa, T. Gamo., J.K. Ladha and I. Watanabe, 1991. Nitrogen accumulation and changes in natural ^{15}N abundance in the tissues of legumes with emphasis on N_2 fixation by stem nodulating plants in upland and paddy fields. **Soil Sci. Plant Nutr., 37:** 75-82.

CHAPTER - 16

EFFECT ON BIOFERTILIZERS WITH GRADED LEVELS OF NITROGEN ON LOWLAND RICE AND ITS CARRYOVER EFFECT ON SUCCEEDING CROP OF RAPESEED (*Brassica campestris* var. Toria)

L.N. Singh

ABSTRACT

A field experiment using BGA and *Azolla* in integration with graded levels of N fertilizer in rice followed by rapeseed was conducted during 1997-'98 and 1998-'99 to find out suitable integration of N fertilizer and biofertilizer for higher grain yield of rice and to study its carryover effect on rapeseed. Among the various treatments, 80 kg N ha^{-1} in integration with either BGA or *Azolla* or both recorded higher grain yield of rice and statistically at par with that of 100 kg N ha^{-1} (alone), 80 kg N ha^{-1} (alone) and 60 kg Nha^{-1} in integration with either BGA or *Azolla* or both. With regards to the effect on subsequent crop of rapeseed, the treatments containing *Azolla* were found better in yield and the highest was recorded with higher dose of N (80 kg N ha^{-1}) in integration with *Azolla*. When rice-rapeseed sequence was considered as a whole, application of 80 kg N ha^{-1} + *Azolla* or 80 kg N ha^{-1} + *Azolla* + BGA gave the highest return per rupee investment (2.33).

INTRODUCTION

The ever increasing cost of chemical N fertilizer, its short supply and the prevailing low purchasing power among the small and marginal farmers, which constitute the bulk of farming community are the major drawbacks in rice production today in India. In this context both free living and symbiotic forms of N_2 fixers have drawn attention. As a result, use of biofertilizers like BGA and *Azolla* in lowland rice cultivation is getting popular now-a-days. Many reports are available about the feasibility of using such biofertilizers for other parts of the country (Mahapatra and Sharma, 1998 and Gangwane, 1992) but little is known about its usefulness in the north eastern part of the country which is quite distinct from agroclimatic point of view from the rest of the country. The present investigation was therefore, undertaken to study the feasibility of using BGA and *Azolla* in rice cultivation and its residual effect on yield of rapeseed.

MATERIALS AND METHODS

The field experiment was conducted at the Research Farm of College of Agriculture, Central Agriculture University, Imphal during 1997-'98 and 1998-'99. The soil of the experimental site was clay in texture with acidic reaction (pH 5.4) and moderate fertility level (organic carbon 0.87%, low in available N, medium in available P and rich in available K). The experiment comprised of 15 treatment combinations (Table1), which were laid out in Randomised Block Design and replicated 3 times. All the 15 treatments were applied to the rice crop and their carryover effect was observed in succeeding crop of rapeseed.

Azolla pinnata @ 200 g sq. m^{-1} was inoculated as dual crop on the 7th day after transplanting rice and incorporated into soil after 25 days. The left over *Azolla* was incorporated again 25 days after first incorporation. Cyanobacteria was also inoculated @ 10 kg flakes ha^{-1} 7th day after transplanting. All the treatments received equal quantities of P and K @ 40 kg P_2O_5 and 30 kg K_2O ha^{-1} as basal dose. N was applied in 3 splits i.e. 1/3 as basal, 1/3 at active tillering and 1/3 at panicle initiation stage of rice crop.

Rapeseed variety M-27 was grown in *rabi* season with a common nutrient dose of 40:20:20 kg N : P_2O_5 : K_2O ha^{-1} and all of which were applied as basal dose.

RESULTS AND DISCUSSION

Effect on rice crop

The yield contributing characters except test weight and yield of rice were increased significantly in all the treatments over control (Table1). Application of 100 kg N ha^{-1} could not brought further significant increase in yield over that of 80 kg N ha^{-1} and it remained statistically at par with that of application of 60 kg N ha^{-1} in integration with either BGA or *Azolla* or both and 80 kg N ha^{-1} in integration with BGA or *Azolla* or both. Thus inoculation of BGA or *Azolla* or both increased the grain yield of rice significantly when treated alone or in combination with N fertilizer upto the level of 60 kg N ha^{-1}. More or less similar trend was observed in straw yield also. It was also further observed that combined application of both the biofertilizers could not or *Azolla* in all the levels of N fertilizer. The results are in conformity with the findings of Betty *et al.* (1987), Mahapatra and Sharma (1988), Dar and Zardar (1989) and Patra *et al.* (1989).

Table 1. Effect of BGA and *Azolla* with graded levels of Nitrogen fertilizer on yield parameters and yield of rice (pooled data of 1997 and 1998)

Treatment	No. of effective tillers hill^{-1}	No. of filled grains panicle $^{-1}$	Test weight (g)	Grain yield (q ha^{-1})	Straw yield (q ha^{-1})
T_0 Control	4.80	99.6	23.66	27.9	39.5
T_1 N_{60}	5.25	128.5	24.55	39.7	58.1
T_2 N_{80}	5.50	131.6	24.49	44.7	65.1
T_3 N_{100}	6.10	130.9	24.45	45.9	74.0
T_4 N_{40} + BGA	5.60	113.4	24.65	36.8	54.1
T_5 N_{40} + *Azolla*	5.75	112.1	24.98	38.1	56.1
T_6 N_{40} + BGA + *Azolla*	6.00	118.3	24.69	40.4	60.4
T_7 N_{60} +BGA	6.02	126.5	24.68	44.6	59.0
T_8 N_{60} + *Azolla*	6.18	119.7	24.84	44.0	58.9
T_9 N_{60} + BGA + *Azolla*	6.19	128.2	24.75	45.1	68.8
T_{10} N_{80} + BGA	6.26	130.7	24.60	45.8	65.4
T_{11} N_{80} + *Azolla*	6.25	129.8	24.75	45.9	66.1
T_{12} N_{80} + BGA + *Azolla*	6.25	130.4	25.04	46.2	72.2
T_{13} BGA	5.46	114.0	24.50	33.4	51.0
T_{14} *Azolla*	5.47	113.2	24.49	34.3	49.4
C.D. 0.05	0.68	10.2	NS	4.87	7.93

N = Nitrogen BGA = Blue Green Algae

Carryover effect on rapeseed

Azolla inoculation in rice crop benefited the subsequent crop of rapeseed with regards to number of siliqua plant, number of seeds siliqua^{-1} and grain yield (Table 2).

The highest grain yield (8.39 q ha^{-1}) was associated with the treatment 80 kg N ha^{-1} + BGA + *Azolla* but remained statistically at par with that of 80 kg N ha^{-1} + *Azolla*, 60 kg N ha^{-1} + BGA + *Azolla*, 60 kg N ha^{-1} + *Azolla* and 40 kg N ha^{-1} + BGA + *Azolla*. Thus, irrespective of N level, *Azolla* application in rice crop increased the yield of rapeseed but, BGA alone failed to bring such significant influence. This might be due to the residual effect of organic matter added to the soil due to *Azolla* incorporation to the preceeding rice crop. Similar result of benificial effect of organic matter to the succeeding crop of mustard was reported by Rakesh *et al.* (1999).

Table 2. Effect of BGA and *Azolla* with graded levels of nitrogen on the yield of subsequent rapeseed crop (pooled data of 1997-'98 and 1998-'99)

Treatment	No. of siliqua plant^{-1}	No. of seeds siliqua^{-1}	Test weight (g)	Grain yield (q ha^{-1})
T_0 Control	68.88	11.33	2.52	7.02
T_1 N_{60}	71.19	12.84	2.59	7.12
T_2 N_{80}	70.23	12.87	2.62	7.12
T_3 N_{100}	74.66	12.96	2.58	7.42
T_4 N_{40} + BGA	71.44	12.39	2.61	7.09
T_5 N_{40} + *Azolla*	75.12	13.04	2.67	8.06
T_6 N_{40} + BGA + *Azolla*	76.01	13.72	2.62	8.30
T_7 N_{60} +BGA	72.01	12.88	2.55	7.15
T_8 N_{60} + *Azolla*	75.45	13.90	2.58	8.34
T_9 N_{60} + BGA + *Azolla*	75.90	13.78	2.66	8.41
T_{10} N_{80} + BGA	74.99	12.84	2.60	7.27
T_{11} N_{80} + *Azolla*	76.08	14.14	2.55	8.28
T_{12} N_{80} + BGA + *Azolla*	80.03	14.00	2.59	8.39
T_{13} BGA	70.20	12.81	2.65	7.14
T_{14} *Azolla*	74.97	13.74	2.58	7.38
C.D. 0.05	2.47	0.18	NS	0.24

Economics of the rice-rapeseed sequence

The highest gross return (Rs. 40,937 ha^{-1}) was recorded with the application of 80 kg N ha^{-1} + BGA +*Azolla* in rice crop and closely followed by (Rs. 40,599 ha^{-1}) with the application of 80 kg N ha^{-1} + *Azolla* in rice (Table 3).

As the increase in cost of production in the said treatments over others were minimal, the net return and per rupee investment were found higher in the treatments mentioned above. Such a beneficial effect on total productivity and economics of cropping sequence in treatments which received higher fertility dose preferably with organic matter in *Kharif* crop were reported by Jana and Ghosh (1996) and Raksh *et al.* (1999). The lowest net return and return per rupee investment were observed in control plots.

On the basis of above investigation it can be concluded that *Azolla* can be integrated with 80 kg N ha^{-1} in lowland rice cultivation in order to obtain higher yield in rice and the subsequent crop rapeseed. It is also found economically feasible as it gave the highest return per rupee investment in the rice rapeseed sequence.

Table 3. Effect of BGA and *Azolla* with graded levels of nitrogen on the economics of rice-rapeseed crop sequence

Treatment	Cost of production (Rs. ha^{-1})			Gross return (Rs. ha^{-1})			Net Return (Rs. ha^{-1})	Return per Rupee investment
	Rice	Rapeseed	Total	Rice	Rapeseed	Total		
T_0	10256	5908	16164	18135	9126	27261	11097	1.69
T_1	10890	,,	16798	25805	9243	35048	18250	2.09
T_2	11044	,,	16952	29055	9373	38428	21476	2.27
T_3	11377	,,	17285	228535	9646	39481	22136	2.28
T_4	10934	,,	16842	23920	9217	33137	16295	1.97
T_5	11186	,,	17094	24765	10478	35243	18149	2.06
T_6	11381	,,	17289	26260	10790	36738	19449	2.12
T_7	11085	,,	16993	28990	9295	38285	21292	2.25
T_8	11336	,,	17244	28600	10842	39442	22198	2.29
T_9	11531	,,	17439	29315	10933	40248	22809	2.31
T_{10}	11239	,,	17147	29770	9451	39221	22074	2.29
T_{11}	11490	,,	17398	29835	10764	40599	23201	2.33
T_{12}	11685	,,	17593	30030	10907	40937	23344	2.33
T_{13}	10451	,,	16359	21710	9282	30992	14633	1.89
T_{14}	10703	,,	16611	22295	9594	31889	15278	1.92

Calculation was based on local market price of rice @ Rs. 6.50 kg^{-1} and rapeseed @ Rs. 13.00 kg^{-1}

REFERENCES

Betty, R.A., K.N. Devaraju and S. Lingaraju. 1987. Effect of *Azolla* and green leaf manuring with and without nitrogenous fertilizers on irrigated rice. **Oryza, 24 (3)**: 268-269.

Dar, G.H. and M.Y. Zardar. 1989. Cyanobacterial contribution to rice (*Oryza sativa* L.) in Kashmir. **Indian J. Agric. Sci., 59(9):** 566-569.

Gangwane I.V. 1992. Biofertilizer technology: An agrobased industry for farmers. In: **Biofertilizer Technology Transfer** (eds.) L.V. Gangwane. Associated Publishing Co. New Delhi pp. 21-27.

Jana, M.K. and B.C. Ghosh. 1996. Integrated nutrient management in rice (*Oryza sativa* L) – rice crop sequence. **India J. Agron., 41(2):** 183-187.

Mahapatra, G.S. and G.L. Sharma. 1988. Effects of *Azospirillum*, cyanobacteria and *Azolla* biofertilizers on productivity of lowland rice. **Indian J. Agron., 33(4):** 368-371.

Patra, S.K., A.K. Padhi and S.N. Mishra. 1989. Effect of BGA at varying levels of nitrogen on paddy in north eastern ghat region of Orissa. **Environ. Ecol., 7(2) :** 271-274.

Rakesh, K.S., V.K. Shrivasatva, S.T. Suresh, P.N. Tiwari and R.P. Yadav. 1999. Nutrient management in soybean – mustard crop sequence. **Indian J. Agron., 44(3):** 493-497.

CHAPTER - 17

EFFECT OF INTEGRATED NUTRIENT MANAGEMENT IN RICE (*Oryza sativa*) – WHEAT (*Triticum aestivum*) CROPPING SYSTEM IN WESTERN MAHARASHTRA

S. G. Gholve, S. K. Kamble and S.N. Shinde

ABSTRACT

Application of 50% RD of N + 10 tonnes of Glyricidia / ha realised highest paddy yield (51.19 q/ha) which indicated the possibility of substituting 50% N of Paddy crop through Glyricidia green manuring @ 10 tonnes / ha. Similarly, the yield obtained due to application of 75% R. D. of N + BGA @ 20 kg/ha^{-1} was statistically equal to that of 100% R.D. of N thereby indicating the efficient role of BGA for substitution of nitrogen N up to 25%. While in succeeding wheat crop N level responded upto 100% RD of N. This was also confirmed by the assessment of the system in terms of monitory returns.

INTRODUCTION

Rice-Wheat sequential cropping is the most dominant cropping system in rice growing areas of Western Maharashtra. Since both these crops are exhaustive in nature, this system needs optimum application of fertilizers to meet its nutritional need. However, very meagre research information is available about the nutritional needs of this system as a unit. Integrated nutrient supply system which involves the conjunctive use of fertilizers alongwith organic sources, may be manures or biofertilizers, assuring great importance in recent years due to inadequate availability of fertilizers and the need to sustain productivity. Integrated nutrient management technology plays a vital role in getting maximum net returns without impairing the soil health. Therefore, it becomes imperative to develop economically viable, technically, feasible and sustainable integrated nutrient management technology for rice – wheat sequence. With this background in view the present study was undertaken.

MATERIALS AND METHODS

A field experiment was laid out in medium deep block soil at Agrl. Res. Station, Vadgaon Maval (M. S) during 1998-99 and 1999 – 2000 with Randomized Block Design (Factorial) and three replications. The soil of the experimental field was low in available Nitrogen (153 kg ha^{-1}) medium in available Phosphorus (15.4 kg ha^{-1}), high in available Potassium (505 kg ha^{-1}) and having Organic Carbon 0.64%. The experiment comprised three combination of integrated nutrient management for rice (Variety Pawana) *viz.*, T1 –100% RD of N, T2-50% RD of N + 10 tonnes Glyricidia / ha^{-1}, T3 – 75% RD of N + BGA cultures @ 20 kg / ha^{-1} and 3 nitrogen levels for wheat (Variety HD 2189) *viz.*, F1-100% F2-100% and F3-125% RD. The recommended dose of fertilizer for both the crops was 100, 50 and 50 kg N,P & K ha^{-1}, out of which N dose was applied as per treatments and P, K applied uniformly as a basal dose in both the crops. Incorporation of Glyricidia just before pudding and application of BGA eight days after transplantation of rice seedlings was undertaken in respective treatments. Need based cultures operations were practiced during both the years.

RESULTS AND DISCUSSION

Data on grain yield of both the crops are presented in Table 1, while rice grain equivalent and various aspects of economics *viz.,* cost of cultivation, gross and net monetary returns, B: C ratio are presented in Table 2.

A. Effect of nutrient management for rice

Rice yield

The data presented in Table 1 revealed that application of 50% R. D. of N + 10 tonnes Glyricidia ha^{-1} (T2) to rice crop production highest yield during both the individual years as well as in pooled data (51.60, 50.78 and 51.19 q ha^{-1} respectively). This treatment was found significant during both the year and in pooled data over the treatment 100% R. D. of N (T1) as well as 75% R. D. of N + BGA @ 20 kg ha^{-1} (T3). Tripathi (1999) as well as Hiremath and Patel (1998) also observed more or less similar trend. The response of Glyricidia was because of quick decomposition and mineralization for consequent easy availability of nutrient to rice crop due to succulence of leaves. Moreover, the combination of green manuring of Glyricidia with N fertilizer was of added advantage due to continuos supply of N. Treatment T1 and T3 were found to be at par to each other. This indicates the efficacy of BGA @ 20 kg ha^{-1} in rice crop for substitution of 25% N.

Wheat yield

As regards residual effect of nutrient management for rice on succeeding wheat crop grown in sequence, more or less, similar trend as in rice yield was observed during individual years as well as in pooled results. Application of 50% R. D. of N alongwith 10 tonnes Glyricidia ha^{-1} (T2) produced maximum wheat yield which was significantly superior over T1 and T3 during the year 1998-99 and in pooled data. It was also observed that wheat yields of T1 and T3 were also at par with each other. This clearly indicates the beneficial residual effect of BGA on wheat yield due to application of BGA @ 20 kg ha^{-1} to rice (T3).

Rice grain yield equivalent

Data in Table 2 revealed that application of 50% R. D. of N alongwith Glyricidia @ 10 tonnes ha^{-1} (T2) produced highest rice grain yield equivalent values during individual years as well as in pooled results also (98.05, 102.76 and 100.41 q ha^{-1} respectively) which were also significantly superior over T1 and T3. This again indicated the overall advantageous effect of 50% N through Glyricidia for rice crop in rice –wheat sequence. Similarly treatment T1 and T3 were at par with each other. This indicates the importance of application of biofertilizer i.e BGA for rice –wheat cropping system and thereby substitution of 25% N through organic source i.e BGA.

INM for rice shear cropping system

Table 1. Grain yield of rice and wheat as influenced by INM

Treatment	Rice grain (q ha^{-1})			Wheat grain (q ha^{-1})		
	1998-99	1999-2000	Pooled	1998-99	1999-200	Pooled
A. Nutrient management for rice						
T_1-100% RDF of N	46.34	41.45	43.40	30.57	38.94	34.76
T2- 50% RDF of N + 10 tonnes Glyricidia / ha	51.60	50.78	51.19	34.55	40.36	37.46
T3-75% RDF of N + BGA @ 20 kg / ha	47.39	37.47	42.43	32.42	37.51	34.97
SE ±	1.41	2.52	2.03	0.41	1.41	0.79
C. D. at 5%	4.15	7.29	5.88	1.61	N.S.	2.43
B. Nutrient management for wheat						
F1-75% RD of N	48.48	39.99	44.24	27.49	36.80	32.15
F2-100% RD of N	48.91	45.53	47.22	34.12	39.77	36.94
F3- 125% RD of N	47.53	42.64	45.09	35.94	40.24	38.09
S. E. ±	2.39	1.87	1.40	0.82	0.93	1.29
C. D. at 5%	NS	5.39	NS	2.52	2.87	3.97
Interaction	NS	NS	NS	NS	NS	NS

Economics

Under this, cost of cultivation, gross as well as net monetary returns and B:C ratio was determined and the data are presented in Table 2 . Data revealed that the cost of cultivation was highest (Rs.35967 ha^{-1}) in treatment T2 – wherein 50% R. D. of N + Glyricidia @ 10 tonnes ha^{-1} was used. This is because of the fact that collection of green biomass of Glyricidia and its application in rice field is labourious and hence, treatment T2 was expensive as compared to treatment T1 i.e 100% R . D. of N (Rs.34363 ha^{-1}).

Table 2. Rice grain equivalent and its Economics

Treatment	Rice grain equivalent			Economics			
	1998-99	1999-2000	Pooled	Total monetary returns pooled (Rs. ha^{-1})	Cost of cultivation (Rs. ha^{-1})	Net Returns (Rs. ha^{-1})	B:C Ratio
A. Nutrient management for Rice							
T_1-100% RDF of N	86.15	89.46	87.81	52,680	34,363	18,317	1.53
T2- 50% RDF of N + 10 tonnes Glyricidia/ ha	98.05	102.76	100.41	60,246	35,967	24,279	1.68
T3-75% RDF of N+BGA @ 20 kg/ ha	90.35	85.00	87.68	52,608	34,198	18,410	1.54
SE ±	2.68	2.82	3.81	2,286			
C. D. at 5%	8.24	8.49	11.71	7,026			
B. Nutrient management for wheat							
F1-75% RD of N	84.93	87.24	86.08	51,648	34,138	17,560	1.51
F2-100% RD of N	94.46	95.17	94.82	56,892	34,363	22,529	1.66
F3- 125% RD of N	95.15	95.21	95.18	57,108	34,588	22,520	1.65
S. E. ±	2.82	1.62	1.57	942			
C. D. at 5%	8.70	4.98	4.82	2,892			
Interaction				NS			

As regards gross monetary returns, it revealed that nutrient management in rice crop in rice-wheat crop sequence affected the gross monetary returns significantly. Significantly higher value (Rs. 60246 ha^{-1}) was observed when 50% R.D. of N + 10 tonnes of Glyricidia ha^{-1} (T2) was applied for rice in rice –wheat cropping sequence. It is very important to note that in treatment T1 and T2 the values of gross monetary returns were more or less similar. The trend for net returns as well as B: C ratio (1.68) was observed in treatment T2 (50% R. D. of N + 10 tonnes Glyricidia ha^{-1}). In both these characters, the values were second highest in treatment T3 (75% R.D.

of N + BGA @ 20 kg ha^{-1}). This indicates the beneficial substitution of 25% R.D of N through use of biofertilizer i.e. BGA for rice. In other words, this indicates the efficient role of organic sources *viz.* Glyricidia and BGA for substitution of inorganic N upto 50 and 25% of recommended dose respectively for rice crop in rice-wheat cropping sequence. These results are in conformity to the finding by Reddy and Reddi (1997) who reported that depending on favourable condition, the amount of atmospheric N fixed by BGA ranges from 15-40 kg ha^{-1} in low land rice crop.

Soil fertility status

Organic carbon content found to be increased only due to adoption of Glyricidia green manuring (T2), while, increase in available N content was observed both due to use of Glyricidia green manuring (T2) as well as BGA (T3).

B. Effect of nutrient management for wheat

Rice yield

The data in Table 1 revealed that the effect of fertilizer management for wheat on rice yield in rice –wheat cropping sequence was found to be non significant. It was as per expectations during first year. However, this trend was also observed in pooled data indicating that there were no residual effects of nutrient management for wheat on succeeding rice crop during next year.

Wheat yield

Data revealed that during both the individual years as well as in pooled data the nutrient management for wheat affected the wheat yield significantly. Application of 125% R. D. of N (T3) produced highest wheat yield (35.94, 40.24 and 38.09 q ha^{-1} during first, second year and pooled data respectively), however, it was on par with 100% R. D. on N (F2). This clearly indicates the suitability of prevailling recommended dose of N to wheat crop even after adoption of integrated nutrient management approach in rice – wheat cropping. As reported by Thakur *et al.* (1999) this might be due to only marginal residual effect of green manuring as well as biofertilizers used in preceding rice crop.

Rice grain yield equivalent

Data in Table 2 revealed that as that of wheat yield, more or less similar trends were obtained in rice grain equivalent. Though highest yield – equivalent was obtained at application of 125% R.D. of N to wheat crop, it was at par with the application of 100% R.D.of N.

Economics

Various aspects *viz.*, cost of cultivation, gross and net monetary returns and B:C ratio were studied. The gross monetary returns were affected significantly due to varying N levels for wheat. The highest gross monetary returns (Rs. 57,108 ha^{-1}), though at par with F2 (100% R.D. of N) were obtained in F3 (125% R.D. of N). However, highest net returns (Rs. 22,529 ha^{-1}) as well as B : C ratio (1.66) was observed due to application of 100% R. D. of N. i.e. recommended dose for wheat crop.

Soil fertility status

Noticeable change was not observed in organic carbon content, however, available Nitrogen content status was found to be improved in the order of level of Nitrogen dose applied to the wheat crop.

Table 3. Effect of INM for rice – wheat cropping system on soil fertility status

Treatment	Organic carbon (%)		Available Nitrogen (kg ha^{-1})	
	After 1st year	After 2nd year	After 1st year	After 2nd year
A. Nutrient management for Rice				
T1-100% RDF of N	0.62	0.61	156	158
T2- 50% RDF of N + 10 tonnes Glyricidia / ha	0.66	0.67	158	159
T3-75% RDF of N + BGA @ 20 kg / ha	0.63	0.63	155	156
B. Nutrient management for wheat				
F1-75% RD of N	0.63	0.63	155	156
F2-100% RD of N	0.63	0.64	156	157
F3- 125% RD of N	0.64	0.63	158	160
Interaction	0.64		153	

REFERENCES

Chandrasekharan, V., G.S. Thangamthu and P. Balsubramanyam. 1988. Effect of biofertilizer and urea on growth and yield of low land rice. **Madras Agric. Journal. 75 (9)**218-322.

Hiremath, S.M. and Z.G. Patel. 1988. Effect of winter green manuring and nitrogen application on summer rice. **Indian Journal of Agronomy. 43 (1):** 71-76.

Mitra, B.N. and R.S. Tripathi. 1999. Effect of organics and chemical fertilizers on rice chickpea cropping system. **Indian Journal of Agronomy. 44(2):** 223-227.

Reddy, T.Y. and G.H.S. Reddi. 1997. **Principles of Agronomy. pp.** 251 – 252.

Thakur, R. B., S. K. Chaudhary and G. Jha. 1999. Effect of combined use of green manuring crops and nitrogen on productivity of rice – wheat system. **Indian Journal of Agronomy. 44(4):** 664-668.

CHAPTER - 18

IMPACT OF *AZOLLA* BIOFERTILIZER ON RICE YIELD IN HIRAKUD COMMAND AREAS OF ORISSA

K.C. Samal

ABSTRACT

Rice cultivars of about 1001 was investigated continuously for 2 years in Hirakud command areas. A comparison of yield data indicated that the combined inoculation of *Azolla* (10t/ha) with 60 Kg N/ha gave the maximum grain yields of 49.24 q/ha with an increase of 56 per cent over the uninoculated control (31.52 q/ha). The average number of productive tillers as well as panicle per meter square was much higher in the same treatment when compared to that of uninoculated control. Again consecutive application of *Azolla* for 2 years improves soil fertility and organic matter status of soil. In 3rd year, *Azolla* was not inoculated to rice field, but *Azolla* mat appeared automatically in wetland rice field during the growth of rice. This was possible because of sporulation of *Azolla microphylla* under stress condition during previous winter seasons when the temperature drops below 12°C.

INTRODUCTION

The combined application of organic and inorganic fertilizer for increasing crop productivity and maintaining soil health is being increasingly practised in recent years, particularly in the context of sustainable agriculture. Of the various biofertilizers, water fern *Azolla* with its cyanobacterial symbiont, *Anabaena azollae* is found to be very effective in paddy field and can contribute 40-60 Kg N/ha (Kannaiyan, 1985). It is used as green manure before transplanting and grown along with rice as dual cropping after transplanting, the latter method being more practicable and economical. Orissa is a state where rice is a major crop. Sambalpur, the rice bowl of Orissa, is an intensively rice cultivating area which is located under Hirakud command area. The productivity of rice in Hirakud command areas is steadily decreasing owing to the indiscriminate application of synthetic fertilizer and pesticides. In this context, a strategy has been evolved for the alleviation of chemicals in agriculture by the utilisation of biofertilizer for sustainable development and clean ecosystem.

MATERIALS AND METHODS

Field experiments were conducted at the Krishi Vigyan Kendra, Sambalpur for four consecutive seasons. The experiment was laid out in randomised block design with 7 treatments and 3 replications. Rice variety MTU 1001 was transplanted in the 5x3 m plots in 15x10 cm spacing. Phosphorus and potassium were applied at 30 Kg /ha as basal dressings. Nitrogen was applied at 3 equal spilt doses during transplanting time, tillering stages and panicle initiation stage. *Azolla* was inoculated at 1000 Kg/ha on 7 days after transplanting (DAT). The observations like number of tillers, grain yield and straw yield were recorded at the time of harvest.

RESULTS AND DISCUSSION

The application of *Azolla* increased the tillering capacity significantly over control. The application of *Azolla* increased the number of tillers with different levels of nitrogen. The percentage increase in the number of tillers was higher in the *Azolla* inoculated plots than uninoculated plots. Plots where *Azolla* was inoculated along with 60 kg N/ha gave the result equal to the plots where 90 kg N/ha alone was applied (Table 1).

Application of *Azolla* along with 60 kg N/ha has recorded the highest grain yield (49.2 q/ha), with a 56.2 per cent increase over control. In the plots where *Azolla* was inoculated at the rate of 1000 kg /ha along with 30 kg N/ha, the grain yield was 43.4 q/ha which was at par with the plot inoculated with 60 kg N/ha alone (44.22 q/ha). Again in the same experiment, when *Azolla* was inoculated with 60 kg N/ ha, the grain yield was equal to the plots where 90 kg N/ha alone was applied (Table 1). The *Azolla* inoculation with different levels of nitrogen increases the straw yield significantly over uninoculated plots. The straw yield was maximum with 90 Kg N/ha without *Azolla* (66.8/ha). Plots inoculated with 30 Kg N/ha along with *Azolla* had recorded the same straw yield as that of the plots inoculated with 60 Kg N/ha alone. (Table 1).

Consecutive use of *Azolla* along with 60 kg N/ha resulted in the highest beneficial effect, which was even more than 90 kg N/ha treatment. From this experiment it was found that the use of *Azolla* as dual crop has contributed 30-40 kg N/ha to rice crop.

Besides nitrogen contribution, weed suppression in wetland rice by *Azolla* is an added economic advantage to rice growers (Kannaiyan, 1984). Thus the use of *Azolla* and inorganic nitrogenous fertilizers in rice cultivation could economise nitrogen use and serve as a technology adoptable by farmers of Hirakud command areas.

Table 1. Effect of *Azolla* with different levels of nitrogen on the number of tillers, grain yield and straw yield of rice. Mean rice yield of four consecutive seasons

Levels of nitrogen (Kg/ha)	No of tillers	Grain yield (Q/ha)	Straw yield (Q/ha)
Control	8.4	31.5	54.7
Azolla alone	10.2	38.2	58.1
30 Kg N	10.7	37.6	57.8
30 Kg N + *Azolla*	13.6	43.4	62.3
60 Kg N	13.2	44.2	62.9
60 Kg N + *Azolla*	15.8	49.2	65.4
90 Kg	16.1	48.9	66.8
CD (P=0.05)	1.6	4.2	2.3

From the field experiment, it was also observed that in the 3rd year, *Azolla* was not inoculated to any plots of transplanted rice crop. But spontaneously *Azolla* mat appeared in rice field after 3 weeks of transplanting. This may be due to the germination of sporocarp of *Azolla*, which is formed during stress period. From the (Table 2) it was also observed that in winter season temperature goes below 9^{0}C in the months of December and January. In summer season, the temperature rises above 40^{0}C in the month of April, May and June. This extreme temperature condition induces sporulation in *Azolla*. Sporulation is often associated with beginning and end of a period of environmental stress. Kannaiyan and Rains (1985) have shown the positive influence of low temperature stress (20+1^{0}C / 15 + 1^{0}C day / night cycle) on sporocarp formation in *Azolla mexicana*. Generally sporocarps are formed during winter month in India and China (Peters and Meeks, 1989). Also, sporocarp formation due to over crowding, decomposition of lower layer of mat and lower temperature are known (Singh *et al.,* 1984, 1987).

Table 2. Mean maximum and mean minimum temperature from 1980 to 1999 at Chiplima under Hirakud command areas

Month	Mean of Highest Temperature of the Month	Mean of Lowest Temperature of the Month	Temperature ^{0}C	
			Mean Maximum	Mean Minimum
January	30.9	9.6	27.7	13.8
February	34.3	12.3	30.4	16.0
March	38.7	15.8	34.1	19.1
April	42.2	19.5	38.7	22.8
May	44.1	21.7	40.2	26.0
June	42.5	22.8	36.4	26.3

July	36.2	23.4	32.0	25.4
August	34.5	23.1	31.5	25.2
September	34.7	23.0	32.2	24.9
October	34.1	18.2	32.0	21.9
November	34.9	12.3	29.8	16.9
December	30.1	9.5	27.6	13.0

In the present investigation, it was observed that inoculation of *Azolla* to rice field increases the organic carbon (%) of soil significantly. (Table 3) From the initial level of 0.4 and 0.5 percent, the organic carbon was increased to 0.8 and 0.9 percent after inoculation of *Azolla* for four consecutive seasons.

Table 3. Effect of inoculation of *Azolla* on soil organic carbon

Levels of Nitrogen (Kg/ha)	Organic carbon (%) before planting of 1st season	Organic carbon (%) after harvesting of 4th season
Control	0.4	0.5
Azolla alone	0.4	0.8
30 Kg N	0.4	0.5
30 Kg + Azolla	0.4	0.9
60 Kg N	0.5	0.5
60 Kg N + Azolla	0.4	0.8
90 Kg N	0.4	0.4
CD (P= 0.05)	NS	0.2

CONCLUSION

Azolla is a free-floating water fern that floats in water and fixes atmospheric nitrogen in association with blue green algae *Anabaena azollae.* In the present experiment, it was found that consecutive application of *Azolla* in rice field ecosystem increases grain yield of rice (about 55 percent) over control. Inoculation of *Azolla* to rice field increases not only grain and straw yield of rice but also improves soil health by adding organic matter. *Azolla* undergoes sporulation during stress period and perpetuates itself during favourable conditions. Hence, *Azolla* is a renewable source of biofertilizer which adds not only 30-40 Kg N/ha but also helps for improvement of soil health in a natural manner.

REFERENCE

Kannaiyan S. 1984. Studies on the *Azolla* biofertiliser for rice production in Tamil nadu, In: **Proc. Inte. Workshop to assess the potential of *Azolla* use in Tropical Asia**. NIFTAL PROJECT, Bankhen, Bangkok, Thailand, pp. 1-40

Kannaiyan S. 1985. Use of *Azolla* in India, In: **Inte. Workshop on *Azolla* use.** Fujian Acad. Sci Fuzhou, Fujian, China.

Peters, G.A. and G.C. Meeks 1989. The *Azolla-Anabaena* symbiosis. Basic biology. ***Ann. Rev.*Plant Physiol. Plant Mol. Biol., 40:** 193-210.

Singh P.K., R.N. Patra and S.K. Nayak 1984. Sporocarp germination, cytology and mineral nutrition of *Azolla.* In: **Practical application of *Azolla* for rice Production,** (eds.) Silver W. and E.C. Schroder, Dr.W. Junk Publ., Martinus, The Netherlands p.55.

Singh, P.K., D.P. Singh and K.D. Pandey 1987. The influence of fertilizer on sporocarp formation on *Azolla pinnata.* **Proc. Indian Acad. Sci. (Plant Sci.). 97(3):** 223-226.

CHAPTER - 19

EFFECT OF *AZOLLA* BIOFERTILIZER ON THE GROWTH AND YIELD OF RICE IN COASTAL KARNATAKA

N.A. Janardhana Gowda, A.S. Kumaraswamy, T.R. Guruprasad and P.S. Herle

ABSTRACT

Mangalore is a coastal district receiving 4000 mm precipitation received from June through the September. Soil of the experimental site is taxonomically characterized as fine clay, mixed lethic Isohyperthermic family of union dysteropents practiced by the farmers. Due to the continuous use of high yielding varieties with high cropping intensity the yield of rice is drastically declining. The applied chemical fertilizers particularly nitrogenous fertilizer is lost through flooding, leaching and evaporation losses. In order to over come the stability of rice yields, through substitute biological nitrogen the experiment was conducted during 1993 to 1994 both the rainfed and irrigated conditions at the Agricultural Research Station, Mangalore to know the effect of *Azolla* biofertilizer on the growth and yield of rice combination with inorganic fertilizers souces. The experiment consists of eight treatments including control. Rice variety Phalguna and Jaya were used in *kharif* and *rabi* respectively. The experimental results indicatied that the application of 10 tons fresh *Azolla* + 30 Kg N per hectare and 90 Kg N per hectare recorded significantly maximum yield of 4719 kg / ha, height, tillers per hill and panicles per. sq. mt. during *kharif* as well as in *rabi* season.

INTRODUCTION

Rice (*Oryza sativa*) is the major food crop of the zone covering an area of 0.225 million ha and 52% of the net cropped area. The stagnation of yield are mainly due to low recovery of applied nitrogen fertilizer by rice ranging from 19.1 to 39.8 percent (Meelu, 1980). And it is deterrent to getting the potential yields from the high yielding varieties. The poor utilization of nitrogen fertilizer by rice crop under flooded condition is due to finite supply and escalating cost of nitrogen fertilizer in recent years and to economise its use without any adverse effect of productivity. The various approaches has been undertaken with *Azolla* bio-fertilizer in combination with chemical nitrogen fertilizer during *kharif* and *rabi* season.

MATERIALS AND METHODS

The field experiment was conducted at the Agricultural Research Station, Kankanady, Mangalore during the year 1993 and 1994 by using completely randomised design with three replication. The experiment consists of eight treatments are given in Table 1 & 2. The testing varieties used Phalgu and Jaya during *kharif* and *rabi* season respectively. The seedlings were transplanted at 25th day after sowing (DAS) in the plot size of 15 m^2. The crop was applied uniformly with nitrogen as per treatments and P_2O_5 and K_2O applied @ 30 kg / ha each for all the treatments except control. The cultural practices like irrigation, weeding etc were attended as per package.

RESULTS AND DISCUSSIONS

The data on the influence of *Azolla* bio-fertilizer on the growth and yield parameters of rice is shown in Table 1. The growth parameters showed the significant response with the application of *Azolla* with and without inorganic source of nitrogen fertilizer. The maximum plant height of 88.1 cm is obtained with the application of 10 t *Azolla* + 30 kg N/ha. during *kharif* while the highest plant height of 86.6 cm is recorded with 90 kg N/ ha and these two treatments are significantly on par. Significantly highest response on the production of tillers and panicles / m^2 of 8.8 tillers / hill and 10.8 tillers / hill and 440 panicles per m^2 and 540 panicles per m^2 during *kharif* and *rabi* respectively compared to the control which recorded 6.2 tillers / hill and 7.8 tillers / hill during *kharif* and *rabi* while panicles per m^2 was 311 and 390 panicles per m^2 during *kharif* and *rabi* respectively. Similar trends were observed by Manjappa (1987), Kannaiyan (1992) and Krishnaiah (1984). Further, the significant response was observed with respect to grain yield. The maximum mean grain yield of 4875 kg/ha. was recorded with the application of 90 kg N / ha during *kharif* while in *rabi*, application of 10 t / ha *Azolla* + 30kg N /ha recorded significantly highest grain yield of 4904 kg/ha and these two treatments are comparable with that of 10 t/ha. *Azolla* alone and in combination with 5 t *Azolla* + 30 kg N/ha compared to control which recorded lowest grain yield of 3451 kg/ha in both the seasons. Singh (1979) reported that grain, straw yield thousand and grain wt. increased by application of 30 kg / ha *Azolla* as basal dose. Similar trend was noticed by Marimuthu and Prasad (1983). Kannaiyan (1992) also reported the similar instances with respect to grain increase of rice due to *Azolla* application.

Performance of *Azolla* bio-fertilizer on growth and yield of rice during *kharif* and *rabi* seasons under rice – rice crop rotation (1993 to 1994)

Table 1. Growth parameters of rice.

Treatments	Mean growth parameters of rice					
	Plant ht. (cm)		Tillers / hill (no)		Panicles / m^2	
	Kharif	*Rabi*	*Kharif*	*Rabi*	*Kharif*	*Rabi*
Control	80.6	79.0	6.2	7.8	311	390
5 t fresh *Azolla* before planting	85.5	83.2	8.2	9.8	410	490
10 t fresh *Azolla* before planting	87.0	84.6	8.4	9.6	420	480
5 t fresh *Azolla* + 30 Kg N / ha	86.8	84.8	8.1	10.2	405	510
10 t fresh *Azolla* + 30 Kg N /ha	88.1	85.0	8.8	10.8	440	540
30 Kg N / ha	85.3	82.1	7.6	8.8	380	440
60 Kg N / ha	86.8	84.4	7.6	8.2	380	410
90 Kg N / ha	87.1	86.6	8.6	9.2	430	460
Sem. ±	1.0	0.5	0.2	0.2	9.9	9.9
C.D (0.05%)	3.10	1.4	0.6	0.6	30.1	29.9
C.V %	2.1	0.9	4.1	3.7	4.4	3.7

Table 2. Yield parameters of rice (Mean of 1993 & 1994)

Treatments	Mean grain yield of rice (Kg / ha)			
	Kharif 1993	*Kharif* 1994	Mean	*Rabi* 1993
Control	3644	3257	3451	3480
5 t fresh *Azolla* before planting	4379	4206	4293	4310
10 t fresh *Azolla* before planting	4795	4581	4688	4641
5 t fresh *Azolla* + 30 Kg N / ha	4565	4262	4414	4650
10 t fresh *Azolla* + 30 Kg N /ha	4810	4627	4719	4904
30 Kg N / ha	4011	3279	3645	2843
60 Kg N / ha	4018	4128	4073	3846
90 Kg N / ha	4825	4925	4875	4330
Sem. ±	63.5	27.5		30.4
C.D (0.05%)	192.5	82.3		92.2
C.V %	2.5	1.1		1.2

REFERENCES

Kannaiyan, S. 1992. ***Azolla* bio-fertilizer technology for rice crop.** TNAU, Coimbatore, pp. 54.

Krishnaiah 1984. **Effect of *Azolla* on paddy and on the succeeding maize – soybean intercrops in conjunction with straw incorporation.** M.Sc. (Agri.) Thesis. UAS, Bangalore.

Manjappa, R. 1987. **Effect of combined use of USG, phosphorus and *Azolla* green manuring on growth and yield of rice.** M. Sc. Thesis, UAS, Bangalore, pp. 164.

Marimuthu, P. and Prasad, N. N. 1983. **Effect of *Azolla* application at graded levels of nitrogen on rice.** Sixth S. other Regional Conf. Microbial Inoculants, Sept. 22-23, 1983, Proc. and Abs., Dept. Microbiology, UAS, Bangalore.

Meelu, O.P. 1980. Efficiency of nitrogen in Indian rice culture. **Fert. News, 25 (10):** 27-30.

Singh, P. K. 1979. Use of *Azolla* in rice production in India. **Nitrogen and Rice, Los Banos,** Philippines, pp. 407 – 418.

CHAPTER - 20

PERFORMANCE OF CYANOBACTERIAL AND *AZOLLA* BIOINOCULANTS IN RICE IN COIMBATORE AND ERODE DISTRICTS OF TAMIL NADU

P. Yasotha, K. Kumar and S. Kannaiyan

ABSTRACT

The lowland rice eco-system largely favours the growth and development of bio fertilizers like *Azolla* and cyanobacteria. The use of these biofertilizers largely contributes towards sustaining the fertility of the tropical rice fields. Besides saving fertilizers N costs and improving plant yield they also add considerably to the soil organic pool and influence the soil physical characteristics, thereby aiding the soil health maintenance. Experiments were conducted during the K*harif* and *Rabi* 1996-97 to study the effect of BGA and *Azolla* inoculants respectively in rice fields of Coimbatore and Erode districts of Tamil Nadu. Yield increase of 13.0% and 12.9% in the rice varieties *viz.,* ADT 36 and IR 50 were obtained by the cyanobacterial inoculation in Coimbatore and Erode districts respectively. In the *Rabi* season, dual cropping and incorporation of *Azolla* as a biofertilizer recorded a yield increase of 10.2% and 8.9% in ADT 36, IR20 and Entry 2 rice varieties in these two districts. The positive influence of these biofertilizers, *viz., Azolla* and cyanobacteria on the growth and yield of rice was well established by these on farm trials conducted in farmers fields and their potential needs to be exploited fully.

INTRODUCTION

The lowland rice ecosystem provides a conducive environment for the growth and development of bioinoculants like cyanobacteria and *Azolla.* Besides cutting down fertilizer costs, they also contribute to the long term fertility of the rice fields by improving soil structure, soil organic matter and microbial activity of the soil. The identification of suitable bioinoculants for the *Kharif* and the *Rabi* seasons will help in promoting the use of the bioinoculants in a long way. To demonstrate the benefits of cyanobacterial and *Azolla* biofertilizers to the farmers, onfarm trials on cyanobacterial (310 trials) and *Azolla* biofertilizers (221 trials) were conducted for the *Kharif* and *Rabi* seasons respectively in the rice growing areas of Erode and Coimbatore districts of Tamil Nadu.

MATERIALS AND METHODS

A series of experiments were conducted in the farmer's fields in the rice growing areas of Erode and Coimbatore districts of Tamil Nadu (Table 1).

Table 1. Performance of *Azolla* and BGA biofertilizers in the demonstration trials conducted in the farmer's holdings in Tamil Nadu.

Sl. No.	Name of the districts	No. of trials conducted	Season	Variety	Average grain yield (t ha^{-1})		Per cent increase over control
					Uninoculated	Inoculated	
I	Azolla Biofertilizer						
	(i) Coimbatore	97	Rabi	IR 50/IR 20/ ADT 36	5.60	6.20	10.2
	(ii) Periyar	124	Rabi	ADT36/ Entry 2	6.50	7.08	8.9
II	BGA Biofertilizer						
	(i) Coimbatore	172	Kharif	ADT 36	5.30	5.98	13.0
	(ii) Periyar	138	Kharif	IR 50/ ADT 36	6.29	7.10	12.9

Bioinoculants *viz.*, cyanobacteria and *Azolla* were used for the *Kharif* and *Rabi* season respectively. Cyanobacterial biofertilizers were applied @ of 10 kg ha^{-1}, while *Azolla* was inoculated @ 500 kg ha^{-1} as dual crop and incorporated at the time of first and second weeding during the *Rabi* season. The rice varieties used for the studies were ADT 36, IR 20 and Entry 2. The grain yield and straw yield were recorded after harvest of rice in the onfarm trials.

RESULTS AND DISCUSSION

The beneficial effects of cyanobacterial inoculation during *Kharif* season has been demonstrated in the Coimbatore and Erode districts where yield increases of 13.0 and 12.9 % were observed due to the inoculation. The yield increase due to *Azolla* incorporation was 10.2 % in Coimbatore district and 8.9 % in Erode districts. The effects of cyanobacterial inoculation have been proved in several field experiments conducted in different parts of Tamil Nadu. (Kannaiyan, 1998). The abundant sunlight available during the *Kharif* season largely favours the growth of cyanobacteria. A bloom of cyanobacteria may contribute 10-15 kg N ha^{-1}. During the *Rabi* season, the growth of *Azolla* is favoured (Kannaiyan, 1992). The beneficial effects of *Azolla* incorporation has been well documented (Kannaiyan, 1993). Besides, the capacity of *Azolla* to check weeds is also well established. Hence it can be profitably exploited as a biofertilizer for the rice crop raised during the *Rabi* season.

REFERENCES

Kannaiyan, S. 1992. ***Azolla* biofertilizer technology for rice crop**. Tamil Nadu Agricultural University, Coimbatore. p. 54.

Kannaiyan, S. 1993. **Biofertilizers for rice**. Tamil Nadu Agricultural University, Coimbatore. p. 60.

Kannaiyan, S. 1998. **Cyanobacterial biofertilizer for rice crop**. Tamil Nadu Agricultural University, Coimbatore. p. 247.

CHAPTER - 21

A NEW METHOD FOR INTERCROPPING OF *AZOLLA* IN COCONUT GARDENS

R. Venkitaswamy

ABSTRACT

A new method was developed at TNAU Agricultural Research Station, Aliyarnagar to multiply *Azolla* under coconut shade in coconut gardens. Pit size of 4 x 1 m to a depth of 25 cm was dug in-between 4 coconut palms in 12 years old coconut field planted with VHC – 2 Hybrid seedlings. The soil was completely removed from the pit. The bottom of the pit was completely covered with old fertilizer gunnies to prevent the coconut root growth. The pit was covered with polyethylene sheet (200 gauge thickness) and above the polyethylene sheet the soil was spread upto 10 cm height. The pit is filled with water and brought to puddled condition and water was stagnated upto 10 cm height. Cowdung and super phosphate at the rate of 250 and 5 g m^{-2} were added to the pond. *Azolla microphylla* at 250 g m^{-2} was introduced into the pond. Water was maintained in the pit to a depth of 10 cm. The inoculated *Azolla* formed complete mat in 15 days and ready for harvest. A quantity of 840 g m^{-2} fresh *Azolla* was recorded in fifteen days and the multiplication rate was 3.4 times under Aliyarnagar conditions during July 2001. This method was highly economical since it requires very less water when compared to the traditional method.

INTRODUCTION

Azolla is free floating water fern that floats in water and fixes atmospheric nitrogen in association with symbiotic N_2 fixing cyanobacterium,

Anabaena azollae (Moore, 1969). *Azolla* fronds consists of a floating rhizome and small overlapping bilobed leaves and roots and it is used as biofertilizer for increasing rice yields (Kannaiyan, 1984, Kannaiyan *et al.*,1982). In Pollachi Taluk of Coimbatore district (Tamil Nadu) coconut and paddy are grown in an area of 40000 and 5000 ha respectively. In riverside areas of Aliyar river paddy is grown in low lying areas and coconut is grown in high level fields. A new method was developed at TNAU Agricultural Research Station, Aliyarnagar to multiply *Azolla* under coconut shade in the interspaces of coconut.

MATERIALS AND METHODS

Pit size of 4 x 1 m to a depth of 25 cm was dug in-between 4 coconut palms in 12 years old coconut field planted with VHC – 2 Hybrid seedlings. The bottom of the pit was completely covered with old fertilizer gunnies to prevent the coconut root growth. The pit was covered with polyethylene sheet (200 gauge thickness) and above the polyethylene sheet the soil was spread upto 10 cm height. The pit was brought to puddled condition and water was stagnated upto 10 cm height. Cowdung and super phosphate @ 250 and 5 g m^{-2} were added. *Azolla microphylla* @ 250 g m^{-2} was introduced into the pond. Water level was maintained at 10 cm height. Fresh *Azolla* was harvested after 15 days and thereafter it was harvested once in three days.

RESULTS AND DISCUSSION

The inoculated *Azolla* formed complete mat in 15 days. The mean maximum and minimum temperatures during the growth period were 25.5 and 23.0^{0}C respectively (Table 1). The optimum mean temperature for most of the *Azolla* species is around 25- 30^{0}C and species like *Azolla microphylla* tolerates upto a temperature of 35^{0}C (Datta, 2000). The temperature factor is highly favourable during most of the year under Aliyarnagar conditions. The coconut that is predominantly grown in Pollachi region provides required shade for *Azolla* multiplication. The light transmission was 23.0 % at 8.00 hrs and it increased to 51.1 % at 12.00 hrs and it deceased to 28.0 % at 16.00 hrs. The water requirement for multiplication of *Azolla* comes to 93 litres m^{-2} during the month whereas the average water requirement under puddled condition was 200 litres m^{-2}. The use of polyethylene sheet saves about 107 litres of water per month and the saving works out to 53.5 %. (Table 2). The provision of polyethylene sheet prevents total downward and horizontal movement of water due to leaching, seepage etc., A quantity of 840 gm^{-2} fresh *Azolla* was recorded in fifteen days and the initial multiplication rate was 3.4 times under Aliyarnagar conditions during July 2001(Table 3). Thereafter a quantity of 250 g m^{-2} was harvested once in three days. Quantity of *Azolla* that can be harvested is estimated as 30 kg m^{-2} year $^{-1}$ and it is equal to 60 gms of N. A quantity equal to 600 gms of N can be harvested in 10 m^2 between four palms in one year. It is possible to harvest fresh *Azolla* equal to 25.8 kg N ha $^{-1}$ year $^{-1}$ in coconut garden.

CONCLUSIONS

- *Azolla* can be successfully multiplied under coconut shade with the use of polyethylene sheet. The use of polyethylene sheet saves about 107 litres of water per month and the saving works out to 53.5 %.

Table 1. Weather data during the period

Parameter		Max. 14.22 hrs		Min.7.22 hrs	
Mean temperature		25.5 ^{0} C		23.0 ^{0} C	
Relative Humidity		89.1 %		62.1 %	
Light transmission (%)	8 00 hrs	10 00 hrs	12 00 hrs	14 00 hrs	16 00 hrs
	23.0	48.0	51.1	38.0	28.0
Total Pan evaporation in 15 days				75.6 mm	
Total rainfall in 15 days				7.8 mm	

Table 2. Water requirement per month

Details	Quantity of water m^{-2}
Requirement under puddled condition without polyethylene sheet	200 litres
Requirement with the use of polyethylene sheet	93 litres
Saving of water	107 litres
Percentage of saving	53. 5 %

Table 3. *Azolla* harvested and N_2 fixation per year

Details	Quantity
1.Quantity of fresh *Azolla* multiplied in 15 days m^{-2}	840 gms
2.Quantity of fresh *Azolla* harvested once in three days m^{-2}	250 gms
3. Quantity of fresh *Azolla* harvested in one month m^{-2}	2500 gms
4.Quantity of fresh *Azolla* to be harvested during one year m^{-2}	30000gms
5.Dry weight of *Azolla* (5 % of fresh weight) m^{-2}	1500gms
6. Nitrogen content of *Azolla*	4.00 %
7.Quantity of N fixed year m^{-2}	60 gms
8.Qunatity of N fixed in 10 m^2 between 4 palms	600 gms
9. Quantity of N fixed by *Azolla* in 1 ha coconut garden	25.8 Kg

- A quantity of 840 g m^{-2} fresh *Azolla* was recorded in fifteen days and the initial multiplication rate was 3.4 times under Aliyarnagar conditions during July 2001.

- A quantity of 250 g m^{-2} of fresh *Azolla* was harvested once in three days which is equivalent to 25.8 kg N ha $^{-1}$ year $^{-1}$.

ACKNOWLEDGEMENT

The author is highly grateful to Prof. Dr. S. Kannaiyan, Vice-Chancellor, Tamil Nadu Agricultural University, for nominating the author for the Workshop on *Azolla* and Cyanobacterial Biofertilizers Technology for rice, held at TNAU during March, 2001. The above work was based on the experience gained from the workshop.

REFERENCES

Datta, S.K. 2000. *Azolla* and its utilisation for rice production in West Bengal. In: **Biofertilizers Blue green algae and *Azolla*.** pp.122-141.

Kannaiyan, S. 1984. Studies on *Azolla* biofertilizer for rice production in Tamil Nadu. In: **Proc. Inte. Workshop to assess the potential of *Azolla* use in Tropical Asia**, NIFTAL Project, Thailand.

Kannaiyan, S., M. Thangaraju and G. Oblisami. 1982. *Azolla* a potential biofertilizer for rice production. In: **Int. Workshop on practical application of *Azolla* for rice production**, University. Pureto Rico, Haggaguez campus, Pureto Rico.

Moore, A.W. 1969. *Azolla*: Biology and agronomic significance. **Bot. Rev., 35**:17-35.

CHAPTER - 22

AZOLLA - A SUSTAINABLE FEED SUBSTITUTE FOR LIVESTOCK WITH SPECIAL EMPHASIS ON SILPAULINE BASED PRODUCTION AND FEEDING TECHNIQUE

P. Kamalasanan Pillai, S. Rajamony and S. Premalatha

ABSTRACT

With the change in food habit due to Western influence and better purchasing power, the demand for milk and meat has increased many fold. So Animal Husbandry as such is expanding and is becoming a profitable occupation. There is a substantial decrease in fodder production from traditional source due to the decreasing area under forest and grassland. The fodder availability from cultivated crops also have decreased substantially due to the introduction of high yielding dwarf varieties of crops like paddy, wheat etc.

The shortage of fodder is compensated with grain based commercial feed which increased the cost of production of meat and milk. It is estimated that for the production of 1 kg meat one has to use atleast 10 kg cereals. So if people convert 2.5% of their total in take of food from vegetarian into non – vegetarian, there will be atleast 25% increase in cereal consumption. This will be another threat to our national food security in the long run. More over farmers mix urea and other hormones along with feed to increase milk. This will affect both the quality of milk and the longevity of animal and it is not at all sustainable. An alternate sustainable feed substitute for live–stock especially dairy cattle, poultry, piggery and fish could be *Azolla*.

Normally *Azolla* is grown in paddy field as a bio-fertilizer for wetland paddy. In paddy field the biomass production was less than 50 g/ sq. m / day and hence not cost effective and profitable for feed production. The *Azolla* grown in paddy field is also not hygienic as it contains dirt and other injurious organisms which may be harmful to live- stock. Series of trials conducted at NARDEP-Vivekananda Kendra as part of a project sponsored by Department of Bio-technology, Govt., of India, we could standardize a package of practices for *Azolla* to suit farmers homesteads. Here the *Azolla* was grown in artificial silpauline based lined water bodies with required nutrients. *Azolla* could be harvested daily and could be directly fed to cattle without any processing. The daily production of biomass ranges from 300 – 350 g / sq. m / day, which may be perhaps the highest in any vascular plant of this kind.

The cost of production of *Azolla* in this method comes to less than 50 Ps / Kg and could be grown in homesteads with waste-water from kitchen by ordinary housewives.

Analytical reports from renowned institution like CFTRI shows that *Azolla* contains 30-40% protein on dry weight basis, almost all essential amino acids, almost all essential minerals, many vitamins – especially B12, β carotene. Moreover *Azolla* is easy to digest and assimilate and it is no way harmful to live – stock.

These rare contributions of protein, minerals, vitamins along with rapid biomass production make *Azolla* a potential and sustainable feed substitute for live – stock particularly Dairy cattle, poultry, pig and fish.

INTRODUCTION

With the change in food habits due to Western influence and better purchasing power, the demand for milk and meat has increased many fold. So, Animal husbandry, as such, is expanding and is becoming a profitable occupation. There is a substantial decrease in fodder production from traditional source due to the decreasing area under forest and grassland. The fodder availability from various crops has also decreased largely due to the introduction of high yielding dwarf varieties of paddy, wheat etc.

The shortage of fodder is compensated with commercial feed, which, however, has increased the cost of production of meat and milk. It is estimated that for production of 1 kg meat, one has to use at least 10 kg cereals by traditional method. Use of steroids and other innovation like by-pass protein technology have still reduced the ratio. So if an average Indian converts 2.5% of his total food requirement from vegetarian to non-vegetarian, the annual additional demand of cereals will be around 25%. This will lead to another threat to our National food security in the long run. More over, feed manufacturers mix urea and other artificial milk boosters along with commercial feed. This has a serious deleterious effect on the quality of milk produced and the longevity of animal and is not at all sustainable. This will also lead to degenerative diseases like cancer and heart problems in human beings.

Search for an alternative took us to a wonderful plant "*Azolla*" which holds the promise for a sustainable feed substitute for livestock especially Dairy cattle, sheep, poultry, pig and fish.

Azolla is a floating fern, I which belongs to the family of *Azolla*ceae. The fern *Azolla*, hosts a symbiotic blue green alga, *Anabaena azolla*e which is responsible for the fixation and assimilation of atmospheric nitrogen (Kannaiyan, 1992). *Azolla* in turn provides the carbon source and favorable environment for the growth and development of the BGA symbiont. In *Azolla*, the endo-symbiont, *A. azollae* is even carried through the sexual reproductive phase, perhaps the only one of this kind in the plant kingdom. It is this unique symbiotic relationship, which makes *Azolla* a wonder plant.

MATERIALS AND METHODS

Based on the initial feed back of milk increase received from farmers who were advised to feed *Azolla* along with regular feed, detailed studies were carried out to ascertain the exact bio-chemical constitution of *Azolla* and its role in increasing the quality and quantity of milk.

Studies carried out by NARDEP in association with other research institutions like CFTRI showed that *Azolla* is rich in protein, almost 25 – 35% on dry weight basis. It is also found to contain essential minerals like iron, calcium, magnesium, potassium, phosphorous, copper, manganese etc., apart from appreciable quantities of vitamin A precursor and Vitamin B12. It is also found to contain almost all the essential amino acids, many probiotics, biopolymers and β- carotene. The above mentioned bio-chemical constitution along with the rapid multiplication rate makes *Azolla* an ideal organic feed substitute for livestock, apart from its utility as a bio-fertilizer for wetland paddy. Livestock can easily digest *Azolla* due to its high protein and low lignin content. Trials carried out by NARDEP, Vivekananda Kendra, with *Azolla* as feed substitute at Kanyakumari and Southern district of Kerala demonstrated an overall increase of milk by 15 – 20% with 1.5 – 2kg of *Azolla* along with regular feed. Commercial feed specially oil cake could also be reduced to 15 – 20% by supplementing feed with the same quantity of *Azolla* on dry weight basis without affecting milk production. (Pillai *et al.*, 2001). It is also found that *Azolla* feeding improves the quality of milk, and the health and longevity of animals. *Azolla* feeding also improves the weight of broiler chicken and increase the egg production of layers. *Azolla* can also be fed to sheep, goat, pig, and rabbit as feed substitute. In China cultivation of *Azolla* along with paddy and fish is reported to have increased the rice production by 20% and fish production by 30%. Trials conducted at Kanyakumari with *Azolla* as "fish feed substitute" are also found to be encouraging.

Azolla production for feed purposes

Azolla was found to be a very nutritive and cheap organic feed substitute for dairy cattle, poultry, pig and fish. *Azolla* when grown for feed purpose should be hygienic and there should be optimum production of biomass, preferably near the homestead (Pillai *et al.*, 2000). In paddy fields the bio-mass efficiency will be less i.e. less than 50 gm/sq.m./day against the possible optimum production of 300 gm./ sq. m. /day. The possibility of contamination by competitive algae is also more if grown in paddy fields. It is also observed that most of the dairy farmers do not have their own paddy fields and if at all they have it will be far off where daily supervision becomes difficult. All these problems can be solved if it can be grown in homesteads.

During our different trials in our campus as well as in farmers fields it was found that *Azolla* grew very well in an artificial Silpauline lined water body in homesteads preferably under a tree.

Nardep method

Natural Resources Development Project (NARDEP) an organ of Vivekananda Kendra, to address developmental and environmental issues, has been working on *Azolla* for the last 3 – 4 years. Finding its potential as a feed substitute, cost effective methods for the mass multiplication of *Azolla* in farmers homesteads were perfected.

In NARDEP method, *Azolla* multiplication plots preferably under the shade of a tree, with the help a Silpauline sheet of size 2.8 mts x 1.8 mts a pit of the size of 2 meter length, 1 meter width and 20 cm depth should be dug as the first step. This pit is then covered with second hand plastic gunnies to prevent the growth of roots of the nearby trees, which may pierce the silpauline sheet. The silpauline sheet is spread over the plastic gunnies without any fold. About 10 – 15 kgs sieved fertile soil is uniformly spread over the silpauline sheet. Two kg. Cow-dung and 30 gm of super phosphate made into slurry in 10 litres water is poured into the sheet. More water is poured to make the water level at about 10 cm. About 500 gms – 1kg of fresh and pure culture of *Azolla* is inoculated in the pit. *Azolla* growth will cover the pit within 10 – 15 days and about 500 gms – 600 gms of *Azolla* can be harvested daily thereafter. About 20 gms of super phosphate and about 1kg of cow-dung should be given once in 5 days to keep the *Azolla* in rapid multiplication phase and to maintain the daily yield of 500 gm./pit. Micro-nutrient mix containing magnesium, iron, copper, sulphur etc. can also be added at weekly intervals to enhance the mineral content of *Azolla*. (Special manipulation technique also has been developed to enhance a particular mineral or amino acid content in *Azolla*). In this method the cost of production of *Azolla* is less than 65 ps./kg.

Cultural practices for optimum biomass

1. It is important to keep *Azolla* at the rapid multiplication growth phase with the minimum doubling time.
2. Plant should not be allowed to enter maturative stage or sporulation stage by periodic application of cow dung slurry, super phosphate and other macro and micro nutrients except nitrogen.
3. The temperature should be retained below 25^{0}C. In case the temperature goes up, the light intensity should be maintained below 40 Klux by providing shade net or other devices.
4. Biomass should be removed every day or on alternate days to avoid over crowding.
5. pH should be tested periodically to see that it never goes below 5.5 & above 7.

Precautions

1. A shady place, preferably under a tree, with sufficient sunlight should be chosen for the *Azolla* production unit. A place of direct sunlight should be avoided.
2. All corners of the pit should be of the same level so that the minimum 10cm water level can be maintained uniformly.
3. *Azolla* biomass @ 300gms – 350gms/sq.meter/day should be removed daily to avoid over crowding and for keeping the fern at rapid multiplication phase.
4. Suitable nutrient should be supplied as and when deficiency is noticed.
5. Plant protection measures against pests and diseases should be taken as and when required.
6. About 5kg soil should be taken out from the bed and another 5kg of fresh soil are to be spread into the bed once in 30 days to avoid nitrogen build up and prevent micro-nutrient deficiency for optimisation of yield.
7. ¼ to 1/3 water also needs to be removed once in 10 days to prevent nitrogen build up in the bed and replaced with fresh water.
8. Fresh inoculation of *Azolla* after removing soil and water should be made at least once in six months repeating the whole procedure afresh.
9. A fresh bed has to be prepared and inoculated with pure culture of *Azolla* when contamination of lemna and other competitive algae, pest and diseases are observed.

Manipulation with nutrients to enhance protein, vitamins and mineral content

Azolla is a versatile fern capable of slightly altering the secondary and tertiary synthetic pathway as per the nutritional status and environmental condition. So the plant can be easily manipulated to synthesize more quantities of desired secondary metabolites like amino acids, vitamins, anthocyanines and even mineral compounds by the proper manipulation with nutrients and environmental factors (Subba Rao *et al.*, 1993; Subudhi and Singh, 1979 and Watanabe *et al.*, 1980). Adding the respective salts in the medium can increase the much needed iron, copper and calcium content of *Azolla* biomass. Anthocyanine and β- carotenoids can be enhanced by subjecting the plant to periodic temperature and light stress, and by adjusting shade. Further trials are needed to standardise the nutrient mix for a specific class of compounds.

Maintenance of pure culture

Azolla is very sensitive to pest and diseases. So it is important to keep seed materials free from pest and diseases. Chance for contamination with pest and disease is more if *Azolla* is grown in open fields. So it is always advisable to maintain a seed stock separately treated with pesticide and fungicide. For this, two or more plastic basins filled with soil mixture containing super phosphate and cow-dung can be inoculated with "disease – pest free healthy *Azolla*" and kept in farm house or verandhas. This can be used as a pure culture when pest or disease affects the *Azolla* in the main field.

Plant protection

Though there is no large-scale incidence of pests and diseases, the following pest and disease problem were noticed during intensive cultivation.

a) **Pests:** Severe pest problems are not noticed in Silpauline based production system (NARDEP method). If at all there is a pest attack it can be controlled by the application of Furadon @ 10gram/sq.meter. Biomass collected from field applied with pesticide should not be used as a feed for livestock. In case of severe pest attack, the entire bed should be cleaned and a fresh bed may be laid out in a different location. It is, however, found that pest attack is almost nil in Silpauline based production system if pest free seed material is used.

b) **Diseases:** Black rot disease and powdery mildew caused by fungus are the only serious diseases affecting *Azolla* in Silpauline base production system. It is always better to collect and remove the *Azolla* affected by fungus from the bed and bury it as and when disease is noticed. Treating *Azolla* with Bavistin 2% solution can control most of the fungal diseases. The treated *Azolla* should not be fed to livestock (Kannaiyan and Kumar, 1993). Whenever there is an occurrence of pest and disease, it is better to have a fresh inoculation after pesticide / fungicide application.

Feeding *Azolla*

Azolla should be harvested with a plastic tray having holes of 1 sq.cm mesh size to drain the water. The tray along with *Azolla* should be kept in a bucket half filled with water. Then washing of *Azolla* is to be done in the tray to get rid of the smell of cow-dung. Washing like this will also help the small plantlets to get out of the tray. The plantlets along with water in the bucket can be poured back in to the original bed. Fresh *Azolla* thus collected may be mixed with commercial feed in 1:1 ratio and fed as such to dairy cattle, sheep and pig. For poultry, *Azolla* as such can be fed both for layers and broilers. However, it advisable to mix *Azolla* in regular feed in 1:1 ratio in the beginning for one week. It is found that after a fortnight of feeding on *Azolla* mixed with regular feed, Dairy cattle, Goat, Poultry etc. feed *Azolla* directly with out the addition of regular feed material. Medium sized herbivorous fish, like Grasscarp, Silvercarp etc. also take *Azolla* directly when broadcast in to the fish tank.

RESULTS AND DISCUSSION

Nutritive value of *Azolla*

Azolla is very rich in protein and almost all essential amino acids, vitamin A precursor, carotene, several growth promoter intermediaries, minerals like calcium, phosphorous, potassium, ferrous, copper, magnesium etc. The protein content is 25 – 35% on dry weight basis, mineral content 10 – 15%, amino acids and bio-active substances and bio-polymers constitute 7 – 10%. Carbohydrate and oil content in *Azolla* is very low. Analysis report received from CFTRI, Mysore shows that it also contains appreciable quantities of ß carotene and vitamin B12. The above combination of protein, minerals, vitamins and amino acid make *Azolla* one of the most economic and efficient feed substitutes for livestock (Tables 1 and 2).

The nutrient constitution of *Azolla* is found to be almost similar to that of commercial poultry feed, except that the protein content is high and calcium content is slightly low. Feeding trials carried out in poultry with different combination of commercial feed and *Azolla* showed that 20 – 25% of commercial feed could be replaced by supplementing it with fresh *Azolla* in diet. Birds with 75% of the regular feed with 12.5% diet in the form of *Azolla* weighed almost equal to the birds with 100% feed. The birds receiving normal diet with 5% extra *Azolla* diet grew faster than the bird with 100% feed alone and there was 10 – 12% increase in the total body weight (Table 3).

Table 1. Comparison of biomass and protein content of *Azolla* with different fodder

S. No	Item	Annual production of Bio-mass MT	Dry matter content MT	Protein content MT
1.	Hybrid Napier	250	50	4
2.	Kolakattai grass	40	8	0.8
3.	Lucerne	80	16	3.2
4.	Cowpea	35	7	1.4
5.	Subabul	80	16	3.2
6.	Sorghum	40	3.2	0.6
7.	Jowar (Cambu)	35	3	0.6
8.	*Azolla*	730	56	20

Yield of *Azolla* 730 MT/ hectare @ of 200gm./sq.m/day

Table 2. Nutrient content of *Azolla*

Sl. No.	Parameters	Result	Test method AOAC 12th END.1975
1.	Calcium, mg/100g	67	Sec.14.014
2.	Iron, mg/100g	7.3	Sec.2.096
3.	Magnesium, mg/100g	43.7	Sec.2.096
4.	Copper, ppm	0.9	Sec.25.068&2.096
5.	Zinc, ppm	4.1	Sec.2.096
6.	Total carotene, mcg/100g	4475	AOAC 11th Edn.1970

7.	B-Carotene mcg/100g	1380	JAOAC, 1980, 63:8-94
8.	Protein, (N x 6.25)%by wt.	3.9	AOAC 12th Edn.1975 Sec.47.021
9.	Vitamin B12, mg/100g	1.19	HPLC

Source: Analytical report from CFTRI, Mysore

Table 3. Result of feeding trials conducted for poultry

S. No.	Mean weight / bird / week in grams				
	Age of birds in weeks	100 % feed	50 % feed + 16% *Azolla*	75 % feed + 12.5% *Azolla*	100 % feed + 5% *Azolla*
1.	7	155	120	120	149
2.	8	193	151	149	188
3.	9	234	179	181	245
4.	10	278	201	252	335
5.	11	320	231	312	398
6.	12	373	264	416	480
7.	13	433	309	454	544
8.	14	494	344	518	614

The trials conducted at Kanyakumari also showed that there is a substantial increase in the weight to broiler chicken. The number of eggs per bird and quality of eggs (the yellow yolk portion of egg being more prominent and yellowish) was also better than the control. The detailed study is still in progress.

In Dairy cattle also there was a substantial increase in the production of milk when *Azolla* was fed along with commercial feed. It is found that *Azolla* feeding improves the health of the animal and the quality of milk produced. It is found that the increase in the quantity of the milk produced on the basis of nutrient was higher than the quantity of *Azolla* fed. Hence it is presumed that more than the carbohydrate, protein content, other components like carotenoids, bio-polymers, probiotics etc. may be contributing in the over all increase in the production of milk. It was found that there was a clear cut 10-15% increase in milk production, which even went up to 20% during summer month from February to May.

Economics of *Azolla* in Nardep method

The cost of production of *Azolla* comes to 65 paise per kilogram in a three kilo per day silpauline based production unit developed by VK-NARDEP as part of DBT project is as detailed below:

1. Cost of 120 gauge silpauline 2.8mts x 1.8mts		
(Production in 1 sq.meter/day is 300 grams)		
Cost of 4 units Silpauline Rs.80 x 5	-	400.00
2. Labour charges for bed preparation	-	100.00
3. Cowdung 1 x 4 = 4kg Rs.2 x 73	-	146.00
4. Superphosphate 25 x 4 =200g x 73 =1,400 grams	-	7.50
5. Magnesium sulphate 1 kg	-	4.00
6. Micro-nutrients 73 x 5 =365grms.	-	15.00
7. Cost of inoculation material 200 x 8 = 1,600grms.	-	5.00

		Total 677.50

Total cost of production	=	Rs.677.50
Total production 350 days in 4 units	=	1050 kgs/annum
(4kg/day for 350 days for one year)		
Cost of production	=	Rs.677/1050kgs = 65ps./kg.

*Application of cow-dung and Super phosphate once in 5 days. i.e. 73 time in a year

CONCLUSION

Azolla can be used as an ideal feed substitute for cattle, fish, pig and poultry, apart from its utility as a biofertilizer for wetland paddy. It is popular and cultivated widely in other countries like China, Vietnam, Philippines etc. The technology is yet to be taken up by the Indian farming community, in a big way, as the production technology has not yet been fully standardised to the diverse and different Agro-climatic zones of the country.

Dairy farmers of South Kerala through Milma and few farmers of Kanyakumari district have taken up the low cost production technology developed by VK-NARDEP as part of the project funded by Department of Bio-technology, (Govt. of India) and initial reports are satisfactory. In fact, it is the report of the dairymen in Kanyakumari, about the increase in milk production, that motivated us to work on this line to popularise *Azolla* as a cattle feed substitute. We sincerely hope that in the coming days, dairy farmers especially those who experience scarcity of land for growing fodder will take up the *Azolla* technology in a big way.

ACKNOWLEDGEMENT

The authors acknowledge with thanks the financial assistance granted by Department of Biotechnology (Govt. of India) for carrying out the research on the propagation of BGA and *Azolla* in Nardep, Vivekananda Kendra, Kanyakumari. We also acknowledge the help and support provided by *Azolla* laboratory, Tamil Nadu Agricultural University, Coimbatore.

REFERENCE

Kannaiyan, S. 1992. *Azolla* Bio-fertilizer Technology for Rice. Tamil Nadu Agrl. University, Coimbatore, Tamil Nadu, India, p. 54.

Kannaiyan, S. and K. Kumar. 1993. Experimental techniques of *Azolla* biofertilizer. Tamil Nadu Agri. University, Coimbatore. p.93.

Pillai P.K., S. Rajamony and S. Premalatha. 2000. Mass Multiplication of *Azolla* in homesteads – **"Training Manual for training Officers of MILMA**, State Milk Union, Kerala.

Pillai P.K, S. Rajamony and S. Premalatha. 2001. *Azolla* as an organic feed substitute for Dairy cattle – **Proceedings of the National workshop on organic farming in MCRC**, Chennai.

Subba Rao, N. S, Venkataraman G.S. and Kannaiyan, S. 1993. **Hand book on Nitrogen fixation.** Indian Council of Agrl. Research, New Delhi. P.111.

Subudhi, B.P.R. and P.K. Singh 1979. Effect of phosphorous, nitrogen on growth, chlorophyll and algal heterocysts of water fern *Azolla pinnata*. **Biol. Plantarum. 21:** 401-406.

SECTION – III

AZOSPIRILLUM BIOFERTILIZER

CHAPTER – 23

STUDIES ON *Azospirillum* ASSOCIATED WITH RICE VARIETIES

K. Govindarajan and K. Kavitha

INTRODUCTION

The recognition in the mid seventies that strains of *Azospirillum* are abundantly associated with the roots of cereals and millets and help fixing substantial amounts of nitrogen led to the opening of the new line of work, the associative symbiosis. The concept of "associative symbiosis" has widened our knowledge of free-living nitrogen fixing microorganisms. Almost all crop plants harbour this bacterium in their roots. These bacteria colonizing the roots not only remain on the root surface but also sizeable populations penetrate into the root tissues and live in harmony with the plant. Of late, studies on specificity of *Azospirillum* to host plants has gained importance, since specific strains if developed will be highly useful to increase the productivity of selected crops. Hence investigations were taken up with the main objective of studying the diazotrophic association and host specificity of *Azospirillum* with different paddy varieties.

MATERIALS AND METHODS

Isolation of *Azospirillum* cultures

The root samples of the paddy varieties were used for isolation of *Azospirillum* as per the root enrichment technique described by Day and Dobereiner (1976). Seeds of the following paddy varieties obtained from Paddy Breeding Station, Tamil Nadu Agricultural University, Coimbatore were used for the study.

1. ASD 18 5. CO RH2 hybrid A line
2. ADT 36 6. CO RH2 hybrid B line
3. CO 47 7. CO RH2 hybrid R line

4. CO 47

Synthesis of Indole Acetic Acid (IAA) by *Azospirillum* isolates

The *Azospirillum* isolates were grown in Okon's broth supplemented with and without tryptophan. At desired intervals of time cell free culture filtrate was taken, pH adjusted to 2.8 and IAA was extracted with diethyl ether. The quantity of IAA produced was estimated using salper's reagent (Gorden and Paleg, 1957).

Selective colonization of *Azospirillum* isolates in the root region of rice varieties

Fahraeus nutrient solution (half strength) was taken in one litre bottles and an aluminium mesh was placed inside the bottle just above the solution level as a support to the seedlings and the entire set was sterilized. Forty eight hour old seedlings of the seven paddy varieties were aseptically transferred to the wire mesh in the bottles in such a way that the roots alone were dipping into the nutrient solution. The bottles were inoculated with 0.5 ml of the isolated *Azospirillum* culture and kept under sunlight for 2 weeks. Separate bottles were kept for each *Azospirillum* culture. After 2 weeks the seedlings were removed carefully, shoot portion was discarded and the organisms adhering to the roots alone were dislodged by shaking it in 100ml of sterile water. Serial dilutions were made and the population of *Azospirillum* was estimated.

RESULTS AND DISCUSSION

Seven *Azospirillum* cultures were isolated from the selected rice varieties following the enrichment culture technique (Table 1). The formation of white subsurface undulating fine pellicles and change of colour of the N free semisolid malate medium from green to blue revealed the presence of *Azospirillum*. Moreover, typical small dense white colonies formed by all the isolates on N free medium supplemented with traces of yeast extract further confirmed the isolation of *Azospirillum*.

Table 1. *Azospirillum* cultures isolated from different rice varieties

S. No.	Rice varieties	*Azospirillum* cultures
1.	ASD 18	AGK 1
2.	ADT 36	AGK 2
3.	CO 46	AGK 3
4.	CO 47	AGK 4
5.	A line of hybrid CO RH 2	AGK 5
6.	B line of hybrid CO RH 2	AGK6
7.	R line of hybrid CO RH 2	AGK 7

The indole acetic acid production was more in all the isolates in the presence of tryptophan in the medium. Among the isolates, AGK7 recorded the highest quantity of 2.90 g ml^{-1} without tryptophan and 11.65 g ml^{-1} with tryptophan. Isolate AGK1 recorded the least quantity of 1.80 g ml^{-1} without tryptophan and 7.20 g ml^{-1} with tryptophan (Table 2).

Table 2. Production of indole acetic acid (IAA) by *Azospirillum* cultures isolated from rice varieties

S. No.	*Azospirillum* cultures	IAA production (g ml^{-1})	
		Without tryptophan	With tryptophan
1.	AGK 1	1.80	7.20
2.	AGK 2	2.52	11.28
3.	AGK 3	2.85	11.52
4.	AGK 4	1.95	7.80
5.	AGK 5	2.70	10.85
6.	AGK6	2.72	10.96
7.	AGK 7	2.90	11.65

(Data represents mean of three determinations)

Azospirillum is known to produce plant growth promoting substances and the hormones produced by *Azospirillum* significantly enhanced growth parameters (Tien *et al,* 1979). The possibility of production of IAA from the precursor tryptophan occurring in the root exudates was reported by Scott (1972). Reynders and Vlassek (1979) observed the conversion of tryptophan to IAA by *A. brasilense* in the culture medium. Prinsen *et al.* (1993) postulated different pathways for IAA biosynthesis like tryptophan dependent and tryptophan independent pathways in *Azospirillum*. In the present studies also supplementation of tryptophan in the medium helped the organism to produce more IAA *in vitro* and the increased synthesis of IAA confirmed the earlier findings for the requirement of tryptophan and tryptophan mediated pathways for increased IAA biosynthesis. (Bar and Okon, 1992).

Studies on the colonization of *Azospirillum* isolates in the root region of rice varieties indicated the preferential colonization of the *Azospirillum* in the roots of different rice varieties (Table 3). The population of the isolate AGK1 isolated from the variety ASD18 varied from 39.82 to 1.92 x10^5 among the different rice varieties tested. Maximum colonization was recorded in the variety ASD18 followed by ADT 36.Similarly the homologous isolates of the different varieties colonized their respective host in significantly more numbers. Baldani and Dobereiner (1980) observed that homologous strains could establish well with in the roots than the non-homologous strains. In the present study also it was observed that the colonization of *Azospirillium* was greater when the homologous strains were inoculated to the paddy varieties under axenic conditions.

Table 3. Selective colonization of *Azospirillum* isolates in the root region of rice varieties grown under axenic conditions

S. No.	*Azospirillum* cultures	Population of *Azospirillum* (x 10^5)						
		ADT 18	ADT 36	CO 46	CO 47	A line	B line	R line
1.	AGK 1	39.82	34.22	32.93	19.87	3.81	2.46	1.92
2.	AGK 2	37.29	54.38	43.01	1.28	2.54	3.69	4.48
3.	AGK 3	34.13	38.49	53.09	33.97	5.72	3.69	1.28
4.	AGK 4	27.18	21.99	23.52	58.97	3.81	3.08	2.56
5.	AGK 5	10.11	7.33	9.41	5.77	50.71	37.52	37.76
6.	AGK6	3.16	4.89	4.03	5.77	33.66	44.28	40.96
7.	AGK 7	1.26	6.72	8.06	10.89	24.13	31.5	57.6
	SEd	1.12	0.94	0.05	0.08	0.06	0.07	0.60
	CD (5%)	2.40	0.20	0.11	0.17	0.13	0.15	0.12

Expressed g^{-1} of root tissue on dry weight basis

REFERENCES

Baldani, V.L.D and J.Dobereiner.1980.Host plant specificity in the infection of cereals with *Azospirillium* sp. **Soil Biol.Biochem.**,12:433-439.

Bar, T and Y. Okon.1992.Tryptophan conversion to indole –3 acetic acid via indole –3 acetamide in *Azospirillium brasilense* P7.**CanadianJ.Microbiol.**,39:81-86.

Day, J.M and J. Dobereiner 1976. Physiological aspects of nitrogen fixation by a *Spirillum* from Digitaria roots. **Soil Biol. Biochem.**, 8:45-50.

Dobereiner and H. De-Polli 1980. Diazotrophic rhizocoenosis. In: **Nitrogen fixation** (eds.), W. D. P. SteWart and J. R. Gallon Academic Press, New York. pp.303-333.

Gorden, S.A and L.G. Paleg. 1957. Quantitative measurement of indole acetic acid. **Physiol. Plant**, 10:347-348.

Prinsen, E., A. Castacurta, K. Michiels, J. Vanderleyden and H.V Ouckelen. 1993. *Azospirillium brasilense* indole-3-acetic acid biosynthesis: evidence for a non-tryptophan dependent pathway. **Mol. Plant Microbe interact.**, 6:609-615.

Reynders, L and K.Vlassak.1979. Use of *Azospirillium brasilense* and biofertilizer in intensive wheat cropping. **Plant Soil,** 66 : 212-223.

Scott, T.K. 1972. Auxin and roots. **Ann. Rev. Plant Physiol.**, 23 : 235-258

Tien, T. M., H. G. Diem, M. H. Gaskins and D. H. Hubbell 1979. Plant growth substances produced by *Azospirillium brasilense* and their effect on the growth of pearl millet (*Pennisetum americanum* L.), **Appl. Environ. Microbiol.**, 37:1016-1024.

CHAPTER – 24

EFFECT OF *Azospirillum* INOCULATION ON RICE GROWN IN ACIDIC SOILS OF KERALA

M. Govindan and C.K. Yamini Varma

ABSTRACT

Azospirillum was found to be very common diazotrophic bacteria in soil and root environment of rice. Their population was more in surface soil than deeper layers. The Kypad soil which is saline acidic also harbors *Azospirillum* in a sizeable manner. Soil population of *Azospirillum* was more in the month of September. During dry periods their population declined drastically. Most of the isolated strains obtained from rice root tissues/soil were found to be *Azospirillum lipoferum*. Thirteen strains were selected and used for pot culture experiments. Local isolate S-10 and a commercial strain performed better. The population of *Azospirillum* in the root zone of rice grown in field conditions was studied both in rhizosphere and endorhizosphere. Inoculation increased the MPN count of *Azospirillum*. The rhizosphere population was higher than endorhizosphere population. This shows the colonisation of inoculated strain in root environment and this might have helped in better plant growth. Effect of *Azospirillum* inoculation on growth and yield of rice was studied over three seasons. The results established that *Azospirillum* inoculation enhanced crop yield significantly over uninoculated controls. The present results suggest that by selecting suitable strains, it is possible to increase the yield and supplement the nitrogen requirement of the crop grown in acidic soil conditions of Kerala.

INTRODUCTION

Biological nitrogen fixation is a simple and low cost technology, which has a tremendous role in the immediate future of agriculture in Kerala. Rice is the staple food of Kerala. The advancing crop production technology, development of high fertilizer responsive crop varieties and pressure for intensive farming, demand huge inputs of which nitrogenous fertilizer take a major share. A suitable bio-fertilizer like *Azospirillum* can become a promising alternative for the above problems. Soil pH is one of the major environmental constraints governing the establishment and performance of *Azospirillum*. The soils of Kerala are having low pH, high aluminium and iron. Hence an attempt has been made to evaluate the native population of *Azospirillum* and study its biocoenosis with the rice to exploit its potentialities as a microbial inoculant.

MATERIALS AND METHODS

Population of *Azospirillum* was enumerated from the soil samples collected from paddy fields of Regional Agricultural Research Station, Pilicode and Kypad soils. To determine the seasonal variation of bacterial population, samples were collected during the months of September 1995, December 1995 and April 1996 from different depths of 0 - 10 cm, 11 – 20 cm and 21 – 30 cm. MPN technique was used for the enumeration of *Azospirillum* (Okon *et al.*, 1977; Hegazi *et al.*, 1979) and expressed per gram dry weight of soil. For the isolation of *Azospirillum* from soil/root tissues of rice, enrichment culture technique was used (Day and Dobereiner 1976).

Samples were collected from Kasaragod, Kannur, Kozhikode, Thrissur, Wayanad, Ernakulam and Kollam Districts of Kerala. Various morphological and biochemical characters like growth in Congo Red medium, motility, Gram reaction, presence of PHB, Pleomorphism on alkaline medium, utilisation of malate, succinate, lactate, glucose, ketoglutamate, mannitol, sucrose, acidification of peptone broth, biotin requirement, nitrate reductase, denitrification and nitrogen fixation were studied for tentative identification and *in vitro* selection of isolates.

To select the strains with plant growth promoting effect, a pot culture experiment was conducted using sterilised laterite soil (pH 5.0) and an uninoculated control was also maintained. Thirteen strains were used for the study. Ten replicates were used for each treatment in a completely randomised blocks design. *Azospirillum* cultures were grown in malate broth and lignite based inoculum was prepared. Paddy seeds (Variety: Jyothi) were surface sterilized and treated with the carrier based inoculum, air dried for one hour and sown in plastic pots. One month after sowing shoot length, root length and biomass were recorded.

To assess the potential of *Azospirillum* as a microbial inoculant for rice grown in acidic soil and to select a suitable strain, field experiments were conducted for three seasons in the paddy fields at Regional Agricultural Research Station, Pilicode. A commercially available strain of *Azospirillum* (CS) and the selected local strain S 10 were used for comparison. Inoculations were given to the seeds as well as seedlings. The seedlings inoculated and uninoculated were transplanted and raised to maturity following standard cultural practices.

The fertilizers and organic manures were applied as per the package of practices recommended by Kerala Agricultural University. The NPK fertilizers recommended are 90,45,45 kg/ha. Nitrogen fertilizers were given in three levels, *viz.,* 100%, 75% and 50%. nitrogen was given as urea and applied in three doses. Inoculations were given in all the three levels of nitrogen. The treatments were replicated three times in a randomized block design in plots of 30 m.sq.

In order to study the root colonization and establishment of inoculated strains, plant samples were collected 10 DAT and 40 DAT. Then rhizosphere and endorhizosphere population was enumerated following standard procedures. Periodic observations on plant height, number of tillers and other biometric observations were recorded. After harvest, grain weight at 12% moisture and the dry weight of straw were recorded. The data collected over a period of three years were pooled and analysed.

RESULTS AND DISCUSSION

The population of *Azospirillum* was maximum during September in the soils of R.A.R.S., Pilicode (7.97×10^5 cells) and Kypad soils (4.8×10^5 cells) in 10 cm depth (Table 1). The deeper layers of 20 cm and 30 cm recorded higher counts during April. Adequate organic matter and soil moisture on top layers during September and December might have influenced the bacterial population. During April surface layers of field were dry and the population was drastically reduced, but the deeper layers offered some protection to bacteria possibly by mulching effect.

Table 1. Population of *Azospirillum* in the soils of Pilicode and Kypad during wet and dry season

Sl. No.	Months	*Azospirillum* x 10^5	
		R.A.R.S. Pilicode soil	Kypad soil
1	September 95		
	10 cm	7.97	4.80
	20 cm	6.84	3.40
	30 cm	6.40	3.10
2	December 95		
	10 cm	23.00	2.14
	20 cm	17.00	3.50
	30cm	15.00	1.20
3	April 96		
	10 cm	15.00	3.00
	20 cm	45.00	5.00
	30 cm	30.00	4.00

The present findings describe the isolation and selection of *Azospirillum* strains which are effective in association with rice plants in acidic soils. The pH of the soils from where the isolates were obtained varied from 4 to 6.5. From the soil/root samples of rice, 25 *Azospirillum* isolates were obtained from Kasaragod, Kannur, Kozhikode, Thrissur, Wayanad, Ernakulam and Kollam districts of Kerala. Based on the morphological and biochemical characters, isolates were tentatively identified and 20 of them were *Azospirillum lipoferum* and five *Azospirillum brasilense*. Ladha *et al.* (1987) found that 84% of isolates from rice roots were *Azospirillum lipoferum*.

Pot culture studies with selected thirteen cultures showed that the effect of inoculation with *Azospirillum* was variable with different strains (Table 2). The variation ranged from 10.34% to 59.19% increase over uninoculated control with respect to the shoot length and 14.14% to 73.94% in the case of root length. This indicates that the effect is more pronounced on root development. The shoot biomass also recorded 6.75% to 42.08% increase over control. Based on the data it is found that inoculation with isolate S-10 enhanced maximum root length and plant biomass and this isolate was selected for field experiment (Table 2).

Table 2. Response of rice seedlings to inoculation with different isolates

Isolate	Shoot length (cm)	Root length (cm)	Biomass (mg) (Dry weight)
S1	29.3 (42.85)	7.12 (21.29)	178.7 (34.06)
S2	31.4 (53.10)	7.6 (29.47)	158.2 (18.68)
S3	28.2 (37.49)	6.9 (17.55)	173.1 (29.86)
S4	26.3 (28.23)	7.01 (19.42)	142.3 (6.75)
S5	27.5 (34.08)	8.61 (46.68)	179.1 (34.36)
S6	31.3 (52.61)	9.2 (56.73)	167.3 (25.51)
S7	23.2 (13.11)	6.7 (14.14)	180.1 (35.11)
S8	24.98 (21.79)	8.1 (37.99)	169.3 (27.01)
S9	21.2 (10.34)	7.1 (20.95)	173.1 (29.86)
S10	32.65 (59.19)	10.21 (73.94)	189.4 (42.08)
S11	24.1 (17.50)	7.3 (22.66)	171.3 (28.51)
S12	26.5 (29.21)	8.61 (46.68)	163.2 (22.43)
S13	29.2 (42.37)	9.4 (60.14)	178.2 (33.68)
Control DW	20.51	5.87	33.3

(Data in parenthesis indicate percentage increase over controls)

Both on 10 DAT and 40 DAT, rhizosphere population was higher than endorhizosphere population. The MPN count was increased which might be due to the inoculation of the bacteria. These observations clearly indicated that the inoculated strain colonized the root environment and resulted in better plant growth. To assess the potential of *Azospirillum* as a crop inoculant, it is important to determine the extent to which the introduced bacteria become established in the rhizosphere. Results from the present field trial showed that there is substantial increase in bacterial colonization due to inoculation. (Table 3). The bacterial numbers increased from 10th DAT to 40 DAT which may be due to better competitive ability of the selected strain to establish and multiply in the host environments. This is in congruent with the observations of Rao and Venkateshwarlu (1985) that the introduced *Azospirillum* become established and continue to multiply on roots. When different levels of fertilizers were applied, the bacterial counts varied. It was found that the growth in endorhizosphere and rhizosphere population was maximum for 100 per cent nitrogen + inoculation.

Table 3. Population* of *Azospirillum* in the root zone of rice

Treatments	10 DAT		40 DAT	
	Rhizosphere	Endorizosphere	Rhizosphere	Endorizosphere
T1 : 100% N	5.40	0.27	42.00	2.89
T2 : 100% + S-10	91.80	0.59	220.00	158.10
T3 : 75% N	4.48	0.21	36.00	2.00
T4 : 75% N+ S-10	66.80	0.49	157.00	73.56
T5 : 50% N	4.30	0.17	12.80	1.26
T6 : 50% N+ S-10	35.06	0.28	60.10	50.56

* Population expressed in 10^6 g^{-1} root/soil ; DAT – Days After Transplanting

The experiment on effect of *Azospirillum* on grain and straw yield of paddy was conducted in paddy field at R.A.R.S., Pilicode in three seasons. Biometric observations like root length, shoot length, root dry weight and shoot dry weight recorded after transplanting showed that 100% N + local strain inoculation had increased root length, shoot length, root and shoot dry weight (Table 4). Rao *et al.* (1987) reported that yield response to inoculation is most pronounced at moderate levels of fertilizer nitrogen. Thus inoculation with nitrogen fixing bacteria combined with fertilizer nitrogen could increase the yield.

The pooled data showed that the field response of rice to *Azospirillum* inoculation was positive. 17.75 per cent increase in grain yield was recorded in T6 due to inoculation (75% N+*Azospirillum* S10) where as the commercial strains resulted in only 10.58 % yield increase. Grain yield due to 75% N+*Azospirillum* S10 was on par with the treatment T1 (100%N). This indicates that *Azospirillum* inoculations is effective at 50%, 75% and 100% nitrogen levels. However, its effect was more prominent at 75% N level. Kumar and Balasubramanian (1989) reported that 25-50 kg nitrogen/hectare can be saved by *Azospirillum* inoculation in rice cultivation. Seed treatment and soil application of *Azospirillum* increased total biomass and grain yield of a number of rice varieties.

Rao *et al.* (1983) and Subba Rao *et al.* (1979) reported positive yield response to inoculation with *Azospirillum* species at different levels of nitrogen. As per the observations by Tien *et al.* (1979), the increased growth and yield characters might be due to the production of auxins, gibberellins and cytokinins in the pure cultures of *Azospirillum*. Kundu and Sangwan (1992) also established the role of phytohormones produced by *Azospirillum*. All the above observations showed that *Azospirillum* enhanced the growth parameters and yield significantly in the rice crop.

Table 4. Effect of *Azospirillum* inoculation on plant growth and yield

Treatments	Root* length (cm)	Shoot* length (cm)	Root* dry Wt (g)	Shoot* dry Wt (g)	Productive tillers	Total tillers	Grain yield (ton/ha)	Straw yield (t/ha)
100% N	21	60.33	0.72	4.18	8.2	11.1	3.053	2.522
T1+ CS	23	62.66	0.93	4.78	9.1	12.1	3.130	2.740
T1+ S-10	26	68.10	1.14	5.54	9.5	12.6	3.467	3.232
75% N	18	69.30	0.615	4.01	7.3	10.3	2.777	2.281
T4+ CS	22	74.66	0.75	5.10	8.5	11.8	3.071	2.588
T4+ S-10	23	70.66	1.08	5.31	9.0	12.1	3.270	2.671
50% N	15	60.33	0.511	3.21	7.1	10.2	2.428	2.130
T7+ CS	17	69.00	0.68	2.99	7.5	10.6	2.573	2.324
T7+ S-10	22	71.66	0.98	4.10	7.8	10.9	2.800	2.370
CD (P=0.05)							0.36	0.391

* Observations recorded 45 days after transplanting

REFERENCES

Day, J.M. and J. Dobereiner. 1976. Physiological aspects of N_2-fixation by a *Spirillum* from *Digitaria* roots. **Soil Biol. Biochem., 8**:45-50

Govindan, M. and D.J. Bagyaraj. 1995. Response of wetland rice to *Azospirillum* inocurlation **J. Soil Biol. And Ecol., 15**:20-24

Hegazi, N.A., H. Amer and M. Monob. 1979. Enumeration of N_2 fixing spirilla. **Soil Biol. Biochem., 11**:437-438

Kumar, K. and A. Balasubramanian. 1989. Evaluation of two methods of *Azospirillum* biofertilizer application in rice. **Mysore J. Agric. Sci., 23**:1-5

Kundu, B.S. and P. Sangwan. 1992. Role of phytohormones produced by *Azospirillum* in crop responses. **Association of Microbiologists of India, 33rd Annual Conf.** Goa, (Abstr.). 17

Ladha, J.K., A. Tirol-Padre, M.L.G Daroy, G. Punzalan and I. Watanabe. 1987. The effects of N_2 fixation (C_2H_2 reduction), bacterial population and rice plant growth of two modes of straw application to a wet land rice field. **Biol. Fertil. Soil., 5**:106-111

Okon, Y., S.L. Albrecht and R.H. Burris. 1977. Methods for growing *Spirillum lipoferum* and for counting it in pure culture and in association with plants. **Appl. Environ. Microbiol., 33 :** 85-88

Rao, A.V. and B. Venkateswaralu 1985. Most probable number of *Azospirillum* associated with roots of pearl millet. **Plant Soil, 88:**153-158

Rao, V.R., D.N. Nayak, P.B.B.N. Charyulu and T.K. Adhya. 1983. Yield responses of rice to root inoculation with *Azospirillum.* **J. Agric. Sci. Camb., 100:** 689-691

Rao, V.R., P.K. Jena and T.K. Adhya. 1987. Inoculation of rice with nitrogen fixing bacteria – Problems and Perspectives. **Biol. Fertil. Soils., 4:** 21-26

Subba Rao, N.S., K.V.B.R. Tilak, C.S. Singh and M. Lakshmi Kumari. 1979. Response of a few economic species of graminaceous plants to inoculation with *Azospirillum lipoferum.* **Curr. Sci., 48:**133-134.

CHAPTER – 25

Azospirillum – A BIOFERTILIZER FOR RAINFED TRANSPLANTED RICE

K. Manjappa

ABSTRACT

The performance of rainfed transplanted rice as influenced by *Azospirillum* and inorganic nitrogen levels was evaluated at Agricultural Research Station (Paddy), Sirsi, during *kharif* seasons of 1997 and 1998. On an average over two years, the grain yield, straw yield and net returns of rice increased significantly with *Azospirillum* dip compared to no *Azospirillum* treatment. Similarly, the grain yield, straw yield and net returns of rice increased significantly with increased levels of nitrogen. The interaction effect of *Azospirillum* and nitrogen levels on grain yield, straw yield and net returns was also significant. The grain yield recorded with *Azospirillum* + 50% RDN (4802 kg/ha) was on par with that obtained with no *Azospirillum* + 75% RDN (4703 kg/ha). Whereas, the grain yield recorded with *Azospirillum* + 75% RDN (5330 kg/ha) was significantly higher than that obtained with no *Azospirillum* + 100% RDN (5146 kg/ha). The trend was almost similar with respect to gross and net returns also.

INTRODUCTION

Rice is the important food crop of hill zone of Karnataka occupying almost 79 per cent of the net sown area. The yield levels of rice in this zone are decreasing year by year. In order to maintain sustainable yield levels of rice, there is a need to adopt integrated nutrient management system. Recently, *Azospirillum*, a microbial inoculant (biofertilizer) has been tried as one of the components of integrated nutrient management system. Several workers (Gopalaswamy and Vidhyasekaran, 1987; Gopalaswamy *et al.*, 1989; Balasubramanian and Kumar, 1990 and Somani *et al.*, 1990) reported the role of *Azospirillum* culture in fixing atmospheric nitrogen and enhancing rice yields. The nitrogen gains from biofertilizers are highly variable depending on soil environment, nature of native microbial population etc. Hence, an experiment was conducted to know the efficacy of *Azospirillum* in conjunction with inorganic nitrogen on rice yield in rainfed lowland situation in hill zone of Karnataka.

MATERIALS AND METHODS

A field experiment was conducted at Agricultural Research Station (Paddy), Sirsi during *kharif* seasons of 1997 and 1998 to know the efficacy of *Azospirillum* in conjunction with inorganic nitrogen on yield of rainfed transplanted rice. The soil of the experimental site was silty clay loam with low available nitrogen (196 kg/ha), medium phosphorus (20.3 kg/ha) and very low potassium (53.5 kg/ha). The treatments consisted of two levels of *Azospirillum* (no *Azospirillum* and *Azospirillum*) and four levels of nitrogen (0, 50, 75 and 100% recommended dose of nitrogen). The experiment was laid out in factorial randomized block design with three replications. Rice variety Abhilash was used for planting. In treatments having *Azospirillum*, *Azospirillum* slurry was prepared by mixing 200 g of *Azospirillum* culture and 4-5 liters of water. In this slurry, the roots of the seedlings were dipped for 20 minutes and then used for transplanting. Nitrogen was applied at the quantity as per the treatments. Recommended dose of nitrogen (RDN) used was 75 kg/ha. 50 per cent of nitrogen was given at the time of transplanting and remaining 50 per cent was given in two splits (25% at tillering and 25% at panicle initiation stage) as top dressing. Entire quantity of recommended dose of phosphorus and potassium (75:87.5 kg/ha, respectively) was given at the time of transplanting. The crop was grown as rainfed crop. Observation on grain yield was made during both the years. The economics was calculated based the prevailed market prices of inputs and outputs. The data was analyzed as per the factorial randomized complete block design by using M-STAT-C programme.

RESULTS AND DISCUSSION

Effect of *Azospirillum*

The grain yield of rice differed significantly due to *Azospirillum* during both the years. The grain yields were significantly higher with *Azospirillum* (4021, 5892 and 4957 kg/ha, respectively) compared to no *Azospirillum* (3375, 5412 and 4394 kg/ha, respectively) during 1997, 1998 and in pooled data (Table 1). The results are in conformity with those of Gopalaswamy *et al.* (1987). The straw yield was also significantly higher with *Azospirillum* during both the years as well as in pooled data. On an average over two years, there was 12.8 and 14.8 per cent increase in grain and straw yield, respectively, with *Azospirillum* compared to no *Azospirillum*. This better response of rice for *Azospirillum* dip was may be due to increased nitrogen availability through nitrogen fixation by the *Azospirillum* culture.

Gopalaswamy *et al.* (1997) reported that *Azospirillum* inoculation promotes early tillering and reproductive growth of rice and significantly increasing grain filling percentage and grain weight per plant at harvest. The gross and net return was also influenced significantly by *Azospirillum* application. The gross and net return was significantly higher with *Azospirillum* dip compared to no *Azospirillum* dip during both the years of study as well as in pooled data (Table 2).

Effect of Nitrogen

The variation in grain and straw yield due to nitrogen levels was significant during both the years as well as in pooled data (Table 1). The grain and straw yields were increased significantly with increased levels of nitrogen during 1998 and in pooled data. During 1997, the grain yield recorded with 0% RDN (3217 kg/ha) was lowest and was found to be on par with that of 50% RDN (3400 kg/ha). Application of 100% RDN recorded significantly higher grain yield than other nitrogen levels during both the years as well as in pooled data. Similarly, the straw yield was also significantly higher with 100% RDF during both the years and in pooled data. Similar results indicating instances of increased yields of grain and straw with increased level of fertilizer were reported by Dubey *et al.* (1997) and Gopalaswamy *et al.*, (1997). The gross and net returns increased significantly with increased levels of inorganic nitrogen during both the years as well as in pooled data. The highest gross and net returns were realized with application of 100% recommended nitrogen (Table 2).

Interaction effect

The interaction effect of *Azospirillum* and nitrogen levels on grain and straw yield was significant during both the years and as well as in pooled data (Table 1). The grain yield recorded with *Azospirillum* + no nitrogen (3519 kg/ha) was significantly higher compared to that of no *Azospirillum* + 25% RDN (3017 kg/ha) during 1997. During 1998, the grain yield recorded with *Azospirillum* + no nitrogen (4234 kg/ha) and no *Azospirillum* + 25% RDN (3017 kg/ha) was on par. The grain yield recorded with *Azospirillum* + 50% RDN (3783, 5821 and 4802 kg/ha, respectively) was on par with that obtained with no *Azospirillum* + 75% RDN (3558, 5847 and 4703 kg/ha, respectively) during 1997, 1998 and in pooled data. Whereas, the grain yield recorded with *Azospirillum* + 75% RDN (4350 and 5330 kg/ha, respectively) was significantly higher than that obtained with no *Azospirillum* + 100% RDN (4050 and 5146 kg/ha, respectively) during 1997 and in pooled data. During 1998, the grain yield recorded with *Azospirillum* + 75% RDN (6311 kg/ha) was on par with that of no *Azospirillum* + 100% RDN (6241 kg/ha). Similarly, the grain yield recorded with *Azospirillum* + 100% RDN (4392, 6528 and 5460 kg/ha, respectively) was on par with that of *Azospirillum* + 75%.

Table 1: Effect of *Azospirillum* at different levels of nitrogen on grain and straw yield of rainfed transplanted rice

Treatments	Grain Yield (kg/ha)			Straw Yield (kg/ha)		
	1997	1998	Pooled	1997	1998	Pooled
Azospirillum treatment (A)						
A_0 - No *Azospirillum*	3375	5412	4394	4892	6238	5565
A_1 – *Azospirillum*	4021	5892	4957	5943	6835	6289
S.Em±	49.54	61.66	39.55	65.94	79.72	51.73
C.D. at 5%	150.27	187.03	167.52	200.02	241.79	106.47
Nitrogen levels (N)						
N_0 - no nitrogen	3217	4528	3872	4563	5164	4863
N_1 - 50% RDN	3400	5618	4509	5175	6370	5772
N_2 - 75% RDN	3954	6078	5017	5446	7052	6249
N_3 - 100% RDN	4221	6385	5303	6488	7559	7023
S.Em±	70.06	87.21	55.93	93.26	112.74	73.16
C.D. at 5%	212.51	264.51	115.11	282.88	341.95	150.57
Interactions (A x N)						
A_0 x N_0	2875	4146	3510	4375	4765	4570
A_1 x N_0	3559	4910	4234	4750	5563	5157
A_0 x N_1	3017	5415	4216	4517	6058	5287
A_1 x N_1	3783	5821	4802	5833	6682	6258
A_0 x N_2	3558	5847	4703	4808	6733	5771
A_1 x N_2	4350	6311	5330	6083	7372	6728
A_0 x N_3	4050	6241	5146	5867	7394	6630
A_1 x N_3	4392	6528	5460	7108	7724	7416
S.Em±	99.09	123.33	79.10	131.89	159.44	103.46
C.D. at 5%	300.34	374.07	162.79	400.05	483.59	212.93

RDN (4350, 6311 and 5530 kg/ha, respectively) during 1997, 1998 and in pooled data. On an average over two years, the extent of increase in grain yield with *Azospirillum* treatment over no *Azospirillum* treatment was 20.6, 13.9, 13.3 and 6.1 per cent at 0, 50, 75 and 100% RDN. The results are in conformity with those of Gopalaswamy *et al.* (1997) who reported that *Azospirillum* inoculation along with 75 kg N/ha significantly increased the grain yield of rice. They also reported that at 100 kg N/ha, *Azospirillum* treatment did not have significant beneficial effect. The trend in straw yield was also almost similar to that of grain yield. The interaction effect of *Azospirillum* and nitrogen levels on gross and net returns was also significant during both the years as well as in pooled data. On an average over two year, the net returns increased significantly with *Azospirillum* dip compared to no *Azospirillum* dip at respective levels of inorganic nitrogen. But, the extent of increase in net returns with *Azospirillum* dip was more at lower levels of nitrogen. The net return realized with *Azospirillum* + 75% RDN (Rs. 22529/ha) was significantly higher when compared to that obtained with *Azospirillum* + 50% RDN ((Rs. 19316/ha) but was found to be on par with that of no *Azospirillum* + 100% RDN (Rs. 21310/ha) and *Azospirillum* + 100% RDN (Rs. 23767/ha). Similarly, the net returns realized with *Azospirillum* + 50% RDN (Rs. 19316/ha) was on par with that of no *Azospirillum* + 75% RDN (Rs. 18185/ha).

Table 2. Effect of *Azospirillum* at different levels of nitrogen on yield and economics of rainfed transplanted rice

Treatments	Gross Returns (Rs./ha)			Net Returns (Rs.ha)		
	1997	1998	Pooled	1997	1998	Pooled
Azospirilum treatment (A)						
A_0 – no *Azospirilum*	23450	36005	29727	10121	22676	16399
A_1 – *Azospirilum Dip*	28058	39243	33651	14665	25849	20257
C.D. at 5%	795	653	697	795	653	697
Nitrogen levels (N)						
N_0 – no Nitrogen	22254	30067	26160	9260	17072	13166
N_1 – 50% RDN	23875	37267	30571	10555	23947	17251
N_2 – 75% RDN	27194	40486	33840	13711	27003	20357
N_3 – 100% RDN	29694	42675	36185	16048	29029	22538
C.D. at 5%	1125	1731	987	1125	1731	987
Interactions (A x N)						
A_0 x N_0	20188[d]	27567[e]	23877[e]	7226[d]	14605e	10915e
A_1 x N_0	24321[c]	32566[d]	28444[d]	11294[c]	19539[d]	15417[d]
A_0 x N_1	21108[d]	35839[c]	28474[d]	7820[d]	22551c	15185d
A_1 x N_1	26642[b]	38695[b]	32669[c]	13289[b]	25342[b]	19316[c]
A_0 x N_2	24379[c]	38891[b]	31636[c]	10928[c]	25441[b]	18185[c]
A_1 x N_2	30009[a]	42081[a]	36045[b]	16493[a]	28565[a]	22529[ab]
A_0 x N_3	28125[b]	41722[a]	34924[b]	14511[b]	28108[a]	21310[b]
A_1 x N_3	31263[a]	43628[a]	37446[a]	17584[a]	29949[a]	23767[a]
C.D. at 5%	1591	2447	1394	1591	2447	1394

The data of present investigation clearly indicates the beneficial effect of *Azospirillum*. Based on the results, it can be concluded that combination of *Azospirillum* + 75% RDN would be sufficient to get higher yields of rice under rainfed lowland situation of Karnataka. This helps in reducing 25% RDN and there by expenditure on nitrogen fertilizer.

REFERENCES

Balasubramanian, A. and K. Kumar. 1990. Influence of *Azospirillum* biofertilizer on rice crop. In: **Biological nitrogen fixation associated with rice.** pp. 267-275. Dutta, S.K. and Sloger, C. (eds.). Oxford and IBH Publishing Company Pvt. Ltd., New Delhi.

Dubey, S.K., R.S. Sharma and S. K. Vishwakarma 1997. Integrated nutrient management for sustainable productivity of important cropping systems in Madhya Pradesh. **Indian J. Agron., 42(1)**: 13-17.

Gopalaswamy, G. and P. Vidhyasekaran. 1987. Efficacy of *Azospirillum brasilense* in increasing rice yields. **Inte.Rice Res. Newsletter, 12(1):** 34.

Gopalaswamy, G., P. Vidhyasekaran and S. Chelliah. 1989. Effect of *Azospirillum lipoferum* inoculation and inorganic nitrogen on wet land rice. **Oryza, 26:** 378-380.

Gopalaswamy, G., S. Anthoni Raj and T. B. Ranganathan. 1997. Biofertilizer application strategy to rice (*Oryza sativa*). **Indian J.Agron., 42(1):** 68-73.

Somani, L.L., S.C Bhandari, S.N Saxena and K.K. Vyas. 1990. **Biofertilizer**. pp. 13-24,25.

CHAPTER - 26

RESPONSE OF WINTER RICE *(Rabi)* TO INORGANICS AND BIOFERTILIZERS APPLICATION

P.M. Shanmugam, A. Balasubramanian and P.C. Prabu

ABSTRACT

Field experiments were conducted at Agricultural College and Research Institute, Madurai during winter (*rabi*) season of 1995-96 and 1996-97 to study the effect of organic and inorganic nitrogen with zinc on physiological parameters and yield of rice. The results revealed that application of either Green Manure (GM) (*Sesbania rostrata* @ 6.25 t ha^{-1}) or Farm Yard Manure (FYM) (12.5 t ha^{-1}) combined with *Azospirllum* (soil application of 2 kg ha^{-1}) significantly increased the growth analysis *viz.*, crop growth rate, relative growth rate and leaf area index than the combined application of either GM or FYM with *Azolla* (@ 1 t ha^{-1} on 3rd DAS) as dual crop. Among the 3 levels of Nitrogen (N) (112.5, 150.0 and 187.5 kg ha^{-1}) with or without 25 kg Zinc Sulphate ($ZnSO_4$) ha^{-1}, combined application of 187.5 kg N with 25 kg $ZnSO_4$ ha^{-1} recorded significantly the highest Relative Growth Rate(RGR), Crop Growth Rate (CGR) and Leaf Area Index (LAI) and its effect was comparable with 150 kg N + 25 kg $ZnSO_4$ ha^{-1}. Application of GM with *Azospirllum* has significantly shortened the period of 50 per cent flowering and highest grain yield (5282 and 5218 kg ha^{-1} during 1995-96 and 1996-97 respectively) in both years of study. The highest level of N (187.5 kg ha^{-1}) with Zn application recorded significantly shorter days for 50 per cent flowering and grain yield (5526 and 5376 kg ha^{-1} during 1995-96 and 1996-97 respectively), and its effect was on par with 150 kg ha^{-1} with Zn application.

INTRODUCTION

The unfettered use of inorganics was justified by the fact that new crop varieties and hybrids require readily available nitrogen for their fast growth.

Application of organic manures increases the nutrient status, microbial activities and physical properties of soil and thus favouring the crop growth. The complimentary use of organic manures and chemical fertilizers will augment the individual efficiency for better growth and production of rice. The synergistic effect between applied nitrogen and zinc in rice was reported by Vyas *et al.* (1990). Keeping this in view, an attempt has been made to find out the influence of combined application of organic manures with biofertilizers and inorganic nitrogen with zinc on physiological parameters and yield of rice.

MATERIALS AND METHODS

Field experiments were conducted at Agricultural College and Research Institute, Madurai during winter (*rabi*) season of 1995-96 and 1996-97 to study the effect of organic and inorganic nitrogen with zinc on physiological parameters and yield of rice. The experiments were laid out in split plot design with three replications. The main plot treatments consisted of application of either GM (*Sesbania rostrata* @ 6.25 t ha^{-1}) or FYM (12.5 t ha^{-1}) in combination with biofertilizers (soil application of *Azospirllum* @ 2 kg ha^{-1} or *Azolla* as dual crop by inoculating 1 t ha^{-1} on 3 DAS). The subplot treatments comprised of three levels of N (112.5, 150.0 and187.5 kg ha^{-1}) with or without 25 kg $ZnSO_4$ ha^{-1}. The soil of the experimental field was sandy clay loam with a pH of 7.6, having low in available N, medium in available P and high in available K. the medium duration (125 days) rice variety ADT 39 was used with spacing of 20 x 10 cm. The data on leaf area index (at flowering), CGR and RGR (during tillering to flowering and flowering to harvest) and yield were recorded and staistically analysed.

RESULTS AND DISCUSSION

The results obtained from the experiments are discussed hereunder.

Crop Growth Rate (CGR)

Application of organic manures, biofertilizers, inorganics with Zn significantly influenced the physiological parameters (Table 1). Addition of GM + *Azospirillum* (M_2) recorded highest CGR and it was on par with FYM + *Azospirillum* (M_4) in both the stages of observation. Similar trend was observed in 1996-97 also. The effect of inorganics on CGR was significant during 1995-96 only at both the stages of sampling. Addition of 187.5 kg N + 25 kg $ZnSO_4$ ha^{-1} (S_6) registered maximum CGR which was on par with 150 kg N + 25 kg $ZnSO_4$ ha^{-1} (S_4).

Relative Growth Rate (RGR)

Application of GM + *Azospirillum* also recorded significantly the higher values of RGR (30.80 and 30.71 gg^{-1} day^{-1} in 1995-96 and 1996-97 respectively) and was comparable with FYM + *Azospirillum.* Significant response of inorganics and zinc on RGR was observed at the early stages in both the years of study and only at latter stages in first year of study.

Application 187.5 kg N + 25 kg $ZnSO_4$ ha^{-1} registered significantly higher RGR values from tillering to flowering in both the years and it had comparable performance with S_4 and S_5 at vegetative phase.

Table 1. Influence of organics, inorganic N and Zn on physiological parameters of rice

Treatments		1995-96				1996-97			
		Tillering to flowering		Flowering to harvest		Tillering to flowering		Flowering to harvest	
		CGR (gm^{-2} day^{-1})	RGR (gg^{-1} day^{-1})	CGR (gm^{-2} day^{-1})	RGR (gg^{-1} day^{-1})	CGR (gm^{-2} day^{-1})	RGR (gg^{-1} day^{-1})	CGR (gm^{-2} day^{-1})	RGR (gg^{-1} day^{-1})
Organics and biofertilizers									
M_1	GM + *Azolla*	11.2	27.6	14.6	19.2	10.7	26.9	14.9	18.5
M_2	GM + *Azospirillum*	12.7	30.8	18.6	19.9	12.2	30.7	17.9	19.0
M_3	FYM + *Azolla*	10.2	26.2	13.7	17.7	10.6	29.4	14.7	17.8
M_4	FYM + *Azospirillum*	12.0	30.7	17.3	19.3	11.6	30.2	17.6	18.7
	SEd	0.49	0.78	1.02	1.03	0.32	0.55	1.14	1.00
	CD(P=0.05)	1.24	1.88	2.5	NS	0.79	1.34	2.79	NS
Inorganics									
S_1	112.5 kg N ha^{-1}	9.5	27.5	12.9	17.2	10.4	28.2	14.7	17.6
S_2	112.5 kg N + 25 kg $ZnSO_4$ ha^{-1}	9.5	27.6	14.8	17.9	10.5	28.6	15.6	18.2
S_3	150 kg Nha^{-1}	11.0	28.0	15.0	18.5	10.9	29.0	16.6	18.3
S_4	150 kg N + 25 kg $ZnSO_4$ ha^{-1}	13.1	29.5	17.8	20.8	11.0	29.5	16.7	19.0
S_5	187.5 kg N ha^{-1}	12.2	29.3	16.9	18.5	11.0	30.0	17.2	20.1
S_6	187.5 kg N + 25 kg $ZnSO_4$ ha^{-1}	13.9	31.1	18.8	21.2	11.2	32.4	16.9	20.3
	SEd	0.82	1.22	1.18	1.33	0.66	1.51	1.07	1.34
	CD(P=0.05)	1.65	2.46	2.39	2.69	NS	3.04	NS	NS

The beneficial effect of organic manures on CGR and RGR might be due to the advancement of time and proper availability of nutrients by release of the required N more easily and slowly to match the crop demand (Buresh *et al.,* 1993).

Leaf Area Index (LAI)

Though the organic manures did not influence the LAI significantly, inorganic N and Zn application significantly affected the LAI. Application of 187.5 kg N + 25 kg $ZnSO_4$ ha^{-1} (S_6) registered significantly higher LAI in 1995-96 and was comparable with 150 kg N + 25 kg $ZnSO_4$ ha^{-1} (S_6) during 1996-97. The higher LAI under high level of N application might be due to the effect of N on protein synthesis and meristematic growth through hormonal synthesis would have helped leaf production and leaf expansion (Kannan *et al.,* 1982).

Days to 50 per cent flowering

Application of GM + *Azospirillum* took significantly shorter period to 50 per cent flowering (82 and 84 days respectively during 1995- 96 and 1996-97) and was on par with M_4 in both the years (Table 2). Early flowering was observed with S_6 and it was comparable with S_5 and S_4 in both the years of study. Earliness in flowering under the above treatment was attributed to vigorous growth and leaf area expansion leading to accumulation of more photosynthates, which would have caused early flowering (Virdia *et al.,* 1993).

Grain yield

The effect of organics combined with biofertilizers with various levels of N with Zn on rice grain yield was well pronounced during 1995-96 and 1996-97 (Table 2). Application of GM + *Azospirillum* significantly recorded higher grain yield (5282 and 5218 kg ha^{-1} respectively during 1995-96 and 1996-97) which was on par with FYM + *Azospirillum* application. Regarding inorganics, application of 187.5 kg N + 25 kg $ZnSO_4$ ha^{-1} registered significantly higher grain yield of 5516 and 5376 kg ha^{-1} and was comparable with 150 kg N + 25 kg $ZnSO_4$ ha^{-1} application which registered 5414 and 5210 kg ha^{-1} during 1995-96 and 1996-97 respectively. The favourable and significant influence of organic manures with biofertilizers might be due to enhanced growth characters, increasing rate of N, which supply nutrients at a steady rate for longer period from the decomposing green manure and synchronised the rice crop demand. Combined application of N and Zn increased the grain yield owing to better utilisation of N and P for grain production through metabolic activities (Vyas *et al.,* 1990).

Table 2. Influence of organics, inorganic N and Zn on LAI, days to 50 per cent flowering and grain yield of rice

Treatments		LAI at Flowering		Days to 50 per cent flowering		Grain yield (kg ha^{-1})	
		1995-96	1996-97	1995-96	1996-97	1995-96	1996-97
Organics and biofertilizers							
M_1	GM + *Azolla*	5.4	5.5	86	87	5072	4800
M_2	GM + *Azospirillum*	5.6	5.6	82	84	5282	5219
M_3	FYM + *Azolla*	5.4	5.4	90	91	4897	4712
M_4	FYM + *Azospirillum*	5.4	5.5	83	85	5204	5102
	SEd	0.09	0.84	1	1	55	51
	CD (P=0.05)	NS	NS	3	2	135	123
Inorganics							
S_1	112.5 kg N ha^{-1}	5.1	5.4	88	90	4683	4503
S_2	112.5 kg N + 25 kg $ZnSO_4$ ha^{-1}	5.2	5.4	87	91	4875	4743
S_3	150 kg N ha^{-1}	5.4	5.4	85	87	4887	4774
S_4	150 kg N + 25 kg $ZnSO_4$ ha^{-1}	5.5	.5.5	83	85	5414	5210
S_5	187.5 kg N ha^{-1}	5.5	5.6	84	84	5308	5136
S_6	187.5 kg N + 25 kg $ZnSO_4$ ha^{-1}	5.7	5.8	82	84	5516	5376
	SEd	0.09	0.09	2	1	91	115
	CD (P=0.05)	0.18	0.18	3	2	185	238

CONCLUSION

Thus, it is evident from the study that oombined application either GM (*Sesbania rostrata* @ 6.25 t ha^{-1}) or FYM (12.5 t ha^{-1}) combined with soil application of *Azospirillum* (2 kg ha^{-1}) and application of 150 kg N with 25 kg $ZnSO_4$ ha^{-1} maximised the physiological parameters and grain yield of *rabi* rice.

REFERENCES

Buresh, R.J., T. T. Chua, E.G. Castillo, S.P. Liboon and D.P. Garrity. 1993. Fallow and Sesbania effects on soil nitrogen dynamics in lowland rice based cropping system. **Agron. J. 85:** 316-321.

Kannan, V. 1982. Studies on the effect of *Azolla* and organic manures for N economy in rice (CO 37). M.Sc. (Ag.) Thesis. Tamil Nadu Agricultural University.

Viridia, H.M., H.S. Patel and M.P. Patel. 1993. Spacing and nitrogen requirements of rice varities. **GAU Res. J., 19(1):** 133-135.

Vyas, M.K., A.K. Palival, S.B. Gupta, M.L. Adil and S.S. Senegar. 1990. Response of rice to zinc application on production, N utilization and quality of irrigated rice. *Narendradva* **J. Agric. Res., 4(1):** 11-14.

SECTION – IV

RHIZOBIUM BIOFERTILIZER

CHAPTER – 27

SYMBIOTIC NITROGEN FIXATION IN LEGUMES ASSOCIATED WITH RICE CROPPING SYSTEMS

J. Prabakaran and M. Rangarajan

INTRODUCTION

It has been the practice from time immemorial that legumes are grown in different cropping systems associated with rice. Promising cropping system includes the raising of pre –rice legumes and post-rice legumes. Besides the above two systems, legumes are grown along with rice as bund, border and *in situ* crops. Several multipurpose cover legumes as short duration crops are grown after the harvest of the second rice crop of the first year and before the planting of the first crop of the subsequent year. Grain legumes admirably fit into the system in which several pulses such as green gram, black gram, soybean etc. which not only yield grains but add nitrogen to the soil, are included. Leguminous crops which may yield forage or green manure, may also be grown. The legumes associated with rice cropping systems, by their ability to fix atmospheric nitrogen in the root nodules have the potential to accumulate nitrogen varying from 10 to 200 kg N/ha (Becker and Johnson,1999) and even upto 450 kg N/ha.

LEGUMES GROWN AS PRE AND POST RICE CROPS

Different legumes based on their utility as forages, grains, or green manures, are listed here under:

(i) Forage legumes

Siratro (*Macroptilium atropurpureum*), Clitoria (*Clitoria ternata*), Desmanthus (*Desmanthus virgatus*) and Lablab bean (*Lablab purpureus*).

(ii) Grain legumes

Greengram or Mungbean (*Vigna radiata),* Cowpea (*V. unguiculata),* Groundnut (*Arachis hypogea*), Pigeon pea or Redgram (*Cajanus cajan),* Urd or Black gram (*V. mung),* Soybean (*Glycine max*), Chickpea or Gram (*Cicer arietinum),* Lentil (*Lens esculenta)* and Horse gram (*Dolichos uniflorus*)

(iii) Green manure legumes

Sesbania (*Sesbania speciosa, Sesbania rostrata*), Indigo (*Indigofera tinctoria*) Kolingi (*Thephrosia purpurea*) Pillipesra (*Phaseolus aconitifolius*) and Sunn hemp (*Crotalaria juncea*)

ROLE OF LEGUMES IN RICE BASED CROPPING SYSTEMS

Various legumes mentioned above associated with rice cropping systems bring about the following benefits

(a) Soil fertility enhancement

Because of addition of organic matter through incorporation of above and below ground portion of the legumes, soil fertility status including organic fraction is enhanced.

(b) Nitrogen addition

Legumes by their potentiality to fix nitrogen from atmosphere add nitrogen to benefit the standing crop and also the succeeding crop.

(c) Increased N uptake in rice

When legumes are grown in rice solely for green manure production and incorporated, the succeeding rice crop takes up more nitrogen and reduces requirement of the nitrogen for rice. There are several reports that rice produced more grains after a legume than after a non legume or after a fallow, only due to increased nitrogen uptake.

(d) Suppression of insect pests, diseases and weeds

By raising a legume in between two rice crops or before a rice crop in the succeeding year, building up of pests and diseases of rice is checked. Moreover, the weed growth in the soil is also arrested.

CROP SEQUENCES IN RICE BASED CULTIVATION

There are several crop sequences in which rice and legumes are included. The important legumes in rainfed condition involving rice in kharif crop are listed in the Tables 1 and 2.

Table 1. Prominent crop sequences with pulses as component crop under rain fed condition

Region	Kharif (irrigated)	Rabi (rainfed)
Jhansi	Rice	Chickpea
Varanasi	Rice	Chickpea /Lentil
Bhubaneshwar	Rice	Horse gram
Ranchi	Rice	Chickpea / Lentil

Table 2. Prominent cropping system with pulses in different states of India

States	Kharif	Rabi	Summer
Uttar Pradesh	Rice	Mustard	Black gram
	Rice	Gram/Lentil	--------------
Madhya Pradesh	Rice	Rice	Green gram
	Rice	Linseed	Cowpea
Bihar	Rice	Wheat	Green gram
	Rice	Mustard	Black gram
Punjab	Rice	Wheat	Green gram
Haryana	Rice	Wheat	Green gram
Andhra Pradesh	Rice	Wheat	Black gram
	Rice	Ground nut	Black gram
Assam	Rice	Wheat	Cowpea / Green gram Green gram
	Rice	Mustard	
Tamil Nadu	Rice	Rice	Green gram
			Black gram, Kolingi
			Daincha, Pillipesra

PROMISING CROPPING SYSTEMS IN DIFFERENT STATES OF INDIA AND THEIR YIELD POTENTIALITY

The need based cropping with summer pulses followed in different regions of India along with yield potentials are indicated in Table 3.

Table 3. Cropping sequences in different parts of India and their yield potential.

States	Crop sequences			Yield (q/ha)		
	Kharif	Rabi	Summer	Kharif	Rabi	Summer
Uttar	Rice	Mustard	Black gram	50.0	1.4	8.7
Pradesh	Rice	Wheat	Green gram	45.8	34.0	0.8
Madhya	Rice	Wheat	Green gram	25.1	10.4	5.4
Pradesh	Rice	Linseed	Cowpea	56.7	3.2	10.5
Andhra	Rice	Groundnut	Green gram	61.4	24.6	11.9
Pradesh	Rice	Wheat	Black gram	44.5	18.1	10.2

Tamil Nadu	Rice	Rice	Green gram	63.9	39.8	5.4
	Rice	Rice	Cowpea	49.9	21.6	8.9
Kerala	Rice	Wheat	Green gram	25.0	9.3	5.9
	Rice	Mustard	Green gram	23.2	2.1	5.3
Assam	Rice	Wheat	Green gram	50.1	5.0	8.4
Gujarat	Rice	Wheat	Cowpea	47.2	15.2	10.6

LEGUMES AND THEIR NITROGEN FIXING POTENTIALS

The nitrogen fixing capacities of the several legumes vary depending upon the kind and genotype of legume, soil type and other environmental conditions under which they are grown. The amount of nitrogen fixed ranged from 0-450 N kg/ha (Table 4). Estimated Ndfa (nitrogen derived from atmosphere) for green manure legumes and forage legumes is generally higher than that of grain legumes. Among the green manure legumes, *Sesbania rostrata* fixed maximum quantity of nitrogen ranging from 70-458 kg N/ha. Among the grain legumes, soybean and pigeonpea fixed more nitrogen than other legumes (Ladha and Kundu, 1997).

Table 4. Nitrogen fixing capacity of legumes in rice field

Legumes	N Fixation Ndfa %	Amount kg N /ha
Forage / Green manure		
Clitoria	78-79	200-240
Sesbania rostrata	68-94	70-458
Indigofera	70	78
Siratro	69	91-132
Desmanthus	80	120
Aschenomene	68-94	110-145
Grain legumes		
Pigeon pea	71-77	111-167
Cowpea	48-56	60-78
Mungbean	45-56	61 -90
Soy bean	69-78	26-176

CONTRIBUTION OF SYMBIOTICALLY FIXED NITROGEN TO RICE

Growing legumes and incorporating their biomass in soil generally improve soil fertility which results in the increased yield of subsequent rice crop and reduces the requirement of fertilizer nitrogen. Many studies revealed that rice crop yielded more after a legume. The results of the trial conducted in Sri Lanka indicated clearly that rice crop after legumes such as mungbean or crotalaria gave higher grain yield than after a fallow. Similar trial in India also revealed that growing any legume before rice is better than growing rice before rice (Table 5) (Ladha and Kundu, 1997).

Table 5. Effect of preceding crops on grain yield of rice

Location	Length of pre rice season	Pre rice crop	Grain yield (kg/ha)
1. Sri Lanka	2 months	Fallow	2951
		Mung bean	3205
		Crotalaria	4256
		Bush bean	3384 LSD (p: 0.05) 114
2. India	4 months	Rice	4510
		Chickpea	5120
		Mung bean	4980
		Cowpea	5230 LSD (p: 0.05) 470

NITROGEN REMOVAL WITH LEGUME HARVEST

When grain legumes are grown, the quantity of nitrogen left in the above ground residue following grain or pod harvest, depends on the total nitrogen accumulation of the legume and the harvest index for of N. George *et al.* (1995) examined the effect of growing soybean, cowpea, black gram and mung bean for 70-84 days on N balance in a Philippines low land rice (Table 6). The N balance was negative with soybean and cowpea and positive with mungbean and black gram. Most farmers in the tropics remove the haulms or legume beside pods from their fields. In such cases the soil N balance would always be negative (Ladha and Kundu, 1997).

Table 6. Sources and balances of above ground N in grain legumes and their yield (Legume seeds sown on 20th Feb, pods picked from 30th Apr to14 th May)

Legumes	Legume N accumulated (kg N /ha)			Legume grain yield (t/ha)	N removed in pods (kg N /ha) (b)	N Balance (kg N /ha) (a-b)
	Total	Soil N	N fixed (a)			
Soy bean	217	68	149	2.63	186	-37
Cowpea	124	64	60	2.59	65	-5
Mung bean	147	57	90	1.75	80	+10
Black gram	150	60	90	1.8	85	+5

NITROGEN FERTILIZER EQUIVALENCE [NFE] OF LEGUME FIXED NITROGEN : FALLOWS

Many studies revealed an indirect measurement of the legume N contribution to rice by comparing rice yield response to chemical N fertilizer, with and without legume residues (Ladha and Kundu, 1997). Nitrogen fertilizer equivalence of a legume residue is calculated as the quantity of chemical N fertilizer to attain grain yield equal to that obtained with legume residue and no chemical N fertilizer. Nitrogen fertilizer equivalence values for common green manure legumes and grain legumes are given in Table 7.

Table 7. Nitrogen fertilizer equivalence of common legumes

Legume species	Dry mass in residues (t/ha)	Legume N incorporated (kg/ha)	N fertilizer equivalence (kg/ha)
Green manure legumes			78
Sunn hemp	Direct incorporation of the entire legume crop	78	50
Sesbania aculata		57	80
S. rostrata		131	100-120
S.cannabina		98-147	
Grain legumes			
Cowpea	2.7 to 4.0	53-54	37-50
Green gram	4.6	101	100

EFFECT OF PRE–RICE LEGUME MANAGEMENT ON N-MINERALIZATION

Incorporation of green manure legumes or above ground residues of dual purpose legumes after grain/forage harvest adds legume fixed N to soil, which in turn becomes available to a succeeding rice crop. Results of an experiment on mineral N release from cowpea green manure and cowpea residues in a low land rice are given in Table 8. Within 15 days of incorporation, the green manure cowpea released about 28 kg mineral N/ha, while the cowpea residue supplied 20 kg mineral N/ha. However, the below ground portion of cowpea residue after the harvest of grain and straw, supplied only about 4 kg mineral N /ha (Ladha and Kundu, 1997).

Table 8. Effect of pre-rice legume management of mineral N availability in soil in low land

Pre rice low land Pre rice legume management	Mineral N (kg/ha in 0-30 cm soil layer)	
	Before rice sowing	15 days after rice sowing
Control	10.1	16.5
Cowpea incorporated in soil at flowering	19.5	44.4
Cowpea residue incorporated after grain harvest	16.5	36.0
Cowpea grain and straw removed from plot	13.0	20.8

GRAIN LEGUMES IN RICE FALLOWS

Growing grain legumes in rice fallows is being followed by farmers for long time. It is usually grown wherever single or two rice crops are cultivated. The grain legumes such as black gram or green gram are sown by broadcasting in the standing rice crop 15-20 days before harvest. The crop thus sown survives entirely on the residual moisture and fertility. Cost of cultivation is rather low. Net returns are high but the scope of manipulation is limited. Nodulation under rice fallow is good. Native rhizobia are equally efficient and no advantage is noticed by inoculation with good rhizobial strain developed elsewhere. Screening of 38 genotypes of black gram in lowland Lam farm in Andhra Pradesh had indicated the variation in nodulation depending upon the genotypes (Table 9).

Table 9. Genotypic differences in black gram for nodulation

Genotypic characters	Range (38 Genotypes)
Nodules/plant	7.3-29.5
Nodule dry weight (g/plant)	21-96
Grain yield (g/plant)	5- 15.4

NODULE OCCUPANCY BY INOCULANT STRAIN IN GRAIN LEGUME UNDER RICE FALLOW CONDITION

Effective nodules are formed by the competitive strains of rhizobia which enter the soil through inoculation. Survival and formation of effective nodules by the inoculant strains in legumes depends upon the several ecological factors including the host, bacterium and environmental conditions. These have been studied in rice fallow pulses by serological tests including immunofluorescent technique (IF). The IF technique estimates accurately the extent of nodule occupancy by the inoculated rhizobial strain in pulses such as green gram (Thamil Selvi and Rangarajan, 1986) The result of a field testing of inoculated rhizobial strains on green gram under rice fallow system at Aduthurai indicated that a strain of *Rhizobium* VRS-6 competed with the resident rhizobia and occupied nodule ranging from 60 to 70 percent (Rangarajan and Purushothaman, 1987).

MOLECULAR BIOLOGY OF RHIZOBIUM –LEGUME ASSOCIATION

Molecular basis of nodule forming bacterium-legume plant interaction is known. The bacterium contains nodulation (*nod or nol)* genes. Most of these genes are inactive in the absence of the plants. However the root exudate compounds such as flavonoids and nucleotides activate the positive regulatory protein, *Nod*D (Phillips *et al.,* 1977). This results in the transcription of inducible nod genes which are required in the biosynthesis of series of compounds such as *nod* factors or lipo chitin oligosaccharides (LCOs). These host specific LCOs induce the formation of nodule primordia. Other genes such as *nif* or *fix* genes present in the bacterium are also involved in the nitrogen fixation process. Host plant plays important role in the synthesis of nodulins which are nodule specific proteins. The synthesis of some of these nodulins can also induced by purified LCOs. The specific LCOs are major determinants of the host range of nodulation (Spaink *et al.* 1987).

In contrast, usually flavonoids do not act as host range determinants. During evolution the nod D gene seems to have fine-tuned its interaction with the flavonoid inducers or its host plant. Another important factor which is crucial in the host range is plant lectin. Diaz *et al.* (1989) showed that transfer of pea lectin gene to hairy roots of white clover resulted in nodulation and nitrogen fixation of the transferred clover roots by *R. leguminosarum pv viciae* (pea rhizobia) which normally are unable to nodulate white clover.

CONCLUSION

Rhizobium species and strains have been used as legume inoculants to supply fixed nitrogen for various legumes. But resident rhizobia exert significant effect on inoculant strains in field. Some times resident rhizobia may be less effective or even ineffective in nitrogen fixation, in which case inoculant strain may perform very well resulting in the increased nodulation and yield. Therefore it is essential to test the efficacy of introduced strain by assessing its nodule occupancy under controlled condition and later under field condition before we recommend suitable *Rhizobium* strain for any legume associated with rice cropping system.

REFERENCES

Becker. M and D.E. Johnson. 1999. Role of legumes in fallows intensified upland rice based systems in West Africa. **Nutr. Cyc. Agroecol., 53(1)** : 71-81.

Diaz, C.L, L.S. Melchers, P.J.J. Hooykass, B.J.J. Lughanberg and J.W. Kijne.1989. Root lectin as a determinant of host plant specificity in the *Rhizobium*-legume symbiosis. **Nature. 338:** 579.

George, T., J.K. Ladha, D.P. Garrity and R.O. Torres. 1995. Nitrogen dynamics of grain legumes weedy fallow rice sequences in the tropics. **Agron. J., 87**:1-6.

Ladha, J.K and D.K. Kundu. 1997. Legumes for sustaining soil fertility in lowland rice. In: Extending Nitrogen Fixation research in farmer's field; Proceeding of an International Workshop on Managing legume nitrogen fixation in the cropping systems of Asia, on 20-24, August 1996. (eds.) O.P. Rupela, C. Johansen and D. F. Herridge) ICRISAT, Hyderabad, India, pp. 76-112.

Phillips, D.A., C.M. Joseph and P.R. Hirsh. 1977. Occurrence of flavonoids and nucleotides in agricultural soils. **Appl. Env. Microbiol., 63:** 4573-4577.

Rangarajan, M. and A.S. Purushothaman. 1987. Successful inoculation of *Vigna radiata* (L) Wilczek. grown in rice fallow soils containing indigenous cowpea rhizobia. **FEMS Microbiol Ecol., 46**: 85-88

Spaink, H.P. C.A. Vijfelman, E. Pees, R.J.H. Okker and B.J.J. Lugtenberg. 1987. *Rhizobium* nodulation gene *Nod* D as determinant of host specificity. **Nature, 328:** 337.

Thamil Selvi, T. and M. Rangarajan. 1986. Promiscuity in infection and nodule occupancy of rhizobial strains in certain heterologous hosts. **National Seminar on Microbial Ecology.** Held at TNAU, Jan 23-24, 1986. p. 8. Abst.

CHAPTER - 28

INDUCTION OF DESICCATION TOLERANCE IN GREENGRAM RHIZOBIA

K. Kumutha, J. Sempavalan and K. Vijila

ABSTRACT

For exploiting the maximal benefits of *Rhizobium* in the field, it would be benefiting to consider the ecological constraints it faces. *Rhizobium* is a poor survivor in the soil and it is sensitive to various abiotic and edaphic factors. Soil desiccation which prevails during drought situation affects the survival of the organism resulting in the failure of nodulation of crops grown there on. To circumvent this problem in inoculation, a study was conducted for inducing desiccation tolerance in *Rhizobium*. Sixteen isolates of *Rhizobium* were obtained from the nodules of green gram collected at different locations of Tamil Nadu. By growing the isolates on osmotic substrate poly ethylene glycol, the osmotic pressure (OP) level was enhanced in the cells. Of the sixteen isolates tested the isolates CRM 11 and CRM 15 tolerated higher OP levels. The ability of these isolates to grow under soil moisture levels of even 5% was observed. Hence such isolates which are able to survive under higher OP levels may tolerate less water potentials of soils.

INTRODUCTION

Legume – rhizobium interaction is a highly regulated process that results in a complex symbiosis. Numerous host and bacterial factors influence rhizobial infection and host proliferation (Wilson *et al.*, 1995). *Rhizobium* is a poor survivor in the soil and it is sensitive to various abiotic and edaphic factors. Drought is the major abiotic stress that affect the survival of rhizobia in legume rhizosphere which resulted in poor or even failure of nodulation. Prevailence of desiccation during drought is the cause for its failure of inoculation response. To circumvent this problem in inoculation, a study was conducted for inducing desiccation tolerance in greengram rhizobia.

MATERIALS AND METHODS

Sixteen rhizobial isolates were obtained from the nodules collected from different locations of Tamil Nadu. These isolates were subjected to induction of drought tolerance using osmotic substrate Poly Ethylene Glycol (PEG). PEG is employed for altering osmotic pressure (OP) for similarity drought conditions (Parner and Moore, 1968). The isolates were grown in YEMA broth with three levels of osmotic pressure (OP) *viz.*, 3,5 and 6 for 30 days. Simultaneously the log phase culture of the isolates were inoculated in acid washed sand and the moisture levels were maintained to 10, 20 and 30 per cent and incubated for 30 days. After the incubation period, the samples were plated and the population was estimated.

RESULTS AND DISCUSSION

The results indicated the failure of many isolates in different OP levels. CRM 8, 9, 11 and 15 survived well at all OP levels *viz.*, 3, 5 and 6. The reduction in population was noticed with the increase in OP concentration. Maximum population was observed with CRM 11 and CRM 15 (Table 1).

Table 1. Population of *Rhizobium* at different OP concentrations and at different moisture levels

Strains	PEG Concentrations (population x 10 ^{5} cfu /ml)			Moisture levels (population x 10 ^{5} cfu /g)			
	3 OP	5 OP	6 OP	5 %	10 %	20 %	30 %
CRM 3	-	-	-	-	-	2	2
CRM 4	-	-	-	-	-	1	2
CRM 5	-	-	-	-	-	2	3
CRM 6	-	-	-	-	-	2	2
CRM 7	-	-	-	-	-	1	2
CRM 8	62	55	30	-	1	2	4
CRM 9	56	51	40	-	-	3	6
CRM10	-	-	-	-	-	1	1
CRM11	71	63	53	1	3	17	58
CRM12	-	-	-	-	-	1	2
CRM13	-	-	-	-	-	1	3
CRM14	-	-	-	-	-	2	4
CRM15	60	48	40	2	5	9	81
CRM16	-	-	-	-	-	1	3
COG15	-	-	-	-	-	1	3
CRM5 TT	-	-	-	-	2	3	5

The population at moisture stress conditions showed that all the isolates survived well at 20 and 30 per cent moisture. When coming to the lower levels at moisture, almost all the isolates failed to survive except CRM 8, 11, 15 and CRM 5. At 5 per cent moisture, still more reduction in population was observed. Only two isolates CRM 11 and CRM 15 survived at 5 per cent moisture level.

By growing the isolates in osmotic substrates osmotic pressure level in the cells was enhanced and may have the impact under desiccation. Cowpea rhizobia isolated from dry areas are having the tolerance towards desiccation and temperature (Hartel and Alexander, 1984), which may be the indication for drought tolerance. Hence the isolates which were able to survive at high O.P levels and low temperature may have the tolerance towards drought situation and the isolates CRM 11\CRM 15 may have positive response under drought situation to enhance the survival and nodulation. Further, it is reported that the ability under moisture stress may be due to plasmids (Baldani and Werver, 1992). Hence it is essential to characterize the isolates genetically and field-tested for the performance.

REFERENCES

Baldani, J.I and R.W. Weaver, 1992. Survival of clover rhizobia and their plasmid used derivatives in soil under heat and drought stress. **Soil Biol. Biochem., 24(8)** : 737-742.

Hartel, P.G. and M. Alexander, 1984. Temperature and desiccation tolerance of cowpea rhizobia. **Can. J. Microbiol., 30** : 820-823.

Parmar, M. T. and R.P. Moore, 1968. Carbowax 6000, Mannitol and sodium chloride for simulating drought conditions in germination studies of corn (*Zea mays* L.) of strong and weak vigour. **Agronomy Journal, 60**: 192-195.

Wilson, K.J., M.B. Peoples and R.A. Jefferson, 1995. New techniques for studying competition by Rhizobia and for assessing nitrogen fixation in the field. **Plant Soil, 174**: 241-253.

CHAPTER – 29

FORMULATION OF STRESS TOLERANT BIO-INOCULANTS FOR RAINFED CROPPING SYSTEM

S.B. Gupta, K. Tedia, A. Singh, Anurag, N. Lakpale, Anjali Pal and P.K. Chhonkar

ABSTRACT

Identified 9 thermal tolerant (55 C), 8 moisture stress tolerant (% soil moisture < 3.5), 8 acidity tolerant (pH 4.5) and 11 salt tolerant (40,000 ppm) *Rhizobium* isolates. During second part of this study, mung plant has been identified as a macro symbiont of recently identified one of the stress tolerant *Rhizobium* isolates, which was isolated during summer season, 2000 from top soil sample (0-5 cm depth) of Chhattisgarh region. This studies is being continued for identification of macro symbionts of other stress tolerant especially *Rhizobium* isolates. Further, effectiveness of these isolates will be compared with that of available national and international mesophilic checks of specific crops.

INTRODUCTION

Newly formed Chhattisgharh state is bigger than many states of the country. Its economy mainly based on agriculture / forest. Soils of this region very rich with natural resources. In spite of this facts, people of Chhattisgarh region has been under constant threat of poverty, unemployment and hunger that often forced rural poor people in particular to leave their home to find job elsewhere especially after every *Kharif* season because Chhattisgarh is predominantly monocropped, mostly with rainfed rice. Productivity of rice per unit area of this region is about half to the average productivity of the nation. In the Rabi / Summer season, when one can take more reliable crops, in this region, very less areas is cultivated, that too with crops like Lathyrus, linseed, gram, lentil, mung, etc. with productivity of less than half a tonne per hectare, making such farming virtually uneconomic. The situation is becoming worse day-by-day with mounting population pressure and shrinking rate of job opportunities and income for illiterate resource poor farmers. The green revolution phenomenon that transformed agriculture elsewhere has not been able to make the desired impact here and the transformed agriculture seem to realise the need of more massive support and commitment of state, national, international and other agencies for the agricultural development of this region. Numerous research proposal submitted for consideration to various funding agencies by the Indira Gandhi Agricultural University, Raipur have commonly voiced the need of increased agricultural production of not only cereals but also pulses and oilseeds in the region on a sustainable low cost input basis. Inherent low levels of soil fertility status (Sinha and Gupta, 1985). Crop beneficial soil microbial population including soil organic matter content (Gupta *et al.*, 1992, Gupta *et al.*, 1995a, 1995b, Katre *et al.*, 1997 and Gupta, 2000a) and resources scarce situations are now recognized among the basic causes of low crop yield in this region.

Out of 135.00 lakh hectares (1.35 lakh sq. km) area of Chhattisgarh, about 46.02 lakh hectares is under *Kharif* mostly with rice crop, while about 12.45 lakh hectares only under *Rabi* crops. About 30.00 lakh hectares are remaining fallow during *Rabi*. Lathyrus is an important pulse taken as bonus crop in rice belt of central and eastern parts of India. The total cultivated area of this crop is more than 0.89 million hectatre out of this and nearly 0.61 million hectares are situated in Chhattisgarh during *Rabi* season representing the major Lathyrus belt in the country under rice based relay (Utera) cropping system. This crop is very popular among the marginal and poor farmers which includes tribals and backward population of Chhattisgarh. The important Lathyrus growing districts in Chhattisgarh are Raipur, Durg, Dilaspur and Rajandgaon with production is 224.8 thousand tonnes with a productivity 4.8 quintal ha^{-1}. The productivity per unit area of this crop is also very low mainly because of sub optimal nutrition.

Chhattisgarh is not just predominantly rural in character, it also has substantial tribal population. A large proportion of this tribal population continue to live in abysmal poverty. Appropriate technologies, affordable and suitable to socioeconomic conditions of the tribes and other people who inhabit the region need to be developed without any further loss of time. Extension of low cost technologies including popularization of location specific effective bio-fertilizers under INM to the door steps of the farmers to increase production and productivity of their lands, thereby ensuring food and nutritional security, should be the core of our developmental efforts. To achieve all this, no other single institution is better equipped than an Agricultural University.

Agriculture, in a broad sense, is not an enterprise which leaves everything to nature without intervention. Rather it is a human activity in which the farmer attempts to integrate certain agroecological factors and production inputs for optimum crop and livestock production. Thus, it is reasonable to assume that farmers should be interested in ways and means of controlling beneficial soil microorganisms as an important component of the agricultural environment. Agricultural crop growth and development are closely related to the nature of the soil microflora, especially those in close proximity to plant roots, i.e the rhizosphere. Thus, it will be difficult to overcome the limitations of conventional agricultural technologies without controlling soil microorganisms. This particular tenet is further reinforced because the evolution of most forms of life on earth and their environments are sustained by microorganisms. Most biological activities are influenced by the state of these invisible, minuscule units of life. Presently an area that appears to hold the greatest promise for technological advances in crop / tree production, protection and natural resource conservation is that of beneficial, effective and stress tolerant microbial strains applied as soil, plant and environmental inoculants. Therefore, to significantly increase sustainable agricultural production, with the help of low cost environmental friendly microbial inoculants, it is essential to identify / search / develop microbial strains with improved genetic capabilities and with a higher level of environmental competitiveness, particularly under stressed conditions. Hence, this study is planned with the following objectives

Objectives

- Isolation of crop beneficial from rainfed stressed environments
- Selection of efficient isolates capable to survive under stressed environments in order to formulate bio inoculants for rainfed areas like Chhattisgarh
- Creation of national level respectively on microbes from rainfed stressed environments.

MATERIALS AND METHODS

Field survey cum preliminary microbiology laboratory studies were conducted for population dynamics and isolation of soil microflora in order to identify stress tolerant beneficial microbes (*Rhizobium*, *Azospirillum*, *Azotobacter*, PSB and PGPRs etc) from soil samples of Chhattisgarh region, which is exposed to extreme dry and hot climate during summer.

Collection and preservation of soil samples

A total of 687 soil samples from different depth (0.5, 5-10, 10-15 cm) were collected mostly at 10 days interval from different parts of Chhattisgarh during summer and monsoon seasons (Name of villages, Pirda, Dharampura, Serikheri, Labhandi, Phundar, Nakti, Suddu, Termri, Nakta, Ahiwara, Semair, Kurud, Damiya, Tanakhar, Bansari, Kapnavapara, Kotma, Murga, Salhi, Ajirima, Amapara, Menpat, Nagadand, Navparakala, Sapandar, Ropakhat, Tiger point, Kotmernapur, Nandanakot, Jajag, Fukena, Bagudega, Boirdader, Badpali, Tilga, Bhagora, Chotebhandar, Ambagarh, Basna *etc.*). About 150 –200 gm soil was kept in polythene bags to prevent the moisture loss. These samples were properly tagged, sealed and stored in refrigerators for the isolation and identification of stress tolerant *Azospirillum*, PSB, *Rhizobium*, *Azotobactor*, PGPRs *etc.* These samples were also preserved for determinations of their pH and other physico-chemical characteristics.

Isolation of crop beneficial microbes

In addition to already available 3408 representative soil samples of Chhattisgarh, 687 soil samples were recently collected from this region during summer and monsoon seasons for isolation and identification of stress tolerant crop beneficial effective microbes. For isolation of microbe, serial dilution was prepared by taking 10 gm of soil in 90 ml of sterilized water (Cappuccino and Sherman, 1992). Plating was done separately for each micro-organism by using specified selective medium like YEMA (Fred *et al.* 1932) for *Rhizobium*, Malate medium (Okon *et al.* 1977) for *Azospirillum*, Jenson's medium (Jenson, 1954) for *Azotobacter* and Pikovaskya medium (Pikovaskya, 1948) for PSB and Thronton's agar medium (Okon *et al.*, 1977) for *Azospirillum*, Jenson's medium (Jenson, 1954) for *Azotobacter* and Pikovaskya medium (Pikovaskya, 1948) for PSB and Thronton's agar medium for PGPRs *etc.* Aliquots of soil microbes – water suspension were inoculated on the agar plate containing selective medium. After inoculation, Petri plates were incubated in BOD incubator at 28 ± 2°C for 4 -15 days depending upon the appearance of new specific bacterial colonies. Number of pure colonies were counted individually for each microorganism especially for their isolation and population density under stressed environments prevailing in rainfed Chhattisgarh region. Further, isolated crop beneficial microbes were tested for their stress tolerant behaviour.

Studies under stressed environments

Rhizobium isolates from high temperature and low moisture affected top soil (0.5 cm depth) were tested for their growth / survival at high temperature, low pH and high salt concentration in order to select microbial strains capable to work in stress environments of rainfed regions like Chhattisgarh.

(a) Thermal tolerance

Sixty two *Rhizobium* sp. of different legumes (broth cultures) were subjected to 45,50 and 55° C for 10- 30 minutes at water bath. After thermal shocks when broth culture cool down up to room temperature, each isolate was (loopful) inoculated into petri plates containing specified medium with four replication. Control was also maintained for comparison. All the inoculated petri plates were incubacted in BOD incubator at 28± 2°C for 2-10 days (Benson, 1990). Observations were recorded for survival and / or growth of inoculum. Promising isolates were repeated for their confirmation.

(b) Moisture tolerance

Isolates obtained from top soils of Chhattisgarh region which is exposed to extreme dry and hot climate during summer (Table 1) can be considered as temperature and moisture stress tolerant isolates for further detailed studies.

Table 1. Weather conditions at Raipur during summer and monsoon seasons

Weeks	Weekly average weather**			
	Soil Temp. at 5cm depth	Soil Moisture (%) of top 5 cm depth		Rainfall (mm)
		Light soil	Heavy soil	
1-7 April 2 K	43.87	-	-	0.00
8-14 April 2K	46.53	-	-	0.00
15-21 April 2K	46.23	1.19-3.22	1.54-3.45	0.00
22-28 April 2K	47.43	0.71-2.15	0.87-2.82	0.00
29-5 May 2K	47.50	0.60-1.15	0.73-1.44	0.00
6-12 May 2K	44.84	1.07-3.50	1.11-4.00	2.40
13-19 May 2K	46.60	0.69-2.30	0.80-2.17	0.00
20-26 May 2K	46.86	0.60-1.13	0.66-1.40	0.26
27 May – 3 June 2K	49.97	-		0.00
4-10 June 2K	39.02	-	-	5.37
1-7 July 2K	39.05	-	-	5.72
8-14 July 2K	32.48	-	16.18-23.35	11.32
15-21 July 2K	28.78	17.01-26.51	22.35-29.00	16.75
22-28 July 2K	32.38		-	12.60

** Data presented in Table indicate that very high degree of soil exposure to extreme dry and hot climate during summer resulting in severe loss of mesophilic crop beneficial microbes. Hence, identification of stress tolerant microbes is certainly useful in order to formulate those cultures which are able to survive / persist for longer periods in such conditions.

(c) Acidity

Isolates of *Rhizobium* were inoculated on specific agar medium having adjusted the pH to 4.5, 5.0, 5.5, 6.0 and 6.5. Four replications were maintained along with control. Observations for survival and / or growth of inoculum were started after 48 hrs of incubation. Promising isolates were repeated for their confirmation.

(d) Salt tolerance

Isolates of *Rhizobium* were inoculated on specific agar medium containing 10,000, 20,000, 30,000, 40,000 and 1,00,000 ppm salt concentration (NaCl). Four replications of the plates for each isolate were maintained along with control. After 48 hrs of incubation, observations for survival and / or growth of inoculum were recorded (Benson, 1990). Promising isolates were repeated by tested for their confirmations.

Surface sterilization of seeds

Uniform size seeds of different legumes were rinsed with 95% ethanol and then immersed in 0.1% mercuric chloride solution. The Seeds were, then washed thoroughly with double distilled sterilized water for at least five times before sowing along with selected bacterial isolates for different types of testing.

Cross inoculation

Different leguminous plants were raised from seeds inoculated with broth cultures of recently isolated stress tolerant *Rhizobium* isolates by using N free sand culture technique (Vincent, 1970 and Subba Rao, 1988) for identification of their homologous macro-symbiont (leguminous plants). Timely irrigation to all the cups was given by N free Mcknight seedling nutrient solution as and when required.

During second part of this study, moong plant has been identified as a macro symbiont of one of the recently identified stress tolerant *Rhizobium* isolates, which was isolated during summer season, 2000 from top soil sample (0.5 cm depth) of Chhattisgarh region. This studies is being continued for identification of macro symbiont of other stress tolerant especially *Rhizobium* isolates. Further, effective isolates will be compared with that of available national and international mesophilic checks of specific crops.

RESULTS AND DISCUSSION

The present investigation is an important part of study which is being carried out in the Department of Soil Science, IGKV, Raipur in order to formulate those crop beneficial bioinoculants (cultures) which can give more survival and effectiveness under climatic conditions of rainfed regions like Chhattisgarh which is exposed to extreme dry and hot climate during summer (Table 1).

(a) Microbial population

During isolation of soil microbes, population density of crop beneficial bacteria (*Rhizobium*, *Azospirillum*, *Azotobacter* and PSB) was found to be much lower in pre monsoon (summer) soil samples than population in soil samples of monsoon season. On the basis of qualitative observations during isolation work under this study, it can be concluded that variation in population density of symbiotic microbes (Soil *Rhizobium*) with respect to seasons much higher than non symbiotic microbes. Some of the soil samples are to be analysed at lower dilutions to obtain colonies of above refered beneficial bacteria for their isolation, quantitative estimation and testing under stressed environments.

(b) Behaviour of newly isolated *Rhizobium* isolates

Study related to stress tolerant behaviour of crop beneficial microbes revealed that among recently isolated *Rhizobium* isolates from temperature and moisture affected top soils (0.5 cm depth), 9 showed thermal tolerance

(55°C : Table 2) and 8 showed acidity tolerance (pH 4.5 : Table 3). Most of the above referred temperature tolerant isolates can also be considered moisture stress tolerant isolates because they were isolated from extremely dry and hot soil conditions (soil moisture level below 3.5% and temperature above 45°C) as mentioned in Table 1,2,4. In addition to above, 11 *Rhizobium* salt tolerant isolates have also been identified and observed their survival at 40,000 ppm under controlled conditions (Table 5).

(c) Efforts to create repository of crop beneficial microbes related to stressed environments of rainfed region.

Efforts are being continued for creation of a national level repository on crop beneficial microbes from stressed environments. In this connection, recently added germplasm of about 1500 bacterial isolates including *Rhizobium*, *Azospirillum*, *Azotobacter*, PSB and PGPR etc. in already available huge germ plasm of crop beneficial microbes including *Rhizobium*.

Table 2. Growth of temperature tolerant *Rhizobium* isolates after thermal shocks for different duration

Sl. No.	Isolates No.	Temperature (°C)												Control
		45				50				55				
		10 min		30 min		10 min		30 min		10 min		30 min		
		1	2	1	2	1	2	1	2	1	2	1	2	
1.	3450	+	+	+	+	+	+	+	+	+	+	+	+	+
2.	3468	+	+	+	+	+	+	+	+	+	+	+	+	+
3.	3475	+	+	+	+	+	+	+	+	+	+	+	+	+
4.	3486	+	+	+	+	+	+	+	+	+	+	+	+	+
5.	3498	+	+	+	+	+	+	+	+	+	+	+	+	+
6.	3518	+	+	+	+	+	+	+	+	+	+	+	+	+
7.	3521	+	+	+	+	+	+	+	+	+	+	+	+	+
8.	3535	+	+	+	+	+	+	+	+	+	+	+	+	+
9.	3654	+	+	+	+	+	+	+	+	+	+	+	+	+

Table 3. Growth of acidity tolerant *Rhizobium* isolates at low pH

S. No.	Isolate No.	pH									
		4.5		5.0		5.5		6.0		6.5	
		1	2	1	2	1	2	1	2	1	2
1.	3460 R-A	+	+	+	+	+	+	+	+	+	+
2.	3475 R-A	+	+	+	+	+	+	+	+	+	+
3.	3486 R-A	+	+	+	+	+	+	+	+	+	+
4.	3513 R-A	+	+	+	+	+	+	+	+	+	+
5.	3514 R-A	+	+	+	+	+	+	+	+	+	+
6.	3518 R-A	+	+	+	+	+	+	+	+	+	+
7.	3520 R-A	+	+	+	+	+	+	+	+	+	+
8.	3653 R-A	+	+	+	+	+	+	+	+	+	+

Table 4. Survival of *Rhizobium* isolates at high soil temperature and low moisture environments of Chhattisgarh

S. No.	Isolates No	Date of soil sample collection	Moisture (%) in soil sample during the isolation	Soil temp. at 5 cm depth (°C)
1.	3450 R-A	27.4.2000	0.71	49.0
2.	3468 R-A	1.5.2000	0.73	45.5
3.	3475 R-A	2.5.2000	0.63	49.8
4.	3486 R-A	8.5.2000	3.08	43.0
5.	3498 R-A	12.5.2000	2.61	44.4
6.	3518 R-A	20.5.2000	1.10	46.4
7.	3521 R-A	20.5.2000	1.12	46.8
8.	3535 R-A	23.5.2000	1.02	45.6

Table 5. Survival of *Rhizobium* isolates at 40,000 ppm salt concentration

S. No.	Isolate No.	Salt concentration (ppm)										Control
		10,000		20,000		30,000		40,000		1,00,000		
		1	2	1	2	1	2	1	2	1	2	
1	3437 R -A	+	+	+	+	+	+	+	+	-	-	+
2	3438 R -A	+	+	+	+	+	+	+	+	-	-	+
3	3439 R -A	+	+	+	+	+	+	+	+	-	-	+
4	3460 R -A	+	+	+	+	+	+	+	+	-	-	+
5	3486 R -A	+	+	+	+	+	+	+	+	-	-	+
6	3494 R -A	+	+	+	+	+	+	+	+	-	-	+
7	3495 R -A	+	+	+	+	+	+	+	+	-	-	+
8	3508 R -A	+	+	+	+	+	+	+	+	-	-	+
9	3509 R -A	+	+	+	+	+	+	+	+	-	-	+
10	3510 R -A	+	+	+	+	+	+	+	+	-	-	+
11	3652 R -A	+	+	+	+	+	+	+	+	-	-	+

Similar observations were recorded by Segura (1995) screening of acidity tolerant *Rhizobium* strains. He observed out of fifty, five strains of *Rhizobium* tolerance to acidity up to pH 4.5 in liquid and solid culture, while, Mpepereki *et al.* 1997, observed their growth at pH 4.0 and salt tolerance (NaCl) up to 55,000 ppm, Surange *et al.* 1977 they isolated *Rhizobium* strain from *Albizia lebbeck* which showed its survival at 50°C while *Rhizobium* strains isolated from *Sesbania formosa*, *Acacia farnesiana* and *Dalbergia sissoo* were well adapted to grow on pH 12.0. All the *Rhizobium* strains tolerate salt concentrations up to 5.0% (50,000 ppm).

Findings of this investigation is also close to observations of Higa (1991) who clearly supported the view of this study that "Once the stress tolerant new microbial strain(s) is established and stabilized, the desired effects will continue indefinitely and no further applications are necessary unless organic amendments cease to be applied, or the soil is subjected to severe drought of flooding." Hence, identification of stress tolerant crop beneficial microbes is certainly useful in order to formulate those cultures which are able to survive/persist for longer period and work more efficiently under climatic conditions of Chhattisgarh plains.

REFERENCES

Benson, H. J. 1990. Temperature lethal effects. In : **Microbiological applications.** pp.125 –126.

Benson, H. J. 1990. Osmotic pressure and bacterial growth. In: **Microbiological applications.** p.127.

Cappuccino, J. G. and N. Sherman. 1992.Microbial population in soil. Enumeration. In: **Microbiology A Laboratory manual,** pp.315-317.

Guptha, S. B. and P. K. Chhonkar. 1995. Effective utilization of phosphours in rice wheat cropping system in a Vertisol through VA – mycorrhiza and phosphorus solubilizer. **Proc. of the 60 Annual Conversion of the ISSS,** held during 02-05 Nov. 95 at PAU Ludhiana, p. 82.

Guptha, S. B., R. K. Katre, P. K. Chhonkar, M. K. Guptha and M. L Adil.2000. Significant research and development with reference to biofertilizers in Chharrisgarh. **Fert. News** 45 (11) : 35-40.

Higa,T. 1991. Effective microorganisms. A biotechnology for mankind. In: **Proce. of the First International Confront of Kyuse Nature Farming.** USDA, Washington, D. C., USA, pp. 8-14.

Katre, R. K., M. L. Adil and S. B. Guptha. 1997. Department of Biotechnology, New Delhi sponsored project report submitted by Department of Soil Science, I. G. A. U. Rajpur.

Kumar, A. and M. Srivastava. 1994. Survey of local rhizobial isolates for their efficiency in nitrogen fixation and biomass production in *Vigna mungo* (L.) **Indian J. Plant Phyiol. Vol XXXVII. No. 4** pp.264 – 255.

Mpepereki, S., F. Makonese and A. G. Wollum.1977. Physiological characterization of indigenous rhizobia nodulating *Vigna unguiculata* in Zimbabwean soils. **Symbiosis – Rehot. 1997, 22:3,** 275 – 292.

Seguta, R. C. 1995. Selecting *in vitro* of *Rhizobium* for tolerance to acidity and aluminium. **Agronomia- Colombiana, 1995. 12:2,** 142-148.

Sinha, S. B. and G. P. Gupta. 1985. Soils of Mathya Pradesh and their management. In: **Soils of India,** Fertilizer Association of India, New Delhi. pp.225-249.

Surange, S., A. G. Wollum, N. Kumar and C. S. Nautiyal. 1977. Characterization of *Rhizobium* from root nodules of leguminous tress growing in alkaline soils. **Canadian Journal of Microbiology, 43:** 9,891-894.

Vincent, J. M. 1970. **A manual for the practical sutdy of the root nodule bacteria.** IBP. Hand Book No. 15. Blackwell Scientific Publication, Oxford and Edinburgh.

<u>CHAPTER – 30</u>

FIELD EVALUATION OF LIQUID AND CARRIER BASED RHIZOBIAL INOCULANTS IN REDGRAM

S. Gunasekaran, D. Balachandar and K. Mohana Sundaram

ABSTRACT

A field experiment was conducted to study the influence of the commercially available carrier based rhizobial inoculants and liquid rhizobial inoculant (UAS, Bangalore) on nodulation and grain yield of redgram (Vamban 1). The results of the experiments clearly indicated that inoculation of TNAU rhizobial culture CRR 6 had given the maximum nodules (14 nodules / plant), plant dry weight and grain yield (875 kg/ha), which was 28.48 per cent higher over the uninoculated control (681 kg/ha). Among the commercially available rhizobial cultures, the inoculant supplied by the Department of Agriculture, Govt. of Tamil Nadu had given a grain yield of 763 kg/ha but was on par with other commercially available cultures where as liquid inoculant had recorded a grain yield of 735 kg/ha only.

INTRODUCTION

The modern day intensive crop cultivation requires the use of chemical fertilizers but fertilizers are not only in short supply but they are becoming expensive. Therefore chemical fertilizers are being supplemented with biofertilizers of microbial origin. The role of legumes in enriching the fertility of soil was known through centuries. Pulse crops have been an important component of agriculture since ancient times. These leguminous plants in association with symbiotic bacterium, *Rhizobium* forms nodules which fixes the atmospheric nitrogen, providing an alternative energy to the chemical nitrogenous fertilizers.

The quality of rhizobial inoculants is great importance in ensuring the field performance of inoculant industry. Basically quality means the presence of right type of microorganisms in active form and desired numbers (Hiltbold *et al.*, 1980), to survive and form effective nodules for nitrogen fixation.

Hence a field experiment was conducted with different rhizobial inoculants obtained from various manufactures along with liquid inoculant to study the quality of the cultures in their effectiveness in nodulation, nitrogen fixation and grain yield of redgram.

MATERIALS AND METHODS

A field experiment was conducted at National Pulses Research Centre, Vamban with the following treatments.

1. Uninoculated control
2. Carrier based rhizobial inoculant TNAU culture CRR – 6.
3. Carrier based commercial inoculant – NIT
4. Carrier based commercial inoculant – Dept. of Agriculture, Govt. of Tamil Nadu.
5. Carrier based commercial inoculant – FACT
6. Liquid inoculant (UAS-Bangalore)

All the commercial inoculants were obtained from the local market. The details of the field experiment are furnished below.

Design	:	RBD
Replication	:	Four
Crop	:	Redgram – Vamban 1
Spacing	:	45 x 20 cm
Basal dressing	:	50 kg P_2O_5 / ha.

Before sowing, the seeds were treated uniformly with the rhizobial inoculants and shade dried. For liquid inoculant, the redgram seeds were soaked in the rhizobial broth for 30 min, dried and sown. The initial population of *Rhizobium* in the carrier based inoculant and liquid inoculant were X 10^8 cfu/g and X 10^8 cfu /ml respectively. The nodule number, nodule dry weight, plant dry weight were recorded at 45 DAS and grain yield was recorded at harvest.

RESULTS AND DISCUSSION

The results indicated that there was significant differences among the treatments. The uninoculated control had recorded the lowest yield (681 kg/ha) and did not form any nodules. The TNAU rhizobial inoculant had recorded the maximum nodules (5.6 nodules/plant), nodule dry weight (14.36 mg/pl) and grain yield 875 kg/ha which was 28.48 per cent higher than the uninoculated control. Among the commercially available rhizobial inoculant, there were no significant differences among the cultures, although the rhizobial culture obtained from the Dept. of Agriculture had given the higher grain yield (763 kg/ha) and nodulation (Table 1 and Fig. 1).

Table 1. Evaluation of liquid and carrier based *Rhizobium* inoculants in redgram (var. Vamban) under field condition

Sl. No.	Treatments	Nodule No./ plant	Nodule dry weight (mg/pl.)	Plant dry weight (g/pl.)	Plant height (cm/ pl)	Grain yield (kg/ha)	Per cent increase over control
1.	Uninoculated control	-	-	14.31	68.1	681	-
2.	Carrier based inoculant of local best *Rhizobium* (CC 1)	5.60	14.36	18.00	85.6	875	28.48
3.	Commercial carrier based *Rhizobium* – (Dept. of Agriculture)	3.20	10.28	15.50	80.1	763	12.04
4.	CCR 1 – FACT *Rhizobium*	2.10	8.21	14.36	68.7	715	4.99
5.	CCRI – NIT – *Rhizobium*	1.00	2.10	14.10	72.7	759	11.45
6.	Liquid inoculant (UAS – Bangalore)	2.33	3.66	14.00	70.9	735	7.92
	CD (P=0.05)	NS	NS	0.23	8.03	66.72	-

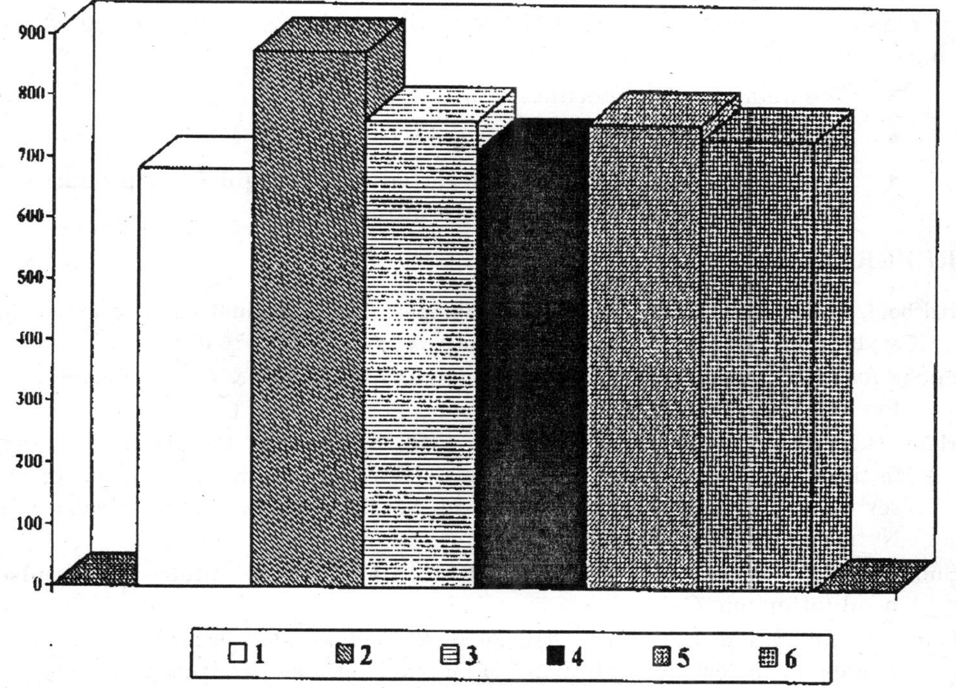

Fig. 1. Evaluation of liquid and carrier based *Rhizobium* inoculants on the grain yield in redgram

The liquid inoculant had given the lowest yield (735 kg/ha) than the carrier based inoculants. The plant dry weight and nodulation was also poor compared to other inoculants. When broth is directly used for inoculation large quantities of fresh broth is necessary to obtain good nodulation comparable to that of peat based inoculants. Secondly, the agar and broth cultures are inferior to peat based inoculants because peat provides the natural protection to rhizobia and thus enhances its survival on seeds (Subba Rao, 1988).

The quality of cultures vary between individual manufacturers. Many of the manufactures are not strictly following the BIS specifications. Failure to obtain the desired response may be due to (1) presence of native ineffective strains which could not be displaced by the introduced strains (2) presence of antagonistic bacteria which minimizes the number of rhizobia in the rhizosphere and (3) the availability of soil conditions which limit symbiosis caused by acidity, alkalinity *etc.* (Hem *et al.*, 1976, Subba Rao and Tilak, 1977).

The TNAU rhizobial inoculant CRR-6 has been found to form effective nodules in redgram at variying agroclimatic zones (Gunasekaran and Balachandar, 2000). Keyser *et al.* (1992) reviewed the characters required for *Rhizobium* to be used as inoculant.

Considerable input of biologically fixed N can be achieved in all agricultural ecosystems through the activity of symbiotic associations. For increasing pulse crop yield through biofertilizers the following strategies are suggested (Kannaiyan *et al.*, 2000).

- The quality of the inoculants
- Effective inoculant delivery systems.
- For success of biofertilizer, concerted efforts right from production, demonstration to distribution is required.

REFERENCES

Hiltbold, A.E., D.L. Thurlow and H.D. Skipper. 1980. Evaluation of commercial soybean inoculants by various techniques. **Agron. J., 72**: 674-681.

Subba Rao, N.S. 1988. **Biofertilizers in Agriculture.** Oxford & IPH publishing Co., Pvt. Ltd., New Delhi.

Hem, G.E., R. J. Lawn and W. A. Brun. 1976. Influence of inoculation, nitrogen fertilizer and photosynthetic sources – Sink manipulation of field grown soybeans. pp. 239-253. In: **Symbiotic N_2 fixation in Plants**, (eds.) P.S. Nutman, Combridge Univ. Press.

Subba Rao, N.S. and K.V.B.R. Tilak. 1977. **Rhizobial cultures – their role in pulse production.** pp. 31-34.

Gunasekaran, S. and D. Balachander, 2000. *Rhizobium* and phosphobacteria – An avenue for increasing productivity of pulses. In: **Pulse production strategies in Tamil Nadu**. (eds.) Prof. Dr. S. Kannaiyan, *et al.*, TNAU, CBE. p. 103-122.

Kannaiyan, S., K. Govindarajan, K. Kumar and K. Chendrayan, 2000. Use of biofertilizer for increasing pulse production. In: **Pulse production strategies in Tamil Nadu** (eds.) Prof. Dr. S. Kannaiyan, *et al.*, TNAU, CBE. p. 8-23.

CHAPTER - 31

STUDIES ON SYNERGISM BETWEEN *Rhizobium*, PLANT GROWTH PROMOTING RHIZOBACTERIA (PGPR) AND PHOSPHATE SOLUBILIZING BACTERIA IN BLACKGRAM

S. Gunasekaran, D. Balachandar and K. Mohana Sundaram

ABSTRACT

A field trial was conducted to study the synergism between *Rhizobium* (COC.10), Plant growth Promoting Rhizobacteria (PGPR) (*Pseudomonas* KB-133) and phosphate solubilizing bacteria (*Bacillus megaterium*) on nodulation and yield of blackgram (Vamban 1). Inoculation of *Rhizobium* alone had given a grain yield of 655 kg ha^{-1} whereas uninoculated control had recorded 472 kg ha^{-1} only. Combination of all the three organisms (*Rhizobium* + PGPR + PSB) had recorded the maximum nodules, plant biomass and grain yield (760 kg ha^{-1}) which was 61.0 per cent higher than individual inoculation of *Rhizobium* alone.

INTRODUCTION

Pulse crops are an important component of agriculture since ancient times. These leguminous plants on symbiotic association with *Rhizobium* fixes atmospheric nitrogen, accounted for about 40 per cent of the 175 million tones of N/year of the annual average global nitrogen fixation. It is clearly revealed that application of *Rhizobium* as seed inoculant is the most essential component of in the INM system of pulses.

The phosphate solubilizing bacteria play a major role in the solubilization and uptake of native and soil phosphorus. It accelerates tillering, flower initiation, good pod and seed setting. The bacteria *Bacillus megaterium*, *B. polymixa*, *Pseudomonas striata*, *Streptomyces* are the few important phosphobacteria commonly available in soil.

The plant rhizosphere is an important soil ecological environment for plant microbe interaction and involves colonization by a variety of microbes in and around the roots. These root colonizing bacteria that stimulate plant growth are generally referred as Plant Growth Promoting Rhizobacteria (PGPR) or Plant Health Promoting Rhizobacteria (PHPR) or Yield Increasing Bacteria (YIB) (Piao *et al.*, 1992). The PGPR includes, the genera of *Acetobacter, Azotobacter, Azospirillum, Bacillus, Bradyrhizobium, Pseudomonas, Rhizobium, Serratia* and *Xanthomonas* (Kloeppe *et al.*, 1980).

Nitrogen, phosphorus and other nutrients are needed for increased growth and better nodulation in pulses. Hence, an experiment was designed to investigate the influence of the synergism between the *Rhizobium*, PGPR and PSB on pulses, in increasing the nutrient availability and grain yield.

MATERIALS AND METHODS

A field experiment was conducted at National Pulses Research Centre, Vamban with the following treatments.

1. Uninoculated control
2. *Rhizobium* (COC. 10)
3. PSB (*Bacillus megaterium*)
4. PGPR (*Pseudomonas* KB 133) obtained from CCSHAU, HISAR.
5. *Rhizobium* + PSB
6. *Rhizobium* + PGPR
7. PGPR + PSB
8. *Rhizobium* + PSB + PGPR

The treatments were replicated thrice with randomized block design. A basal dressing of 40 kg P_2O_5/ha was supplied. The plots size were 4 x 3 m having a spacing of 30 x 10 cm. The nodule number, nodule dry weight, plant height, plant biomass and grain yield were recorded. The blackgram variety Vamban 1 was chosen for their experimental study. The soil pH was 5.8.

RESULTS AND DISCUSSION

The results of the study clearly indicated that combined inoculation of all the three organisms, *Rhizobium* PSB and PGPR was significantly different from other treatments. Inoculation of COC.10 rhizobial strain had significantly enhanced the grain yield (655 kg ha^{-1}), nodule number (14.0 no./plant) and nodule dry weight (83.6 mg/plant). Inoculation of *Rhizobium* with PSB or PGPR had also increased the nodule number, nodule dry weight, plant biomass and grain yield (612 kg/ha with PSB and 702 kg / ha with PGPR) significantly over individual inoculation of *Rhizobium* alone, where as the combination of all the three organisms had given the maximum nodules, biomass (28.3 / pl.) and grain yield (760 kg/ha) (Table 1 and Fig. 1).

Table 1. Synergism between *Rhizobium,* PGPR and PSB organisms

Sl. No.	Treatments	Plant height (cm/pl)	No. of nodules / plant	Nodule dry weight (mg/pl)	Plant biomass (g/pl)	Grain yield (kg/ha)	Per cent increase over control
1.	Uninoculated control	32.4	10	69.4	13.3	472	-
2.	*Rhizobium* (COC 10)	29.5	14	83.6	19.3	655	38.7
3.	PSB (*B. megaterium*)	31.9	13	80.4	18.3	483	2.3
4.	PGPR (Pseudomonas KB 133)	29.3	14	81.2	28.3	616	30.5
5.	*Rhizobium* + PSB	29.8	13	79.8	25.0	612	29.6
6.	*Rhizobium* + KB 133	30.9	14	83.2	15.0	702	48.7
7.	PSB + KB 133	31.8	12	76.2	25.0	483	2.3
8	*Rhizobium* + PSB + KB 133	37.3	19	91.4	28.3	760	61.0
	CD (P=0.05)	2.3	14	2.8	2.3	36.6	

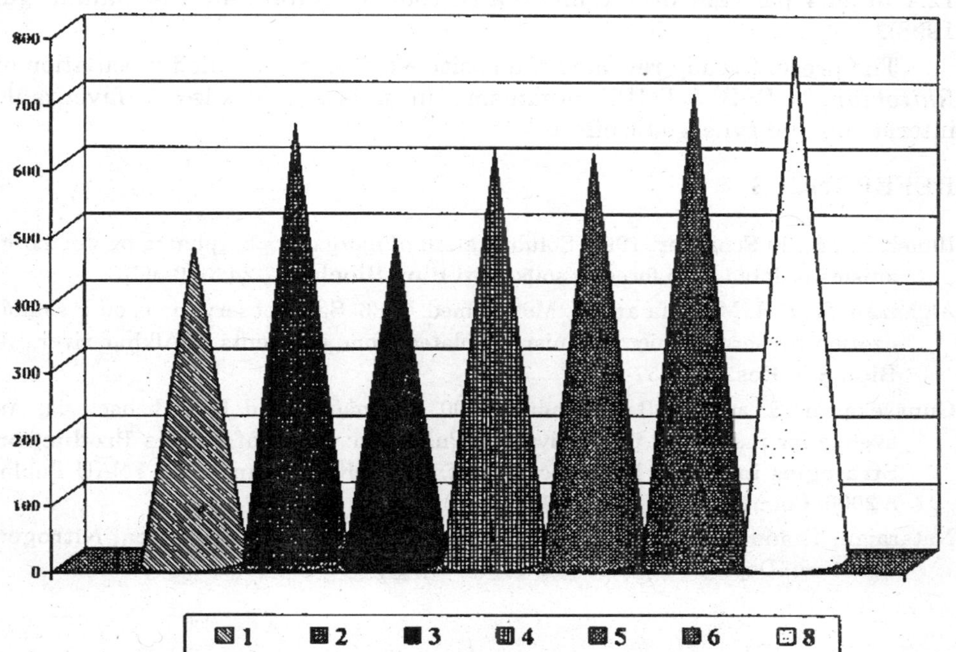

Fig. 1. Synergism between *Rhizobium,* PGPR & PSB organisms on the grain yield in blackgram

Trials conducted at various stations / places clearly indicated that combined inoculation of *Rhizobium* and PSB enhanced the grain yield of pulses (Gunasekaran and Balachandar, 2000). The production of organic acids such as citric, fumaric, malic etc. (Illmer and Schinner, 1992) and also the phosphatase enzyme activity accounted for the solubilization of P (Alghazali *et al.,* 1986), i.e the unavailable form of P is converted released and available form of P. The phosphorus is required for active of nitrogen fixation by the nitrogenare enzyme. Apart from the positive influence on N_2 fixation by the *Rhizobium,* P is also needed for flower initiation and pod setting.

The combination of all the three organisms had a cumulative influence on the plant. Among the PGPR organism, *Pseudomonas* are the early root colonizers, which by producing variety of its secondary metabolites reported to stimulate the growth of cultivated crops.

Through biological nitrogen fixation by *Rhizobium* and solubilization of insoluble phosphates by phosphate solubilizers, more nutrients are made available for crop plants by the root associated organisms. The compatible existence of nitrogen fixing, P solubilizing and plant growth promoting organisms (produce IAA, Gibberellins, Cytokinins and cause reduction of plant diseases) enhanced the nodulation, growth, nitrogen fixation and nutrient uptake of pulses. The field trials conducted revealed that seed inoculation of different varieties of soybean with efficient strains of *Rhizobium / Brady rhizobium* and PSB or PGPR enhanced the seed yield by 12.4 to 34.4 per cent over uninoculated control (Natarajan and Ilamurugu, 1998).

The present study revealed the positive effect of combined inoculation of *Rhizobium* / PSB / PGPR organisms in pulses and also a favourable interaction and synergistic effect.

REFERENCES

Illmer, P. and F. Schinner. 1992. Solubilization of inorganic phosphates by microorganisms isolated from forestry soils. **Soil Biol. Biochem., 24**: 389-395.

Alghazali, R., S.H. Mustafa and K. Mohammed. 1986. Same observations on P solubilization by aerobic microorganisms isolated from sediments of Alkhar river., **J. Biol. Sci. Res., 47:** 157-172.

Gunasekaran, S. and D. Balachandar. 2000. *Rhizobium* and Phosphobacteria: An avenue for increasing productivity in Pulses. In: **Proc. of Pulses Production Strategies in Tamil Nadu.** (ed.,) Prof. Dr. S. Kannaiyan *et al.*, TNAU Publn. 8/2000. Coimbatore 3. pp. 103-122.

Natarajan, T. and K. Ilamurgu, 1998. **Annual Report of the Biological Nitrogen Fixation**. Dept. of Agricultural Microbiology, TNAU. Coimbatore-3.

SECTION – V

ENDOPHYTIC DIAZOTROPHS

CHAPTER – 32

HARNESSING STEM NODULE NITROGEN FIXING SYSTEM FOR LOWLAND RICE

M.D. Sundaram

ABSTRACT

Aeschynomene and *Sesbania rostrata* are better candidates for wet land rice cultivation. The *in situ* incorporation of *S. rostrata* after a period of 60 days of growth contributes 267 kg N ha^{-1} of which one-third is transferred to rice crop and the remainder is left behind. The practice of green manuring with stem nodulating legumes possesses additional advantage over nodulating legumes in their twin abilities to harness atmospheric nitrogen in both stem as well as root nodules.

At the level of fundamental research, the stem nodules are convenient system to investigate the interrelationship between photosynthesis and biological nitrogen fixation. The cortex of the stem nodule is green due to chlorophyll II and the bacteroid tissue is pink possibly due to leghaemoglobin. The transfer of carbon skeletons from the cortex bacteroid tissue for the process of nitrogen fixation can easily be traced and measured with accuracy by the use of labelled carbon.

INTRODUCTION

Rice based cropping system include pulse legumes like black gram, green gram and cowpea; oil seed legumes like groundnut and soybean and non-legumes like sesame and sunflower; green manure legumes like sesbania, daincha, sunnhemp, wild indigo and Indigo. The legume BNF system in the rice based cropping system has got potential in recycling not only nitrogen but also on the dissolution of various plant nutrients.

Cauvery deltaic region of Tamil Nadu comprises of Thanjavur, Nagapattinam and parts of South Arcot district and is well known as rice bowl of the state. There are about 2.0 M ha where rice is grown under monoculture succeeded by legume cultivation. The nitrogen is the most limiting nutrient for rice production under lowland rice ecosystem.

Stem nodule legumes are potential candidates of BNF system for nitrogen enrichment in the Cauvery delta. Harnessing of stem nodule nitrogen fixing system meets 25-50 per cent of the nitrogen requirement of rice.

Legume - *Rhizobium* symbiosis and the resultant N_2 fixation are generally associated with root nodules. But there are a few legumes like *Sesbania, Aeschynomene* and *Neptunia* where nodules are formed on the stem also and N_2 fixation occurs.

Fifteen species of *Aeschynomene* are known to develop stem nodules (Alazard, 1985). *Sesbania rostrata* (Dreyfus and Dommergues, 1981) and *S. punctata* are also known to produce stem nodules. The biological and agricultural significance of *Neptunia* has to be studied in detail. *N. oleraceae* L. commonly found in lakes and swamps of deltaic regions of Tamil Nadu (Sundaram, 1991).

MATERIALS AND METHODS

The performance of *Aeschynomene indica, Sesbania rostrata* and *Neptunia oleraceae* under rice field condition was thoroughly studied for its effect on plant growth and yield and for soil enrichment (Dreyfus and Dommergues; 1984, Latha and Garcia, 1989 and Sundaram, 1991).

RESULTS AND DISCUSSION

Stem nodule legumes have distinct advantage over other legume where the competition with other rhizobia (native) may be either low or non exist. Unlike root nodules, N_2 fixation in stem nodules is not affected by increased concentration of applied fertilizer nitrogen. Further, number of nitrogen fixing nodules can also be increased by repeated application of *Rhizobium* culture. *Sesbania rostrata* is known to fix as much as 250 kg N ha^{-1} in just 60 days time in rice fields. Its high nitrogen fixing ability and succulent nature of the plant tissue at flowering makes it a very ideal green manure crop for rice fields.

Aeschynomene spp. and *Sesbania rostrata* are better candidates in wet land rice cultivation. *S. rostrata* contributes 267 kg N ha^{-1}, about 100 kg N ha^{-1} is utilised by rice crop and the remainder is left in the soil for subsequent utilization and a part is lost by leaching etc.,

Stem nodulating rhizobiaL isolates formed effective nodules on both roots and stems. In *S. rostrata* suited to rice field under aquatic and terrestrial environment, swabbing method of inoculation produced highest stem nodules.

Biomass producing potential, nodule number, nitrogenase activity and nitrogen content of *S.rostrata* was the maximum followed by *S. aculeata* and *S. speciosa*. Chlorophyll content and biomass content was more with *S. rost-*

rata, nitrogen content ARA activity and leghaemoglobin was higher with *N. oleraceae*.

Performance of *S.rostrata* was compared with the *S.rostrata* TSR-1 mutant. The mutant did not flower under Annamalainagar condition. The stem nodule production in the mutant TSR-1 was restricted to its own isolate TAR-1. Application of K upto 60 kg ha^{-1} increased the nitrogenase activity.

Rock phosphate application and inoculation of PB to *S.rostrata* enriched the N and P status of soil and increased the rice yield.

The root nodule isolate of *S. rostrata*, stem and root nodule isolate of *Aeschynomene aspera*, *A.indica* and stem nodule isolate of *Neptunia oleraceae* were characterised as *Rhizobium* sp. The stem nodule isolate of *S.rostrata* was characterised as *Azorhizobium caulinodans*.

Stem nodule rhizobia survive in both the phyllosphere and rhizosphere of root and stem nodule legumes *viz.*, *S.rostrata*, *A.aspera*, *N.oleraceae*, *S.speciosa* and *S.aculeata*. However, *S. rostrata* harboured the maximum population.

The pesticides *viz.*, endosulfan, methyl demeton, metasystox and sevin, the fungicides *viz.*, allochlor 1.25%, oxyfluorfen 0.15% and nitrofen 2.5% inhibited the growth of *Azorhizobium caulinodans*.

The incorporation of *S.rostrata* significantly improved the rice yield and has the capability of supplying all the fertilizer N requirement of rice crop when grown as dual crop with *Azolla microphylla*. The incorporation of *Azolla* and *S.rostrata* had a gradual N release, so that 70 per cent of their N content became available.

Field trials conducted at Annamalai University Experimental Farm, have shown that the stem and root nodulating *S.rostrata* was superior than the root nodulating *S. aculeata* and *S. speciosa* (Table 1) by recording higher biomass, providing large input of nitrogen to the soil within a period by 60 days (Shivakumar and Sundaram, 1989).

The growth and yield of ADT 37 rice at 80 kg N as urea + *S. rostrata* incorporation is comparable with 100 kg N as urea alone and as well as with 100 kg N as urea + *S.rostrata* incorporation (Table 2). Thus a saving of 40 kg inorganic nitrogen is obtained with *S.rostrata* incorporation (Shivakumar and Sundaram, 1989).

Table 1. Studies on the performance of *S. rostrata* in comparison with other species of *Sesbania*

Treatments	Height (cm)	Biomass (g $plant^{-1}$)	Dry matter (g $plant^{-1}$)	N content (%)	Nodule no. $plant^{-1}$		Leghaemoglobin (mg/g nod.)		Nitrogenase activity*	
S. rostrata	200.7	79.2	9.34	3.54	Root	Stem	Root	Stem	Root	Stem
					110	31.0	1.61	1.11	301	266
S. aculeata	182.2	78.8	8.94	2.61	63.8		1.16		275	
S. speciosa	153.0	67.5	7.42	2.20	61.3		1.12		270	

* Micromoles of ethylene produced/g of nodule/hr.

Table 2. Effect of *S. rostrata* incorporation on the growth and yield of ADT 37 rice

Treatments	Height (cm)	Dry matter (g plant^{-1})	N content (%)	Grain yield (kg ha^{-1})	Straw yield (kg ha^{-1})	1000 grain weight (g)
0 kg N	68.2	25.04	1.74	2388	6833	20.14
0 kg N + *S. rostrata*	72.4	27.30	1.79	2555	7050	20.16
20 kg N	68.6	26.20	1.80	2750	7220	20.19
20 kg N + *S. rostrata*	74.8	27.45	1.81	2800	8380	20.45
40 kg N	70.6	26.40	1.83	3080	7450	20.29
40 kg N + *S. rostrata*	74.5	27.45	1.81	3800	8600	20.54
60 kg N	73.0	26.85	1.85	3340	7720	20.33
60 kg N + *S. rostrata*	75.4	27.00	1.85	4555	9000	20.66
80 kg N	73.6	26.85	1.8	3445	7900	20.66
80 kg N + *S. rostrata*	77.2	27.94	1.86	4720	9300	20.96
100 kg N	74.0	26.91	1.85	3680	8225	20.33
100 kg N + *S. rostrata*	75.1	27.00	1.85	4100	8750	20.66

The incorporation of stem nodulating legumes *viz., Neptunia oleraceae, S. rostrata* and *Aeschynomene indica* each separately with 20 kg N ha^{-1} as urea was superior to incorporation of stem nodulating legumes alone as well as 40 kg N ha^{-1} as urea alone in increasing the growth and yield parameter of ADT 36 rice. However, the incorporation of *N. oleraceae* was better than the other two. On the other hand, the dual cropping of *N. oleraceae* with rice at different levels of nitrogen was found to be competitive and reduced the growth and yield parameters of ADT 36 rice (Ranganathan and Sundaram, 1992).

REFERENCES

Alazard, D. 1985. Stem and root nodulation in *Aeschynomene* spp. **Appl. Environ. Microbiol., 50**: 732-734.

Dreyfus, B., D. Alazard and Y.R. Dommergues. 1984. Stem nodulating rhizobia. In: **Current Perspectives in Microbial Ecology**. (eds.) M.C. Klug and C.E.Reddy American Society for Microbiology, Washington, D.C. p. 161-169.

Ladha, J.K., Miyans and M. Garica 1989. *Sesbania rostrata* as green manure for lowland rice: growth, N_2 fixation, *Azorhizobium* sp. inoculation and effects on succeeding crop yields and nitrogen balance. **Biol. Fertil. Soils., 7**: 191-197.

Ranganathan, T.V. and M.D. Sundaram 1992. Comparative performance of stem nodulating legumes on the growth and yield of ADT 36 rice. **Paper presented in the 32nd Annual Conference of Association of Microbiologists of India** held at MKU Madurai, p.30.

Shivakumar, K. and M.D. Sundaram 1989. Studies on certain aspect of *Sesbania rostrata* - *Azorhizobium* symbiosis and effect of *S. rostrata* incorporation on the growth and yield of ADT 37 rice. **Paper presented in International Symposium on Bioinoculants** held at Sri Pushpam College, Poondi, Thanjavur, Tamil Nadu, India, p.25.

Sundaram, M.D. 1991. Studies on *Sesbania rostrata* - *Azorhizobium* symbiosis. **Paper presented in the 31st Annual Conference of AMI** held at TNAU, Coimbatore, p. 125.

CHAPTER – 33

INDUCTION OF LATERAL ROOTLET FORMATION, SYSTEMIC TRANSLOCATION AND ENDOPHYTIC COLONIZATION OF *Azorhizobium caulinodans* ORS 571 – IK – SK M4 BY ITS COMBINED INOCULATION WITH *Agrobacterium rhizogenes*

I. Kannan and S. Kannaiyan

ABSTRACT

The endophytic diazotroph *Azorhizobium caulinodnas* gains entry in to rice through the cracks, created in epidermis during emergence of lateral roots. One of the strategies for extending this approach is to facilitate the congenial conditions for the easy entry of diazotrophic endophytes. *Agrobacterium rhizogenes* hither to known to form tumors in dicots is found to facilitate the smooth entry of *A. caulinodans* in rice. Inoculation of *Agrobacterium rhizogenes* either alone or in combination with *A. caulinodnas* ORS-571 IK-SK-M_4 with triple antibiotic marker induced the formation of lateral roots. Significant increase in the number of lateral roots was seen in the combination of *Agrobacterium rhizogenes* and *A. caulinodans* ORS-571-IK-SK-M_4 compared to individual inoculation of *A. caulinodans*. This had a positive influence on root length, shoot length than any other treatment. Interestingly combined inoculation of *Agrobacterium rhizogenes* and *A. caulinodans* ORS 571 IK-SK-M_4 registered a significantly increased endophytic population of the triple antibiotic resistant marker strain, *A. caulinodans* ORS 571 – IK- SK-M_4 in root, shoot and leaves of ADT 36 rice.

INTRODUCTION

The nitrogen fixed by the conventional BNF systems *viz.*, free living and diazotrophic system is often subject to losses. Hence, an alternative would be the introduction of a viable nitrogen fixing system inside the rice plant. It will be more advantageous than the conventional BNF system, because the nitrogen fixed will be directly available to plants without any losses, which is more economic (Kannaiyan, 2001). This is made possible by the discovery that *Azorhizobium caulinodans* forms nodule like structures in rice roots (Christiansen-Weniger, 1996). The root cracks caused by the splitting of the epidermis at the point of the emergence of the young lateral roots has been found to be the point of entry of endophytic diazotroph like *A. caulinodans* (Cocking *et al.*, 1994). Later findings have shown that the xylem vessels are the sites of colonization by the introduction bacteria (Gopalswamy *et al.*, 2000). So, one of the approaches to increase the endophytic population in rice is to enhance the lateral root formation. This could be made possible by co-inoculating *Agrobacterium rhizogenes* with *A. caulinodans*. *A. rhizogenes* is a members of the family Rhizobiaceae that is known to cause the hairy roots of dicots and its not known to infect monocots. The introduction of *rol* genes of *A. rhizogenes* cause the tillering and dwarf stature of rice. The present investigation was carried out to study the effect of inoculation of the triple antibiotic resistant markers strain *A.c* ORS 571-IK-SK-M_4 and *Agrobacterium rhizogenes* on growth of ADT 36 rice seedlings under hydroponic conditions and systemic translocation and colonization by the introduced endophytic bacterium.

MATERIALS AND METHODS

Method of growing rice seedlings

The seeds of rice (cv. ADT 36) were soaked overnight in water, surface sterilized with 0.1% (w/v) mercuric chloride for 3 min. and washed 4-5 times with sterile distilled water. Surface sterilized paddy seeds (25 nos) were transferred to sterile petridishes containing sterile filter paper that was moistened with sterile distilled water and kept in the dark for germination. Seeds which sprouted well on the 3rd day were placed over a nylon net of hydroponic system (Uma and Kannaiyan, 1996) in 250 ml beakers containing 245 ml of IRRI$^{(-)}$ NO_3 medium (Watanabe, 1977). After 7 days, the roots of the seedlings were immersed in enzyme solutions containing cellulase 1% (w/v), pectinase 0.1% (w/v) either individually or in combination for 10 min at 22°C. For microbial treatment, the roots were immersed in a solution containing equal volume of 25% (w/v) poly ethylene glycol (PEG) with 50 mM $CaCl_2$ and 48 h old cultures of *Agrobacterium rhizogenes* MTCC 532 and the triple antibiotic resistant mutant marker strain *A. caulinodans* ORS 571-IK-SK-M_4 individually and in combination for 25 min at 22°C. The treated seedlings were taken out and the roots of the individual seedlings were washed twice with 8% sterile mannitol solution at one min intervals and transferred to the same hydroponic system. The growth characters of the seedlings such as root length, shoot length, number of lateral rootlets and the endophytic population of antibiotic resistant marker strain *A. caulinodans* ORS 571-IK-SK-M_4 in different parts of the rice plant *viz.*, root, shoot and leaves were estimated on 7,14 and 21 days after inoculation and recorded.

Estimation of endophytic population of antibiotic resistant marker strain *A. caulinodans* ORS 571-IK-SK-M_4 in various plant parts.

One gram of fresh root, stem and leaf samples were collected from the hydroponic system and were surface sterilized with 0.1% (w/v) mercuric chloride for 3 min. and washed repeatedly (3-4 times) with sterile distilled water. The surface sterilized samples were transferred separately to a test tube containing 10 ml of sterile distilled water and crushed by using a sterile blunt end glass rod, and serially diluted upto 10^{-4} dilution. From the appropriate dilution, 0.5ml suspension was poured in sterile petridish under aseptic conditions. Then 15-20 ml of freshly prepared cooled molten YEMA medium containing triple antibiotic *viz.*, chloramphenicol, streptomycin and tetracycling (each 100 μg/ml) was poured and the plates were rotated clockwise and anticlockwise directions for uniform spreading of medium. After solidification, the plates were inverted and incubated at room temperature (28±1°C) for 48 h and the number of colonies were recorded.

RESULTS AND DISCUSSION

Among the different treatments, inoculation of *Agrobacterium rhizogenes* alone recorded the maximum number of lateral rootlets followed by the combined inoculation of *A. rhiozgenes* and *Azorhizobium caulinodans* (Table 1). This could be attributed to the inherent nature of *A. rhizogenes* to cause lateral rootlet proliferation at the point of infection. The subsequent root cracks could serve as points of entries for the *Azorhizobium caulinodans*.

The inoculation of *A. caulinodans* ORS 571-IK-SK-M_4, alone recorded the maximum root length and shoot length (Table 2 and 3). This could be due to production of plant growth promoting substances like IAA by the bacterium. Similar observations have been made by other workers (Cocking *et al.*, 1994; Amutha and Kannaiyan, 2000). The results also indicated that the enzyme treatment decreased the root length and increased the shoot length. This result indicates that the enzyme treatment creates some shock to the rice seedlings. Similar observations have already been recorded (Senthilkumar, 2000; Buvana, 2001).

The combined inoculation of *Agrobacterium rhizogenes* and *Azorhizobium caulinodans* ORS 571-IK-SK-M_4 showed the highest endophytic population of antibiotic resistant markers strains in all parts of the rice seedlings *viz.*, root, stem and leaves and in all the day interval at which observations were recorded (Fig. 1,2 and 3). This could be due to the fact that the induction of more number of lateral roots by these two bacterium results in more number of epidermal cracks. This attracts the free – living bacteria for better colonization and facilitates the entry of more number of endophytic bacteria into the rice seedlings *A. caulinodans* entered into rice seedlings through natural fissures crated during emergence of lateral roots (O'Callaghan *et al.*, 1997).

Table 1. Effect of combined inoculation of *Azorhizobium caulinodans* ORS 571-IK-SK-M_4 and *Agrobacterium rhizogenes* with enzyme mixture on the number of lateral roots of ADT 36 rice

Treatments	Days after treatment					
	7 days		14 days		21 days	
	Rootlet number	Per cent increase (+) or decrease (-) over control	Root let num-ber	Per cent increase (+) or decrease (-) over control	Root let number	Per cent increase (+) or decrease (-) over control
Uninoculated control	12.33	-	37.66	-	68.66	-
Agrobacterium rhizogenes MTCC 532 alone	14.66	(+)18.89	62.66	(+) 64.63	96.33	(+) 40.30
A. c ORS 571-IK-SK-M4 alone	14.00	(+)13.54	43.33	(+) 15.05	78.00	(+) 13.60
A. rhizogenes + *A. c* ORS 571 -IK-SK-M4	14.33	(+)16.22	54.66	(+) 45.14	92.33	(+) 13.40
A. c ORS 571 - IK-SK-M4 + cellulase	11.33	(-) 8.10	36.33	(-)3.53	67.33	(-) 1.93
A. c ORS 571 - IK-SK-M4 + pectinase	11.66	(-) 5.43	36.33	(-) 3.53	67.33	(-) 1.95
A. c ORS 571 - IK-SK-M4 + cellulase + pectinase	12.00	(-) 2.67	36.66	(-)2.65	67.33	(-) 1.93

	SEd	CD (0.05)
Days (D)	2.49	5.02
Treatment (T)	3.80	7.67
D x T	6.58	13.2

Table 2. Effect of combined inoculation of *Azorhizobium caulinodans* ORS 571-IK-SK-M_4 and *Agrobacterium rhizogenes* with enzyme mixture on the root growth of ADT 36 rice

Treatments	Days after treatment					
	7 days		14 days		21 days	
	Root length (cm)	% increase (+) or decrease (-) over control	Root length (cm)	% incre-ase (+) or decrease (-) over control	Root length (cm)	% increase (+) or decrease (-) over control
Uninoculated control	8.80	-	10.10	-	11.31	-
Agrobacterium rhizogenes MTCC 532 alone	8.33	(-) 5.34	9.51	(-) 5.842	10.33	(-) 8.66

A.c ORS 571-IK-SK-M4 alone	9.88	(+) 12.77	11.81	(+) 16.93	13.54	(+) 19.71
A. rhizogenes + *A. c* ORS 571 + IK-SK-M4	9.11	(+) 3.52	10.42	(+) 3.16	11.98	(+) 5.92
A. c ORS 571 + IK-SK-M4 + cellulase	7.90	(-) 10.22	8.91	(-) 11.78	10.11	(-) 10.6
A. c ORS 571 + IK-SK-M4 + pectinase	8.10	(-) 7.95	9.23	(-) 8.61	10.57	(-) 6.54
A. c ORS 571 + IK-SK-M4 + cellulase + pectinase	8.03	(-) 8.75	9.14	(-) 9.50	10.32	(-) 8.75

	SEd	CD (0.05)
Days (D)	0.20	0.58
Treatment (T)	0.44	0.89
D x T	0.77	1.54

Table 3. Effect of combined incoulation of *Azorhizobium caulinodans* ORS 571-IK-SK-M_4 and *Agrobacterium rhizogenes* with enzyme mixture on the shoot growth of ADT 36 rice

Treatments	Days after treatment					
	7 days		14 days		21 days	
	Shoot length (cm)	% incre-ase (+) or decrease (-) over control	Shoot length (cm)	% incre-ase (+) or decrease (-) over control	Shoot length (cm)	% increase (+) or decrease (-) over control
Uninoculated control	15.51	-	19.31	-	20.41	-
Agrobacterium rhizogenes MTCC 532 alone	15.43	(-) 0.51	18.73	(-) 3.00	19.62	(-) 3.87
A. c ORS 571-IK-SK-M4 alone	16.72	(+) 7.80	22.34	(+) 15.69	23.93	(+) 17.24
A. rhizogenes + *A. c* ORS 571 -IK-SK-M4	16.25	(+) 4.77	21.52	(+) 11.44	22.46	(+) 10.04
A. c ORS 571 - IK-SK-M4 + cellulase	14.81	(-) 4.31	18.46	(-) 4.40	19.55	(-) 4.21
A. c ORS 571 - IK-SK-M4 + pectinase	15.13	(-) 2.45	18.97	(-) 2.76	19.93	(-) 2.85
A. c ORS 571 - IK-SK-M4 + cellulase + pectinase	14.96	(-) 3.54	18.61	(-) 3.62	19.62	(-) 3.87

	SEd	CD (0.05)
Days (D)	0.50	1.01
Treatment (T)	0.77	1.55
D x T	1.33	2.68

The results showed that enzymatic treatment increased the endophytic population of antibiotic resistant marker strain in roots, stem and leaves of the rice seedlings. This could be due to softening of the cell wall by the cell wall degrading enzymes, which enables the entry of antibiotic resistant marker strain. Similar observations were made in an earlier study (Al-Mallah *et al.,* 1987).

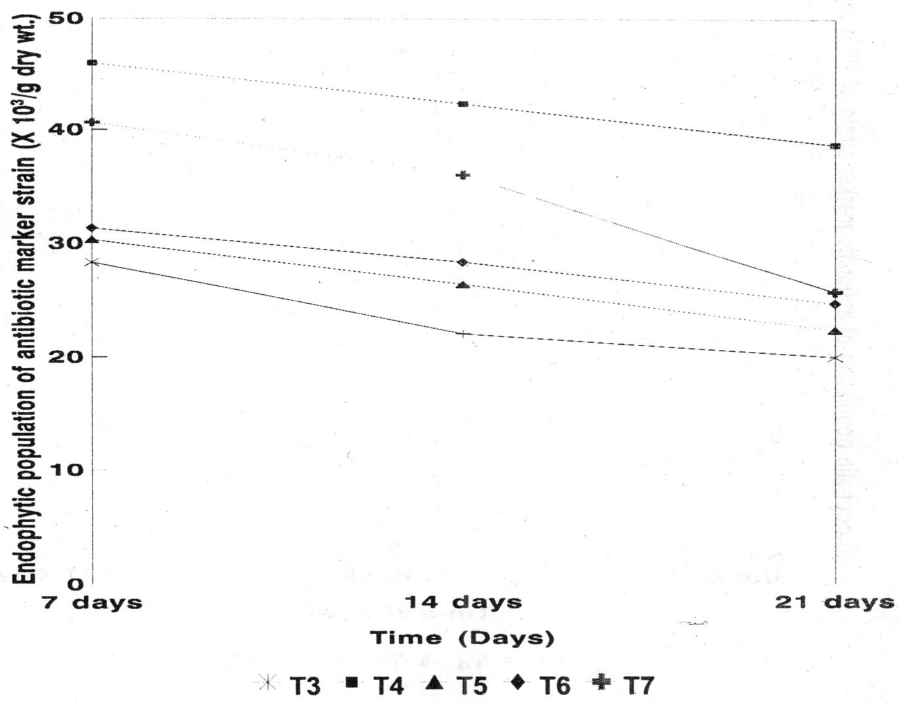

T_1 - Uninoculated control

T_2 - *Agrobacterium rhizogenes* MTCC 532 alone

T_3 - *A.c* ORS 571 + IK-SK-M_4 alone

T_4 - *A. rhizogenes* + *A.c* ORS 571-IK-SK-M_4

T_5 - *A. rhizogenes* + *A.c* ORS 571-IK-SK-M_4 + cellulase

T_6 - *A. rhizogenes* + *A.c* ORS 571-IK-SK-M_4 + pectinase

T_7 - *A. rhizogenes* + *A.c* ORS 571-IK-SK-M_4 + cellulase + pectinase

Fig. 1. Effect of combined inoculation of *Azorhizobium caulinodans* ORS 571-IK-SK-M4 and *Agrobacterium rhizogenes* with enzyme mixture on the endophytic population of antibiotic marker strain (*A.c* ORS 571-IK-SK-M_4) of ADT 36 rice roots.

The antibiotic resistant marker strain *A.c* ORS 571–IK-SK-M_4 was translocated systemically in the upward direction *viz.,* from root to leaf of rice plant system under hydroponic conditions. This is evident from the results obtained from the present study that the endophytic population of *A. c* ORS-571–IK-SK-M_4 in root was highest on 7th day after treatment and it showed a gradual decreasing trend with increasing day intervals at which observations were taken. Similarly the endophytic population of *A. c* ORS 571 – IK-SK-M_4

in shoot was found to be very less on 7th day after treatment and showed a gradual increase upto 14th day after treatment and then it showed a declining trend on 21st day after treatment. The endophytic population of *A.c* ORS 571 SK-M_4 in leaves showed a gradual increase with the sampling intervals.

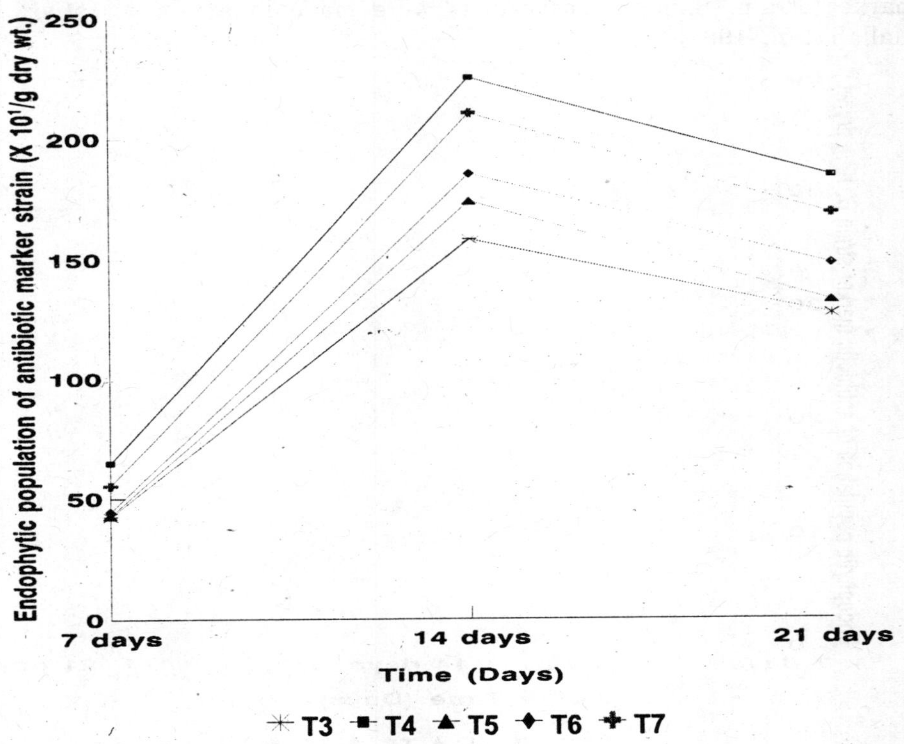

T_1 - Uninoculated control

T_2 - *Agrobacterium rhizogenes* MTCC 532 alone

T_3 - *A.c* ORS 571 + IK-SK-M_4 alone

T_4 - *A. rhizogenes* + *A.c* ORS 571-IK-SK-M_4

T_5 - *A. rhizogenes* + *A.c* ORS 571-IK-SK-M_4 + cellulase

T_6 - *A. rhizogenes* + *A.c* ORS 571-IK-SK-M_4 + pectinase

T_7 - *A. rhizogenes* + *A.c* ORS 571-IK-SK-M_4 + cellulase + pectinase

Fig. 2. Effect of combined inoculation of *Azorhizobium caulinodans* ORS 571-IK-SK-M_4 and *Agrobacterium rhizogenes* with enzyme mixture on the endophytic population of antibiotic marker strain (*A.c* ORS 571-IK-SK-M_4) in the stem of ADT 36 rice.

The endophytic nature of *Azorhizobium caulindans* has been well established and the xylem vessels have been shown as the pockets of colonization (Gopalswamy *et al.*, 2000). Such a stable association between the diazotroph would be more beneficial than the free-living and associative symbionts. The upward movements of the endosymbionts during the course of the investigation could be attributed to its xylem colonization and the subsequent translocation by the plant system. The establishment of such stable associations and development of suitable tracking methods will play a large role in the development of nitrogen fixing endophytic system in non – legumes.

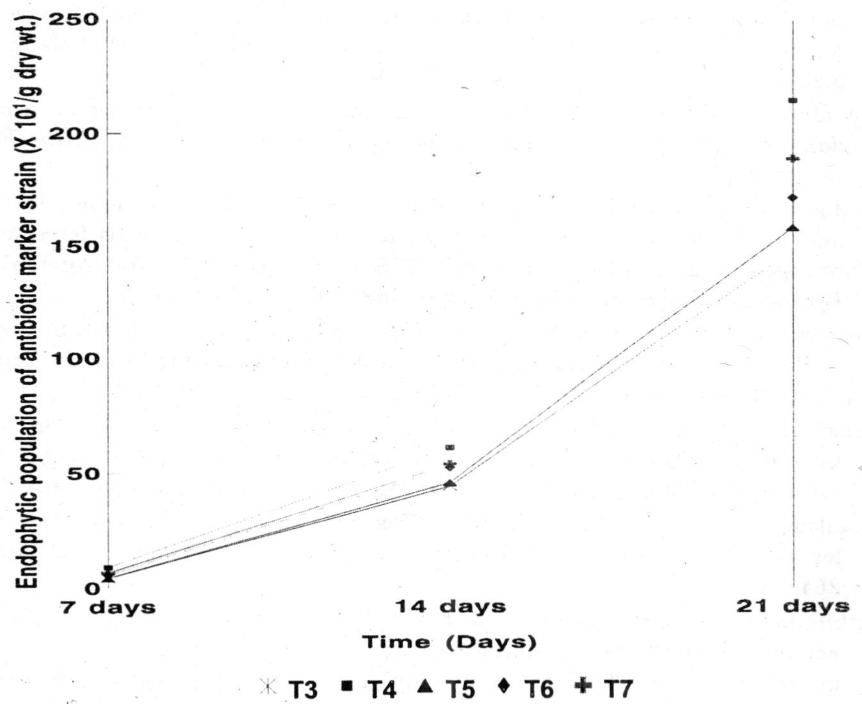

T_1 - Uninoculated control

T_2 - *Agrobacterium rhizogenes* MTCC 532 alone

T_3 - *A.c* ORS 571 + IK-SK-M_4 alone

T_4 - *A. rhizogenes* + *A.c* ORS 571-IK-SK-M_4

T_5 - *A. rhizogenes* + *A.c* ORS 571-IK-SK-M_4 + cellulase

T_6 - *A. rhizogenes* + *A.c* ORS 571-IK-SK-M_4 + pectinase

T_7 - *A. rhizogenes* + *A.c* ORS 571-IK-SK-M_4 + cellulase + pectinase

Fig. 3. Effect of combined inoculation of *Azorhizobium caulinodans* ORS 571-IK-SK-M_4 and *Agrobacterium rhizogenes* with enzyme mixture on the endophytic population of antibiotic marker strain (*A.c* ORS 571-IK-SK-M_4) in the leaves of ADT 36 rice.

REFERENCES

Al-Mallah, M.K., M.R. Davey and E.C. Cocking 1987. Enzymatic treatment of clover root hairs removes a barrier to *Rhizobium* host specificity. **Biotechnol., 5 :** 1319-1322.

Amutha, K., and S. Kannaiyan, 2000. Effect of kinetin and inoculation of *Azorhizobium caulinodans* on induction of lateral rootlets and paranodules in rice. **J. Microb. World. 2(1)**. 1-7.

Buvana, R. 2001. Studies on the effect of naringenin and inoculation of stem nodulating nitrogen fixing bacterium, *Azorhizobium caulinodans* on the induction of paranodules and endophytic colonization in the root systems of rice and maize. Ph.D. Thesis, Tamil Nadu Agric. Univ., Coimbatore, Tamil Nadu India p.235.

Christiansen-Weniger, C. 1996. Endophytic establishment of *Azorhizobium caulinodans* through auxin-induced root tumours of rice (*Oryza sativa* L.). **Biol. Fertil. Soil, 21(4):** 293-302.

Cocking, E. C., C. A. Batchelor, S. L. Kothari, S. Jain, G. Webster, J. Jones, J. Jotham and M. R. Davey. 1994. The interaction of rhizobia with cereals for symbiotic nitrogen fixation. In: **Proceedings Second Asia Pacific Agricultural Biotechnology Symposium,** Madras, India, pp. 281-286.

Gopalaswamy, G., S. Kannaiyan, K.J. O' Callaghan, M.R. Davey and E.C. Cocking. 2000. The xylem of rice (*Oryza sativa*) is colonized by *Azorhizobium caulinodans*. **Proc. R. Soc. Lond., B267**: 103-107.

Kannaiyan, S. 2001. Endophytic Association of *Azorhizobium caulinodans* for Nodulation and Nitrogen Fixation in Cereal crops In: **Agriculture 2001** (eds.) S. Kannaiyan and C. Ramasamy, Tamil Nadu Agrl. University, Coimbatore . pp.251-269.

O' Callaghan, K.J., M.R. Davey and E.C. Cocking. 1997. Xylem colonization of the legume *Sesbania rostrata* by *Azorhizobium caulinodans*. **Proc. R. Soc. Lond., B 264**: 1821-1826

Senthilkumar. M. 2000. Studies on the induction of paranodules by the stem nodulating *Azorhizobium caulinodans* and its nitrogen fixing activity in rice and maize. M.Sc. (Ag.) Thesis, Tamil Nadu Agric. Univ., Coimbatore. Tamil Nadu India p. 120.

Uma, D. and S. Kannaiyan. 1996. Effect of systemic fungicide Bavistin on the nitrogen status of cyanobacteria under immobilized state in polyrethane foam. **S. Afr. J. Bot., 62(3):** 127-132.

Vincent, J.M. 1970. A Manual for the Practical Study of Root Nodule Bacteria. International Biological Programme Hand book 15, Oxford-Black well Scientific Publ., New Delhi.

Watanabe, I. 1977. *Azolla* utilization in rice culture. **Inter. Rice Res. Newslett., 2(3):** 10.

CHAPTER – 34

EFFECT OF SEED INOCULATION OF *Azorhizobium caulinodans, Azospirillum brasilense* AND *Pseudomonas fluorescens* ON THEIR SURVIVAL IN THE SPERMOSPHERE AND RHIZOSPHERE OF CEREAL CROPS

K. Amutha and S. Kannaiyan

ABSTRACT

In recent years, use of microbial inoculants to enrich N and P nutrients in soil has become a practice in most of the countries, as far as economical and environmental view points are concerned. Seed treatment with bioinoculants increases the plant growth and also increases the colonization of beneficial microorganisms in root region of the plants. Hence, an investigation was carried out to study the colonization of bioinoculants such as *A. caulinodans* ZB-SK-5, *A. brasilense* SP-7 and *P. fluorescens* PF-1 in the spermosphere and rhizosphere of cereal crops. The seeds of rice, maize, sorghum, cumbu and ragi were treated with the carrier based bioinoculants at 200g /10 kg of seeds and sown in pots filled with unsterilized field soil. The spermosphere (3 days old sprouted seeds) and rhizosphere population were enumerated. Among the three bioinoculants tested, the maximum spermosphere colonization was observed in rice seeds treated with *A. brasilense* SP-7 followed by *A. caulinodans* ZB-SK-5 and *P. fluorescense* PF-1. In maize, sorghum, cumbu and ragi seeds. The maximum colonization was observed in *A. caulinodans* ZB-SK-5 treated seeds. The rhizosphere population was maximum in *A. caulindoans* ZB-SK-5 treated plants, compared to *A. brasilense* SP-7 and *P. fluorescense* PF-1 treated plants. The rhizosphere colonization of all the three organisms was higher on 30th day, which decreased from the 45th day onwards and was lowest on 60th day. It is clear from the results that the survival of the inoculated rhizobacteria in the rhizosphere was maximum on 30th day after inoculation which decreased thereafter.

INRODUCTION

In recent years, use of microbial inoculants to enrich N and P nutrients in soil has become a practice in most of the countries, as for as economical and environmental view points are concerned. Seed treatment with bioinoculants increases the plant growth and also increases the colonization of beneficial microorganisms in root region of the plants. Survival of *Azorhizobium* in peat was highest compared to vermiculite and different combination of peat and vermiculite (Mekhemar *et al.*, 2000). The inoculation of *Azorhizobnium* to field crops such as wheat, rice and millets have shown a significant increase in the growth parameters, dry matter accumulation and grain yield (Boddey *et al.*, 1986).

The mixed inoculation of *Azospirillum brasilense* and *Azotobacter chroococcum* in maize and sorghum increased dry matter production (Tilak *et al.*, 1982). The *Astragalus sinicus* rhizobia survived well under flooded condition, with a population of 10^5 g^{-1} (dry weight) of soil in a field continuously under *A.sinicus* (Tuzimura and Watanabe 1959). Hence, an investigation was carried out to study the colonization of bioinoculants such as *A.caulinodans* ZB-SK-5, *A.brasilense* SP-7 and *P. fluorescens* PF-1 in spermosphere and rhizosphere of cereal crops.

MATERIALS AND METHODS

The healthy seeds of rice (ADT 36), maize (CO 1), *Sorghum* (CO 26), cumbu (CO 7) and ragi (CO 12) were soaked overnight in water. The 48 hours old standard cultures of *A.caulinodans* grown in yeast extract mannitol broth, *A.brasilense* N – free malic acid broth and *P. fluorescens* King's B broth were taken and blended with sterile neutralized carrier, and the seeds were treated with carrier based preparations of the above bioinoculants at 200 g /10 kg of seeds and sown in pots filled with unsterilized field soil. The spermosphere population was enumerated in 3 days old sprouted seeds. The rhizosphere population was enumerated at 15 days interval on the 15, 30, 45 and 60 days after sowing.

RESULTS

Among the three bioinoculants tested, the maximum spermosphere colonization was observed in rice seeds treated with *A. brasilense* SP–7 followed by *A. caulinodans* ZB-SK-5 and *P. fluorescens* PF–1. In maize, sorghum, cumbu and ragi seeds, the maximum colonization was observed in *A. caulinodans* ZB–SK–5 treated seeds.

The rhizosphere population was maximum in *A. caulinodans* ZB-SK-5 treated plants compared to *A.brasilense* SP-7 and *P. fluorescens* PF-1 treated plants (Table 1 to 3). The rhizosphere colonization of all the three organisms were significantly higher on 30th day after which the rhizosphere population dropped from the initial level. The population was comparatively less on 45th day and was lowest on 60th days after sowing.

Table 1. Colonization of *A. caulinodans* ZB-SK-5 in the rhizosphere of rice (ADT 36), maize (CO 1), sorghum (CO 26), cumbu (CO 7) and ragi (CO 12)

Crop	Population - *A. caulinodans* (ZB-SK-5) (x10^6)			
	15 DAS	30 DAS	45 DAS	60 DAS
Rice	45.51	62.63	40.08	19.00
Maize	36.04	47.50	35.34	15.35
Sorghum	22.47	29.83	16.17	5.73
Cumbu	21.77	38.34	21.80	4.63
Ragi	39.30	50.50	36.00	10.17
SEd	0.49	0.44	0.99	
CD	1.00	0.89	2.00	

Table 2. Colonization of *A. brasilense* SP 7 in the rhizosphere of rice (ADT 36), maize (CO 1), sorghum (CO 26), cumbu (CO 7) and ragi (CO 12).

Crop	Population - *A. brasilense* (SP- 7) (x 10^6)			
	15 DAS	30 DAS	45 DAS	60 DAS
Rice	38.55	48.70	21.23	5.77
Maize	28.54	39.40	19.59	3.30
Sorghum	25.37	32.00	15.11	1.77
Cumbu	17.97	29.35	17.80	1.44
Ragi	19.83	33.33	18.13	10.26
SEd	0.32	0.29	0.64	
CD	0.65	0.58	1.31	

Table 3. Colonization of *P. fluorescens* PF- 1 in the rhizosphere of rice (ADT 36), maize (CO 1), sorghum (CO 26), cumbu (CO 7) and ragi (CO 12)

Crop	Population –*P. fluorescens* (PF-1) (x 10^6)			
	15 DAS	30 DAS	45 DAS	60 DAS
Rice	29.83	55.33	23.33	8.55
Maize	18.93	28.33	13.00	6.66
Sorghum	9.67	19.67	10.50	4.43
Cumbu	6.83	16.66	9.23	1.67
Ragi	6.00	11.67	6.66	2.33
SED	0.72	0.64	1.44	
CD	1.46	1.31	2.92	

DISCUSSION

Plant associated nitrogen fixation involves complex interaction among the rice plants, microbial population and environment. Nitrogen fixing microorganisms in association with the roots of wetland rice provide a part of their fixed nitrogen to the rice plant.

Becker *et al.* (1990) reported that the epiphytically growing rhizobia were washed with rain water then it goes down the stem and there by infects the nodulating sites The survival of indigenous and introduced strains of *A.caulinodans* in flooded soil and in the rice rhizosphere was reported by Ladha *et al.* (1989). They also found that the incorporation of fresh biomass of *S. rostrata* before rice planting has resulted in harbouring a large population of azorhizobia in rice rhizosphere. Chandrasekar and Kannaiyan (1995) studied the presence of *A. caulinodans* in the rice rhizoplane region and also in rice soils. It is clear from the results of the present study that the survival of the inoculated organisms was maximum in the rhizosphere.

The survival of *Azospirillum* in many of the carrier was studied by Tilak *et al.* (1979), who confirmed the fact that soil and farmyard manure gave higher counts of *Azospirillum* than other carriers. Seed inoculation increased the yields of straw and grain of rice and barley (Subba Rao *et al.,* 1979).

REFERENCES

Becker, M., J.K. Ladha and J.C.G. Ottow. 1990. Growth and nitrogen fixation of stem nodulating legumes and their effect as green manure on low land rice. **Soil Biol. Biochem., 22 (8):** 1109 – 1119.

Boddey, R. M., V. L. D. Baldani, J. I. Baldani and J. Dobereiner. 1986. Effects of inoculation of *Azospirillum* spp. on nitrogen accumulation by field grown wheat. **Plant Soil, 90**: 303-304.

Chandrasekar, M.V. and S. Kannaiyan .1995. Induction of nodular structures by *Azorhizobium.* **Tamil Nadu Agric. Univ. Newslett., 24** (12): 3

Ladha, J.K., M.Garcia, S. Miyan, A.T. Padre and I. Watanabe. 1989. Survival of *A.caulinodans* in the soil, rhizosphere of wetland rice under *S.rostrata* rice rotation. **Appl. Environ. Microbiol., 55**: 454 – 460.

Mekhemar, G. A. A., A. A. Abo EL- Soud, A. A. Rag, F. T. Mikhaeel, S. R. S. Sabry, S. A. Saleh and E. C. Cocking . 2000. Survival of *Azorhizobum* and *Azospirillum* in different carrier materials: Inoculation of wheat and *Sesbaniba rostrata.* In : **International symposium on Nitrogen Fixation with non - legumes.** (eds.) I.R. Kennedy and L. Copeland. The Univ. Sydney, NSW, Austtralia, p.51 (Abstr.)

Tilak, K.V.B.R., M. Lakshmi Kumari and C. Nautiyal. 1979. Survival of *Azospirillum brasilense* in different carriers. **Curr. Sci., 48**: 412 – 413.

Tilak., K. V. B. R., C. D. S. Singh, N. K. Roy and N. S. Subba Rao. 1982. *Azospirillum brasilense* and *Azotobacter chrococcum* inoculum. Effect on yield of maize (*Zea mays* and *Sorghum*) (*Sorghum bicolor*). **Soil Biol. Biochem., 14**: 417 – 418.

Tuzimura, K. and I. Watanabe. 1959. Nodulation– dilution frequency method to determine number of root nodule bacteria in soils and its application. **J. Soc. Sci. Soil Manure** Jpn. **30:** 292 –296.

CHAPTER - 35

INDUCTION OF NODULATION IN RICE USING *Azorhizobium caulinodans* UNDER POT CULTURES CONDITIONS

K. Amutha and S. Kannaiyan

ABSTRACT

Rice (*Oryza sativa L.*) is the staple food crop of the world. Nitrogen nutrition is most essential for rice crop. Though nitrogen is the most abundant element present in the atmosphere comprising around 80 per cent, its availability to rice plant is very less. The continuous application of inorganic nitrogenous fertilizers increased the cost of production of rice besides affecting the soil health. In recent years, several novel approaches have been developed to nodulate non- legumes. These include engineering of new rhizobial inoculants, inoculation of rhizobial isolates in *Parasponia* with growth regulators including the 2,4-D. To prove this an investigation was carried out to assess the induction of nodulation in rice (*Oryza sativa*) under pot culture condition. The seeds of rice (cv. ADT 36) were soaked overnight in water and treated with 2,4-D at 0.5, 1.5, 3.0 ppm concentrations, followed by treatment with carrier based 48 hrs old culture of *Azorhizobium caulinodans* ZB-SK-5. The treated seeds were sown in pots containing sterile as well as unsterile soils. After 28 days, the nodule like structure formation was observed in the roots as well as crown region of the seedlings, but the nodule like structures were not observed in control pots and pots in which 2,4-D alone was applied. The treatment with *Azorhizobium caulinodans* ZB-SK-5 and 2,4-D (3.0 ppm) recorded maximum number of nodule like structures compared to other treatments. This experiments clearly showed that the inoculation of *Azorhizobium caulinodans* ZB-SK-5 with growth regulator 2,4-D has got significant effect in inducing the nodulation in rice under pot culture condition.

INTRODUCTION

Rice (*Oryza sativa L.*) is the staple food crops of the world. Nitrogen nutrition is most essential for rice crop. One of the major limitation of associative nitrogen fixation is that diazotrophic bacteria in the rhizosphere of plants utilize the products of nitrogen fixation for their own growth, but release little while they are alive (Alazard, 1990). Several approaches to achieve a more direct transfer of biologically fixed nitrogen to rice plants have been proposed, best approaches are the indentification of stably maintained diazotrophic endophytic bacteria in rice tissues (Ladha *et al.*, 1997) and induction of nodulation in non -legumes by engineering of rhizobial inoculants, inoculation of rhizobial isolates in *Parasponia* with growth regulator 2,4-D. To prove this an investigation was carried out to study the induction of nodulation in rice under pot culture condition.

MATERIALS AND METHODS

The seeds of rice (cv. ADT 36) were soaked overnight in water and treated with 2,4-D at 0.5, 1.5, 3.0 ppm concentrations. The 48 hrs old standard culture of *A. caulinodans* ZB-Sk-5 grown in yeast Extract Mannitol broth was blended with sterile neutralized carrier, and this carrier based preparation was used for seed treatment at 200 g / 10 kg of seeds and sown in pots filled with sterilized as well unsterilized wetland soil. After 15 days of sowing the growth parameters ware recorded, the nodule like structures formation was observed in root and crown region of the seedlings 28 days after sowing.

RESULTS

The growth parameters root growth, shoot growth, biomass dry weight and number of nodule like structures (Table 1) were recorded maximum in the treatment with inoculation of *A. caulinodans* and *A. caulinodans* with 2,4-D at various concentrations compared to 2,4-D alone at various concentration. Inoculation of *A. caulinodans* alone enhanced the growth of the seedlings without any inhibition due to 2,4-D. But inoculation of *A. caulinodans* ZB-SK-5 with 2,4-D at 0.5, 1.5 and 3.0 ppm concentration produced nodule like structure. Increased levels of 2.4-D with inoculation of *A. caulinodans* proportionately increased the nodule like structure formation. There was no nodulation like structures formation in rice seedlings when 2,4-D was applied separately and also uninoculated control.

Table 1. Effect of 2,4-D and inoculation of *A. caulinodans* ZB-SK-5 on the root, shoot length, total biomass, dry weight and number of nodule like structures of rice (ADT 36) under pot cultures condition

Treatments	Rice (ADT 36)				
	Root length (cm)	Shoot length (cm)	Total biomass (g)	Dry weight (g)	Nodule (No.plant^{-1})
Uninoculated	8.34	14.50	0.118	0.035	-
A. caulinodans (*A.c*)	12.38	18.10	0.213	0.064	3.83
2,4-D (0.5 ppm)	8.17	13.51	0.179	0.031	-
2,4-D (0.5 ppm) + *A.c*	9.05	15.67	0.190	0.057	9.03
2,4-D (1.5 ppm)	7.47	12.33	1.161	0.027	-
2,4-D (1.5 ppm) + *A.c*	8.50	16.52	0.170	0.044	10.33
2,4-D (3.0 ppm)	7.33	10.37	0.132	0.019	-
2,4-D (3.0 ppm) + *A.c*	8.20	14.27	0.138	0.032	16.00
SEd	0.78	0.84	0.002	0.002	0.53
CD	1.65	1.78	0.003	0.005	1.12

DISCUSSION

The nitrogen fixing organisms possess the enzyme nitrogenase which catalyse the conversion of molecular nitrogen to ammonia. Many plants including Rice do not have the capacity to fix nitrogen. Sabry *et al.* (1997) showed the endophytic establishment of *A. caulinodans* in wheat. To establish an endophytic nitrogen fixing interactions of *A. caulinodna* with wheat, plant growth regulator 2,4-D was added to induce nodular structures. The use of 2,4-D and polyethylene glycol is loosening the cell wall of roots which helps the entry of rhizobia into the roots and form nodule like structures (Cocking *et al.*, 1995 and Christiansen – Wengier, 1996).

In the present study, an attempt was made to induce nodule like structures in rice with different concentration of 2,4-D and subsequent seed treatment of *A. caulinodans* ZB-SK-5 culture. This treatment was found to induce the formation of nodule like structure in rice both sterile and unsterile soil under pot culture conditions. Panwar *et al.* (2000) reported the presence or absence of nitrogen fixing baceria, different plant hormones like 2,4-D, benzylamino purine and zeatin have been tried but use of 2,4-D has been reported to be the best in inducing nodular out growth.

REFERENCES

Alagard, D. 1990. Nitrogen fixation in pure culture by rhizobia isolated from stem nodules of tropical *Aeschynomene* species. **FEMS Microbiol. lett., 68(12):** 177-182.

Christiansen-Weniger, C.1996. Endophytic establishment of *A. caulinodans* through auxin induced root tumors of rice (*Oryza sativa* L.) **Biol. Fertil. Soils, 21(4):** 293-302.

Cocking, E.C., S.L. Batchelor, S. Kothari, S. Jain, G. Webster, J. Jones and M.R. Davey 1995. **The interaction of rhizobia with cereals for symbiotic nitrogen fixation.** In: Proc. 2nd Asia Pacific Agrl. Biotechnol. Symp., Madras, India.

Ladha, K.K. F.J. de Bruijn and K.A. Malik. 1997. Introduction: assessing opportunities for nitrogen fixation in rice - a frontier project. **Plant Soil, 194**: 1-10.

Panwar, J.D.S. 2000. Nodulation and nitrogen fixation achieved in cereals. A Science and Technology. News letter, IARI, New Delhi. pp. 13-14.

Sabry, S.R.S., S.A. Saleh, C.A. Batchelor, J. Johes, J. Jothao, G. Webster, S.L. Kothari, M.R. Davey and E.C. Cocking. 1997. Endophytic establishment of *A. caulinodans* in wheat. **Proc. Roy Soc. London. 264** : 341–346.

CHAPTER - 36

ENDOPHYTIC NITROGEN FIXATION BY *Gluconacetobacter diazotrophicus* IN SUGARCANE

M. Thangaraju

INTRODUCTION

Although a number of microorganisms belonging to different genera have been identified to be associated with sugarcane and fix atmospheric nitrogen, only microorganisms, which belong to the genera *Azospirillum* and *Azotobacter*, have been considered to be of economical importance. In recent years, explanation for large scale contribution of biological nitrogen fixation in crops like sugarcane and non-legumes has become much easier by description of newer obligate endophytic bacterium *Gluconacetobacter diazotrophicus* (Cavalcante and Dobereiner, 1988; Yamada *et al.*1997; 1998), which get colonized within roots, stems and leaves of certain plant genotypes. In addition to *Gluconacetobacter diazotrophicus*, *Herbaspirillum seropedicae* (Baldani *et al.*, 1986), *Herbaspirillum rubrisubalbicans* (Baldani *et al.*, 1996) and *Azoarcus* (Reinhold-Hurek *et al.*, 1993) have also been reported as other endophytic bacteria. However physiological and nutritional studies with sugar rich crops *viz.*, sugarcane, sweet potato, sweet sorghum and coffee plants revealed the importance of nitrogen fixing bacterium, *Gluconacetobacter diazotrophicus.* In addition to nitrogen fixation, these organisms also produce indole-3-acetic acid, which serve as a growth promoter.

OCCURRENCE OF *Gluconacetobacter diazotrophicus*

Gluconacetobacter diazotrophicus was found to occur in the root, stem, leaf tissues of sugarcane (Cavalcante and Dobereiner, 1988; Gillis *et al.*, 1989), rhizosphere soil and even in cane juice (Muthukumarasamy *et al.*, 1994) in appreciable number. In addition, it could be isolated from Cameroon grass (*Pennisetum purpureum* cv. Cameroon) and sweet potato (Paula *et al.*, 1989; Dobereiner *et al.*, 1993) as well as from different genera of mealy bugs associated with sugarcane plants (Boddey *et al.*, 1995).

GEOGRAPHICAL DISTRIBUTION OF *Gluconacetobacter diazotrophicus*

This bacterium was found in all major sugarcane growing areas of Brazil as well as in Australia, Mexico, Cuba and India (Cavalcante and Dobereiner, 1988; Li and Mac Rae, 1992; Fuentes Ramirez *et al.,* 1993; Muthukumarasamy *et al.,* 1994; Jayakumar and Thangaraju, 1996; Thangaraju, 2000).

SALIENT FEATURES OF *Gluconacetobacter diazotrophicus*

- Acid tolerant nitrogen fixing bacterium associated with sugarcane.
- An endophytic organism colonizes the roots, stems, buds and leaves of sugarcane
- It is located in the apoplastic fluid of sugarcane stems and to a lesser extent in xylem vessels.
- Capable of growing and fixing nitrogen at 30 % sucrose concentration and optimum sucrose concentration for better nitrogen fixation is 10%
- Growth and nitrogen fixation observed even at pH 3. 0
- The optimum pH is 5.5
- An average of 200 kg N ha^{-1} would be fixed by using 30 per cent of the total sugar produced by a sugarcane crop.
- IAA production is also noticed
- Tolerant to higher level of combined nitrogen.
- Mode of entry is through VA mycorrhizae, mealy bug and mechanical injury of plant parts.
- Exhibiting antibiotic resistance.
- It is a member of alpha group of Proteobacteria
- In terms of 16s rDNA based phylogeny *Gluconacetobacter* clusters with the genera *Rhodopila, Acidomonas, Acidiphilium* and *Gluconobacter*
- In terms of *nif H* sequence, it is more related to *Rhizobium* sp. and *Azospirillum brasilense* than to other diazotrophs.

NITROGEN FIXATION BY *Gluconacetobacter diazotrophicus*

Dobereiner *et al* (1988) first reported the nitrogen fixing ability of *Gluconacetobacter diazotrophicus.* However, it was confirmed by Cavalcante and Dobereiner (1988), who reported that these organisms produced approximately 240 n moles C_2H_4 h^{-1} mg of cell $protein^{-1}$. This was later conformed by Gillis *et al.* (1989). Boddey *et al.* (1991) reported that due to the absence of nitrate reductase, nitrogen fixation was not affected by high levels (25 mM) of nitrate.

Based on the laboratory estimates, Dobereiner (1992) reported that an average of 200 kg N ha^{-1} would be fixed by using 30 per cent of the total sugar produced by a sugarcane crop. James *et al.* (1994) reported that *G. diazotrophicus* produced more than 150 n moles C_2H_4 h^{-1} 15 days after inoculation of micro propagated sugarcane plantlets at dilution of 10^{-6} to 10^{-7}. They also suggested that the xylem vessels in the dense shoots of mature plant are possible sites of nitrogen fixation by *G. diazotrophicus.* Jayakumar and Thangaraju (1996) estimated nitrogenase activity of 220 n moles C_2H_4 produced h^{-1} mg of cell $protein^{-1}$ in the *G. diazotrophicus* culture isolated from sugarcane root/stem.

PLANT GROWTH PROMOTING SUBSTANCES RELEASED BY *Gluconacetobacter diazotrophicus*

Libbert and Risch (1969) showed that epiphytic bacteria could increase the IAA content in the plant. Cavalcante and Dobereiner (1988) and Boddey *et al.* (1991) reported that *G. diazotrophicus* could promote root development and improve sugarcane growth. However, first report on the production of indole-3-acetic acid by these organisms was made by Fuentes-Ramirez *et al.* (1993). They showed that all the strains of *G. diazotrophicus* produced indole-3-acetic acid in defined medium supplemented with tryptophan and the amount of IAA produced was estimated from 0.14-to 2.42-μ ml^{-1} in culture medium. Panneerselvam (1997) reported the ability of local sugar cane isolates of *G. diazotrophicus* to produce IAA and estimated the value as high as 2.24 μg ml^{-1}

ANTIBIOTIC RESISTANCE OF *Gluconacetobacter diazotrophicus*

Ureta *et al.* (1995) suggested that the detection of intrinsic antibiotic resistance will be useful for both genetic studies and distinguishing the isolates of *G. diazotrophicus* genera. The isolates of *G. diazotrophicus* are susceptible to tetracycline (10 μg ml^{-1}), rifampicin (10 μg ml^{-1}), novobiocin (25 μg ml^{-1}), nalidixic acid (15 μg ml^{-1}) and resistant to penicillin (50 μg ml^{-1}), carbencillin (150 μg ml^{-1}) and chloramphenicol (75 μg ml^{-1}).

MODE OF ENTRY AND SURVIVAL OF *Gluconacetobacter diazotrophicus*

Endophytic nature of *Gluconacetobacter diazotrophicus* was reported by Dobereiner *et al.* (1988). These organisms infected sugarcane through damaged tissue (Boddey *et al.*, (1991). Many authors described that *G. diazotrophicus* is spread among cane cultivars by the mealy bugs associated with sugarcane (Ashbolt and Inkerman, 1990), as well as by the spores of the Vesicular Arbuscular Mycorrhizal (VAM) fungus (Paula *et al.*, 1992, Panneerselvam, 1997; Thangaraju, 2000).

Caballero-Mellado and Martinez-Romero (1994) reported that dispersal of *Gluconacetobacter diazotrophicus* by sugarcane seeds may not occur and unable to isolate bacteria from seeds. James *et al.* (1994) reported that *G.*

diazotrophicus first colonized the root and lower stem epidermal surface of sugarcane during the first 15 days after planting and then used root tips and lateral root junctions to enter the sugarcane plants where it was distributed around the plant in the translocation stream.

EFFECT OF *Gluconacetobacter diazotrophicus* ON THE GROWTH AND YIELD OF SUGARCANE

Muthukumarasamy *et al.* (1994) reported that *G. diazotrophicus* increased cane yield by 7 to 10 tonnes/acre with 50 per cent reduction in the recommended chemical nitrogen and also observed an increase in sugar recovery from 0.5 to 1 percent. Jayakumar and Thangaraju (1996) reported that *Gluconacetobacter diazotrophicus* in combination with 50 per cent nitrogen recorded cane yield and sugar yield values on par with 100 per cent nitrogen applied treatments. The growth and cane yield was better for *Gluconacetobacter* than *Azospirillum* + *Azotobacter* and *Gluconacetobacter* + *Azotobacter* combinations (Thangaraju and Jayakumar, 2002). Inoculation of *G.diazotrophicus* strain *G.d.* 5 increased the cane yield up to 15 % than the uninoculated control. The increase was 11.69 % in the reduced fertilizer nitrogen level (75% of the recommended nitrogen) than the full dose of nitrogen (Table 1) indicating a saving of 25% of fertilizer nitrogen (Thangaraju and Govindarajan, 2001).

GENETICS OF *Gluconacetobacter diazotrophicus*

A basic and common set of *nif* genes that are necessary for the structure and function of the nitrogenase enzyme are present in all diazotrophs. In addition all diazotrophs are known to have additional genes to encode products involved in N_2 fixation which may be unique to a particular organism or found in subset of usually related bacteria (Lee *et al.*, 2002a).

Table 1. Response of sugarcane to *Gluconacetobacter diazotrophicus* inoculation

Treatment	Cane yield		Sugar yield
	(t.ha^{-1})	% increase*	(t.ha^{-1})
100% fertilizer N	102.67	-	11.01
75% fertilizer N	95.50	-6.98	9.44
G.diazotrophicus + 100% fertilizer N	118.33	15.25	13.05
G.diazotrophicus + 75% fertilizer N	114.67	11.69	12.44
Azospirillum + 100% fertilizer N	116.67	13.64	12.81
Azospirillum + 75% fertilizer N	113.00	10.06	12.08

* over uninoculated control (Treatment 1) (Thangaraju and Govindarajan, 2001)

The order and arrangement of *nif* and associated genes on chromosomes (or sometimes plasmids) varies tremendously among diazotrophs (Dean and

Jacobson, 1992). In *G. diazotrophicus* PAL5, three overlapping DNA fragments were isolated (Lee *et al.,* 2002 a). All the common *nif* genes and several associated *fix* and *fdx* genes similar to those in other organisms were found in a single cluster of about 32kb in length. The individual *G. diazotrophicus* gene products are generally most similar to those found in other alpha group proteobaceria, with 17 gene products being most like those in members of the Rhizobiaceae and 9 gene products being most closely related to *Rhodobacter capsulatus* proteins (Lee *et al.,* 2002b). In addition, *mod D* and *mcp A* genes were found downstream of *fix ABCD* (Lee *et al.,* 2002a) – gene not known to be directly involved in nitrogenase structure, function or regulation of gene expression- encoding a methyl – accepting transmembrane protein that responds to extra cellular signals for chemo tactic responses (Mowbray, 1999).

The overall arrangement of genes in the *nif /fix* cluster of *G. diazotrophicus* is more like that of *A. brasilense* than that of any other diazotroph although some differences are apparent. Also *G. diazotrophicus* and *A. brasilense* are the only two diazotrophs so for characterized that have a *mcpA* - like gene that is associated with the *nif / fix. Mcp A* protein is involved in chemo taxis in several other organisms via their methylation in response to nutrient and / or other environmental effectors (Lee *et al.,* 2002 a). Since both *G. diazotrophicus* and *A.brasilense* are found naturally associated with monocot plants, it will be interesting to determine whether *mcp A* is responsible for chemo tactic responses to plant exudates for signaling or nutrition. On the other hand the individual gene products of *A. diazotrophicus* are more similar to those of rhizobial species than to *A.brasilense* proteins. The similarity of *G. diazotrophicus* gene products to those of known *nif, fix* and other genes is given in Table 2.

Table 2. Similarity of *G. diazotrophicus* gene products with that of other diazotrophs

Functions	Genes	Most similar organisms	Identity	Product size
Nitrogenase structure	nifH	*R. leguminosorum*	85	31.9
	nifD	*H. seropedicae*	83	55.9
	nifK	*B. japonicum*	76	57.2
Fe Mo Cofactor Synthesis	nifB	*R. capsulatus*	76	51.9
	nifE	*B. japonicum*	59	51.8
	nifN	*R. meliloti*	55	49.5
	nifV	*F. alni*	52	92.2
	nifQ	*R. sp NGR234*	33	22.9
	nifU	*A. chroococcum*	56	32.9
Maturation or activation	nifS	*A. vinelandii*	63	43.0
	nifW	*A. caulinodans*	56	12.0
	nifZ	*R. sp. NGR234*	65	6.9
Electron transfer	fixA	*R. sp. NGR234*	67	30.7

	fixB	*A. caulinodans*	68	38.9
	fixC	*B. japonicum*	62	48.1
	fixX	*R. meliloti*	71	10.7
Transcriptional regulator	nifA	*A. caulinodans*	51	62.2
	nifX	*R. sp NGR234*	64	17.4
Chemotaxis response	mcpA	*C. crescentus*	67	69.3
Molybdenum transport	modA	*R. capsulatus*	42	26.9
	modB	*R. capsulatus*	57	24.7
	modC	*R. capsulatus*	41	36.8
	modD	*R. capsulatus*	47	26.7
	fdxB	*R. capsulatus*	65	9.9
Ferridoxin	fdxN	*R. leguminosorum*	51	6.8
Sigma factor	rpoN	*B. japonicum*	40	51.3
Unknown	orf4	*R. capsulatus*	50	16.9
	orf1	*A. caulinodans*	53	7.6
	orf6	*R. capsulatus*	48	10.9
	ynf1	*R. capsulatus*	41	26.4
	fixU	*R. leguminosorum bv. trifoli*	56	8.6

(Lee *et al.*, 2002b)

(Organisms : *Azotobacter chroococcum; Azorhizobium caulinodans; Azotobacter vinelandii; Bradyrhizobium japonicum; Caulobacter crescentus: Frankia alni; Herbaspirillum seropedicae; Rhizobium leguminosarum; Rhizobium meliloti; Rhizobium sp.NGR 234; Rhodobacter capsulatus***)**

The isolation of *nif / fix* clusters of *G. diazotrophicus* has allowed the construction of defined *nif* mutants which have been extremely important in plant growth experiments to test the hypothesis that *G. diazotrophicus* benefits plant growth by providing fixed nitrogen.

REFERENCES

Ashbolt, N.J. and P.E. Inkerman 1990. Acetic acid bacterial biota of the pink sugarcane mealy bug *Saccharococcus sacchari* and its environs. Appl. Environ. Microbiol., 56 : 707-712.

Baldani, J. I., V. L. D. Baldani, L. Seldin and J. Dobereiner 1986 Characterization of *Herbaspirillum seropedicae* gen.nov., sp.nov., a root associated nitrogen-fixing bacterium. **Int. J. Syst. Bacteriol. 36,** 86–93

Baldani J I, Pot B, Kirchhof G, Falsen E, Baldani V L D, Olivares F L B, Hoste, Kersters K, Harmann A, Gillis M and Dobereiner, J 1996 Emended description of *Herbaspirillum*; inclusion of (*Pseudomonas*) *rubrisubalbicans*, a mild plant pathogen, as *Herbaspirillum rubrisubalbicans* com. nov., and classification of a group of clinical isolates (EF group 1) as *Herbaspirillum* species. **Int. J. syst. Bacteriol. 46,** 802–810.

Boddey, R.M., S. Urquiaga, V.M. Reis and J. Dobereiner 1991 Biological nitrogen fixation associated with sugarcane. **Plant Soil, 137** : 111-117.

Boddey, R.M., de Oliveira, O.C., Urquiaga, S., Reis, V.M. de olivares, F.L., Baldani, V.L.D and Dobereiner, J. 1995. Biological nitrogen fixation associated with sugarcane and rice; contributions and prospects for improvement. **Plant Soil, 174**: 195-209.

Caballero-Mellado, J. and E.Martinez-Romero. 1994. Limited genetic diversity in the endophytic sugarcane bacterium *Acetobacter diazotrophicus*. **Appl. Environ. Microbiol., 60** : 1532-1537.

Cavalcante, V.A. and J.Dobereiner. 1988. A new acid-tolerant nitrogen fixing bacterium associated with sugarcane. **Plant Soil, 108**: 23-31.

Dean, D.R. and M.R. Jacobson. 1992. Biochemical genetics of nitrogenase. **In: Biological Nitrogen Fixation,** (eds.) G. Stacey, H.J. Evans and R.Burris. Chapman & Hall, New York, pp. 763-834.

Dobereiner, J. 1992. Recent changes in concepts of plant bacteria interactions: endophytic N_2 fixing bacteria. **Cienc. Cult., 44** : 310-313.

Dobereiner, J., V. M. Reis and A. C. Lazarine. 1988. A new nitrogen fixing bacteria in association with cereals and sugarcane. In: **Nitrogen Fixation: Hundred years after,** (eds.) H. F. Bothe, F. .J. De Bruijn and W. E. Newton), Gustav Fischer, Stuttqart. pp.717-722.

Dobereiner, J., V.M.Reis, M.A. Paula and F.L. Olivares. 1993. Endophytic diazotrophs in sugarcane, cereals and tuber plants. In: **New Horizons in Nitrogen fixation, Current Plant Science and Biotechnology in Agriculture,Vol. 17,** (eds.) R. Palacios, J. Mora and W.E. Newton), Kluwer Academic Publishers, Dordrecht. pp. 671-676.

Fuentez-Ramirez, L.E., T.Jiminez-Salgado, I.R.Abarca-Ocampo and J. Caballero-Mellado. 1993. *Acetobacter diazotrophicus,* an Indole-3-acetic acid producing bacterium isolated from sugarcane cultivars in Mexico. **Plant Soil, 154**: 145-150.

Gillis, M., B.Kerters, D.J. Hoste, R.M. Kroppenstedt, M.P. Stephen, K.R.S. Teixeira, J. Dobereiner and J. De Ley. 1989. *Acetobacter diazotrophicus* sp. Nov. a nitrogen fixing acetic acid bacterium associated with sugarcane. **Int. J. Syst. Bacteriol., 39** : 361-364.

James, E.K., V.M. Reis, F.L. Olivares, J.I. Baldani and J. Dobereiner. 1994. Infection of sugarcane by the nitrogen fixing bacterium *Acetobacter diazotrophicus*. **J. Exp. Bot., 45** : 757-766.

Jayakumar, P. and M.Thangaraju. 1996. Assessment of nitrogen fixation by *Acetobacter diazotrophicus* in sugarcane. **TNAU Newslett., 26** : 1-2.

Lee, S., A. Reth, D. Meletzus, M. Sevilla and C. Kennedy. 2002 a. Characterization of major cluster of nif, fix and associated genes in the sugarcane endophyte, *Acetobacter diazotrophicus*. **J. Bacteriol.** (In press).

Lee, S., M. Sevilla, D. Meletzus, N. Gunupala and C. Kennedy. 2002 b. Characterization of nitrogen fixation genes and plant growth promoting properties in *Acetobacter diazotrophicus*,, and endophyte of sugarcane. **In: Plant Microbe Interactions. – Vol. 5,** (Ed) G. Stacey and N. Keen, APS press. (In press).

Li, R. and I.C.MacRae. 1992. Specific identification and enumeration of *Acetobacter diazotrophicus* in sugarcane. **Soil Biol. Biochem., 24** : 413-419.

Libbert, E. and H. Risch. 1969. Interactions between plants and epiphytic bacteria regarding their auxin metabolism. V. Isolation and identification of the IAA-producing and destroying bacteria from pea plants. **Physiol. Plant, 22** : 51-58.

Mowbrey, S.L., 1999. Bacterial chemorecepotors: recent progress in structure and function. **Mol. Cell., 9**: 115-118.

Muthukumarasamy, R., G.Revathi and A.R.Solayappan.1994. Biofertilizers – A supplement or substitute for chemical nitrogen for sugarcane crop. **Cooperative Sugar, 25** : 287-290.

Panneerselvam P 1997. Studies on *Acetobacter diazotrophicus* and its interaction with VA-mycorrhizal fungi in sugarcane. M.Sc. (Ag.) thesis, Tamil Nadu Agric. Univ.. Coimbatore. India. 140p

Paula, M.A.,J. Dobereiner and J.O. Siqueira. 1989. Efeio da inoculacao com fungo microrrizico VA e bacterias diazotroficas no crescimento e producao de batata-doce. **22nd Congresso Brasileiro de Ciencia do Solo,** Recife. Programme resumos, pp.109.

Paula, M.A., S.Urquiaga, J.O.Siqueira and J.Dobereiner. 1992.Synergistic effects of vesicular-arbuscular mycorrhizal fungi and diazotrophic bacteria on nutrition and growth of sweet potato (*Ipomoea batatas*). **Biol. Fertil. Soils, 14** : 61-66.

Reinhold-Hurek, B., T. Hurek, M. Gillis, B. Hoste, M. Vancannyeyt, K. Kersters and J. Deley. 1993. *Azoarcus* gen. Nov., a nitrogen fixing proteobacteria associated with roots of Kallar grass (*Leptochloa fusa* (L.) Kunth, and description of two species *Azoarcus inolegens* sp. nov., and *Azoarcus communis* sp. nov.., **Int. J. Syst.Bacteriol. 43,** 574-588.

Thangaraju M 2000 Prospects of biofertilizer usage in sugarcane. In: Proc. of symp. on current status of biofertilizer technology and its promotion in Karnataka, Tamil Nadu, Kerala and Pondicherry. (ed.) R. N. Bisoyi. Regional Biofertilizer Development Centre, Bangalore, India. pp. 202-208.

Thangaraju M and K. Govindarajan 2001 Sugarcane / A new bacterium: Better nitrogen fixing activity. **The Hindu,** April 5, 2001, pp. BS. 8.

Thangaraju M and P. Jayakumar 2002. *Acetobacter diazotrophicus* - A new endophytic nitrogen fixing bacterium associated with sugarcane. Ed. S Kannaiyan, Narosa Pub . House, New Delhi (*In press*).

Ureta, A., B.Alvarez, A.Ramon, M.A.Vera and G.Martinez Dretz. 1995. Identification of *Acetobacter diazotrophicus, Herbaspirillum seropedicae* and *Herbasipirillum* rubrisubalbicans using biochemical and genetic criteria. **Plant Soil, 172** : 271-277.

Yamada, Y., K. Hoshino and T. Ishikawa. 1997. The phylogeny of acetic acid bacteria based on the partial sequences of 16s ribosomal RNA: the elevation of the subgenus *Gluconacetobacter* to the generic level. **Biosci. Biotech. Biochem., 61:** 1244-1251.

Yamada, Y., K. Hoshino and T. Ishikawa 1998. *Gluconacetobacter* nom. Corrig. (*Gluconacetobacter* (sic). In: validation of publication of new names and new combination previously effectively published outside the IJSB, List no. 64. **Int. J. Syst. Bacteriol., 48** : 327-328.

CHAPTER – 37

OCCURRENCE OF PINK -PIGMENTED FACULTATIVE METHYLOTROPHIC BACTERIA IN TROPICAL PLANTS

M. Madhaiyan, Sp. Sundaram and S. Kannaiyan

ABSTRACT

Pink - pigmented facultative methylotrophs (PPFMs) are ubiquitous in nature and found in a variety of habitats. Members of the genus *Methylobacterium* form the major group of PPFMs and occur in the phyllosphere of a variety of plants. An attempt was made in the present study to isolate the PPFMs from phyllosphere of cereals, pulses, oil seeds, vegetables, fruit crops, spices, commercial crops, flower crops, trees and weeds growing under tropical conditions. As such seventy isolates were obtained from the above tropical plants. The populations of the PPFMs occurring in the phyllosphere region were determined. Among the different crops tested, the maximum population of PPFMs were found in the phyllosphere of *Dolichous lablab* (70 square cm^{-1}), *Phaseolus mungo* (67 square cm^{-1}), *Phaseolus aureus* and *Bougainvillea spectabilis* (65 square cm^{-1}), *Hibiscus rosasinensis* (60 square cm^{-1}) and *Pisum sativum* (53 square cm^{-1}). In general, PPFMs were dominating on the phyllosphere of pulses. The occurrences of PPFMs were comparatively less in *Carica papaya* and *Cucumber sativus* which recorded only 2 colonies square cm^{-1}.

INTRODUCTION

Pink-pigmented facultative methylotrophs (PPFMs) coming under the genus *Methylobacterium* are physiologically interesting group of bacteria that preferentially utilize substrates having one carbon compounds like methanol and methylamine as sources of energy and carbon (Patt *et al.,* 1976; Green and Bousifield, 1982 and 1983; Green, 1992). PPFMs have been found in large numbers on leaves of a diversity of plant species, including angiosperms (monocots and dicots), gymnosperms, and even bryophytes and ferns (Basile *et* al., 1969; Austin and Goodfellow, 1979; Corpe and Basile, 1982; Corpe, 1985; Corpe and Rheem, 1989; Hirano and Upper, 1989; Hirano and Upper,

1991) whereas these PPFMs were not detected on leaves of olive plants in southern Italy (Ercolani, 1978; Ercolani, 1991).

Although they are not always present in large numbers, PPFMs frequently constituted upto 79 per cent of the heterotrophic bacteria recovered from leaves of white clover (Corpe and Rheem, 1989) or more than 90 per cent of the bacteria culturable from snapbean (Hirano and Upper, 1991). Methanol concentrations of leaves were measured by Corpe and Rheem (1989) and Nemecek-Marshall *et al.* (1995) and they demonstrated the availability of this C_1 compound on leaf surfaces. It has been suggested that methanol emission from leaves may be a major source of methanol found in the atmosphere (Nemecek-Marshall *et al.*, 1995). Effects of PPFMs on germinating soybean seeds have been suggested to be mediated by cytokinins or other plant growth substances (Holland and Polacco, 1994; Freyermuth *et al.*, 1996). The vitamin B_{12} production by PPFMs was established in liverwort tissue culturing (Basile *et al.*, 1985). They are also known to enhance the germination ability and seedling vigour index of many plants (Holland and Polacco, 1994; 1996). However, literatures are not available on the occurrence of PPFMs in tropical plants. Hence in the present study, an attempt was made to isolate and estimate the quantitative occurrence of PPFMs on various tropical plants ranging from crop plants to weeds and the results are reported hereunder.

MATERIAL AND METHODS

Sample site

Leaf samples of tropical plants listed in Table 1 collected in and around the farm of Tamil Nadu Agricultural University, Coimbatore were used for isolation of PPFMs.

Isolation of PPFMS

As PPFMs have the ability to utilize methanol as sole energy source and their characteristic colony pigmentation, these organisms were isolated using ammonium mineral salts (AMS) medium developed by Whittenbery *et al.* (1970).

NH_4Cl	-	0.50 g
K_2HPO_4	-	0.70 g
KH_2PO_4	-	0.54 g
$MgSO_4.7H_2O$	-	1.00 g
$CaCl_2.2H_2O$	-	0.20 g
$FeSO_4.7H_2O$	-	4.00 mg
$ZnSO_4.7H_2O$	-	100.0 µg
$MnCl_2.4H_2O$	-	30.0 µg
H_3BO_4	-	300.0 µg
$CoCl_2.6H_2O$	-	200.0 µg
$CuCl_2.2H_2O$	-	10.0 µg
$NiCl_2.6H_2O$	-	20.0 µg
$Na_2MoO_4.2H_2O$	-	60.0 µg

Agar	-	15.0 g
Distilled water	-	1000 ml
pH	-	6.8

The AMS medium was sterilized by autoclaving at 121°C for 20 min and cooled to 50°C. Filter sterilized cyclohexamide antibiotic solution and 0.5 percent (v/v) methanol were incorporated into the sterilized AMS medium before plating and transferred into sterile petridishes approximately in 20 ml quantities. Leaf imprinting techniques developed for the isolation of PPFMs by Corpe (1985) was adopted in this study. The upper surface leaf prints of fresh leaf samples were made on the solidified media separately and incubated at 28 ± 2° C for 5 days (Holland *et al.*, 2000). The characteristic pink colored colonies capable of utilizing methanol as a sole carbon source growing over the medium were identified (Corpe and Rheem, 1989).

Quantification of PPFMS population on phyllosphere regions

The phyllosphere population of PPFMs was estimated on upper leaf surfaces and calculated using the formula,

Total PPFMs population = Number of PPFMs square cm^{-1} leaf $^{-1}$ x Leaf area (square cm^{-1}) and expressed as CFU^{-1} square cm^{-1} leaf $^{-1}$.

RESULTS

The pink-pigmented facultative methylotrophs were identified on the surface of AMS agar medium within the leaf imprinted area, based on the characteristic pink-pigmented bacterial colonies having the diameter of 1-10mm. These isolates were further confirmed by the characteristic pink-pigmented surface ring and/or thin pellicle indicating the aerobic nature in AMS liquid media. Seventy PPFMs isolates were obtained including 5 from cereals, 7 from pulses, 4 from oilseeds, 10 from vegetable crops, 6 from fruit crops, 6 from spices, 5 from commercial crops, 9 from flower crops, 8 from forest trees and 10 from weed plants. The quantitative occurrence of PPFMs on the phyllosphere of different tropical plants showed a wide variation ranging from 2 to 70 CFU square cm^{-1}. Among the different tropical crops studied the maximum of PPFMs was recorded in pulse crops *viz.*, 70 (Colong forming units) CFU square cm^{-1} in *Dolichous labalb*, 42 CFU square cm^{-1} in *Glycine max*, 65 CFU square cm^{-1} in *Phaseolus aureus* and 67 CFU square cm^{-1} in *Phaseolus mungo.* Among the cereals crops studied there was no comparable variation of PPFMs population except in *Eluecine coracana*, which recorded 5 CFU square cm^{-1}. Likewise in oilseed crops and spices crops the population variation in phyllosphere was not significant, whereas in vegetable crops the quantitative occurrence varied from 2 CFU square cm^{-1} in *Cucumber sativus* to 30 CFU square cm^{-1} in *Ipomea batatus*; in fruit crops the variation was found to be ranging from 2 CFU square cm^{-1} in *Carica papaya* to 40 CFU square cm^{-1} in *Musa paradisicca*; in commercial crops the variation of PPFMs occurrence ranged from 7 CFU square cm^{-1} in *Gossypium sp.* to 43 CFU square cm^{-1} in *Cocos nucifera.* Similarly the variation of the phyllosphere occurrence of PPFMs on flower crops were ranging from 16 CFU square cm^{-1} in *Tagetus patula* and *Crossandra undulaefolia* to 65 CFU square cm^{-1} in *Bougainvillea spectabilis*; on tree crops 5 CFU square cm^{-1} in *Tamarindus indica* to 45 CFU square cm^{-1} in *Erythrina indica*; 18 CFU square cm^{-1} on *Cyperus rotandus*, 5 CFU square cm^{-1} in *Pothas aureus*, 35 CFU square cm^{-1} in *Agava sessilana* and 5 CFU square cm^{-1} in *Agava americana* (Fig.1).

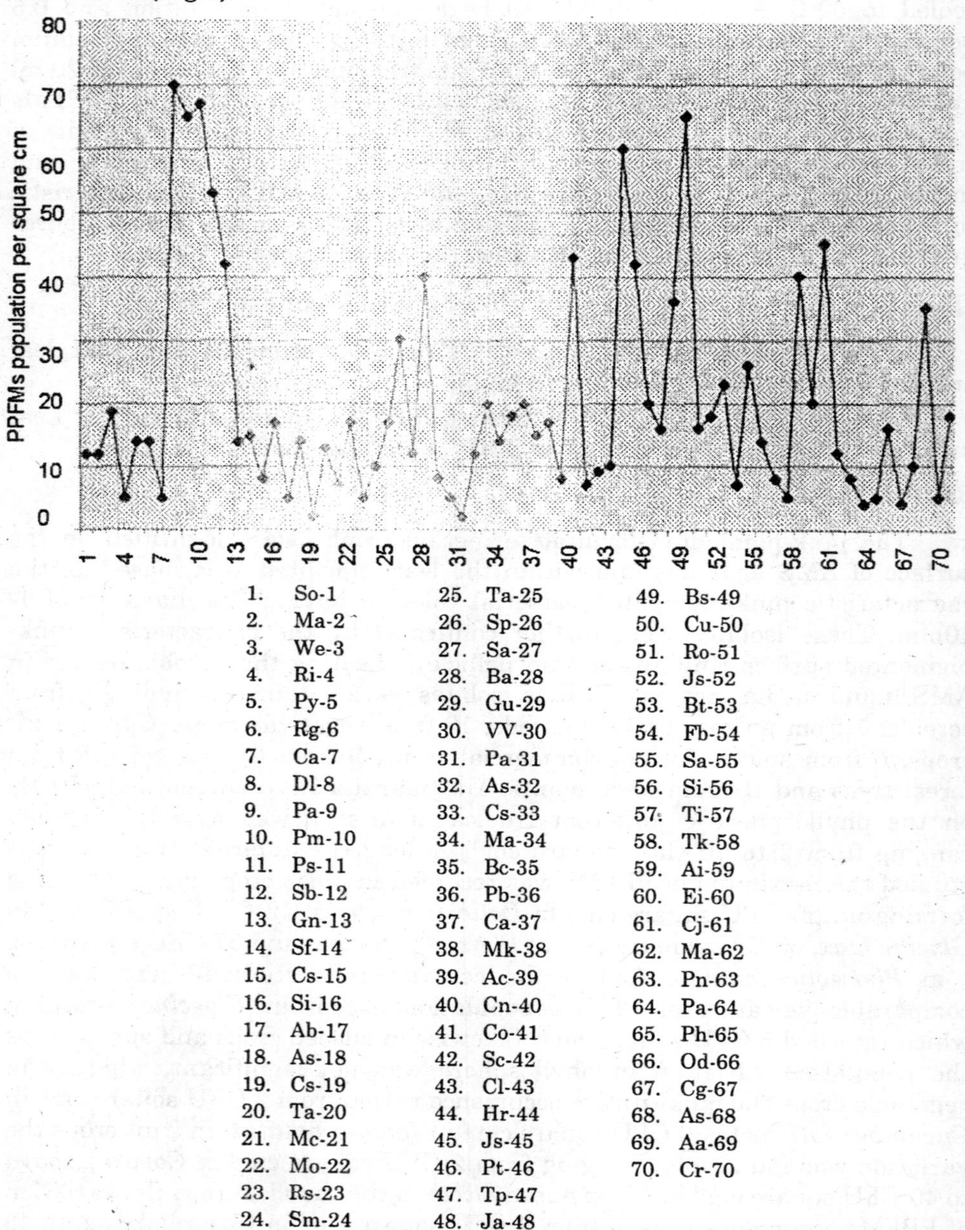

Fig.1. Survival of indigenous methylotrophs population on phyllosphere regions of different tropical plants.

There were many variation and on the occurrence of PPFMs indigenous micro flora on the phyllosphere regions of tropical crops. Total population of PPFMs was maximum in pulse crops -*Dolichus lablab* (89.60 X 10 ^{2} CFU leaf $^{-1}$) followed by flower crops- *Hibiscus rosa- sinensis* (12.72 X 10 ^{2} CFU leaf $^{-1}$), tree crops- *Tectona grantis* (7.40 X 10 ^{2} CFU leaf $^{-1}$), fruit crops- *Achras zapota* (5.72 X 10 ^{2} CFU leaf $^{-1}$), commercial crops- *Curcuma longa* (5.50 X 10 ^{2} CFU leaf $^{-1}$), weed crops- *Agava sessilana* (5.04 X 10 ^{2} CFU leaf $^{-1}$), vegetable crops- *Ipomea batatus* (4.11 X 10 ^{2} CFU leaf $^{-1}$), spice crops- *Murraya koenigii* (2.58 X 10 ^{2} CFU leaf $^{-1}$), oil seed crops- *Sesamum indicum* (2.41 X 10 ^{2} CFU leaf $^{-1}$) and cereal crops- *Oryza sativa* (1.97 X 10 ^{2} CFU leaf $^{-1}$). In general the concentration of pigmented colonies were found to be concentrating along the veins of the leaves studied. However, the density varied between different crops studied (Table 1).

Table 1. Occurrence of pink-pigmented facultative methylotrophic (PPFMs) bacteria in different tropical plants

Sl. No.	Crops	Colonies character				Leaf area (square cm $^{-3}$)	Total population leaf $^{-1}$ (x 10^2)
		Colony morphology	Diameter (mm)	Nature of occurrence	Colour		
	I. CEREALS						
1	*Sorghum bicolor*	Spreaded type	3-4	Along the veins	Pink	5.3	0.64
2	*Zea mays*	Pinpointed colony	1-2	Densely occurred	Pink	3.2	0.28
3	*Triticum aestivum*	Minute colony	2-3	Densely occurred	Pink	5.1	0.95
4	*Elusine coracona*	Minute colony	2-3	Along the veins	Pink	8.5	0.43
5	*Oryza sativa*	Pinpointd colony	1-2	Densely occurred	Pink	14.1	1.97
	II. PULSES						
6	*Cajanus cajan*	Minute colony	2-3	Densely occurred	Pink	12.7	1.78
7	*Cicer areitinum*	Minute colony	2-3	Densely occurred	Pink	14.2	0.71
8	*Dolicus lablab*	Pinpointed colony	1-2	Along the veins	Pink	12.8	89.60
9	*Phaseolus aureus*	Minute colony	2-3	Densely occurred	Pink	18.1	11.77
10	*Phaseolus mungo*	Spreaded colony	2-4	Densely occurred	Pink	12.2	8.17
11	*Pisum sativum*	Spreaded colony	2-4	Densely occurred	Reddish pink	13.2	7.00
12	*Glycine max*	Minute colony	2-3	Densely occurred	Pink	12.4	5.21
	III. OIL SEEDS						
13	*Arachis hypogaea*	Spreaded type	3-4	Along the veins	Pink	8.4	1.18
14	*Helianthus annuus*	Pinpointd colony	1-2	Densely occurred	Pink	2.0	0.30
15	*Ricinus communis*	Minute colony	2-3	Densely occurred	Pink	4.0	0.32
16	*Sesamum indicum*	Minute colony	2-3	Along the veins	Pink	14.2	2.41
	IV. VEGETABLES						
17	*Abelmoschus* spp.	Pinpointd colony	1-2	Along the veins	Reddish pink	28.2	1.41
18	*Amarathus spinosus*	Minute colony	2-3	Densely occurred	Pink	13.2	1.85

19	*Cumcumber sativus*	Minute colony	2-3	Few colony on veins	Pink	18.2	0.36
20	*Lycoperscium solanacearum*	Minute colony	2-3	Along the veins	Pink	7.2	0.94
21	*Momordia charantia*	Minute colony	2-3	Along the veins	Pink	12.4	0.87
22	*Moringa oleifera*	Spreaded colony	3-4	Densely occurred	Pink	3.0	0.51
23	*Rephanus sativus*	Minute colony	2-3	Densely occurred	Reddish pink	14.3	0.72
24	*Solanum melongena*	Pinpointd colony	1-2	Densely occurred	Pink	23.5	2.35
25	*Trichosanthus anguina*	Minute colony	2-3	Densely occurred	Pink	15.2	2.58
26	*Ipomea batatus*	Minute colony	2-3	Along the veins	Pink	13.7	4.11
	V. FRUITS						
27	*Achras zapota*	Pinpointed colony	1-2	Along the veins	Pink	9.4	1.13
28	*Musa paradisicea*	Spreaded colony	3-4	Densely occurred	Pink	14.3	5.72
29	*Psidium guajava*	Minute colony	2-3	Along the veins	Pink	13.0	1.04
30	*Vitis vinifera*	Minute colony	2-3	Densely occurred	Pink	40.2	2.01
31	*Carica papaya*	Pinpointed colony	1-2	Along the veins	Reddish pink	20.7	0.41
32	*Anona squmosa*	Minute colony	2-3	Densely occurred	Pink	14.3	1.72
	VI. SPICES						
33	*Corander sativum*	Minute colony	2-3	Densely occurred	Pink	6.1	1.22
34	*Mentha arvensis*	Pinpointed colony	1-2	Along the veins	Reddish pink	11.2	1.57
35	*Brassica campestris*	Minute colony	2-3	Densely occurred	Pink	14.2	2.56
36	*Piper betle*	Spreaded colony	2-4	Along the veins	Pink	8.2	1.64
37	*Capsicum annum*	Spreaded colony	2-4	Densely occurred	Pink	13.6	1.04
38	*Murraya koenigii*	Minute colony	2-3	Along the veins	Pink	15.2	2.58
	VIII. COMMERCIAL CROPS						
39	*Areca catechu*	Spreaded colony	2-4	Along the veins	Pink	8.4	0.67
40	*Cocos nucifera*	Minute colony	2-3	Along the veins	Pink	12.7	5.46
41	*Gassypium* sp.	Pinpointed colony	1-2	Densely occurred	Pink	20.3	1.42
42	*Saccharum officianarum*	Spreated colony	2-4	Densely occurred	Pink	50.5	4.55
43	*Curcuma longa*	Minute colony	2-3	Densely occurred	Pink	55.0	5.50

DISCUSSIONS

The PPFMs bacteria are strict aerobes and can be isolated from almost all the environment including fresh water where some dissolved oxygen exists (Graf and Bauer, 1973), where they produce pink ropy masses of growth (Hanson, 1980) and occupy a special niche. Although Corpe (1985) produced evidence to suggest that the PPFM bacteria present on mosses and liverwort producing growth stimulatory substances for these lower plants, there is presently no further evidence to support this theory of symbiotic association.

The isolation of many facultative methylotrophic bacteria on the phyllosphere of plants in recent years have been made (Stocks and McCleskey, 1964; Kouno and Okaki, 1975). Their occurrence was reported to be abundant accounting more than 90 per cent of the phyllosphere microflora in *Phaseolus vulgaris* throughout the growing season (Hirano and Upper, 1992). Holland (1997) reported 10^5 to 10^9 CFU g^{-1} of fresh weight of various plant tissues studied. The occurrence has also been reported in dry seeds of different plants. Routine sterilization techniques carried out in tissue culture preparation does not always remove the PPFMs (Holland and Polacco, 1992).

In particular, the PPFMs refer to the species *Methylobacterium* that is seed transmitted and is regularly associated with plants (Basile *et al.*, 1985; Dunleavy, 1988a; 1988b). Although microbiologists have known the bacterium for some time, its relationship to plants is relatively unknown among plant scientists. This is distributed ubiquitously, since the bacterium has been isolated from plants including monocots and dicots, angiosperms and gymnosperms, and vascular and non vascular plants (Corpe 1985; Corpe and Basile, 1982; Dunleavy, 1988a). Significantly, all plant tissues analyzed to date have been colonized with the bacteria (Corpe and Rheem, 1989; Dunleavy, 1988a; Dunleavy, 1990; Holland and Polacco, 1992). We reported seventy strains isolated from the agricultural, horticultural, ornamental, commercial crops and weed plants established the occurrence of PPFMs in a wide variety of tropical plants tested. All the isolates were easily identified from the other bacteria based on the production of pink pigment, a carotenoid which is a product of isoprenoid metabolism. Further investigations on characterization and testing their ability on crop improvement will be much useful in tapping their potential.

ACKNOWLEDGEMENT

We thank Dr. M.A. Holland (Department of Biology, Salisbury University, Salisbury, MD 21801, USA) for his excellent guidance and support in identification of the *Methylobacterium* sp.

REFERENCES

Austin, B. and M. Goodfellow. 1979. *Pseudomonas mesophilica,* a new species of pink bacteria isolated from leaf surfaces. **Int. J. Syst. Bacteriol., 29:** 373

Basile, D. V., L. L. Slade, and W. A. Corpe. 1969. An association between a bacterium and a liverwort, *Scapania nemorosa.* **Bull. Torrey Bot. Club, 96:**711-714.

Basile, D.V., M.R. Basile., Q.Y. Li., and W.A. Corpe. 1985. Vitamin B_{12} stimulated growth and development of *Jungermannta leiantha* Grolle and *Gymnocolea inflata* (Huds.) Dum. (Hepatiaceae). **Bryologist, 88(2):** 77-81.

Corpe, W.A. 1985. A method for detecting methylotrophic bacteria on solid surfaces. **J. Microbiol. Methods, 3:** 215-221.

Corpe, W.A. and D.V.Basile.1982. Methanol utilizing bacteria associated with green plants. **Dev. Indust. Microbiol., 23:** 483-493.

Corpe, W.A. and S.Rheem.1989. Ecology of the methylotrophic bacteria on living leaf surfaces. **FEMS Microbiol. Ecol., 62:** 243-250.

Dunleavy, J.M. 1988a. *Curtobacterium plantarum* sp.nov. is ubiquitous in plant leaves and is seed transmitted in soybean and corn. **Int. J. Syst. Bacteriol., 39:** 240-249.

Dunleavy, J.M. 1990. Urease production by *Methyloabcterium mesophilum,* a seed transmitted bacterium ubiquitous in soybean. Presented **at 3rd Biennial Conf. Mol. Cell. Biol. Soybean**, Ames. Iowa, July 23-25.

Dunleavy, J.M.1988b. *In vitro* expression of the cellulose gene in *Methylobacterium mesophilum,* a seed transmitted bacterium ubiquitous in soybean. Presented at **2nd Biennial Conf. Mol. Cell. Biol. Soybean**, Ames, Iowa, July 25-27.

Ercolani, G. L. 1978. *Pseudomonas savastanoi* and other bacteria colonizing the surface of olive leaves in the field. **J. Gen. Microbiol., 109:**245-257.

Ercolani, G. L. 1991. Distribution of epiphytic bacteria on olive leaves and the influence of leaf age and sampling time. **Microb. Ecol., 21:**35-48.

Freyermuth, S.K., R.L.G. Long, S. Mathur, M.A. Holland, T.P. Holtsford, N.E. Stebbins, R.O. Morris, J.C. Polacco 1996. Metabolic aspects of plant interaction with commensal methylotrophs. In: **Microbial Growth on C1 compounds,** (eds.). M.E. Lidstorm and R.F. Tabita, pp. 277-284, Kluwer, Dordrecht, Netherlands.

Graf, W., and L. Bauer. 1973. Red bacterial growth (*Corynebacterium rubrum*) in tap water systems. **Zentralbl. Bakeriol., (**Abteilung II) **73:** 74-96.

Green, P.N. 1992. The genus Methylobacterium, p. 2342-2349. In : **The prokaryotes,** 2nd ed. (eds.), A. Balows, H. G. Trüper, M. Dworkin, W. Harder, and K.-H. Schleifer vol. III. Springer-Verlag, New York, N.Y.

Green, P.N. and I.J. Bousfield, 1982. A taxonomic study of some Gram negative facultatively methylotrophic bacteria. **J. Gen. Microbiol., 128:** 623.

Green, P.N. and I.J. Bousfield, 1983. Emendation of *Methylobacterium* Patt, Cole and Hanson 1976, *Methylobacterium rhodinum* (Heumann 1962) comb. Nov.corig.; *Methylobacterium radiotolerans* (Ito and Iizuka 1971), Comb.nov.corrig., and *Methylobacterium mesophilicum* (Austin and Goodfellow 1979) comb.nov. **Int. J. Syst. Bacteriol., 33:** 875.

Hanson, R.S. 1980. Ecology and diversity of methylotrophic organisms. **Adv. Appl. Microbiol., 26 :** 3-39.

Hirano, S.S. and C.D. Upper. 1989. Diel variation in population size and ice nucleation activity of *Pseudomonas syringae* on snap bean leaflets. **Appl. Environ. Microbiol., 55:623**-630.

Hirano, S. S., and C. D. Upper. 1991. Bacterial community dynamics, p. 271-294. In: **Microbial ecology of leaves.** (eds.) J.H. Andrews, and S.S. Hirano, Springer-Verlag, New York,

Hirano, S.S. and C.D. Upper. 1992. Bacterial community dynamics. In: **Microbial Ecology of leaves.** (eds.) J.H. Andrews and S.S. Hirano, pp. 271-294. Springer-Verlag, New York.

Holland, M. A. and J. L. Polacco. 1994. PPFMs and other covert contaminants: Is there more to plant physiology than just plant?. **Ann. Rev.Plant Physiol. Plant Mol.Biol.,45:** 197-209.

Holland, M. A. and J. L. Polacco. 1996. US Patent No. 5,512,069 " **Seeds coated or Impregnated with a PPFM".**

Holland, M.A. 1997. *Methylobacterium* and plants. **Rec. Res. Dev. in Plant Physiol., 1:** 207-213.

Holland, M.A. and J.C. Polacco 1992. Urease null and hydrogenase null phenotypes of a phylloplane bacterium reveal altered nickel metabolism in two soybean mutants. **Plant Physiol., 98:** 942-948.

Holland, M.A., R. Davis, S. Moffitt, K.O' Laughin, D. Peach, S. Sussan, L. Wimbrow and B. Tayman 2000. Using "Leaf prints" to investigate a common bacterium. **The American Biology Teacher, 62(2) :** 128-131.

Kouno, K., and A. Okaki. 1975. Distribution and identification of methanol utilizing bacteria, p. 11-21. In: **Microbial growth on C1 compounds** (eds.) G. Terui *et al.*, Society of Fermentation Technology, Tokyo, Japan.

Nemecek-Marshall, M., R. C. MacDonald, J. J. Franzen, C. L. Wojciechowski, and R. Fall 1995. Methanol emission from leaves. **Plant Physiol., 108 :** 1359-1368.

Patt, T.E., G.C. Cole, and R.S. Hanson 1976. *Methylobacterium,* a new genus of facultative methylotrophic bacteria. **Int. J. Syst. Bacteriol., 26:** 226-229.

Stocks, P. K., and C. S. McCleskey 1964. Identity of the pink-pigmetned methanol-oxidising bacteria as *Vibrio extoquens.* **J. Bacteriol., 88:** 1065-1070.

Whittenbury, R.S., S.L. Davies, and J.F. Davey 1970. Exposures and cysts formed by methane utilizing bacteria. **J. Gen. Microbiol., 61:**219-226.

CHAPTER – 38

STUDIES ON PINK-PIGMENTED FACULTATIVE METHYLOTROPHIC BACTERIA ON *Glycine max*

M. Madhaiyan, Sp. Sundaram and S. Kannaiyan

ABSTRACT

The plant canopy provides an excellent niche for the establishment and the multiplication of a wide range of microorganisms. Among the bacterial genera on the plant surface, the pink- pigmented facultative methylotrophs (PPFMs) associate with seeds and leaves of all the tropical and temperate crops. They are present ubiquitously on the plant canopy and play role in the production of cytokinins and vitamins. An attempt was made to study effect of inoculation of ten PPFM strains on germination and the growth of soybean. The soybean seeds were treated with PPFMs strains and tested under pot culture condition. Results suggested that the *Methylobacterium sp* (Sb 34) isolates from soybean increased the shoot length (26.4 cm plant^{-1}), root length (25.01 cm plant^{-1}), plant biomass (11.25g fresh weight plant^{-1} and 2.05 g dry weight plant^{-1}), and nodulation efficiency (13.5 percent) followed by *Methylobacterium sp* (Bg 19) obtained from blackgram recording shoot length (25.4 cm plant^{-1}), root length (23.22 cm plant^{-1}), plant biomass (10.25 g fresh weight plant^{-1} and 2.01 g dry weight plant^{-1}), and nodulation efficiency (12.5 percent) in soybean. The increased plant parameters may be attributed to the production of cytokinins and vitamins by pink- pigmented facultative methylotrophic bacteria.

INTRODUCTION

Pink-pigmented facultative methylotrophic bacteria (PPFMs) have been found in large numbers on leaves of a diversity of plant species, including angiosperms, gymnosperms, and even bryophytes and ferns. PPFMs are bacterial symbiont of plants, shown previously to participate in plant metabolism. This organism is likely the descendants of seed borne in nature and enhances the rate of seed germination under *in vitro* conditions. They provide exogenous source of plant growth regulator cytokinin, vitamin B_{12} and play significant role in nitrogen metabolism by inducing urease activity.

Although they are not well known, these bacteria are co-evolved, interacting partners in plant metabolism. Evidence exists for a complex interaction between the PPFMs and its plant host (Basile *et al.*, 1969; Corpe and Basile, 1982; Holland and Polacco, 1992, Holland and Polacco, 1994).

The genus *Methylobacterium* is a physiologically interesting group of bacteria that preferentially utilize substrates having one carbon compounds like methanol and methylamine as sources of energy and carbon. Because of this unusual ability to grow utilizing methanol or other one carbon compounds, they are easily isolated from plant tissues using selective media containing methanol as sole carbon source (Corpe, 1985). Their occurrence was reported to be abundant accounting more than 90 per cent of the phyllosphere micro flora in *Phaseolus vulgaris* throughout the growing season (Hirano and Upper, 1992). Holland (1997) reported 10^5 to 10^9 CFU g^{-1} of fresh weight of various plant tissues studied. However, there was no research work available as the beneficial role of PPFMs on the crop plants grown under tropical conditions. Hence the present study was conducted to know the effect of PPFMs in enhancing the productivity of soybean by spraying PPFMs on a plant.

MATERIALS AND METHODS

Source of PPFMs

The following PPFMs bacterial strains viz., *Methylobacterium* sp (Sb-34), *Methylobacterium* sp (Bg-19), *Methylobacterium* sp. (Mb-41), *Methylobacterium extorquens* (Tk-33), *Methylobacterium* sp. (Re-32), *Methylobacterium* sp. (Dt-34), *Methylobacterium* sp. (Pa-38), *Methylobacterium* sp. (Ma-42), *Methylobacterium* sp. (Co-45), and *Methylobacterium* sp. (Tk-39) which were already isolated and identified based on the carbon utilization test from our laboratory were used in the present study.

PPFMs Treatment to soybean seeds

PPFMs were grown in AMS liquid medium (Whittenbury *et al.,* 1970) at 30^0C for 5 days till it reached 10^8 cells ml^{-1}. The soybean seeds were surface sterilized with 5 per cent sodium hypochlorite solution for 2-3 min and 70 per cent alcohol for 2 minutes and repeatedly washed in sterile distilled water for 6 times and the soybean seeds were imbibed with PPFMs strains for 12 h (Holland and Polacco, 1994).

Growth parameter and germination test

Twenty-five seeds treated with PPFMs culture were kept in a petridish containing the sterilized moisturized filter paper and kept undisturbed for germination in dark for three days. After three days the number of seeds germinated in each treatment was recorded and the germination percentage was calculated. The other growth parameters like root and shoot length and plant biomass were observed by growing the germinated seeds using paper towel method.

In this method the sterile pouches made of polypropylene (16x18 cm) were used. Inside the pouches, paper towel was kept in such a way that it supports the growth of the seeds by folding the paper towel to form a small fold and holes for placement of the seeds. Thirty ml of sterilized plant nutrient solution was taken in each pouch and the three days old germinated seeds were kept in a manner that the radicle is inserted inside the hole for its growth. After 15-20 days the growth of seedlings, nodule efficiency, seedling length and plant biomass was estimated (Somasegaran and Hoben, 1994).

RESULTS

Germination test

All PPFMs treated seeds exhibited significantly higher germination percentage over uninocualted control. Among the PPFMs treated seeds the maximum germination percentage of 99 was obtained in the seeds treated with *Methylobacterium* sp (Sb- 34) followed by *Methylobacterium* sp (Bg-9), which recorded 98 per cent where as the uninocualted control recorded only 91 per cent (Table 1).

Plant growth

All the treatments recorded significantly higher root and shoot growth except *Methylobacterium* sp. (Ma-42), which recorded a lower shoot growth (22.98 cm plant^{-1}). However, *Methylobacterium* sp. (Bg-19) recorded significantly higher growth (25.40 cm plant^{-1}) than uninocualted control. The maximum plant growth was noticed in *Methylobacterium* sp (Sb-34) treated soybean plants which recorded 26.40 and 25.01 cm plant^{-1} of respective shoot and root growth (Table 1).

Plant biomass production

Methylobacterium sp (Sb-34) recorded the maximum plant biomass of 11.25g fresh biomass and 2.05g dry biomass 10 seedlings^{-1}, which is significantly higher than the other inoculated treatments as well as uninocualted control. *Methylobacterium* sp. (Bg-19) followed next recording 10.25g fresh biomass and 2.01g dry biomass 10 seedlings^{-1}. While the uninoculated control recorded only 9.23g fresh biomass and 1.82g dry biomass 10 seedlings $^{-1}$ (Table 1).

Table 1. Effect of pink- pigmented facultative methylotrophs on the growth of *Glycine max*

Sl. No	Treatments	Host	Plant growth (cm plant^{-1})*		Plant biomass (gm plant^{-1})*		Germination (%)	Nodulation efficiency (%)
			Shoot length	Root length	Fresh weight	Dry weight		
1	Uninocualted control		22.01	20.82	9.23	1.82	91.0	11.15
2	*Methylobacterium* sp (Sb-34),	Soybean	26.40	25.01	11.25	2.05	99.0	13.50
3	*Methylobacterium* sp (Bg-19),	Blackgram	25.40	23.22	10.25	2.01	98.0	12.00
4	*Methylobacterium* sp.(Mb-41),	Mulberry	23.12	23.12	10.01	1.98	97.0	11.50
5	*Methylobacterium extorquens*	Teak	24.06	22.82	10.22	1.96	97.0	11.98
6	(Tk- 33)	Rose	24.32	22.48	9.48	1.97	96.0	11.42
7	*Methylobacterium* sp. (Re-32),	Agava	24.40	21.07	9.52	2.00	95.0	11.83
8	*Methylobacterium* sp. (Dt-34),	Papaya	25.02	23.18	10.03	1.99	94.0	11.92
9	*Methylobacterium* sp. (Pa-38),	Maize	22.98	22.98	10.24	1.95	97.0	11.34
10	*Methylobacterium* sp.(Ma-42),	Cotton	25.12	23.01	10.19	1.83	97.0	11.62
11	*Methylobacterium* sp. (Co-45), sp. (Tk-39)	Teak	24.53	23.15	9.92	1.87	96.0	11.35
	SEd		0.424	0.091	0.078	0.036	0.020	0.036
	CD (0.05%)		0.879	0.187	0.162	0.074	0.042	0.074

* Average of 10 seedlings

Nodulation efficiency

Results suggested that the pink pigmented facultative methylotrophic bacterial strains obtained from soybean (Sb-34) has the maximum nodulation efficiency of 13.5 percent followed by blackgram isolates (Bg-19) with nodulation efficiency of 12.5 percent in soybean over the uninocualted control having 11.15 percent (Table 1).

DISCUSSION

Methylobacterium sp. is abundant, non-pathogenic bacteria ubiquitously distributed on plants. Although they are not well known, these are co-evolved, interacting partners in plant metabolism in a number of important ways including, production of enzymes (Holland and Polacco, 1992), stimulation of germination (Holland and Polacco, 1994), enhancement of growth (Basile *et al.*, 1969; Holland, 1997), production of cytokinins (Freyermuth *et al.*, 1996) and production of vitamin B_{12} (Basile *et al.*, 1985). The present investigation provides a method for increasing the productivity of a plant by seed imbibitions on soybean plants. The study also relates to increasing productivity of a plant under stress conditions by applying PPFMs to a plant and subsequently applying an aqueous solution-containing methanol to the plant. Pink-pigmented facultative methylotrophic bacteria are symbiotic, seed transmitting on plants and enhance germination rate of seeds (Holland, 1999).

PPFMs stimulate germination in almost all crops, which synthesize cytokinin plant growth regulators and vitamin B_{12}. PPFMs are mostly isolated from phyllosphere regions (Corpe and Basile, 1982; Corpe and Rheem, 1989) and they are tentatively identified as the genus *Methylobacterium* sp. (Green and Bousifield, 1982). Foliar application of PPFMs during the seed set in soybean lowered the incidence of embryo abortion and resulting in higher yields (Munsanje, 1999). The stimulation of growth and development of soybean germination and yield by PPFMs suggests that they produce plant growth regulators. Our results suggested that the imbibitions of soybean seeds into the culture media (PPFMS) has increased the productivity, germination percentage, nodulation efficiency, total soluble sugar and reducing sugar. From these result, it is evidently clear that the novel PPFMs bacteria can be used as a potential bioinoculant for soybean.

ACKNOWLEDGEMENT

We thank Dr. M.A. Holland (Department of Biology, Salisbury University, Salisbury, MD 21801, USA) for his excellence guidance and support in identification of the *Methylobacterium* sp.

REFERENCES

Basile, D. V., L. L. Slade, and W. A. Corpe. 1969. An association between a bacterium and a liverwort, *Scapania nemorosa*. **Bull. Torrey Bot. Club, 96:**711-714.

Basile, D.V., M.R. Basile, Q.Y. Li., and W.A. Corpe. 1985. Vitamin B12 stimulated growth and development of *Jungermannta leiantha* Grolle and *Gymnocolea inflata* (Huds.) Dum. (Hepatiaceae). **Bryologist, 88(2):** 77-81.

Corpe, W.A., and D.V. Basile 1982. Methanol-utilizing bacteria associated with green plants. **Dev. Ind. Microbiol., 23:** 483-493.

Corpe, W. A. and S. Rheem. 1989. Ecology of the methylotrophic bacteria on living leaf surfaces. **FEMS Microbiol. Ecol., 62:** 243-250.

Corpe, W.A. 1985. A method for detecting methylotrophic bacteria on solid surfaces. **J. Microbiol. Methods, 3:** 215-221.

Freyermuth, S.K., R.L.G. Long, S. Mathur, M.A. Holland, T.P. Holtsford., N.E. Stebbins, R.O. Morris, J.C. Polacco. 1996. Metabolic aspects of plant interaction with commensal methylotrophs. In: **Microbial Growth on C1 compounds,** (eds.) M.E. Lidstorm and R.F. Tabita, *pp.* 277-284, Kluwer, Dordrecht, Netherlands.

Green, P. N. and I. J. Bousfield. 1982. A taxonomic study of some gram-negative facultatively methylotrophic bacteria. **J. Gen. Microbiol., 128:** 623-638.

Hirano, S.S. and C.D. Upper. 1992. Bacterial community dynamics. In: **Microbial Ecology of leaves,** (eds.) J.H. Andrews and S.S. Hirano, pp. 271-294. Springer-Verlag, New York.

Holland, M. A. and J. C. Polacco. 1992. Urease-null and hydrogenase-null phenotypes of a phylloplane bacterium reveal altered nickel metabolism in two soybean mutants. **Plant Physiol., 98:**942-948.

Holland, M.A and J.L Polacco. 1994. PPFMs and other covert contaminants: Is there more to plant physiology than just plant? **Ann. Rev.Plant Physiol. Plant Mol.Biol.,45:** 197-209.

Holland, M.A. 1997. *Methylobacterium* and plants. **Rec. Res. Dev.in plant Physiol. 1:** 207-213.

Munsanje, E. M. 1999. **Potential of a cytokinins-secreting *Methylobacterium* as a biofertilizer in soybean production**. Ph.D.dissertation. University of Maryland Eastern Shore, Princess Anne, MD.

Somasegaran, P. and H.J. Hoben. 1985. **Methods in legume - *Rhizobium* technology.** NifTAL Project and MIRCEN. University of Hawaii, Honolulu, pp. 1-367.

Whittenbury, R.S., S.L. Davies and J.F. Davey. 1970. Exposures and cysts formed by methane- utilizing bacteria. **J. Gen. Microbiol., 61:** 219-226.

CHAPTER – 39

EFFECT OF PINK- PIGMENTED FACULTATIVE METHYLOTROPHIC BACTERIAL ISOLATES ON THE COTTON SEED QUALITY

M. Madhaiyan, Sp. Sundaram and S. Kannaiyan

ABSTRACT

Pink pigmented facultative methylotrophic (PPFM) bacterial strains belonging to the genus *Methylobacterium*, normally associated with seeds and leaves of tropical plants were tested to study their influence on the cotton seed quality. Ten PPFMs strains were grown in ammonium mineral salt (AMS) liquid medium for 4 to 5 days at room temperature and the cotton seeds were treated with PPFM strains and sown in germination trays. The germination test was conducted in a seed germination room at a temperature of 20-25^{0}C and relative humidity 80-90 per cent for 12 days. Samples were taken after incubation periods and analyzed for seedling length, fresh weight and dry matter content. Results suggested that *Methylobacterium* sp. (CO-47) obtained from cotton increased the shoot length (13.60 cm plant^{-1}), root length (13.75 cm plant^{-1}), plant biomass (7.50 g fresh weight plant^{-1}and 1.26 g dry weight plant^{-1}), germination percentage (98 per cent) and vigour index (2653) followed by brinjal isolates recording the shoot length (13.12 cm plant^{-1}), root length (13.21 cm plant^{-1}), plant biomass (7.16 g fresh weight plant^{-1} and 1.23 g dry weight plant^{-1}), germination percentage (97 per cent) and vigour index (2617) in cotton when compared with uninocualted control. The initial enhanced seedling vigour may be due to the production of plant hormone and vitamins by PPFMs bacteria colonizing on the phyllosphere regions of cotton.

INTRODUCTION

The relationship between microbes and plants is an unappreciated evolutionary story that probably began with the development of multicellularity, cellular differentiation and division of labour among tissues and organs was accompanied by a loss of biochemical versality that characterize unicellular life (Holland, 1997). The plant canopy is a host to a wide array of microorganisms beneficial, harmful and neutralistic. The occurrence of several such relationships often go undetected.

Many of the microbes living on the phylloplane probably lead a saprophytic lifestyle, feeding on materials leached from the leaf.

The association of *Methylobacterium sp.* with plant has been known to the scientific community, but its relationship to plants is unknown. PPFMs are probably distributed ubiquitously on plants of all taxa (Basile *et al*, 1969; Corpe and Basile, 1982; Dunleavy, 1988a; 1988b; 1990). The genus *Methylobacterium* is composed of a variety of pink pigmented facultative methylotrophic (PPFM) bacteria, which are capable of utilizing one carbon compounds such as formate, formaldehyde, and methanol as sole source of carbon and energy. PPFMs enhanced the germination of soybean seeds, which is mediated by cytokinins or other plant growth substances (Holland and Polacco, 1994; Freyermuth *et al.,* 1996). They are also known to enhance the germination ability and seedling vigour index of many plants (Holland and Polacco, 1994).

Though detailed studies are available on the influence of PPFMs on temperate plants, no such reports are available in tropical agriculture. Since the PPFMs seem to be more potential and novel microbial symbiont especially to enhance the germination percentage and increase the seedling vigour, an attempt was made in the present study to evaluate some of the efficient PPFMs strains isolated in our lab on cotton plants.

MATERIALS AND METHODS

Source of PPFMs

The following PPFMs strains screened by us viz., *Methylobacterium* sp. (Pa-38), *Methylobacterium* sp. (Wp-40), *Methylo-bacterium* sp. (Dt-34), *Methylobacterium* sp. (Mb-41), *Methylobacterium extorquens* (Tk-33), *Methylobacterium* sp. (Je-42), *Methylobacterium* sp. (Br-29), *Methylobacterium extorquens* (Co-47), *Methylobacterium* sp. (Tk-39), *Methylobacterium* sp. (Re-32) were utilized in the present study.

PPFMS treatment to cotton seeds

PPFMs bacteria were grown in AMS liquid medium (Whittenbury *et al.,* 1970) at 30^{0}C for 5 days till it reached 10^8 cells ml^{-1}. The delinted cotton seeds were surface sterilized with 5 per cent sodium hypochlorite solution for 2-3 min and imbibed with mass multiplied PPFMs bacteria for 12 hours (Holland and Polacco, 1994).

Germination test

Twenty five cotton seeds treated with PPFMs strains were sown in aluminium trays containing fine sand with moisturized conditions. These trays were kept in a germination room at 20⁰C for 12 days. Every day the germination trays were observed to record the number of seeds germinated and the rate of germination was computed in each treatment. On 12th day all the seedlings were pulled out with entire root system intact and washed with tap water to remove sand particles. After blot drying 10 seedlings were selected at random to record shoot length, root length and plant fresh biomass. After recording these observation the seedlings were kept in an incubator at 80 ± 2⁰C for 48 h till it reached a constant weight and plant dry matter production was recorded. Vigour index was computed by using the formula and expressed as whole number (Abdul Baki and Anderson, 1973).

Vigor index (VI) = germination percentage x seedling length (shoot + root length).

RESULTS

Rate of germination

PPFMs inoculated plants recorded higher germination, plant growth than uninocualted control. However, *Methylobacterium extorquens* (Co-47) and *Methylobacterium* sp. (Br-45) exhibited higher rate of germination recording 8.08 and 8.07 respectively followed by *Methylobacterium* sp. (Mb-41), *Methylobacterium* sp. (Dt-34), *Methylobacterium* sp. (Tk-39) and *Methylobacterium* sp. (Je-42). In general these PPFMs inoculated plants recorded significantly higher germination rate over other PPFMs as well as uninocualted plants (Table 1).

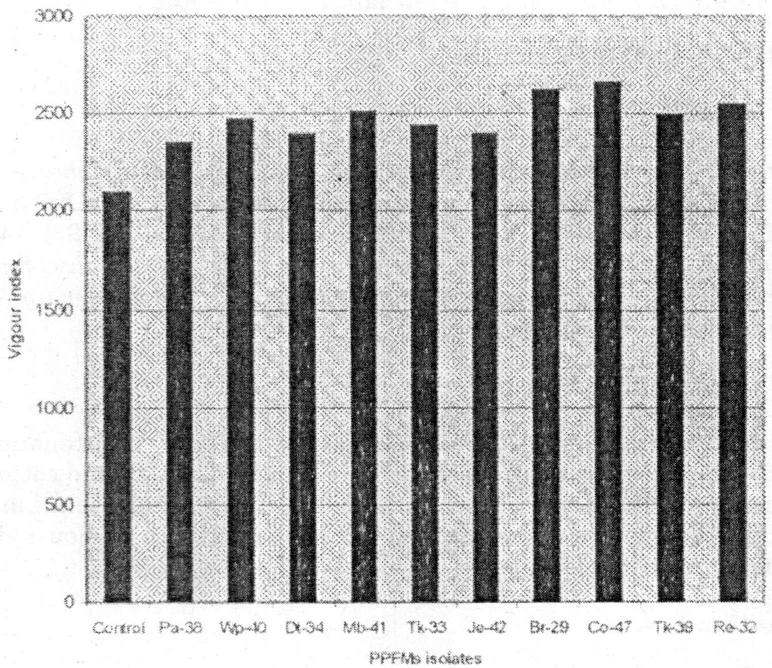

Fig. 1. Effect of PPFMs strains inoculation on the vigour index in cotton

Germination percentage and vigour index

All PPFMs inoculated seeds exhibited significantly higher germination percentage over uninocualted control. The maximum germination (98 per cent) and vigour index (2653) recorded in *M. extorquens* Co-47 inoculated cotton followed by *Methylobacterium* sp. (Br-29) recording vigour index (2617) and germination (97 per cent) whereas uninocualted control recorded the lowest vigour index (2087) and germination (85 per cent) (Fig. 1 and Table 1).

Table 1. Effect of PPFMs strains inoculation on the cotton seed quality.

Sl. No	Treatments	Plant growth (cm^{-1} $plant^{-1}$)*		Plant biomass (g^{-1} $plant^{-1}$)*		Rate of germination	Germination (%)
		Shoot length	Root length	Fresh weight	Dry weight		
1	Uninocualted Control	12.10	11.09	5.88	0.912	7.03	85.0
2	*Methylobacterium* sp. (Pa-38)	12.64	12.08	6.43	1.069	7.56	95.0
3	*Methylobacterium* sp. (Wp-40)	12.95	12.76	7.11	0.983	7.55	96.0
4	*Methylobacterium* sp. (Dt-34)	12.60	12.03	6.77	0.803	7.91	97.0
5	*Methylobacterium* sp. (Mb-41)	12.80	13.05	6.62	0.986	7.99	97.0
6	*Methylobacterium* extorquens (Tk-33)	13.09	12.10	6.34	1.105	7.79	96.0
7	*Methylobacterium* sp. (Je-42)	13.15	11.28	7.06	0.906	7.68	96.0
8	*Methylobacterium* sp. (Br-29)	13.12	13.21	7.16	1.227	8.07	97.0
9	*Methylobacterium extorquens* (Co-47)	13.60	13.75	7.50	1.257	8.08	98.0
10	*Methylobacterium* sp. (Tk-39)	13.04	12.43	6.72	1.093	7.69	97.0
11	*Methylobacterium* sp. (Re-32)	12.90	13.05	6.46	1.107	7.29	97.0
	SEd	0.051	0.045	0.464	0.004	0.034	1.614
	CD (0.05%)	0.105	0.094	0.096	0.009	0.070	3.348

Average of 10 seedlings

Plant growth

All inoculated treatments recorded significantly higher root and shoot growth over uninocualted control. Among the PPFMs tested, *Methylobacterium extorquens* (Co-47) recorded significantly higher shoot (13.60 cm $plant^{-1}$) and root growth (13.75 cm $plant^{-1}$) over other strains followed by *Methylobacterium* sp. (Br-45) inoculated cotton plants recording 13.12 and 13.21 cm $plant^{-1}$ of respective shoot and root growth. *Methylobacterium* sp. (Dt-34), recorded least shoot growth (12.60 cm $plant^{-1}$) and root growth (12.03 cm $plant^{-1}$) among the PPFMs tested (Table 1).

Plant biomass production

Methylobacterium extorquens (Co-47) recorded a highest plant biomass producing 7.50 g fresh biomass and 1.257g dry biomass 10 $seedlings^{-1}$, which is significantly higher than other inoculated treatments as well as uninocualted control. *Methylobacterium* sp. (Br-45) followed next recording 7.16g fresh biomass and 1.227g dry biomass 10 $seedlings^{-1}$. While the uninocualted control recorded only 5.88 g fresh biomass and 0.912g dry biomass 10 seedlings^{-1} (Table 1).

DISCUSSIONS

Methylobacterium sp. a group of bacteria that are familiarly known in the literature as pink-pigmented facultative methylotrophs (Austin and Goodfellow, 1979; Patt *et al.*, 1976; Green and Bousifield, 1982; 1983). All PPFMs tested in the present study gave positive response on the rate of germination, germination percentage vigour index, seedling growth, and plant biomass production in cotton plants. The PPFMs are reported to be closely associated with the seeds and are known to be carried even after surface sterilization (Holland and Polacco, 1996). They also established the importance of PPFMs on seed germination by demonstrating less germination ability of heat-treated soybean seeds at 50^{0}C for 48 h and restoring to normality by PPFMs treatment. Among the methylotrophs tested *Methylobacterium extorquens* (Co-47) and *Methylobacterium* sp. (Br-45) performed better and increased rate of germination, vigour index, plant growth and plant biomass production. *Methylobacterium extorquens* (Co-47) recorded maximum rate of germination (8.08), vigour index (2653) and plant dry matter production (1.257 g 10 seedlings^{-1}) followed by *Methylobacterium* sp. (Br-45) recording rate of germination as 8.07, vigour index (2617), and plant dry matter production (1.227 g 10 seedlings^{-1}). In general, all PPFMs inoculated plants recorded significantly higher rate of germination percentage, vigour index, plant growth, fresh biomass production and dry matter production over uninocualted control. The enhanced vigour of PPFMs treated cotton seedlings may be due to cytokinin produced during the isoprenoid metabolism of the bacterium as well documented by Renoux and Rohmer (1985). Cytokinins production by the PPFMs leads naturally to a consideration of potential of this organism to influence the growth and development of plants (Freyermuth *et al.*, 1996). Holland (1997) has also reported a method for increasing productivity of soybean by spraying pink-pigmented facultative methylotrophic bacteria. Our results suggested that the PPFMs could be possibly used as a bioinoculant to enhance germinability, and vigour of.the cotton seeds, since all the strains tested on cotton in the present study established their beneficial role in cotton seed germination, enhancing vigour and establishment of seedlings.

ACKNOWLEDGEMENTS

We thank Dr. M.A. Holland (Department of Biology, Salisbury University, Salisbury, MD 21801, USA) for his valuable guidance and support in identification of the *Methylobacterium* sp.

REFERENCES

Abdul- Baki, A.A. and J. D. Anderson. (1973). Vigour determination in soybean seed by multiple criteria. Crop Sci., 13: 630-632.

Austin, B. and M. Goodfellow. 1979. *Pseudomonas mesophilica,* a new species of pink bacteria isolated from leaf surfaces. Int. J. Syst. Bacteriol. 29: 373.

Basile, D. V., L. L. Slade, and W. A. Corpe. 1969. An association between a bacterium and a liverwort, **Scapania nemorosa. Bull. Torrey Bot. Club** 96:711-714.

Corpe, W.A. and D.V.Basile.1982. Methanol- utilizing bacteria associated with green plants. Dev. Indust. Microbiol. 23: 483-493.

Dunleavy, J.M. 1988a. *Curtobacterium plantarum* sp.nov. is ubiqutous in plant leaves and is seed transmitted in soybean and corn. Int. J. Syst. Bacteriol. 39: 240-249.

Dunleavy, J.M.1988b. *In vitro* expression of the cellulose gene in *Methylobacterium mesophilum,* a seed transmitted bacterium ubiquitous in soybean. Presented at 2nd Biennial Conf. Mol. Cell. Biol. Soybean, Ames, Iowa, July 25-27.

Dunleavy, J.M. 1990. Urease production by *Methyloabcterium mesophilum,* a seed transmitted bacterium ubiquitous in soybean. Presented at 3rd Biennial Conf. Mol. Cell. Biol. Soybean, Ames. Iowa, July 23-25.

Freyermuth, S.K., R.L.G. Long, S. Mathur, M.A. Holland, T.P. Holtsford, N.E. Stebbins, R.O. Morris, J.C. Polacco. 1996. Metabolic aspects of plant interaction with commensal methylotrophs. In: Microbial Growth on C1 compounds. (Eds.) M.E. Lidstorm and R.F. Tabita, pp. 277-284, Kluwer, Dordrecht, Netherlands.

Green, P.N. and I.J. Bousfield, 1982. A taxonomic study of some Gram- negative facultatively methylotrophic bacteria. J. Gen. Microbiol. 128, 623.

Green, P.N. and I.J. Bousfield, 1983. Emendation of *Methylobacterium* Patt, Cole and Hanson 1976, *Methylobacterium rhodinum* (Heumann 1962) comb. Nov.corig; *Methylobacterium radiotolerans* (Ito and Iizuka 1971), Comb.nov.corrig., and *Methylobacterium mesophilicum* (Austin and Goodfellow 1979) comb.nov. Int. J. Syst. Bacteriol. 33 : 875.

Holland, M.A and J.L Polacco 1994. PPFMs and other covert contaminants: Is there more to plant physiology than just plant? Ann. Rev. Plant Physiol. Plant Mol. Biol.45: 197-209.

Holland, M.A and J.L Polacco 1996. US patent No. 5,512,069 “Seeds Coated or Impregnated with a PPFM”.

Holland, M.A. 1997. *Methylobacterium* and plants. Rec. Res. Dev. in Plant Physiol. 1: 207-213.

Patt, T.E., G.C. Cole, and R.S. Hanson. 1976. *Methylobacterium,* a new genus of facultative methylotrophic bacteria. Int. J. Syst. Bacteriol. 26: 226-229.

SECTION – VI

MICROBIAL SOLUBILIZATION AND MOBILIZATION OF NUTRIENTS

CHAPTER – 40

A NEW MIXED BIOINOCULUM – 'AZOPHOS'

K. Premalatha, P. Subramanian and S. Anthoni Raj

ABSTRACT

A composite biofertilizer inoculum containing both *Azospirillum* and Phosphobacteria was developed. It has been named as AzoPhos. The main advantages of this single biofertilizer packet containing both nitrogen fixer and phosphate solubilizer is, less expensive, easy to use and also better efficiency of both the organisms in mixed culture. After studying their comptability it has been standardized that at 50:50 level of mixing of both *Azospirillum* and phosphobacteria was better in both pot culture and field condition in rice. AzoPhos packets were subjected to survival test and found that *Azospirillum* and phosphobacteria survived in the AzoPhos packets for three months with maximum population of 10^6 cells g^{-1} of inoculum. The field trails conducted in Kuruvai season rice confirmed that AzoPhos inoculation recorded the yield difference which was statistically significant. There was considerable reduction in use of biofertilizer and input cost for biofertilizer purchase as well as labour used for application for large areas. Also the efficiency of *Azospirillum* and phosphobacteria is increased when both are mixed and packed together.

INTRODUCTION

Biofertilizers will have to play a significant role in the present scenario of increased fertilizer cost and indiscriminate use of chemical fertilizer that leads to loss of land vigour in agricultural productivity. With the awakening on organic farming and spiraling of fertilizer cost, farmers are trying to use more and more of biofertilizers. Nitrogen and phosphorus are the two essential nutrients influencing the crop growth and productivity of crops.

At present *Azospirillum* and Phosphobacteria are available in separate packets. Further it is expensive and difficult for the farmers to do the seed treatment with two sets of separate packets. Considering all these aspects, an attempt has been made to formulate a new mixed biofertilizer inoculum containing both *Azospirillum* and phosphobacteria in varying proportions (50:50, 60:40, 70:30, 80:20, 100: 100), to overcome some of the shortocoming mentioned above. The new mixed bioinoculum has been named as 'AzoPhos'.

MATERIALS AND METHODS

The *Azospirillum* and phosphobacteria cultures were mass cultured in nitrogen free malate broth and nutrient broth respectively. The rice mill waste, rice husk ash were used as carrier material for preparing mixed inoculum. The inoculum of *Azospirillum* and phosphobacteria were mixed in different proportions as give below.

1. 50% *Azospirillum* + 50% Phosphobacteria
2. 60% *Azospirillum* + 40% Phosphobacteria
3. 70% *Azospirillum* + 30% Phosphobacteria
4. 80% *Azospirillum* + 20% Phosphobacteria
5. 100% *Azospirillum* + 100% Phosphobacteria

The quantity of broth used for mixing with carrier was adjusted to give 40 per cent moisture content in the carrier.

The shelf-life of AzoPhos was evaluated by enumerating the *Azospirillum* and phosphobacteria population at 1,30,60 and 90 days after packing, using Most Probable Number (MPN) technique for enumerating *Azospirillum* and dilution plate technique for phosphobacteria. The seeds treated with AzoPhos were also examined for the adherence of *Azospirillum* and Phosphobacteria on the seed by adopting the above mentioned technique.

Field experiments were conducted during Kharif season under wetland conditions in plots of 4 x 3m size with 15 x 10 cm spacing. The fertilizer schedule adopted was 75:37.5: 50 kg NPK ha^{-1}. Nitrogen was applied in three split doses at 0,30 and 45 days after transplanting. The entire dose of phosphorus and potassium was applied basally. The water level was maintained at 5-10 cm level throughout the crop growth. The experiment was laid out in randomized block design with the following treatments with three replications.

T_1 - Control

T_2 - AzoPhos (50:50)

T_3 - AzoPhos (60:40)

T_4 - AzoPhos (70:30)

T_5 - AzoPhos (80:20)

T_6 - AzoPhos (100:100)

T_7 - *Azospirillum*

T_8 - Phosphobacteria

The rice husk ash based biofertilizers containing 10^9 cells g^{-1} were used for the experiments. Biofertilizers were applied through seedling root dip (one Kg for seedlings required to plant one hectare) and manifield soil application (Two kg per hactare with ten kg FYM) before transplaning.

Grain and straw yield were recorded at the time of harvest and the yield data were analysed statistically.

RESULTS AND DISCUSSION

The population of *Azospirillum* and phosphobacteria in AzoPhos packet wer found to decline when the storage period increases. On 90th day after storage, the population was 0.9 x 10^6 cells / g. It is evidently clear from the results that AzoPhos (50:50) is equal with AzoPhos (100 : 100) in maintaining the population. In AzoPhos (50:50) less quantity of inoculum has been used. So this treatment is superior to AzoPhos (100 : 100) by taking the cost factor into consideration. (Table 1 & 2). The results are in accordance with the report of Okon and Gonzalez (1994) who confirmed that the biofertilizer inoculum number declined in the carriers to 1 x 10^6 cfu. g^{-1} after three months.

The results of seed coat population showed that paddy seeds treated with AzoPhos (50:50) carried *Azospirillum* (1.30 x 10^4 / seed) and phosphobacteria (1.20 x 10^6 / seed) equal to those with Azophos (100 :100) treatment and also with individual inoculation of *Azospirillum* and phosphobacteria. (Table 3). These findings are in agreement with the earlier results of Kundu and Gaur (1980) who observed that seeds inoculated with mixed cultures of nitrogen fixer and phosphorus solubilizers performed similar to individual inoculation.

The results also confirmed that the inoculation of AzoPhos increased the grain and straw yield of rice. Among all the treatments AzoPhos (50:50) recorded the grain yield of 4.8 t/ha whereas uninoculated control recorded 4.0 t/ha. (Table 4). These findings are in line with the findings of Thamizhvendan and Subramanian (1995) who observed increased rice yield due to combined inoculation of *Azospirillum* and phosphobacteria. Such finding were also reported by Vijaya Nirmala and Sundaram (1996) and Kumar *et al.*, (1998).

Table 1. Studies on the survival of *Azospirillum* in AzoPhos mixed inoculum

Sl. No	Treatments	*Azospirillum population* ($MPNg^{-1}$)			
		0 DAM ($x10^8$)	30 DAM ($x10^8$)	60 DAM ($x10^7$)	90 DAM ($x10^6$)
1.	AzoPhos (50:50)	1.1	1.3	1.1	0.9
2.	AzoPhos (60:40)	0.9	0.4	0.6	0.5
3.	AzoPhos (70:30)	0.3	0.3	0.7	0.4
4.	AzoPhos (80:20)	0.9	0.9	0.6	0.5
5.	AzoPhos (100:100)	1.2	1.2	0.8	0.8

DAM – Days after mixing MPN – Most probable number

Table 2. Studies on the survival of phosphobacteria in AzoPhos mixed inoculum

Sl. No	Treatments	Phosphobacteria Population (cfu.g^{-1})			
		0 DAM ($x10^8$)	30 DAM ($x10^8$)	60 DAM ($x10^7$)	90 DAM ($x10^6$)
1.	AzoPhos (50:50)	2.2	1.2	1.5	1.0
2.	AzoPhos (60:40)	1.7	0.9	1.2	0.8
3.	AzoPhos (70:30)	1.0	0.1	1.0	0.8
4.	AzoPhos (80:20)	1.6	1.1	0.9	1.1
5.	AzoPhos (100:100)	1.8	1.0	1.1	1.1

DAM - Days after mixing MPN - Most probable number

Table 3. Studies on quantification of *Azospirillum* and phosphobacteria adhered on Paddy seeds after AzoPhos Treatment

Sl. No.	Treatments	Azospirillum (MPN- 10^4 / seed)	Phosphobacteria (10^6 cfu/seed)
1.	AzoPhos (50:50)	1.30	1.20
2.	AzoPhos (60:40)	0.37	1.40
3.	AzoPhos (70:30)	1.45	1.30
4.	AzoPhos (80:20)	1.40	1.10
5.	AzoPhos (100:100)	1.40	1.20
6.	Azospirillum	1.40	-
7.	Phosphobacteria	-	1.20

MPN – Most Probable Number

Table 4. Studies on the effect of AzoPhos inoculation of yield of rice (*Kharif* 1999)

Sl. No.	Treatments	Grain Yield (t ha^{-1})	Straw Yield (t ha^{-1})
1.	Control	4.0	9.2
2.	AzoPhos (50:50)	4.8	11.0
3.	AzoPhos (60:40)	4.7	9.6
4.	AzoPhos (70:30)	4.7	9.4
5.	AzoPhos (80:20)	4.4	11.0
6.	AzoPhos (100:100)	4.7	10.0
7.	Azospirillum	4.4	9.7
8.	Phosphobacteria	4.2	11.0
	SED	0.0982	0.1150
	CD	0.2106	0.2466

NPK - 75 : 37.5 : 50 kg ha^{-1}

- Significant at 1% level

REFERENCES

Kumar, C.M.M. Agarwal and B.R. Gupta 1998. *Azospirillum* and its potential as biofertilizer. **Fer. News, 43(11):** 47, 49-50.

Kundu, B. and A.C. Gaur. 1980. Establishment of nitrogen fixing and phosphate dissolving bacteria in rhisosphere and their effect on yield and nutrient uptake of wheat crop. **Plant Soil, 57,** 223-230.

Okon, Y. and C.A.L. Gonzalez. 1994. Agronomic applications of *Azospirillum*, an evaluation of 20 years world wide field inoculation. **Soil Biol. Biochem, 26:** 1591 – 1610.

Thamizhvendan, R. and M. Subramanian. 1998. Phosphorous solubilization and yield increase on rice due to bioinoculants. **J. Ecobiol, 10 (2):** 111 – 116.

Vijiaya Nirmala, G. and M.D. Sundaram 1996. Effect of inoculaion of phosphobacteria with diazotrophs and PGPR on the growth and yield of cumbu varity UCC – 5 at graded level of NPK. In: **Microbiology abstracts XXXVII. Annual Conference of the Association of Microbiologists of India,** IIT, Chennai, December : 4.6. p. 146.

CHAPTER – 41

SOIL SOLUTION PHOSPHORUS STATUS AND MYCORRHIZAL INOCULATION EFFICIENCY OF SELECTED TROPICAL GRAIN LEGUMES IN AN ALFISOL

D. Pragatheswari, A. Manjunath, M. Madhaiyan and K. Kumutha

ABSTRACT

A greenhouse investigation was conducted to determine the Mycorrhizal Inoculation Effect (MIE) of selected tropical grain legumes such as, chickpea, cowpea, frenchbean, greengram, horsegram, peanut and pigeon-pea. These legumes were grown in an unsterile alfisol at established soil solution P levels of 0.01, 0.07 and 0.27 mg P/l, inoculated and uninoculated with a VAM fungus, *Glomus mosseae*. The soil samples were supplied with all other plant nutrients at a level found to be optimum for VAM activity. Based on the MIE values calculated, three distinct MIE categories have been proposed; (I) highly responsive, e.g. cowpea (ii) moderately responsive, e.g., green gram and horse gram and (iii) marginally responsive, e.g. chickpea, french bean, peanut and pigeon pea. The critical soil solution P levels useful for separating host species into different MIE groups were found to be 0.01 and 0.07 mg P/litre. This result exemplifies the importance of defining soil solution P levels while assessing the response of plant species to VAM.

INTRODUCTION

Frank (1885) coined the term 'Mycorrhiza', which means "fungus – root" to describe the association between fungi and higher plants. The term mycorrhiza has since evolved to mean a symbiotic association between fungus and the roots of higher plants (Ainsworth *et al.,* 1971).

Gerdemann (1975) defined Mycorrhizal dependency (MD) as "the degree to which a plant species is dependent on the mycorrhizal conditions to produce maximum growth or yield at a given level of soil fertility". Since then, MD of several plant species has been determined under sterile soil conditions (Plenchette et al., 1983 and Saif, 1987). Mycorrhizal dependency (MD) values determined in sterilized soils are useful in categorizing plant species into distinct mycorrhizal dependency groups (Habte and Manjunath, 1991).

As MD values may not serve as indicators of the performance of the VAM fungi in unsterile soils, because native VAM fungi as well as other soil microflora influence the extent to which a plant species derives benefit from the inoculant strain of VAM fungus (Paulitz and Linderman, 1992). Therefore, the determination of MD of a plant species may not be correct in this condition. So mycorrhizal inoculation efficiency (MIE) values will serve as indicators of the performance of VAM fungus in an unsterile soil (Bagyaraj et al., 1988).

When we attempt to determine MIE values it is essential to define soil phosphorus status as precisely as possible because, among various soil factors, P status of soil regulates the outcome of VAM symbiosis to the greater extent (Mosse, 1981). Further as demonstrated by Habte and Manjunath (1987), such studies should be conducted at defined soil solution P levels since, soil differ in their P retention capacity, plants species differ in their external phosphorus requirement (Fox, 1981) and VAM fungi differ in their P tolerance (Schubert and Hayman, 1986). The determination of MIE values at established soil solution P levels could enable us to decide whether to inoculate a soil with an efficient strain of VAM fungus and /or to raise the soil P level to, optimize plant productivity. Since, external P requirements of legumes are higher than that of other plant species (Munns and Mosse, 1980), they are thought to be more dependent on VAM fungi for P absorption. Hence, this investigation was conducted with an objective to assess the mycorrhizal inoculation efficiency (MIE) of selected tropical grain legumes at three relevant levels of soil solution P in an unsterile soil.

MATERIALS AND METHODS

A greenhouse investigation to determine the effect of inoculation of soil with VAM fungus on selected tropical grain legumes was conducted at the Dept. of Agricultural Microbiology, UAS, Bangalore.

Soil used in this study was a sample collected from the depth of 0 to 43 cm. The soil has been classified as fine, kaolinite isohypothermic, kanhaplustalfs (SMSS Technical Monograph No. 19 USDA, 1990). The soil was ground to pass through a 4 mm sieve and the pH of the soil was adjusted to 6.0 by using calcium hydroxide @ 1.0 g kg^{-1} soil. Soil samples were subjected to three cycles of wetting and drying after addition of $Ca(OH)_2$. The quantity of $Ca(OH)_2$ required to raise the pH of the soil to 6.0 was determined based on the liming curve developed by soil equilibration method (McLean, 1982). Portions (3.66 kg. on dry weight basis) of soil were then transferred into plastic pots (18.0 cm x19.5 cm).

A nutrient solution containing all the nutrients needed for the plant growth was added as a blanket amendment, which has been found to be optimum for VAM activity (Manjunath and Habte, 1990). In addition to native phosphorus concentration in soil solution (0.01mg P/l); two concentrations of soil solution P (0.07 and 0.27mg P/l) were established by adding 100 and 500mg of P as KH_2PO_4.

Seven tropical grain legumes, Chickpea (*Cicer arietinum* L.) cv. annigeri, Cowpea (*Vigna sinensis* Save.) cv C-152, French bean (*Phaseolus vulgaris* L.) cv local, Greengram (*Phaseolus aureus* Roxb.) cv PS-16, Horsegram (*Dolichos biflorus* Roxb.) cv HU-1, Peanut (*Arachis hypogaea* L.) cv TMV-2 and Pigeon pea (*Cajanus cajan* (L). Millsp.) cv. TTB -7 were used as indicator plants in this study.

The air-dried crude inoculum of *Glomus mosseae* Nicol. Gerd. Was uniformly mixed with 3.66 kg of unsterile processed soil at the rate of 60 g / kg soil before planting. Uninoculated controls received the washings of crude inoculum after removal of *G. mosseae* structures by filtration.

The surface sterilized seeds of plants were pregerminated on 0.9% water agar (Habte and Manjunath, 1991). The pregerminated seeds were planted in pots and arranged on greenhouse benches in randomized complete block design. The moisture content of the soil was maintained at 60% of maximum water holding capacity. Plants were harvested after 60 days of growth. Dry weight of shoots and roots were determined after drying them to constant weight at 60^0C.

The values of Mycorrhizal Inoculation Efficiency (MIE) was calculated using the formula suggested by Bagyaraj *et al.* (1988), which is a modification of the original formula suggested by Plenchette *et al.* (1983) to calculate mycorrhizal dependency:

$$\text{MIE}(\%) = \frac{\text{Dry wt. of inoculated plant - Dry wt. of uninoculated plant}}{\text{Dry wt. of inoculated plant}} \times 100$$

The data were subjected to analysis of variance by factorial randomized complete block design and Duncan's Multiple Range (DMRT) Test (Little and Hill, 1978).

RESULT AND DISCUSSION

The dry weight of shoots of all plant species used in this study was increased due to inoculation of soil with VAM fungi, but the difference was significant only in chickpea at the lowest level of soil solution P (Table 1). The extent of increase in dry weight of shoot due to VAM inoculation ranged from 49.13% in pigeonpea to 44.01% in cowpea at this soil P level. At the soil solution P level of 0.07 mg P/l, significant increase in dry weight of shoot due to inoculation of soil with *Glomus mosseae* was noticed in horsegram and at the highest soil solution P these differences were significant with cowpea and horsegram.

Table 1. Influence of soil solution phosphorus levels and inoculation of soil with *Glomus mosseae* on shoot dry weight of selected grain legumes.

Treatment	Shoot dry weight (g/plant)						
	Chick pea	Cow pea	French bean	Green gram	Horse gram	Pea nut	Pigeon pea
P_1 + UN	0.68^{d}	0.30^{de}	1.23^{c}	0.15^{b}	0.10^{e}	1.65^{c}	0.75^{c}
P_1 + IN	1.26^{c}	2.30^{de}	2.68^{c}	0.41^{b}	0.38^{e}	2.43^{bc}	1.20^{bc}
P_2 + UN	1.60^{bc}	4.68^{cd}	4.60^{bc}	0.48^{b}	0.74^{d}	2.45^{bc}	1.42^{bc}
P_2 + IN	2.02^{b}	6.66^{c}	8.33^{b}	0.77^{b}	2.30^{c}	2.66^{b}	1.98^{b}
P_3 + UN	3.36^{a}	16.13^{b}	13.83^{a}	2.30^{a}	5.47^{b}	5.35^{a}	4.00^{a}
P_3 + IN	3.88^{a}	22.02^{a}	16.20^{a}	2.30^{a}	6.37^{a}	5.63^{a}	4.85^{a}

Means followed by the same letter within a plant species do not differ significantly at the 5% level of significance.

P_1 = 0.01 mg P / l

P_2 = 0.07 mg P / l

P_3 = 0.27 mg P / l

UN = Uninoculated

IN = Inoculated with *Glomus mosseae*

At the soil P level of 0.01 mg P / l, significant increase in root dry weight due to VAM fungal inoculation was observed in cowpea and horsegram. But, at the soil P level of 0.07 mg P / l inoculation of soil with *G. mosseae* caused significant increase in root dry weight in chickpea, frenchbean, greengram in addition to horsegram. At the highest level of soil solution P, increase in root dry weight due to inoculation of soil with VAM fungi was noticed in chickpea, cowpea, horsegram and pigeonpea. It is interesting to note that there was significant reduction in dry weight of root of peanut at the highest level of soil solution P, 0.27 mg P / l (Table 2). Such a reduction in dry matter yield was observed by Manjunath and Habte (1992) in a marginally dependent plant species, *Sesbania pachycarpa* at the soil solution P concentration higher than 0.07 mg P / l. Growth depression at high soil P levels have been attributed to the utilization of photosynthates by VAM fungi, if a plant species fail to curtail the level of VAM colonization (Harris and Paul, 1987).

Table 2. Influence of soil solution phosphorus levels and inoculation of soil with *Glomus mosseae* on root dry weight of selected grain legumes

Treatment	Root dry weight (g/plant)						
	Chick pea	Cow pea	French-bean	Green gram	Horse gram	Pea nut	Pigeon pea
P_1 + UN	0.21^{b}	0.21^{b}	0.54^{b}	0.10^{b}	0.08^{f}	0.86^{bcd}	0.40^{c}
P_1 + IN	0.22^{b}	0.50^{b}	0.63^{b}	0.18^{b}	0.14^{e}	0.99^{ab}	0.54^{c}
P_2 + UN	0.13^{c}	0.96^{c}	0.60^{b}	0.17^{b}	0.18^{d}	0.68^{d}	0.43^{c}
P_2 + IN	0.21^{b}	1.00^{b}	0.88^{a}	0.22^{a}	0.27^{c}	0.82^{bcd}	0.55^{c}
P_3 + UN	0.23^{b}	1.56^{b}	0.99^{a}	0.33^{a}	0.39^{b}	1.21^{a}	0.96^{b}
P_3 + IN	0.34^{a}	2.76^{a}	0.82^{a}	0.35^{a}	0.44^{a}	0.94^{bc}	1.15^{a}

Significant increase in total dry matter yield due to inoculation of soil with G. *mosseae* was noticed in cowpea, horsegram and pigeonpea at the highest level of soil solution P (Table 3). In general, inoculation of soil with *G. mosseae* increased total dry matter yield of all legumes at soil P levels of 0.01 and 0.07 mg P / l, but the differences were not significant. Application of P alone, also significantly increased total dry matter yield but the extent of increase was maximum when soil P level was raised from 0.07 to 0.27 mg P /l. Increase in P concentration in plant tissue, increased dry matter yield due to inoculation of soil with *G. mosseae* observed in this study is consistent with many earlier studies conducted using unsterile soils (Bagyaraj and Manjunath, 1980 and Jeffries, 1987).

Table 3. Influence of soil solution phosphorus levels and inoculation of soil with *Glomus mosseae* on total dry matter of selected grain legumes.

Treatment	Total dry matter (g/plant)						
	Chick pea	Cow pea	French bean	Green gram	Horse gram	Pea nut	Pigeon pea
P_1 + UN	0.89[cd]	0.51[de]	1.77[c]	0.25[c]	0.18[e]	2.59[b]	1.15[cd]
P_1 + IN	1.48[d]	2.8[e]	3.31[a]	0.59[C]	0.52[e]	3.42[b]	1.74[d]
P_2 + UN	1.73[b]	5.64[c]	5.2[B]	0.65[c]	0.92[c]	3.13[b]	1.85[c]
P_2 + IN	2.23[bc]	7.66[cd]	9.21[BC]	0.99	2.57[d]	3.48[b]	2.53[cd]
P_3 + UN	3.59[a]	17.69[a]	14.82[a]	2.63[a]	5.86[a]	6.56[a]	4.96[a]
P_3 + IN	4.22[a]	24.78[b]	17.02[a]	2.65[b]	6.81[b]	6.57[a]	6.00[b]

Mycorrhizal inoculation efficiency (MIE) values calculated based on the formula suggested by Bagyaraj *et al.* (1988) are presented in the Table 4. All legumes species tested derived maximum benefit from the inoculation of soil with G. *mosseae* at the lowest level of soil solution P. At this soil solution P, MIE values ranged from 80% (highest) in cowpea and 27% (lowest) in peanut. In general, MIE values decreased as soil solution P level increased. But the significant reduction was noticed in cowpea when soil P level was raised from 0.01 to 0.07 mg P / l. When soil P level was raised from 0.07 to 0.27 mg P /l significant reduction in MIE values was noticed in frenchbean, greengram and horsegram. The changes in soil P levels did not influence MIE values of peanut and pigeonpea.

Table 4. Influence of soil solution phosphorus levels and inoculation of soil with *Glomus mosseae* on mycorrhizal inoculation effect (MIE) of selected grain legumes.

Treatment	Mycorrhizal Inoculation Effect (MIE) (%)						
	Chick pea	Cow pea	French bean	Green gram	Horse gram	Pea nut	Pigeon pea
P_1 + IN	40[a]	80[a]	48[a]	65[a]	65[a]	28[a]	36[a]
P_2 + IN	26[ab]	28[b]	43[a]	44[a]	63[a]	15[a]	26[a]
P_3 + IN	15[b]	30[b]	16[b]	15[b]	20[b]	6[a]	20[a]

Based on the mycorrhizal inoculation efficiency (MIE) calculated at three soil solution P levels (0.01, 0.07 and 0.27 mg P/ l), three MIE categories have been proposed.

Highly Responsive

Species with MIE values of 75% or higher at soil solution P level of 0.01 mg P /l and respond significantly to inoculation of soil with VAM fungus at 0.07 and 0.27 mg P / l in terms of dry weight.

Eg : Cowpea

Moderately Responsive

Species with MIE values of 50 - 75% at soil solution P level of 0.01 mg P/l and no significant decrease in MIE when P concentration is raised to 0.07 mg P / l.

Eg : Greengram and Horsegram.

Marginally Responsive

Species with MIE values of less than 50 at soil solution P level of 0.01 mg P /l and no significant decrease in MIE when P concentration is raised to 0.07 mg P/l.

Eg : Chickpea, Frenchbean, Peanut and Pigeonpea.

These findings clearly support the observations of Habte and Manjunath (1991) that plant species greatly differ in their mycorrhizal dependency (MD) and it will be possible to categorize plant species into distinct MD groups only when the response to VAM colonization is evaluated at different soil solution P levels. Based on the results of the study, the soil solutions levels that could be useful in differentiating legumes tested into distinct MIE groups are 0.01 and 0.07 mg P / l. Although at these P levels VAM fungi function effectively, but the extent of benefit derived from VAM colonization decreases as soil P level increase from 0.01 to 0.07 mg P / l. This is because the optimum soil solution P level for effective VAM symbiosis has been observed to be around 0.02 mg P / l (Habte and Manjunath, 1987).

The result from this study clearly demonstrates that it is essential to match MIE values with soil solution P levels so that it could be possible to determine whether to inoculate the soil with VAM fungus and / or increase the soil solution P level to a level optimum for VAM effectiveness. Further this is required from the point of view of using these fungi in agriculture in order to maximize the productivity of legumes at lower P levels.

REFERENCES

Ainsworth, G.C., P.W. James and D.L. Hawksworth, 1971. **Ainsworth and Bisby's Dictionary of the Fungi.** Commonwealth Mycological Institute, Kew, Surrey, pp. 663 – 664.

Bagyaraj, D.J. and A. Manjunath, 1980. Response of crop plants to VA mycorrhizal inoculation in an unsterile Indian soil. **New Phytol.**, **85**: 33 – 36

Bagyaraj, D.J., A. Manjunath and Y.S. Govinda Rao, 1988. Mycorrhizal inoculation effect on marigold, eggplant and citrus in an Indian soil. **J. Soil Biol. Ecol., 8**: 98 – 103.

Fox, R.L. 1981. External phosphorus requirement of crops. In: **Chemistry in the Soil Environment.** (Eds.) R.H.Dowdy and J.A. Ryan Madison, Wisconsin, pp. 223 – 229.

Frank, A.B. 1985. Uber die au Wurselsymbiont beruhende Ernahung geiwieser Baume durch Uterirdische pilze. **Ber. Drsch. Bot. Ges., 3**: 128 – 145.

Gerdemann, J.W. 1975. Vesicular – Arbuscular Mycorrhizae. In: **The plant root and its environment.** (Eds.) E.W.Carson Academic Press Inc. (London) Ltd. pp. 583 – 584.

Habte, M and A.Manjunath. 1987. Soil solution phosphorus status and mycorrhizal dependency in *Lecucaena leucocephala.* **Appl. Environ. Microbiol., 54**: 701 – 703.

Habte, M and A.Manjunath. 1991. Categories of vesicular –arbuscular mycorrhizal dependency of host species. **Mycorrhiza, 1**:3 –12.

Harris, D and E.A.Paul. 1987. Carbon requirement of vesicular-arbuscular mycorrhizae. In: **Ecophysiology of VA mycorrhizal plants** (Eds.) G. Safir, CRC Press, Boca Raton FL, pp.93 –105.

Jeffries, P. 1987. Use of mycorrihzae in agriculture. **Critical Rev. Biotechnol., 5** : 319 – 357.

Little, T.M. and H.F. Hills. 1978. Mean separation. In: **Agricultural Experimental Design and Analysis**. John Wiley and Sons, U.S.A. pp. 63.

Manjunath, A. and M. Habte. 1990. Establishment of soil solution P levels for studies involving VAM symbiosis. **Commun. Soil Sci. Plant Anal., 21** : 537 – 596.

McLean, E.O. 1982. Soil pH and lime requirement. In: **Methods of Soil Analysis – II. Chemical and Biological properties.** (A.L. Page, R.H. Miller and D.R. Keeney, Eds.), Soil Science Society of America Inc. Publisher, Madison, U.S.A. pp. 199 – 223.

Mosse, B., D.P. Stribley and F. Le Tacon 1981. Ecology of mycorrhizae and mycorrhizal fungi. **Adv. Microbiol. Ecol., 5**: 137-210.

Munns, D.N. and B. Mosse 1980. Mineral nutrition of legume crops. In: **Advances in Legume Sciences.** (Eds.) R.J. Summerfield and A.H.Bunting, University of Reading Press, Reading, UK. Pp.115–125.

Paulitz, T.C. and R.G.Linderman. 1992. Mycorrhizal interaction with soil microorganisms. In: **Handbook of Applied Mycology**. (A.K. Dilip, Bharat Rai, K.G. Mukerji and R.K. Guy Eds.), Marcel Dekker Inc., New York. pp.77 –131.

Plenchette, C., J.A. Fortin and V. Furlan. 1983. Growth responses of several plant species to mycorrhiza in a soil of moderate P fertility. **Plant Soil, 70**: 191 – 209.

Saif, S.R. 1987. Growth response of tropical forage plant species to vesicular arbuscular mycorrhiza. I. Growth, mineral uptake and mycorrhizal dependency. **Plant Soil, 97**: 17 – 26.

Schubert, A. and D.S. Hayman 1986. Plant growth responses to vesicular arbuscular mycorrhiza. XVI. Effectiveness of different endophytes at different levels of soil phosphate. **New Phytol., 103** : 79 – 90.

CHAPTER – 42

EFFECT OF INOCULATION WITH VESICULAR ARBUSCULAR MYCORRHIZAL (VAM) FUNGUS (*Glomus mosseae*) ON SELECTED TROPICAL GRAIN LEGUMES

D. Pragatheswari, A. Manjunath and M. Madhaiyan

ABSTRACT

A greenhouse experiment was conducted to study the effect of inoculation of soil with a VAM fungus, *Glomus mosseae* on selected tropical grain legumes such as, chickpea, cowpea, frenchbean, greengram, horsegram, peanut and pigeonpea at different soil solution P levels (0.01,0.07 and 0.27 mg P/l). Inoculation of soil with *Glomus mosseae* markedly increased the level of VAM colonization in roots of all plant species, but the extent of VAM colonization in roots decreased to 20% at soil solution P level of 0.27 mg P/l and increased in soil solution P concentration when it was 0.01 and 0.07 mg P/l. At the lowest level of soil solution P (0.01 mg P/l), the extent of VAM colonization was more than 70% in chickpea, cowpea, peanut and pigeonpea, while it was lowest (35%) in greengram. In frenchbean and horsegram, the VAM colonization level was 55% and 45% respectively. VAM colonization in tropical grain legumes varied with plant species as well as different soil solution phosphorus levels.

INTRODUCTION

The vesicular arbuscular mycorrhizal (VAM) fungi are unique group of ubiquitous soil microorganisms known to form symbiotic association with roots of economically important crop plants. The importance of this association is well recognized over the last three decades, because of the key role of VAM fungi in enhancing the uptake of diffusion – limited nutrients such as, phosphorus in the soils of low nutrient status. Further, VAM fungi have also been thought to influence the plant growth and development through other mechanism such as, by conferring disease resistance and drought tolerance (Bagyaraj, 1984 and Allen and Allen, 1986), increasing biological nitrogen fixation by synergistic interaction with free- living as well as symbiotic nitrogen – fixing bacteria (Hayman, 1986) and influencing the production of plant harmones (Allen *et al.*, 1986). Considering the influence that VAM fungi exert on plant growth and development, they seem to have great potential as biofertilizers.

The inoculation of unsterile soils with a VAM fungi has resulted in significant increase in nutrient uptake, dry matter yields and grain yield of various plant species (Nalini *et al.*, 1986). Although in all these studies inoculation of soil with VAM fungus has positively influenced plant growth, response trends of a plant species observed in one soil cannot be compared with that of the same plant species in an another soil, since there is no basis for comparison (Stribley, 1987). Therefore, a plant species categorized as highly responsive to VAM colonization may be categorized as less responsive to VAM in another soil. This is mainly because soils differ in their available P status and P retention capacity (Fox, 1981). Further, it is well known that among several soil factors, soil P level influences the out come of VAM symbiosis, to the greatest extent (Manjunath and Habte, 1992). Therefore, it is essential that the levels of soil solution P is precisely defined in studies involving VAM symbiosis. Hence, this investigation was conducted to study the effect of VAM symbiosis with tropical grain legumes at three levels of soil solution P.

MATERIALS AND METHODS

Soil used in this study was a sample collected from the depth of 0 to 43 cm. The soil has been classified as fine, kaolinite isohypothermic, kanhaplustalfs (SMSS Technical Monograph No.19 USDA, 1990). The soil was crushed to pass through a 4-mm sieve and the pH of the soil was adjusted to 6.0 by using calcium hydroxide @ 1.0 g kg^{-1} soil. Soil samples were subjected to three cycles of wetting and drying after addition of $Ca(OH)_2$. The quantity of $Ca(OH)_2$ required to raise the pH of the soil to 6.0 was determined based on the liming curve developed by soil equilibration method (McLean, 1982). Portions (3.66 kg on dry weight basis) of soil were then transferred into plastic pots (18.0 cm x19.5 cm).

A nutrient solution containing all the nutrients needed for the plant growth was added as a blanket amendment, which has been found to be optimum for VAM activity (Manjunath and Habte, 1990). In addition to native phosphorus concentration in soil solution (0.01 mg P/l); two concentrations of soil solution P (0.07 and 0.27mg P/l) were established by adding 100 and 500mg of P as KH_2PO_4.

Seven tropical grain legumes, Chickpea (*Cicer arietinum* L.) cv. annigeri, Cowpea (*Vigna sinensis* Save.) cv C-152, French bean (*Phaseolus vulgaris* L.) cv local, Greengram (*Phaseolus aureus* Roxb.) cv PS-16, Horsegram (*Dolichos biflorus* Roxb.) cv HU-1, Peanut (*Arachis hypogaea* L.) cv TMV-2 and Pigeon pea (*Cajanus cajan* (L). Millsp.) cv. TTB -7 were used as indicator plants in this study.

The crude inoculum of *Glomus mosseae* Nicol. Gerd. was produced by following the method outlined by Sreenivasa and Bagyaraj (1988). The spores of *Glomus mosseae* were isolated from field soil by wet seiving and decanting technique (Gerdemann and Nicolson, 1968). Viable and undamaged spores were picked up using a Pasteur pipette and were surface sterilized using chloromine-T (2% w/v) and streptomycin (200 ppm) for 15 minutes. The surface sterilized spores were inoculated to the root zone of young seedling of Rhode Grass (*Chloris gayana* Kunth.) grown in pots containing sterilized sand: soils mix as growth medium and maintained in a greenhouse. This inoculum contained extramatrical hyphae, spores and infected root bits of Rhodes Grass and was used at the rate of 60 g/kg soil. The air-dried crude inoculum of VAM fungi was uniformly mixed with 3.66 kg of unsterile processed soil before planting. Uninoculated controls received the washings of crude inoculum after removal of *G.mosseae* structures by filtration (Whatman No 1 filter paper).

The surface sterilized seeds of plants were pregerminated on 0.9% water agar (Habte and Manjunath, 1991). The pregerminated seeds were planted in pots and arranged on greenhouse benches in randomized complete block design. The moisture content of the soil was maintained at 60% of maximum water holding capacity. Plants were harvested after 60 days of growth. The extent of VAM colonization in roots was determined by the grid – line intersection method outlined by Giovannetti and Mosse (1980) after staining them with tryphan blue as suggested by Phillips and Hayman (1970).

The data were subjected to analysis of variance by factorial randomized complete block design and Duncan's Multiple Range (DMRT) Test (Little and Hill, 1978) separated treatment means.

RESULTS AND DISCUSSION

Inoculation of soil with *Glomus mosseae* significantly increased the level of VAM colonization in roots of all legumes (Table 1), at the lowest level of soil solution P, the extent of VAM colonization was more than 70% in chickpea, cowpea, peanut and pigeonpea, and it was lowest (35%) in greengram. In frenchbean and horsegram, the VAM colonization level was 55% and 45%, respectively. Application of P at the rate of 100 mg/kg soil (0.07 mg P/l) did not cause significant reduction in VAM colonization level in chickpea, cowpea, frenchbean, peanut and pigeonpea compared to lowest level of (0.01 mg P/l). However, this rate of P application significantly reduced VAM colonization level in greengram and horsegram. At the highest rate of P application of 500 mg/kg soil (0.27 mg P/l), significant reduction in VAM colonization level was noticed in all legumes tested. In general, the level of VAM colonization of plant species grown in uninoculated soil was lower than 20%.

Table 1. Effect of soil solution phosphorus levels and inoculation of soil with *Glomus mosseae* on mycorrhizal colonization level of selected grain legumes.

Treatment	VAM Colonization (%)						
	Chick pea	Cow pea	French bean	Green gram	Horse gram	Pea nut	Pigeon pea
P_1 + UN	25^{bc}	30^{b}	12^{bc}	15^{cd}	10^{bc}	50^{b}	30^{c}
P_1 + IN	78^{a}	75^{a}	55^{a}	35^{a}	45^{a}	90^{a}	70^{a}
P_2 + UN	25^{b}	35^{b}	20^{b}	18^{c}	8^{cd}	25^{c}	40^{b}
P_2 + IN	70^{a}	62^{a}	50^{a}	26^{b}	25^{b}	85^{a}	75^{a}
P_3 + UN	16^{d}	10^{d}	13^{bc}	5^{d}	2^{d}	10^{d}	10^{e}
P_3 + IN	20^{d}	23^{c}	15^{c}	10^{cd}	10^{cd}	26^{c}	25^{d}

Means followed by the same letter within a plant species do not differ significantly at the 5% level of significance.

P_1 = 0.01 mg P / l

P_2 = 0.07 mg P / l

P_3 = 0.27 mg P / l

UN = Uninoculated

IN = Inoculated with *Glomus mosseae*

Inoculation of soil with *Glomus mosseae* markedly increased the level of VAM colonization in roots of all plant species (Table 1). This clearly suggests that either the number of infective propagules in this soil is very low or infectivity of native VAM fungi is lower than that of the inoculant fungus. Such increases in root colonization levels of plant species grown in unsterile soil inoculated with VAM fungus have been observed earlier (Bagyaraj *et al.*, 1988). Further, the extent of VAM colonization in roots decreased to a level lower than 20% at soil solution P level of 0.27 mg P/l, such a decrease in VAM colonization level at higher soil P levels has also been observed (Schubert and Hayman, 1986). Menge *et al.* (1978) attributed this reduction to the decrease in the permeability of plant cell membrane by which VAM fungus was deprived of photosynthates for its development. In general, increase in soil P concentration from 0.01 to 0.07 mg P/l did not cause marked reduction in VAM colonization in roots of legumes tested. This is probably because, even at 0.07 mg P/l, soil solution P level is not high enough to exert any detrimental effect on VAM development.

REFERENCES

Allen, E.B. and M.F. Allen 1986. Water relation of xeric grasses in the field: Interactions of mycorrhizal and competition. **New Phytol., 104** : 559 – 571.

Allen, M.F., T.S. Moore and M. Christensen. 1980. Phytoharmone changes in *Botelus gracilus* infected by VAM. I. Cytokinin increase in the host plant. **Can. J. Bot., 60** : 468- 471.

Bagyaraj, D.J. 1984. Biological interaction with VA mycorrhizal fungi. In: **VA Mycorrhiza.** (Eds.) C.J. Powell and D.J. Bagyaraj, CRC Press Inc., Boca Raton, Florida, pp.131 – 154.

Bagyaraj, D.J., A.Manjunath and Y.S. Govinda Rao, 1988. Mycorrhizal inoculation effect on marigold, eggplant and citrus in an Indian soil. **J. Soil Biol. Ecol., 8**: 98 – 103.

Fox, R.L. 1981. External phosphorus requirement of crops. In: **Chemistry of Soil Environment**. (Eds.) R.H. Dowdy and J.A. Ryan, Madison, Wisconsin, pp. 223 – 239.

Gerdemann, J.W. and T.H.Nicolson. 1968. Spores of mycorrhizal endogone species extracted from soil by wet seiving and decanting. **Trans. Br. Mycol. Soc., 46 :** 235 – 244.

Giovanetti, M. and B.Mosse. 1980. An evaluation of techniques to measure VAM infection in roots. **New Phytol., 84 :** 489 –500.

Habte, M and A.Manjunath. 1991. Categories of vesicular –arbuscular mycorrhizal dependency of host species. **Mycorrhiza, 1**:3 –12.

Hayman, D.S. 1986. Mycorrhizae of nitrogen fixing legumes. **Mircen. J. Appl. Microbiol. Biotechnol., 2** : 121 – 145.

Little, T.M. and H.F. Hills. 1978. Mean separation. In: **Agricultural Experimental Design and Analysis**. John Wiley and Sons, U.S.A. pp. 63.

Manjunath, A. and M. Habte. 1990. Establishment of soil solution P levels for studies involving VAM symbiosis. **Commun. Soil Sci. Plant Anal., 21** : 537 – 596.

Manjunath, A. and M. Habte. 1992. External and internal P requirement of plants species differing in their mycorrhizal dependency. **Arid. Soil Res. Rehabl., 6** : 271 – 284.

McLean, E.O. 1982. Soil pH and lime requirement. In: **Methods of Soil Analysis – II. Chemical and Biological properties.** (A.L. Page, R.H. Miller and D.R. Keeney, (Eds.), Soil Science Society of America Inc. Publisher, Madison, U.S.A. pp.199 – 223.

Menge, J. A., E.L.V. Johnson and R.G. Platt. 1978. Mycorrhizal dependency of several citrus cultivars under three nutrient regimes. **New Phytol., 81** : 553 – 559.

Nalini, P.A., M.S. Byra Reddy and D.J. Bagyaraj. 1986. Selection of an efficient VA mycorrhizal fungus for leucaena – A preliminary report. **Leucaena Res. Rep., 7** : 61 - 62.

Phillips, J.M. and D.S. Hayman. 1970. Improved procedure for cleaning roots and staining parasitic and vascular mycorrhizal fungi for rapid assessment of fungi. **Trans. Br. Mycol. Soc., 55**: 158 – 161.

Schubert, A. and D.S.Hayman. 1986. Plant growth responses to vesicular arbuscular mycorrhiza. XVI. Effectiveness of different endophytes at different levels of soil phosphate. **New Phytol. 103** : 79 – 90.

Sreenivasa, M.N. and D.J. Bagyaraj. 1988. *Chloris gayana* (Rhodes grass) a better host for mass production of *Glomus fasciculatum* inoculum. **Plant Soil, 106**: 100-119.

Stribley, D.P. 1987. Mineral nutrition. In: **Ecophysiology of VA mycorrhizal plants**. (G.R.Safir Ed.,), CRC Press Inc., Boca Raton, Florida, pp. 59 – 70.

CHAPTER – 43

SCREENING OF AM FUNGI FOR EFFICIENCY ON GROWTH AND BIOCHEMICAL CONSTITUENTS OF MULBERRY

K. Kumutha, Sp. Sundaram, J. Sempavalan and P. Santhanakrishnan

ABSTRACT

AM fungi are considered as obligate symbionts of crop plants for better utilization of P and other essential elements. Though the importance of AM fungi in many horticulture and agriculture plants is well documented, studies are lacking in perennial cash crop mulberry. Nursery experiment was conducted to screen the AM fungal isolates on mulberry cultivar MR–2. Three isolates obtained from mulberry rhizosphere namely viz., *Glomus mosseae* MGm–3, *G. fasciculatum* MGf 3 and *Gigaspora* sp. MGi 4 and other two isolates *G. etunicatum* and *Acaulospora laevis* were tested for efficiency. Samples were taken on 60 and 90 DAP and studied for growth, biochemical constituents such as, total chlorophyll, carbohydrate, reducing sugars in leaves, AM fungal colonization and phosphatases activity in roots. N and P contents of the leaves were also estimated. Inoculation of different AM fungal isolates significantly increased the dry matter production over the control. Maximum was registered with inoculation with *G. fasciculatum* and recorded 90% increase in shoot dry matter, 63% increase in root dry matter over control. Total chlorophyll content in leaves was significantly altered by mycorrhizal inoculation and also resulted in 14.6 to 15.7% increase in total carbohydrate content of leaves. Reducing sugar content in leaves was enhanced by 49% due to AM fungal inoculation. Inoculated plants had 2-3 fold increase in phosphatase activities of the root over uninoculated control. The enzyme activities were maximum with *G. fasciculatum* followed by *G. mosseae*. Among all AM fungi tested, *G. fasciculatum* colonized the roots effectively. Significant increase in total N and P content was noticed with *G. fascicualtum* inoculation over uninoculated control.

INTRODUCTION

AM fungi are recognized as playing key role in plant survival and nutrient cycling in the ecosystem. They can be found in nearly all soils of the world. This obligate symbiosis with photosynthetic plants play a key role in both natural ecosystems and man modified systems, mainly because they can carry out the functions for the plants which they are unable to perform themselves (Barea and Azcon-Aguliar, 1983). Generally AM fungi are not host specific, they exhibit some host preferences. Though mulberry plants are intimately associated with AM fungi the effective strains should be identified, and mass multiplied for inoculation. Hence this experiment was conducted to screen the efficiency of AM fungi on growth and biochemical characteristics of mulberry.

MATERIALS AND METHODS

This experiment was conducted under nursery condition (each bed with 1x1 square c.m) in soils having properties of pH 7.4, available N 230 Kg ha^{-1} P_2O_5 10 Kg ha^{-1}, and K_2O 250 Kg ha^{-1}. AM fungal isolates obtained from mulberry rhizosphere *Glomus fasciculatum, Glomus mosseae, Gigaspora* sp. and the other isolates obtained from the Dept. Agricultural Microbiology, Coimbatore (*Glomus etunicatum* and *Acaulospora laevis*) were tested for efficiency. The AM fungal isolates were applied @ 100 g per m^2 in the nursery beds and disease free MR 2 cuttings were planted. Total plant dry weight, and leaf dry weight were taken on 60 and 90 Days after planting (DAP). Total chlorophyll, total carbohydrates and reducing sugars in leaves were estimated. In roots acid and alkaline phosphatase activities and AM fungal colonization were estimated. Apart from this total N and P contents in leaves were also estimated as per the standard procedures.

RESULTS AND DISCUSSION

Leaf dry weight

Significant increase in leaf dry matter was observed with AM inoculation. All the inoculants recorded significantly higher leaf dry matter over uninoculated control. Maximum leaf dry matter was recorded with the inoculation of *G. fasciculatum* MGf 3 (1.18 to 3.86 g leaf dry weight $plant^{-1}$), followed by *G. mosseae* MGm 3 (1.18 to 3.17 g $plant^{-1}$) as against uninoculated control (0.6 to 2.0 g $plant^{-1}$) during the study period. Generally AM fungi inoculation nearly doubled the leaf dry matter production and recorded upto 93.8 per cent increase over uninoculated control.

Shoot Dry Matter Production

All the inoculants except *A. laevis* enhanced the shoot dry matter production significantly over uninoculated control on 60 and 90 DAP. Maximum shoot dry matter production was recorded in *G. fasciculatum* MGf 3 inoculated mulberry (1.46 to 4.25 g $plant^{-1}$), followed by *G. mosseae* MGm 3

(1.26 to 3.68 g plant^{-1}) and the percentage increase by AM fungi inoculation was up to 90.7 over uninoculated control (Table 1).

Table 1. Screening the efficiency of AM fungal isolates on shoot growth of mulberry cultivar MR.2 under nursery condition

No	Treatments	Shoot dry weight (g plant^{-1})			Per cent increase over control	Leaf dry weight (g plant^{-1})			Per cent increase over control
		DAP		Mean		DAP		Mean	
		60	90			60	90		
1	Uninoculated control	0.75	2.25	1.50	-	0.60	2.00	1.30	-
2	*Glomus fasciculatum* MGf 3	1.46	4.25	2.86	90.7	1.18	3.86	2.52	93.8
3	*Glomus mosseae* MGm 3	1.26	3.68	2.47	64.7	1.18	3.17	2.18	67.7
4	*Glomus etunicatum* TNAU	1.20	3.13	2.17	44.7	0.95	2.80	1.88	44.6
5	*Gigaspora* sp. MGi 4	1.10	3.60	2.35	56.7	0.99	3.13	2.06	58.5
6	*Acaulospora laevis* TNAU	1.00	3.10	2.05	36.7	0.84	2.90	1.87	43.8
	SEd	0.14	0.20			0.14	0.28		
	CD (0.05)	0.32	0.43			0.30	0.62		

DAP - Days after planting

The results are in line with Ambika *et al.* (1994). Here among the five isolates, *Glomus fasciculatum* produced 75.2 per cent increase in leaf fresh weight and up to 93.8 per cent increase in leaf dry weight over control. Umakanth and Bagyaraj (1998) reported *G. fasciculatum* recorded 30.7 per cent increase in plant biomass and 57.19 per cent more fresh weight of mulberry than control.

Total Chlorophyll Content

Total chlorophyll content in mulberry leaves was not significantly altered by mycorrhizal inoculation on 60 DAP, whereas at 90 DAP, significant variation was noticed. *G. fasciculatum* MGf 3 was significantly superior (1.0 to 5.28 mg g^{-1} fresh leaf) over other cultures with an average chlorophyll content of 3.14 mg g^{-1} fresh leaf, followed by 2.47 mg g^{-1} fresh leaf in *G. mosseae* MGm 3. These were 85.8 and 46.2 per cent more than the uninoculated control. *A. laevis* inoculated plants had an average chlorophyll content of 2.05 mg g^{-1} fresh leaf. This was minimum, among the inoculants tested (Table 2). Published information on the response of mycorrhizae to photosynthesis, photo respiration and chlorophyll content are very few. Arbuscular mycorrhizal plants showed increased levels of chlorophyll and carotenoids as compared to non mycorrhizal plants of *Ziziphus mauritiana*. However different AM fungi varied in their efficacy to increase chlorophyll and carotenoid contents (Mathur and Vyas, 1995). The increase in chlorophyll content might be due to the greater availability of micronutrients by AM fungi which are most important for chlorophyll synthesis.

Total carbohydrates in leaves showed a significant increase at both stages (60 DAP and 90 DAP) of the experiment and the maximum was registered with *G. fasciculatum* MGf 3 (80.2 and 100.0 mg g^{-1} fresh leaf) as against uninoculated control (46.0 and 69.0 mg g^{-1} fresh leaf) on 60 and 90 DAP. Increase in carbohydrate content by inoculation ranged from 14.6 – 56.7 per cent over uninocualted control (Table 2).

Table 2. Screening the efficiency of AM fungal isolates on total chlorophyll and sugar content of mulberry cultivar MR.2 under nursery condition

Treatments	Total chlorophyll content (mg g^{-1} fresh leaf)			Per cent increase over control	Total carbohydrates (mg g^{-1} fresh leaf)			Per cent increase over control	Reducing sugars (mg g^{-1} fresh leaf)			Per cent increase over control
	DAP		Mean		DAP		Mean		DAP		Mean	
	60	90			60	90			60	90		
Uninoculated control	0.69	2.68	1.69	-	46.0	69.0	57.5	-	10.4	15.1	12.8	-
Glomus fasciculatum MGf 3	1.00	5.28	3.14	85.8	80.2	100.0	90.1	56.7	17.5	20.6	19.1	49.2
Glomus mosseae MGm 3	0.80	4.13	2.47	46.2	60.5	95.1	77.8	35.3	14.0	19.4	16.7	30.5
Glomus etunicatum TNAU	0.70	4.01	2.36	39.6	58.0	83.8	70.9	23.3	12.0	17.1	14.6	14.1
Gigaspora sp. MGi 4	0.97	3.84	2.41	42.6	76.0	82.5	79.3	37.9	14.5	18.3	16.4	28.1
Acaulospora laevis TNAU	0.88	3.21	2.05	21.3	55.3	76.5	65.9	14.6	12.0	17.0	14.5	13.3
SEd	0.15	0.19			1.7	2.0			1.5	1.1		
CD (0.05)	NS	0.42			3.6	4.3			3.2	2.3		

DAP - Days after planting

Reducing Sugars in Leaves

AM fungal inoculants had significant influence on reducing sugar content in leaves on both stages of sampling. *G. fasciculatum* MGf 3 was found to be better (17.5 to 20.6 mg g^{-1} fresh leaf), followed by Gigaspora sp. MGi 4 (14.5 to 18.3 mg g^{-1} fresh leaf) and *G. mosseae* MGm 3 (14.0 to 19.4 mg g^{-1} fresh leaf). The average reducing sugar content of inoculated plants ranged from 14.5 to 19.1 mg g^{-1} fresh leaf as against 12.8 mg g^{-1} fresh leaf in uninoculated plants, and recorded a maximum of 49.2 per cent increase over uninoculated control (Table 2).

In the present study, enhanced chlorophyll contents may contribute for higher sugar content through the enhancement in photosynthesis. This is in conformity with the findings of Mathur and Vyas (1995) and McArthur and Knowles (1993).

Colonisation of AM Fungi

Inoculation of AM fungi significantly increased the percent mycorrhizal root colonisation compared to uninoculated plants. In the present study, inoculation of *G. fasciculatum* MGf 3 caused higher root colonisation (60.3 per cent) than *G. mosseae* MGm 3 (53.2 per cent), *Gigaspora* sp. MGi 4 (50.2 per cent) and *A. laevis* TNAU (40.3 per cent). The relationship between effectiveness of AM fungi to stimulate and its colonisation appears to vary depending on the fungus and host

The reason for the positive response may be due to the host preference of *G. fasciculatum* MGf 3 for mulberry. It has the ability to incorporate external sources of carbon other than photosynthetic assimilates (Losel and Cooper, 1979) when inoculated in mulberry, indicating the reduction of carbon drain by the fungus. This could be the reason for *G. fasciculatum* to adopt well and form aggressive colonization in mulberry than other AM fungi tested.

Uptake of Nitrogen and Phosphorus

Significant increase in N and P uptake due to AM fungi inoculation was noticed on 60 and 90 DAP. Inoculation of *G. fasciculatum* MGf 3, *G. mosseae* MGm 3 and *Gigaspora* sp. MGi 4 recorded significantly higher uptake than uninoculated control. These inoculants either differed significantly or on par with each other. However *G. fasciculatum* MGf 3 recorded comparatively higher uptake values, recording 103.7 per cent increase in N uptake and 131.9 per cent increase in P uptake over uninoculated control plants. All the inoculants significantly differed with each other in terms of P uptake at 90 DAP. Generally, *G. fasciculatum* MGf 3 doubled the N as well as P uptake of plants (on an average) irrespective of days of sampling over uninoculated plants (Table 3).

Table 3. Screening the efficiency of AM fungal isolates on N and P uptake of mulberry cultivar MR.2 under nursery condition

Treatments	N uptake (mg plant^{-1}) DAP			Per cent increase over control	P uptake (mg plant^{-1}) DAP			Per cent increase over control
	60	90	Mean		60	90	Mean	
Uninoculated control	10.23	34.83	22.53	-	1.08	3.90	2.49	-
Glomus fasciculatum MGf 3	21.78	70.00	45.89	103.7	2.49	9.06	5.78	131.9
Glomus mosseae MGm 3	18.04	58.08	38.06	68.9	2.01	7.39	4.70	88.8
Glomus etunicatum TNAU	16.92	49.18	33.05	46.7	2.98	5.94	3.96	59.0
Gigaspora sp. MGi4	16.61	57.78	37.20	65.1	1.90	7.13	4.52	81.5
Acaulospora laevis TNAU	14.85	47.38	31.12	38.1	1.55	5.52	3.54	42.2
SEd	0.50	6.47			0.19	0.14		
CD (0.05)	1.06	13.79			0.40	0.30		

DAP - Days after planting

This is also evident from the studies of Katiyar et al. (1990) and Das et al. (1995) in mulberry who reported that all the three nutrients (NPK) in the saplings were increased due to AM fungi inoculation. A positive correlation was noticed in the present study between per cent AM fungi colonisation and P content as well as per cent AM fungi and leaf yield.

Phosphatase activity as influenced by AM Fungi

A major part of the beneficial effects of AM fungi is attributed to their role in P uptake and translocation and is reasonable to expect an involvement of phosphatases in the transport of phosphorus. In the present study, significant enhancement in acid and alkaline phosphatase activities were observed by inoculation with AM fungi. Maximum activities were recorded with *G. fasciculatum* MGf 3 followed by either *G. mosseae* MGm 3 or *Gigaspora* sp. MGi 4 during the study period. Generally the activity of acid phosphatase was higher compared to alkaline phosphatases and the increase in activity was observed in roots when the crop ages.

Phosphatase enzymes are directly involved in acquisition of phosphorus by plants. Acid phosphatases have been reported in AM fungi and they may be associated with the growth and development of the fungi within the host tissue as well as P acquisition in the rhizosphere. Tarafdar and Marschner (1994) reported acid phosphatase activity was higher than alkaline phosphatase activity and pointed out that phosphatase activity contributed towards phosphorus uptake. Similar results were observed in the present study that increase in acid as well as alkaline phosphatase activity in mulberry roots by inoculating AM fungi. Probably this is the first report about the enhancement of phosphatases by mycorrhizal endophytes in mulberry.

REFERENCES

Ambika, P.K., P.K. Das, R.S. Katiyar and P.C. Choudhury. 1994. The influence of vesicular – arbuscular association on growth, yield and nutrient uptake in some mulberry genotypes (Morus spp.). **Indian J. Seric. 33:** 166-169

Barea, J.M. and C. Azcon-Aguilar. 1983. Mycorrhizas and their significance in nodulating nitrogen-fixing plants. In: **Advances in Agronomy** (Ed.) N. C. Brady, Academic Press, New York. pp. 1-54.

Das, P.K., R.S. Katiyar, M. Hanumantha Gowda, P.S. Fathima and P.C. Choudhury. 1995. Effect of vesicular arbuscular mycorrhizal inoculation on growth and development of mulberry (Morus spp.) saplings. **Indian J. Seric., 34:** 15-17.

Katiyar, R.S., P.K. Das, A. Ghosh, P.C. Choudhury and R.K. Datta. 1990. Interaction between vesicular-arbuscular mycorrhizal colonisation and phosphorus application in mulberry. pp. 79-80. In: **Mycorrhizal Symbiosis and Plant Growth.** (Eds.) D.J. Bagyaraj and A. Manjunath, Mycorrhizal Network Asia and University of Agricultural Sciences, Bangalore.

Losel, D.M. and D.M. Copper 1979. Incorporation of ^{14}C-labelled substrates by uninfected and VA-mycorrhizal roots of onion. **New Phytol., 81:** 105-112.

Mathur, N. and A.Vyas 1995. Influence of VA mycorrhizae on net photosynthesis and transpiration of Ziziphus mauritiana. **J. Plant Physiol., 147:** 328-330

McArthur, D.A.J. and N.R. Knowles. 1993. Influence of vesicular-arbuscular mycorrhizal fungi on response of potato to phosphorus deficiency. **Plant Physiol**. 147-160.

Tarafdar, J.C. and H. Marschner. 1994. Phosphatase activity in the rhizosphere and hydrosphere of VA-mycorrhizal wheat supplied with inorganic and organic phosphorus. **Soil Biol. Biochem. 26:** 387-395.

Umakanth, C.G. and D.J.Bagyaraj. 1998. Response of mulberry saplings to inoculation with VA-mycorrhizal fungi and **Azotobacter. Sericologia, 38:** 669-675.

CHAPTER - 44

EFFECT OF INSOLUBLE PHOSPHATE AND DUAL INOCULATION ON SOYBEAN

K. Kumutha, J. Sempavalan and P. Santhanakrishnan

ABSTRACT

A pot culture experiment was conducted to study the influence of different P sources on the growth and nutrient uptake of soybean (variety, CO1) inoculated with *Glomus fasciculatum* individually and in combination with *Rhizobium japonicum* (SB. 101). Two sources of P viz., super phosphate and Mussoorie rock phosphate were selected. Dual inoculation of both organisms produced higher dry weight, nodule weight, VAM colonization and nutrient uptake of N and P than individual inoculants or control at different sources of P. Mussoorie rock phosphate generally exhibited lesser magnitude of growth response and nutrient uptake due to inoculation of inoculants than superphosphate. However dual inoculation exhibited better nodulation, nutrient uptake and phosphatase activity in Mussoorie rock phosphate over uninoculated control. Thus microbial inoculants provide wide scope for the better utilization of Mussoorie rock phosphate.

INTRODUCTION

Soybean is regarded not only as a high quality protein rich food for man and livestock, but also it is a very cheap source of edible oil. Being a legume crop it fixes atmospheric nitrogen, but needs adequate supply of phosphorus for the same. Mycorrhizal infection might help the plant to obtain the required phosphorus for nodulation. With the increasing cost of chemical sources of P fertilizers, the use of rock phosphate may be an alternative. VAM inoculation helps in better utilization of rock phosphate sources (Morse *et al.*, 1976). This idea was tested in an experiment where in superphosphate and

Mussoorie rock phosphate were compared along with *Rhizobium* and VAM for nodulation, nutrient uptake and phosphatase activity of soybean.

MATERIALS AND METHODS

A pot culture experiment was conducted to study the influence of P fertilizers on the growth and nutrient uptake of soybean (CO1) inoculated with *Rhizobium japonicum* (SB 101) and VAM fungus.

VAM inoculum

Bulk inoculum of *Glomus fasciculatum* was applied (@ 10g/pot contained approximately 50 spores/g) and placed at 2.5 cm below the surface.

Fertilizer application

The recommended dose of N and K was applied through urea and Murate of potash respectively. Superphosphate or Mussoorie rock phosphate was applied to meet the P requirement at the rate of 60 kg P_2O_5/ha (75 % of recommended level).

Rhizobial inoculum

Rhizobium japonicum (SB 101) was applied through seed. Inoculant contained 10^8 cells/g lignite was used.

The experiment consisted of 12 treatments (4 inoculants x 3 P levels) with 3 replication. VAM and *Rhizobium* were inoculated individually as well as in combination along with rock phosphate and super phosphate.

VAM colonization was assessed by root clearing and staining technique (Phillips and Hayman, 1970). Acid and alkaline phosphatase activity was estimated by the method of Plummer (1978). Besides these parameters, plant dry weight and nutrient uptake by soybean plants were recorded on 90th day after sowing.

RESULTS AND DISCUSSION

VAM inoculation either individually or incombination with *Rhizobium*, with the application of P significantly enhanced the plant dry weight. Irrespective of the P sources, dual inoculation was found to be superior to individual inoculation. The results attributed for increase in growth response of VAM and dual inoculation with rock phosphate due to efficient absorption of phosphate ions as they become chemically dissociated from particles of the fertilizers (Bolan, 1991). Further it was mentioned that the formation of mycorrhiza might however make root system more active in the breakdown of complex organic phosphates (Gianinazui-Pearson *et al.*, 1981).

With respect to nodulation it was enhanced through rhizobial inoculation. When it was applied incombination with *Glomus fasiculatum*, the nodulation was triggered to maximum level. Superphosphate applied plant exhibited maximum nodules of 85/plant than rock phosphate.

Coinoculation of VAM enhanced the nodulation of *Rhizobium* by 18% when superphosphate was applied and by 15% when rock phosphate was applied. P deficiency is the principal constraint to nitrogen fixation. Research in past decades has established that VAM can improve plant growth through increased uptake of phosphorus (Gerdemann, 1995) and thus contributing for satisfactory nodulation of legumes. The colonization percentage of VAM was more in superphosphate and it was enhanced to greater extent due to dual inoculation.

Table 1. Influence of *Rhizobium* and VAM inoculation on growth, nodulation and VAM infection of soybean applied with different P sources.

Treatment	Plant dry weight (g/pl)	Nodule number/pl	Nodule dry weight (mg/pl)	%VAM colonization
Control	15.93	24	34.56	9
Rhizobium	18.4	51	109.65	11
VAM	19.82	28	42	50
Rhizobium + VAM	20.82	61	140.3	56
Control + SP	19.5	32	64.17	12
Rhizobium + SP	22.2	72	174.2	15
VAM + SP	25.1	42	88.2	66
Rhizobium + VAM +SP	30.0	85	212.67	79
Control + MRP	17.1	29	49.3	10
Rhizobium + MRP	20.0	59	130.99	15
VAM + MRP	22.18	37	69.09	60
Rhizobium + VAM + MRP	24.85	68	169.32	70
CD at 5%	4.76	4.07	10.48	6.72

VAM – Vesicular arbuscular mycorrhiza

SP – Super phosphate

MRP – Mussoorie rock phosphate

The plant receiving both the inoculants along with superphosphate registered maximum nutrient uptake rather than individual inoculant applied plants. Dual inoculation enhanced the P uptake by 59% in super phosphate and 64% in P unamended plants over uninoculated control. In rock phosphate application P uptake was tremendously enhanced by dual inoculation. The higher acquisition of these nutrients by mycorrhizal plants suggests increased efficiency of mycorrhizal roots in translocation and transportation of nutrients. A similar increase in N content and also P content upon mycorrhizal inoculation have been shown for various crops.(Johansen *et al.*, 1994; Jones *et al.*, 1995).

Acid phosphatase activity was significantly improved due to inoculation of VAM and *Rhizobium* and it was maximum at zero P, rather than with P application. Not much influence on alkaline phosphatase activity was noticed.

The significant increase in enzyme level may be of fungal origin or due to the increased production of enzymes by the induction of fungal hyphae in to the host tissue. However, the activity was independent of phosphorous nutrition of the host plant, since the activity was considerably less in phosphorous supplemented plants (Ubalthose Raja *et al.*, 1989).

Dual inoculation was found to be superior than individual inoculation of either of the inoculants in enhancing the nodulation, drymatter and nutrient uptake of soybean. The effect of super phosphate application and rock phosphate application are of similar trend in growth of soybean. However there are notable quantitative differences between the sources of phosphorous. First the amount of phosphate as rock phosphate which has to be added to soil to achieve maximum response of soybean was greater than the amount of phosphate from super phosphate that required for both mycorrhizal and nonmycorrhizal plants. Secondly the maximum response with rock phosphate was lesser than that with super phosphate application as suggested by Harley and Smith (1983).

Table 2. Influence of *Rhizobium* and VAM inoculation on N and P uptake and phosphatase activity of soybean applied with different P sources.

Treatment	N uptake (mg/pl)	P uptake (mg/pl)	Acid phosphatase activity	Alkaline phosphatase activity
Control	350.64	27.44	1.03	0.46
Rhizobium	381.16	32.06	1.03	0.47
VAM	377.26	37.94	1.61	0.49
Rhizobium + VAM	425.37	45.27	1.83	0.49
Control + SP	431.85	45.23	0.45	0.35
Rhizobium + SP	490.96	53.83	0.49	0.35
VAM + SP	523.3	70.61	0.67	0.37
Rhizobium + VAM +SP	663.3	72.02	0.94	0.38
Control + MRP	373.67	32.23	0.49	0.38
Rhizobium + MRP	419.5	38.16	0.49	0.43
VAM + MRP	440.88	41.22	1.12	0.46
Rhizobium + VAM + MRP	567.03	67.28	1.39	0.47
CD at 5%	12.15	5.11	0.22	NS

REFERENCES

Bolan, N.S 1991. A critical review on the role of mycorrhizal fungi in the uptake of phosphorous by plants. **Plant Soil, 134:** 189-207

Dodd, J.C., C.C. Birton, R.G. Burns and P. Jeffries. 1987. Phosphatase activity associated with the roots and the rhizosphere of plants infected with vesicular-arbuscular mycorrhizal fungi. **New Phytol. 107** :163-172

Gianinazzi-Pearson, V.J.C. Fardeau, S. Asimi and S. Gianinazzi.1981.Source of additional phosphorus absorbed from soil by vesicular-arbuscular mycorrhizal soybeans. **Physiol. Veg., 19**:33-34

Gerdeman, J.W. 1975. Vesicular-arbuscular mycorrhizae. In; **The Development and functions of roots** (Eds) J.G. Torrey and D.T.Clarkson.Academic Press, New York. pp. 575-591

Harley, J.L and S.E. Smith. 1983. **Mycorrhizal symbiosis**. Academic press, London. pp. 1-77

Jones, E.J., J. Magid, T.S. Gahoorial and I. Jakobsen1995. P depletion and activity of phosphatases in the rhizosphere of mycorrhizal and non mycorrhizal cucumber. **Soil Biol. Biochem., 27**:1145-1151

Johansen, A., I. Jakobsen and E.S. Jensen 1994. Hyphal N transport by a vesicular arbuscular mycorrhizal fungus associated with cucumber grown at three nitrogen levels. **Plant soil, 160:**1-9

Kennedy, Z.J.and M.Rangarajan.2000. Biomass production, root colonization and phosphatase activity by six VA-Mycorrhizal fungi in papaya. **Indian Phytopathol., 53**:38-102

Mosse, B., C.L. Powelland D.S. Hayman 1976. Plant growth response to vesicular arbuscular mycorrhiza.1X.Interactions between VA-mycorrhiza, rock phosphate and symbiotic nitrogen fixation. **New Phytol., 76**:331-342

Phillips, J.M. and D.S. Hayman. 1970. Improved process for clearing roots and staining parasite and vesicular arbuscular mycorrhizal fungi for rapid assessment of infection. **Trans. Br. Mycol. Soc., 55**:158-166

Plummer, D.T. 1978. **An Introduction to Practical Biochemistry.** McGraw Hill. UK. pp. 247-254

Ubalthose Raja, N., P. Sundaresan, P. Gunasekaran and M. Lakshmanan. 1989. Involvement of arbuscules and phophatases in P uptake during VA-Mycorrhizal symbiosis. **Indian J. Microbiol., 29** : 303-311.

<u>CHAPTER – 45</u>

DEVELOPMENT OF PELLET FORMULATION OF SULPHUR OXIDIZING BACTERIUM (SOB) FOR GROUNDNUT

R. Anandham and R. Sridar

ABSTRACT

A total of 49 isolates of sulphur oxidizing bacteria was isolated from seven locations. Isolates obtained from Neyveli mine spoil drainage (LCH) which reduced the pH from 8.0 to 2.0 also produced highest amount of sulphate 43.21 mg in 100 ml of Starkey's broth. For developing pellet formulations, TNAU wetland clay soil was collected, powdered, sieved through fine mesh sieve <0.1 mm and sterilized. For one kg of soil, 10g of elemental sulphur powder was mixed and 250 ml of 15 days old culture (LCH) was mixed and pellatised with hand Pellatiser. The pellets were cut into pieces of approximately 1 cm length, air-dried. The pellets contained 25×10^6 CFU (colony forming unit) of sulphur oxidizing bacterium (LCH isolate)/gram). It was tested in groundnut ALR-2 at different doses *viz.*, 20 kg, 40 kg, 60 kg/ha. LCH isolate was identified and found to belongs to the genus *Thiobacillus*. Further studies are needed to confirm the use of same formulation in pulses.

INTRODUCTION

Sulphur has long been recognised as essential element for plant growth. Now a days sulphur deficiency has become wide spread. In 'S' deficient fields, N, P, K alone cannot do the job. Sulphur is now considered as the fourth major plant nutrient after NPK (Katyal *et al.*, 1997).

The microorganisms are mainly responsible to make available sulphate from elemental or reduced forms of sulphur through its oxidation process with the intermediate formation of thiosulphate ($S_2O_2^{2-}$) and tetrathionate

($S_4O_6^{2-}$) in soils. The role of chemolithotrophic bacteria of the genus *Thiobacillus* in this process is usually very much emphasized (Shinde *et al.*, 1996). The microbial inoculant should be effective and farmers should be able to handle it easily. With this objective, an attempt was made to develop a pellet formulation of sulphur oxidizing bacterium using clay soil.

MATERIALS AND METHODS

A total of 12 samples from seven locations was used for isolation of sulphur oxidizing bacteria. For isolation, the procedure described by Starkey (1923) was adopted. Three different selective broth *viz.,* Starkey broth (Starkey, 1923), NCL broth (Waksman, 1922) and Thiosulphate broth (Beijerinck, 1904) were employed for the isolation. The initial pH of all the growth media was adjusted to 8.0 using 0.1 N KOH (Potassium hydroxide). Bromocresol purple served as indicator. The medium was steam sterilized for 3 days consecutively. In Starkey broth and NCL broth, elemental sulphur was dusted on the surface @ 1g for 100 ml of broth. Thiosulphate broth was devoid of elemental sulphur. Twenty ml of the above mentioned liquid media were prepared separately with and without glucose for isolation of heterotrophic and autotrophic 'S' oxidizers respectively. One gram/ml of the sample was added to each tube containing broth under aseptic condition. The tubes were incubated in BOD incubator (CI-125 Renu Ubstrynebts-Bombay) at 32°C for 25 days. The isolates obtained were purified by transferring to fresh broth thrice at fortnightly intervals.

Screening of isolates by pH reduction test

The obtained isolates were screened for their efficacy to reduce initial pH of the growth media. While isolating 'S' oxidising bacteria, pH of the growth media was measured after 25 days of incubation using pH paper. Only those isolates, capable of reducing pH from 8 to 5 or less than 5 were purified and used for further studies.

Estimation of sulphate production

Sulphate production was estimated by following the procedure of Tandon (1995). Fifteen days old cultures were used for estimation of sulphate production. Cultures were centrifuged at 10000 rpm for 15 minutes. A quantity of 0.1 ml of supernatant was taken in a 25ml volumetric flask. The volume was made upto 25 ml with distilled water. A quantity of 2.5 ml of conditioning reagent was added to volumetric flask to reduce the interference of other colouring agent in spectrophotometer reading. Just before taking the reading one gram of barium chloride powder was added to the samples and shaken well. It was kept for one minute. Then reading was taken in Carywin UV optical spectrophotometer at 340 mm wavelength (Varian 8510162500 connected with April, 1999 software imported from Varian Australia Pvt. Ltd. Australia). Standards were prepared with known quantity of potassium sulphate.

Development of formulations of 'S' oxidising bacteria

For developing formulation, wet land clay soil (TNAU) was collected, powdered, sieved through fine mesh sieve < 0.1 mm and sterilised. For one kg of soil, 10 g of elemental sulphur powder was mixed and 250 ml 15 days old culture was mixed and pelletised with hand pelletiser. The pellets were cut into pieces of approximately 1 cm length, air dried and stored for further studies (Table 1).

Table 1. Properties of clay soil used for pellet formulation

1. Physical parameters		Mechanical fractions
	a	Clay - 43.4%
	b	Silt - 13.2%
	c	Coarse sand - 23.9%
	d	Fine sand - 19.2%
	e	Apparent specific gravity 1.38 g cc^{-1}
	f	Absolute specific gravity 2.25 g cc^{-1}
	g	Pore space 52%
2. Chemical properties	a	Available N - 227 kg ha^{-1}
	b	Available P - 12.8 kg ha^{-1}
	c	Available K - 667 kg ha^{-1}
3.Physio-chemical properties	a	Soil reaction - 8.1
	b	EC - 0.60 dSm^{-1}
	c	CEC - 27.1 cmol $p(+)^{-1}$ kg

RESULTS

A total of 49 isolates of sulphur oxidising bacteria was obtained from seven locations (three from paddy rhizosphere soil; three from groundnut field, eight from sewage water, five from sterilite factory drain water, six from biogas slurry and 24 from Neyveli lignite mines) (Table 2).

Table 2. Isolation of sulphur oxidising bacteria from different locations

Source	Locations	No. of isolates obtained
Paddy Rhizosphere soil	Wet land, TNAU	3
Normal acid soil	Groundnut Field, TNAU	3
Sewage water sample (a)	Central farm drainage, TNAU	6
Sewage water sample (b)	Tamizhagam Hostel, TNAU	2
Sterilite factory drainage water sample	Sterlite factory, Tuticorin	5
Biogas slurry	Dept. of Bioenergy, TNAU	6
Neyveli minespoil soil		
Soil I	Neyveli II mines	4
Soil II	Neyveli Lignite Corporation	4
Soil III	Neyveli	4
Soil IV	Neyveli	4
Soil V	Neyveli	4
Neyveli II mines drainage water sample	Neyveli Lignite Corporation, Neyveli	4
	Total	49

Among the 49 isolates obtained from different locations, 15 isolates were selected after screening by pH reduction test. After screening, five efficient isolates from Sewage water, three from Sterilite factory drainage water, one from biogas slurry and six from Neyveli lignite mines were purified and selected for further studies. From the initial pH of 8.0, it was observed some of the isolates could reduce the pH even upto 2.0 (Table 3).

The sulphate production by 15 isolates of sulphur bacteria was estimated. The isolate LCH (Neyveli lignite mines drain water) recorded the highest sulphate production of 43.21 mg/100 ml of broth followed by SWA $_4$ isolate drain water) which registered 43.14 mg/100 ml. The lowest sulphate production of 10.04 mg/100 ml was recorded by heterotrophic isolate J_1G) (Table 4).

Table 3. Screening of sulphur oxidising bacteria by pH*

Source	Starkey broth		NCL broth		Thiosulphate broth	
	With glucose	Without glucose	With glucose	Without glucose	With glucose	Without glucose
Paddy rhizosphere soil	6.0	8.0	6.0	8.0	6.0	8.0
Normal field soil	6.0	8.0	6.0	8.0	6.0	8.0
Sewage water	6.0	2.0[a]	7.0	2.0[aa]	7.0	2.0[aaa]
	5.0[b]	4.0[bb]	8.0	8.0	8.0	8.0
Sterilite factory drainage water sample	6.0	2.0[c]	6.0	2.0[cc]	8.0	4.0[ccc]
Biogas slurry	5.0[d]	6.0	6.0	6.0	7.0	7.0
Neyveli minespoil Soil -I	6.0	5.0[e]	6.0	6.0	8.0	8.0
Neyveli minespoil Soil -II	6.0	5.0[f]	6.0	6.0	8.0	8.0
Neyveli minespoil Soil -III	6.0	4.0[g]	7.0	7.0	8.0	8.0
Neyveli minespoil Soil -IV	6.0	4.0[h]	6.0	6.0	8.0	8.0
Neyveli minespoil Soil -V	6.0	4.0[i]	6.0	6.0	8.0	8.0
Neyveli minespoil drainage water	6.0	2.0[j]	7.0	7.0	8.0	8.0

*Initial pH of the medium = 8.0

a. SWA_4, aa. NWA_4, aaa. TWA_4

b. J_1G, bb. J_2W

c. SWA_5, cc. NWA_5, ccc. TWA_5

d. SGA6, e. LC-I, f. LC-II, g. LC-III, h. LC-IV, i. LC-V, j. LCH

DISCUSSION

A total number of 15 isolates of sulphur oxidising bacteria was selected from a total of 49 isolates obtained from different sources *viz.,* field soil, sewage water, mine spoil and mine drain water. Isolation of sulphur oxidising bacteria from various sources have been attempted by several workers. Kelly and Harrison (1988) isolated sulphur oxidizing bacteria *Thiobacillus thioparus* from mud soil, canal water and other fresh water sources. Barrie Johnson (1992) isolated a *Thiobacillus* sp. from acid streams. *Thiobacillus ferrooxidans* and other *Thiobacillus* sp were isolated from uranium mines by Deborah Berthelot *et al.* (1992). Isolates were screened based on their ability to reduce the pH of the growth medium from 8.0 to 5.0 and less than 5.0 within a span of 15 days. The reduction in pH of the medium is due to production of sulphuric acid. Reduction in pH of the growth medium by sulphur oxidising bacteria was also reported by various authors (Vitolins and Swaby, 1969, Wainwright, 1984; Wainwright and Killham, 1980, and Donati *et al.*, 1994).

Table 4. Sulphate production by sulphur bacteria isolates

Isolates	Sulphate production (mg/100ml)
SWA_4	42.55
SWA_5	43.14
NWA_4	42.44
NWA_5	29.44
TWA_4	42.33
TWA_5	43.33
LC-I	19.70
LC-II	39.39
LC-III	39.05
LC-IV	42.16
LC-V	40.66
LCH	43.21
J_2W	38.96
J_1G	10.04
SGA_6	10.58
NCIM2426*	21.44
Control (uninoculated)	3.16

SEd =2.67 CD (P=0.05) = 5.44

* *Thiobacillus thiooxidans* – standard culture obtained from National Chemical Laboratory (NCL), Pune – Maharashtra, India

In the present study, LCH isolate was found to produce the highest quantity of 43.21 mg of sulphate/100 ml medium. This may be due to the utilization of elemental sulphur as energy source and oxidise the elemental sulphur as sulphate by the isolate. Similar results were reported by various authors (Waksman, 1922, Starkey, 1934; Gommers and Kuenen, 1988).

The pellets contained 25 x 10^6 CFU of sulphur oxidizing bacterium/gram. The pellets when inoculated in the medium @ 10g/100 ml were able to reduce the pH from 8.0 to 4.0 in 15 days. It was tested in groundnut ALR-2 at different doses *viz.*, 20 kg ha^{-1}, 40 kg ha^{-1}, and 60 kg ha^{-1}. It was found that this formulation was effective in oxidizing the elemental sulphur and was easy to handle by the farmers.

REFERENCES

Barrie Jhonshon, D., M.A. Palghauvi and Mazinf Said 1992. Isolation and characterisation of an acidophilic, heterotrophic bacterium capable of oxidizing Ferrous iron. **Appl. Environ. Microbiol., 58:** 1423-1428.

Beijerinek, M.W. 1904. Phenomenes dereduction proguits paries microbes. **Arh. Si. Exateset Nat. Haarlem. Ser., 2:** 9131-9157.

Deborah Berthelot, L., I.G Leduec and G.D. Ferroni 1992. Temperature studies of iron oxidising autotrophs and acidophilic heterotrophs isolated from uranium mines. **Can J. Microbiol., 39:** 384-388.

Donati, E., G. Curutchet, C. Pogliani, and Tedesco P. 1996. Bioleaching of covellite using pure and mixed cultures of *Thiobacillus ferrooxidans* and *T. thiooxidans.* **Pro. Biochem., 31:** 129-134.

Gommers, P.J.F. and J.G. Kuenen. 1988. *Thiobacillus* strain , a chemolithotrophic sulphur bacterium. **Arch. Microbiol., 150:**117-125.

Katayal, J.L., K.L Sharma, and K. Srinivas 1997. Sulphur in Indian Agriculture. In: **Proc. ISI/FAI/IFA-Symposium on sulphur in balanced fertilization.** Feb 13-14 at New Delhi pp.ks.2/1-ks 2/11

Kelly, D.P and A.P. Harrison 1988. Genus *Thibacillus* Berijerink. In: **Bergeys manual of systematic bacteriology Vol. 3** (eds) G.T. Staley, N. Pfennig and J.G. Holt. p. 1842-1871. Williams, Wilkins, Baltimore.

Shinde D.B., P.L. Patil, and B.R. Patil 1996. Potential use of sulphur oxidising microorganisms as soil inoculant. **Crop Res., 11:** 291-295.

Starkey, R.L. 1923. Cultivation of organisms concerned in the oxidation of thiosulphate. **J. Bacteriol., 28:** 365-386.

Starkey, R.L. 1934. Isolation of some bacteria which oxidize thio sulphate. **Soil Sci., 39:** 197-219.

Tandon, H.L.S. 1995. Methods of analysis of soils, plants, water and fertilizers. Fertilizer Development and Consultation Organisation, New Delhi. pp. 28-30.

Vitolins M.I and K.J. Swaby 1969. Activity of sulphur oxidising microorganisms in some Australian soils **Aust. J.Soil Res., 7:** 171-183.

Wainwright, M. 1984. Sulphur oxidation in soils. **Adv. Agron. 37:** 350-392.

Wainwright, M and K. Kilham 1980. Sulphur oxidation by *Fusarium solani.* **Soil Biol. Biochem., 12:** 555-558.

Waksman, S.A. 1922. Microorganisms concerned in the oxidation of sulphur in the soil III Media used for the isolation of sulphur bacteria from the soil. **Soil Sci., 13:** 329-336.

CHAPTER – 46

USE OF SULPHUR BACTERIA FOR INCREASED YIELD AND OIL CONTENT OF GROUNDNUT

R. Anandham and R. Sridar

ABSTRACT

Combined effect of sulphur oxidising bacteria *(Thiobacillus sp.)* and *Rhizobium* was studied in groundnut. Sulphur oxidising bacteria pellets (25 x 10^6 CFU/g) were employed in the study at three different doses *viz.,* 20 kg, 40kg and 60 kg ha^{-1} along with *Rhizobium* application as seed treatment at recommended level. Among different doses 60kg pellets ha^{-1} with *Rhizobium* application recorded the highest nodule number (136.31), nodule dry weight (0.740 g), pod yield (2006 kg ha^{-1}) and oil content (52%). It is concluded that sulphur oxidising bacteria can be used as a biofertiliser for groundnut to meet requirement of sulphur nutrition. It can be safely used in combination with *Rhizobium.*

INTRODUCTION

Sulphur has long been recognized as essential element for plant growth. Now a days sulphur deficiency has become widespread. In 'S' deficient fields, N,P,K, alone cannot do the job. Sulphur is now considered as fourth major plant nutrient after NPK (Katyal *et al.,* 1997). The microorganisms are mainly responsible to make available sulphate from elemental or reduced forms of sulphur through its oxidation process with the intermediate formation of thiosulphate ($S_2O_3^{2-}$) and tetrathionate ($S_4O_6^{2-}$) in soils. The role of chemolithotrophic bacteria of the genus *Thiobacillus* in this process is usually emphasized (Shinde *et al.,* 1996b).

Application of biologically activated sulphur with *Thiobacillus and Aspergillus avamorii* in rice increased yield and grain filling (Muralidharan and Jose, 1993). Inoculation of sulphur oxidising bacteria like *Thiobacillus thiooxidans, T. thioparus* and fungus like *Sclerobasidium constrictum, Aspergillus terreus* increased the grain yield, grain protein content and sulphur uptake in wheat (Shinde *et al.*, 1996a). This study was conducted to assess the individual effect of sulphur oxidising bacteria as well as its combined effect with *Rhizobium* in the nodulation, yield and oil content of groundnut.

MATERIALS AND METHODS

A field trial was conducted in Agricultural Research Station, Aliyar Nagar, TNAU, Coimbatore, Tamil Nadu. The trial was laid in Split Plot Design with three replications. *Rhizobium* was taken in the main plot and in sub-plots four indigenous sulphur oxidising bacterial isolates (SWA_4, SWA_5, LCH and SGA_6) belonging to the *Thiobacillus* genus and one standard strain NCIM-2426 *Thiobacillus thiooxidans* obtained from National Chemical Laboratory-Pune, Maharastra, India were applied in the form of pellets in the rows along with groundnut seeds at different dose *viz.*, 20kg, 40kg and 60kg ha^{-1}. For developing pellet formulation, wet land clay soil was collected, powdered, sieved through fine mesh sieve <0.1mm size and sterilized. For 1 kg of soil, 10g of elemental sulphur powder was mixed and 250ml of 15 days old culture was added mixed and pelletised with hand pellatiser. The pellets were cut into pieces of approximately 1cm length and air dried. The pellets contained 25 x 10^6 CFU (Colony forming unit of sulphur oxidising bacteria) / gram. Hundred gram of lignite based *Rhizobium* inoculum TNAU-14 obtained from Department of Agricultural Microbiology, TNAU was mixed with 150 ml of cooked rice kanji to prepare a slurry. To this slurry, 10 kg of groundnut ALR-2 added and mixed well, then shade dried for 30 minutes and used for sowing.

Nodulation, nodule dry weight were recorded on 80th day after sowing and the pod yield was recorded at maturity. Oil content of groundnut was estimated using NMR analyzer (Oxford 4000 NMR Analyzer, Oxford Analytical Instrument Limited, London).

RESULTS

Effect of sulphur oxidising bacteria and *Rhizobium* on the nodulation

The results revealed that on the 80th day after sowing, the isolate LCH registered the highest mean nodule number of 59.17, which was significantly superior then SWA_5 (65.22), SGA_6 (35.38), NCIM-2426 (35.91), SWA_4 (32.36) and control (26.09). Among the different doses the pellets applied @ 60kg ha^{-1} recorded the highest mean nodule number of (50.89), which was statistically superior than the other doses. The combined inoculation of sulphur oxidising bacteria and *Rhizobium* registered the highest mean nodule number of (57.18), which was significantly superior than sulphur bacterial inoculation alone (27.53). The results revealed from the interaction studies that the isolates LCH@ 60 kg ha^{-1} registered significantly higher nodule number of 136.83, when combined with *Rhizobium* application (Table 1).

Effect of sulphur oxidising bacteria and *Rhizobium* on nodule dry weight

The results revealed that after 80 days of sowing, the isolate SWA_5 registered the highest mean nodule dry weight of 0.302g, which was significantly superior of SGA_6 (0.173g), NCIM-2426 (0.166g), SWA_4 (0.145g) and on par with LCH (0.278g). Among the different doses, pellets applied @ 60kg ha^{-1} recorded the highest mean nodule dry weight of 0.246g, which was statistically superior to other doses. The combined inoculation of sulphur bacteria and *Rhizobium* registered the highest mean nodule dry weight of 0.278g, which was significantly superior than sulphur oxidizing bacteria inoculation alone (0.116g). The results of interaction studies revealed that the isolate LCH@ of 60kg ha^{-1} registered the highest nodule dry weight of 0.740g, when combined with *Rhizobium* application (Table 1).

Effect of sulphur oxidizing bacteria and *Rhizobium* on pod yield

The results revealed that the isolate, LCH applied crop registered the highest yield of 1929.44 kg ha^{-1} which was significantly superior to SWA_4 (1847.78 kg ha^{-1}), NCIM-2426 (1847.78 kg ha^{-1}), SGA_6 (1847.22 kg ha^{-1}) and control (1728.33 kg ha^{-1}) but was on par with SWA_5 (1921.11 kg ha^{-1}). The highest yield of 1874.17 kg ha^{-1} was recorded when pellets were applied @ 20 kg ha^{-1}, which was statistically superior to other doses. The combined inoculation of sulphur bacteria and *Rhizobium* registered the highest yield of 1876.48 kg ha^{-1} which was statistically superior to sulphur oxidising bacteria inoculation alone (1830.74 kg ha^{-1}).

The results of the interaction studies revealed that the LCH isolate, @ of 60 kg ha^{-1} registered the highest mean yield of 2006.67 kg ha^{-1}, when combined with *Rhizobium* application (Table 2).

Effect of sulphur oxidising bacteria and *Rhizobium* on oil content

The results revealed that crop which received the isolate SWA_5, registered the highest oil content of 50% which was on par with LCH (49.57%), but significantly superior to SWA_4 (47.68%), SGA_6 (46.89%), NCIM-2426 (45.57%) and control (44.65%). Among the different doses, pellets applied @ 60 kg ha^{-1} registered the highest oil content 48.04% which was statistically superior to the other doses. The highest oil content of 47.83% was recorded in the combined inoculation of sulphur oxidising bacteria and *Rhizobium*, which was significantly higher than the sulphur oxidising bacteria inoculation alone (46.96%). The interaction studies revealed that the isolate LCH @ of 60 kg pellets ha^{-1} registered the highest oil content of 52.58% when combined with sulphur oxidising bacteria and *Rhizobium* application (Table 2).

Table 1. Effect of combined inoculation of sulphur oxidising bacteria and *Rhizobium* on the nodule number and nodule dry weight in groundnut

Parameters	Treatment	S_1 (SWA$_4$)			S_2 (SWA$_5$)			S_3 (NCIM-2426)			S_4 (LCH)			S_5 (SGA$_8$)			S_6 (No inoculum)		
		D_1	D_2	D_3	D_1	D_2	D_3	D_1	D_2	D_3	D_1	D_2	D_3	D_1	D_2	D_3	D_1	D_2	D_3
Nodule Number	W_1	47.33	6.84	37.67	96.47	58.87	121.67	43.47	52.00	53.80	43.47	58.00	136.83	58.73	38.93	33.80	34.17	35.00	36.00
		(6.91)	(46.33)	(6.18)	(9.85)	(7.70)	(10.90)	(6.62)	(7.23)	(7.36)	(6.23)	(7.63)	(11.72)	(7.69)	(6.28)	(5.85)	(5.88)	(6.00)	(6.04)
	W_2	19.00	20.80	23.00	31.33	36.20	50.10	25.07	25.20	15.93	23.07	35.27	58.40	23.13	28.33	29.33	17.00	16.73	17.43
		(4.41)	(4.62)	(4.85)	(5.63)	(6.05)	(7.09)	(5.05)	(5.00)	(4.04)	(4.85)	(5.98)	(7.67)	(4.83)	(5.36)	(5.46)	(4.18)	(4.17)	(4.23)
Nodule dry weight (g)	W_1	0.213	0.208	0.169	0.436	0.266	0.607	0.195	0.234	0.242	0.195	0.261	0.740	0.264	0.219	0.204	0.184	0.179	0.184
	W_2	0.080	0.093	0.103	0.141	0.163	0.196	0.133	0.113	0.077	0.102	0.143	0.228	0.089	0.127	0.132	0.053	0.058	0.058

	Nodul Number		Nodul dry weight (g)	
	SED	CD (P = 0.05)	SED	CD (P = 0.05)
W	0.08	0.33	0.017	0.040
D	0.09	0.19	0.006	0.01
S	0.13	0.34	0.016	0.08
D x S	0.23	0.45	0.01	0.03
W x S	0.23	0.60	0.01	0.03
W x D x S	0.32	0.64	0.02	0.04

MEAN VALUES

Parameters	S_1	S_2	S_3	S_4	S_5	S_6	D_1	D_2	D_3	W_1	W_2
Nodule Number	32.36	65.22	35.91	59.17	35.38	26.09	38.52	37.66	50.89	57.18	27.53
	(5.63)	(7.87)	(5.88)	(7.41)	(5.91)	(5.08)	(6.04)	(6.07)	(6.78)	(7.40)	(5.19)
Nodule dry weight (g)	0.145	0.302	0.166	0.278	0.173	0.119	0.174	0.172	0.246	0.278	0.116

(Figures in parenthesis are $\sqrt{x + 0.5}$ transformed values)

W_1 - with *Rhizobium* W_2 - without *Rhizobium* D_1 – 20kg of pellets/ha D_2 – 40 kg of pellets /ha D_3 – 60 kg of pellets/ha

Table 2. Effect of combined inoculation of sulphur oxidising bacteria and *Rhizobium* on the yield and oil content in groundnut

Parameters	Treat-ment	S_1 (SWA_4)			S_2 (SWA_5)			S_3 (NCIM-2426)			S_4 (LCH)			S_5 (SGA_6)			S_6 (No inoculum)		
		D_1	D_2	D_3	D_1	D_2	D_3	D_1	D_2	D_3	D_1	D_2	D_3	D_1	D_2	D_3	D_1	D_2	D_3
Yield kg/ha	W_1	1843. 33	1860.00	1873.33	1886.67	1976.67	1996.67	1856.67	1870.00	1900.00	1900.00	1950.00	2006.67	1870.00	1850.00	1860.00	1760.00	1756.67	1760.00
	W_2	1826.67	1840.00	1843.33	1870.00	1880.00	1916.67	1786.67	1823.33	1850.00	1883.33	1896.67	1940.00	1843.33	1830.00	1830.00	1680.00	1700.00	1713.33
Oil Content %	W_1	48.17	48.83	48.60	49.41	49.27	52.33	47.17	46.17	45.17	48.37	47.64	52.58	46.39	48.26	47.20	45.06	45.13	45.13
	W_2	45.12	46.93	48.43	49.29	48.37	51.33	46.47	45.13	43.30	46.73	49.96	52.16	45.21	47.91	46.20	44.73	43.77	44.11

	Yield kg/ha		Oilcontent (%)	
	SED	CD (P = 0.05)	SED	CD (P = 0.05)
W	2.43	10.45	0.02	0.08
D	4.38	8.74	0.32	0.64
S	6.19	12.35	0.45	3.13
D x S	10.72	21.39	0.78	1.56
W x S	10.70	21.35	0.79	5.46
W x D x S				

MEAN VALUES

Parameters	S_1	S_2	S_3	S_4	S_5	S_6	D_1	D_2	D_3	W_1	W_2
Yield kg/ha	1847.78	1921.11	1847.78	1929.44	1847.22	1728.33	1874.17	1852.78	1833.89	1876.48	1830.74
Oil conte %	47.68	50.00	45.57	49.57	46.89	44.65	46.86	47.28	48.04	47.83	46.96

W_1 - with *Rhizobium* W_2 - without *Rhizobium* D_1 – 20 kg of pellets/ha D_2 – 40 kg of pellets /ha D_3 – 60k g of pellets/hass

DISCUSSION

Effect on the nodulation and nodule dry weight

The LCH isolate @ 60 kg ha^{-1} registered the highest nodule number of 36.83 and nodule dry weight of 0.740 g, when combined with *Rhizobium* application. These results suggested that the sulphur oxidising bacteria and *Rhizobium* were campatible with each other. This is similar to the earlier findings of Naidu and Ram (1995) who found sulphur nutrition and *Rhizobium* have synergistic relation. Saber *et al.* (1989) reported that combined application of *Bacillus megaterium* var. *phosphaticum* and *Thiobacillus* increased the nodulation and nodule biomass in soybean.

Effect on pod yield

In field studies, the LCH isolate @ 60 kg pellets ha^{-1} when combined with *Rhizobium* application recorded the highest yield of 2006.67 kg ha^{-1}. Sulphate nutrition through elemental sulphur increased the yield of maize, soybean and wheat (Kabesh *et al.,* 1989; Saber *et al.,* 1989 and Shinde *et al.,* 1996b) as a result of increase in sulphate content of soil through the natural oxidation of elemental sulphur. In the present studies also increase in soil sulphate content due to oxidation of sulphur by sulphur oxidising bacteria might have increased the crop yield.

Effect on oil content

The LCH isolate @ 60 kg ha^{-1} in combination with *Rhizobium* registered the oil content of 52.58%. Application of sulphur increases the soil sulphate which ultimately increases the oil content of the crop (Aulakh *et al.,* 1980; Gupta and Agrawal, 1982; Kumar *et al.*, 1992; Panda *et al.,* 1997). In this study also, increase in oil content was observed. This might be due to increase in soil sulphate by oxidation of sulphur by sulphur oxidising bacteria.

CONCLUSIONS

Based on the above studies interaction between sulphur oxidising bacteria and *Rhizobium* was found to be synergistic. Such an interaction can be efficiently tapped for better crop growth, yield and oil content of groundnut.

REFERENCES

Aulakh, M.S, N.S Pasricha and A.S. Azad. 1980. Phosphorus and sulphur application boosts oil seed production in coarse textured soils. **Indian Farming. 39:** 29-30.

Gupta, H.K. and H.R. Agrwal. 1982. Influence of low grade iron pyrite on the availability of iron, sulphur and phosphorus in alluvial soil and their uptake by wheat. **Indian J. Agric. Chem. 15:** 141-147.

Kabesh, M.O., T.T.G. Behairy and M.S.M. Saber. 1989. Utilization of biofertilizers in field crop production. Effect of elemental sulphur application in the presence and absence of two biofetilizers on growth and yield of maize. **Egypt. J. Agron. 14:** 95-102.

Katyal, J.L., K.L Sharma and K. Srinivas. 1997. Sulphur in Indian Agriculture Proc. ISI/FAI/IFA-Symposium on sulphur in balanced fertilization. Feb. 13-14 at New Delhi. pp. KS 2/1-KS 2/11.

Kumar, V., G.C. Shrtotriya and V. Kaore. 1992. Crop response to sulphur application. IFFCO, New Delhi. pp. 50.

Muralidharan, P. and A.I. Jose. 1993. Effect of application of magnesium and sulphur on the growth, yield and uptake in rice. **J. Trop. Agri. 31:** 24-28.

Naidu, M.V. and H. Ram 1995. Effect of sulphur and *Rhizobium* inoculation on rhizosphere and non rhizosphere microflora. **Agro Pedology. 5:** 53-58.

Panda, S. C., S.S. Nanda, and R.C. Mishra. 1997. Response of winter season groundnt (*Arachis hypogaea*) to methods of sowing and sources of sulphur. **Indian J. Agron. 42:** 680-682.

Saber, M.S.M., F.K. El. Baz and M. O. Kabesh. 1989. The utilization of biofertilizers in increasing field crop production. Effect of elemental sulphur application and/ or biofertilization on the chemical composition of soybean seeds. **Egypt. J. Agron. 14:** 241-251.

Shinde D.B, P.L Patil and K.K. Khade. 1996a. A study on sulphur biofertilization of green gram for yield and quality. **J. Maharastra Agric. Univ. 21:** 365-367.

Shinde, D.B., P.L. Patil and B.R. Patil 1996b. Potential use of sulphur oxidising microorganisms as soil inoculant. **Crop Res. 11:** 291-295.

CHAPTER - 47

SOLUBILIZATION OF SILICATE AND CONCURRENT RELEASE OF PHOSPHORUS AND POTASSIUM IN RICE ECOSYSTEM

S. Anthoni Raj

ABSTRACT

Bacteria are plenty in soil and a few of them have the capacity to solubilize silicate minerals releasing silica. Potassium is also released concurrently when K-bearing silicate minerals are attacked. Natural occurrence of silicate solubilising bacteria (SSB) in rice ecosystem, in agroinputs and in materials of anthropogenic activities is observed. A bacterium (*Bacillus* sp) isolated from the soil of granite cursher yard was found to solubilize silicate minerals more efficiently under *in vitro* conditions and also in incubation studies with different types of soils. Inoculation of SSB to rice augumented available silica in soil, plant silica content and enhanced the grain yield in rice. Dissolution of silicates results in rendering phosphorus available for plant absorption as silicon competes with phosphorus fixation sites. Silica acts like a substitute for phosphorus in plants. SSB inoculation with siliceous rice residues like straw, husk ash and husk @ 5t/ha facilitated rapid breakdown of these residues and also enhanced the grain yield. The organic acid production and polysaccharide production are supposed to create conditions favourable for silica solubilisation.

Several organic siliceous materials like rice straw and sugar cane trash are available in rice ecosystem which can be utilized as manure either directly or after composting. Split rice straw is left in the field and the chopped rice straw left by the mechanical harvester is abounded in the field as collection is laborious and costly. SSB can be utilized for hastening its decomposition. Similarly rice husk ash is available in plenty and about 4 –5 million tonnes is available annually accumulating in rice mills. This contains 65% SiO_2, 25 – 30% carbon and a varying content of trace elements. Its utilization along with the SSB enhances the silica, phosphorus and potassium in soil and plant and also enhances the grain yield. Since SSB has a potential to release three different nutrients, it finds wide application in rice ecosystem.

INTRODUCTION

Silicon, an element 146 times more in earth's crust, though shares many properties of carbon which forms the backbone of most organic molecule, seldom forms an integral part of any cellular components. Si-Si bonds are considerably weaker than C-C bonds and many Si-H bonds are relatively unstable and react with oxygen. The larger size of the Si atom compared to C also renders it unsuitable as a building block despite the fact that it can also form bonds with four other atoms creating a three dimensional net work similar to carbon. But a few organisms like diatoms, chrysophytes, radiolarians, plants, insects and vertebrates absorb and accumulate silica. Among the plants, silica (SiO_2) concentrations are found to be higher in monocotyledons than in dicotyledons and its level increased from legumes< fruit crops< vegetables<grasses<grain crops. Grasses accumulate 2-20% of the foliage dry weight as hydrated polymer or silica gel which is not at all added exogenously as done with N, P and K.

Silica is abundantly available in soil but occurs in a polymerized form. Plants absorb only monosilicic acid and this could arise in soil solution only when depolymerisation occurs in soil silicates. Organic acids resulting from the decomposition of organic matter, metabolites from rhizosphere and non-rhizosphere microorganisms, root exudates etc., might result in the break-down of silicate minerals releasing the ionic component. It is well known that biological reactions hinge around bond making (anabolic) and bond breaking (catabolic) reactions while soil reactions depend on ion exchange, their sorption and desorption. Any biological actions that render ions soluble might influence the release and accumulation of exchangeable ions in soil solution. Bacteria are in plenty in soil and a few of them have the capacity to breakdown silicate minerals releasing silica.

The ability of bacteria to depolymerize crystalline silicates has been demonstrated. The release of Si, Mg and Al by fungal action from rocks was described. The breakdown of natural and synthetic silicates by bacteria was reported by several investigations (Webley *et al.*, 1963; Malinovoskaya *et al.*, 1990). A soil strain of *Pseudomonas* was found to dissolve synthetic silicate of calcium, zinc and magnesium and the minerals of wallostonite, apophyllite and olivine. *Proteus mirabilis, Bacillus caldolacticus, Sarcina ureae, Penicillium notatum* and *P. simplicissimus* were also found to degrade silicates. *Bacillus mucotaginosus* var *siliceous* was also reported to degrade silicate minerals. Several microorganisms like *Aspergillus niger, Bacillus extroquens* and *Clostridium pasteurianum* were found to grow on muscovite, biotite, orthoclase and micas *in vitro* (Reitemeier, 1951).

Aleksandrov (1958) reported that organo-minerals increased the yield by an average of 18% in maize, wheat, potato and tomatoes and an additional 15% increase was obtained by the application of organominerals inoculated with silicate bacteria. Ciobanu (1961) found that in a degraded chernozem soils 'Azotobacterin' and 'Silicabacterin' when applied simultaneously increased the yields of cotton. The effect of soil inoculum with a selected strain of silicate bacteria *Bacillus circulans* on the release of potassium and silicon from different minerals was reported (Zahro *et al.,* 1984). Soil contains silicate minerals that contain potassium in them. Hence when silicates are acted upon potassium will also be released. Silica is considered as a substitute for phosphorus and when silica is in solution it competes with the fixation sites for phosphorus thereby rendering its availability.

MATERIALS AND METHODS

Silica solubilising bacteria (SSB) was isolated after serially diluting the samples and plating 1 ml aliquots of appropriate dilution in Bunt and Rovira (1955) medium containing 0.25 per cent insoluble magnesium trisilicate. The plates were incubated at room temperature ($30 \pm 1^{0}C$). The colonies exhibiting clear zones were counted and the population expressed per gram on oven dry basis. Selected isolates were transferred to nutrient agar slants and maintained.

The bacterial isolate from the soil of granite crusher yard showed a larger clearing zone and identified as *Bacillus* sp and Silicate minerals like feldspar, mica muscovite, biotite kaolin, vermiculite, quartz and black ash (rice husk char) pulverized sieved through 60 mesh BSS and incorporated into 100 ml aliquots of basal medium devoid of dipotassium hydrogen phosphate at 0.25% level. The flasks were sterilized, cooled and inoculated with 1 ml of silicate solubilising bacteria (@10^9 cells/ml). An uninoculated control was also maintained for each mineral after sterilization. After incubation, samples were filtered through Whatman No.1 filter paper and centrifuged to remove the debris and cells. The pH, available silica, phosphorus and potassium contents were determined in the clear supernatant solution.

In another experiment, the bacterial species solubilising predominantly silicate (SSB), phosphorus (PSB) and solubilising both silicate and phosphate (SPB) all belonging to the genus *Bacillus* were inoculated to sterilized soils and incubated to assess the release of Si, P and K in four different soils. Red soil, clayey soil, sand and forest soil were filled in tubular pots (12 cm dia and 18 cm h) at 1 kg per pot excepting forest soil, whereas only 700 g soil could be filled in each pot as the bulk density was low. The soil was moistened and one set was sterilized and the other set was left unsterilized. The bacterial suspension of each of test bacterium grown in nutrient broth (@10^9 cells/ml) was inoculated at the rate of 10 ml per 100 g. The water samples were withdrawn at weekly intervals, filtered and analysed for available silica (Saxena 1989), phosphorus (Olsen *et al.,* 1957) and potassium (Jackson, 1967). Similarly soil samples were also collected initially and on fourth week, dried, powdered, sieved and analysed. The silica in soil was determined employing the method described by Imaizumi and Yoshida (1958).

RESULTS AND DISCUSSION

The bacterial growth on silicate containing agar media resulted in clearing zones indicating solubilization of silicates. Large clearing zones were observed in mica, quartz, silica mineral, and magnesium trisilicate. A clearing zone annule of 19.79 cm^2 was observed in mica and the lowest area of 2.64 cm^2 was observed in feldspar. The available silica increased in culture filtrate suggesting their release (Table 1). Quartz and silica mineral were most solubilized with release of 10.9 mg of SiO_2/1 after 20 days of inoculation. It was also found that autoclaving itself released some amount of silica from these organic siliceous materials. Straw was found to be more degraded releasing more of silica (10 mg/1).

The breakdown of organic siliceous materials also released phosphorus. A higher level of phosphorus was found in straw. There was no phosphorus detected in media containing silicate minerals.

There was a simultaneous release of potassium from silicate minerals and organic siliceous materials by SSB. The release of potassium increased with the incubation period. A higher potassium release was observed in straw. Higher release was also observed in the K bearing silicate minerals like biotite, muscovite, mica and feldspar. After 20 days of incubation as much as about 8-9 ppm increase was found in biotite. In vermiculite, quartz and silica mineral K release was almost negligible.

Soil Incubation Studies

Four different soils inoculated with the bacteria released silica, phosphorus and potassium to different extent. A higher quantity of silica was found in water sample of forest soil followed by clay, sand and red soils (Table 2). SSB inoculation showed higher availability of silica and a higher amount was always observed in unsterile soil than sterile soil (Table 3). Sand showed higher release of silica than other soils.

PSB inoculation showed higher release of phosphorus compared to SSB and it was higher in forest soil followed by clay, red and sand. Similarly higher phosphorus was found in unsterilized soil compared to sterile soil. SSB inoculation simultaneously released the potassium. A higher quantity was found in forest soil followed by clay, red and sandy soils after 4 weeks. A higher release was observed in unsterile set compared to sterilized set (Table 4).

The growth of SSB, *Bacillus* sp. on media supplemented with mica, quartz, silica mineral (sand) and magnesium trisilicate was quick and prominent. Lauwers and Heinen (1974) reported the ability of *Bacillus caldolacticus* and *Proteus mirabilis* to grow and solubilize quartz. Higher amount of silica was released from organic siliceous materials like straw, husk and black ash. This might be due to their higher silica content itself besides they are easily degraded and mineralized in nature in the process of their utilization. The solubilization of silica from silicate might be due to production of carbon di-oxide, organic acids, exopolysaccharide production, alkali production and H_2S production (Ehrlich, 1981; Malinovskaya *et al.*, 1990). Autoclaving for sterilization itself released to some extent silica from both organic siliceous materials and silicate minerals suggesting the role of higher temperature similar to weathering.

The solubilization of P may not be solely due to the direct action of organic acids elaborated by the organisms as the silica solubilized by SSB can reverse P-fixation and keep the P-ion available in solution. The release of phosphate by silicate and the reversal of phosphorus fixation by silica was also reported (Jones and Handreck, 1967). A higher release of phosphorus was noted in soils treated with PSB or SSB. Bassitisse (1947) reported that silica mobilized phosphate anions by limiting the energy of fixation of soil colloids and precipitating the oxides of iron and aluminium in laterite soils. Roy *et al.* (1971) had also shown that silicic acid might compete against phosphate for a place on the hydrated hydroxides. Such a mechanism might have operated in the red soil in the present study.

The breakdown of silicate concurrently released potassium. As the incubation progressed, there was higher release of K from K-bearing silicate minerals like muscovite, biotite and feldspar by SSB. Reitemeier (1951) reported the growth of *Aspergillus niger* on muscovite, biotite, orthoclase and microcline. The release of K observed might arise due to (i) leaching from the organic siliceous residues, (ii) decomposition of the organic matter and / or (iii) breakdown of silicate in these materials.

Soil incubation studies with and without inoculation of SSB, PSB, SSB and PSB showed that all the of them played a varying role in dissolution of silica, phosphorus and potassium and this varied with the nature of soil and the organism. Their survival over a period of time in sterile as well as unsterile soils indicated that the organic matter and the other nutrients available in them might have favoured their survival even in the absence of plants to provide nutrients through root exudates, root decay or leaf fall. The available silica in sand and forest soil was higher as sand is primarily silica and SSB inoculation might have solubilized a larger proportion. Chinnasamy and Chandrasekaran (1976) reported a higher available silica in clay soils than clayey loam and sandy loam. However, Nayar *et al.* (1982) reported that the available silica was greatly influenced by flooding and the magnitude of increase in silica was more in laterite soils than black soil.

A mere submergence of soil released very little silica in sterile soils whereas in unsterile soil it was higher. This revealed the role of micro-organisms in depolymerisation of silicates. This view is also strengthened by the fact that inoculation of SSB augmented its level greatly in all these soils both under sterile and unsterile conditions. It is evident from these studies that SSB inoculation augmented silica content in soil solution and in soil of any type and this could be exploited as a bio-inoculant.

The potassium release was relatively higher in forest soil followed by clay, red and sandy soil due to SSB inoculation. This might be due to variation in K-bearing silicate minerals present in these soils and the organic matter content in the soils which play a major role in the release of potassium (Tandon and Sekhon, 1988). The K may be liberated to the immediately exchangeable form both by decomposition of K bearing feldspar, and by replacement of interlayer K ions of micas. A part of this K may be refixed between the layers of particles of micaceous and montmorillonite minerals. The return of plant residue to soil will result in the liberation of their potassium to soluble and exchangeable ions by way of silicate solubilization. The release of K might be due to (i) Breakdown of K bearing minerals and (ii) Si ions causing changes in the exchangeable ion status of minerals. This can occur in any type of soil where SSB is introduced.

There are a variety of siliceous materials available in rice ecosystem. Rice stubbles (straw and root) left in the field constitute easily 1 to 2 t per ha depending upon the harvest and season. Rice straw rich in silica and potassium, is left in the field uncollected when mechanical harvesting is done. Similarly rice husk ash is available to the tune of 4 – 5 million tones annually. This contains 65% silica 20 to 30% carbon and a variety of micronutrients. Addition of these rice straw and rice husk ash was found to augment grain yield of rice particularly when inoculated with SSB (Table 5). Black ash is available in rice mills scattered all over the rice growing area and is a renewable waste that can be used as a source of silica alongwith SSB. Raw sugarcane trash also can be used in rice culture with SSB so as to provide silica and other nutrients to rice.

REFERENCES

Aleksandrov, V.G. 1958. Organomineral fertilizers and silicate bacteria. **Dokl. Akad-S. Kh. Nauk., 7:** 43 – 48.

Bassitisse K. 1947. Studies on theoretical and practical Conditions in which the availability of phosphoric acid as maintained in lateritic soils. **Annals Agronomy 16,** 463-475.

Chinnasamy K.N. and S. Chandrasekaran. 1976. Effect of Silica gel on rice in three soil types. **Annamalai University Agricultural Annuals, 6:** 3-29.

Ciobanu, I. 1961. Investigation on the efficiency of bacterial fertilizers applied to cotton. **Cent. Exp. Ingras. Bact. Lucrari. Stiint., 3:** 203 – 214.

Ehrlich H.L. 1981. **Geo Microbiology**. p. 393. Marcel Dekker Inc., New York.

Imaizumi K. and S. Yoshida. 1958. Edaphological studies on Silicon supplying power of paddy fields. **Bulletin National Institute of Agricultural Sciences, Japan B8,** pp. 261-304.

Jackson, M.L. 1967. **Soil Chemical Analysis.** Prentice Hall of India Pvt., New Delhi, India. P.498.

Jones, L.H.P. and K.A. Handreck. 1967. Silica in soil, plants and animals. **Advances in Agronomy, 19:** 107-149.

Lauwers, A.M. and W. Heinen. 1974. Biodegradation and utilization of silica and quartz. **Archives Microbiology, 95:** 67-78.

Malinovskaya, I.M., L.V.Kosenko, S.K. Votselko and V.S. Podgorskii. 1990. Role of ***Bacillus mucilaginosus*** polysaccharide in degradation of silicate minerals. **Mikrobiologiya, 59:** 70-78.

Nayar, P.K., A.K.Misara and S. Patnaik. 1982. Silica in rice and flooded rice soils. **Oryza, 19:** 34-40.

Olsen, S.R., C.V.Cole, F.S.Watanabe and L.A. Dean. 1954. Estimation of available phosphorus in soils by extraction with sodium bicarbonate. **US Department of Agriculture Circular No. 939,** Washington DC.

Reitemeier R.F. 1951. Soil Potassium. In: **Advances in Agronomy Vol.III** (eds.)A.G. Norman, Academic Prem Intl. Publ. New York. pp.113-164.

Roy A.C., M.Y.Ali, R.L. Fox and J.A. Silva. 1971. Influence of Calcium silicate on phosphate solubility and availability in Hawaiian Latosols. In: **Proceedings on International Symposium on Soil Fertility Evaluation 1:** 751-765.

Saxena M.M. 1989. **Environmental analysis of water, soil and air.** Agrobotanical Publ. India. P. 786.

Tandon H.L.S. and G.S. Sekhon. 1988. **Potassium Research and Agricultural Production in India.** Fertilizer Development and consultant Organisation, New Delhi, India. P.144

Webley D.M., M.E.K. Henderson and E. Taylor 1963. The Microbiology of rocks and weathered stones. **J. Soil Sci., 14:** 102-114.

Zahro M.K., M. Mohib, S.I. Abdel -Al and H. Heggo. 1984. Significance of inoculation with silicate bacteria. **Zentral. Microbiologiya, 139:** 349-357.

CHAPTER – 48

ISOLATION AND CHARACTERIZATION OF ZINC SOLUBILIZING BACTERIA (ZSB) FROM PADDY SOIL

V.S. Saravanan and S. Anthoni Raj

ABSTRACT

An attempt was made to isolate zinc solubilizing bacterial cultures from soil and ore sources, both by direct plating and by enrichment technique in the modified Bunt and Rovira medium with 0.1 per cent zinc. Three cultures (ZSB- O-1, ZSB-S-2 and ZSB-S-4) were isolated by direct plating and one (ZSB-S-3) by enrichment technique. Among this ZSB-O-1and ZSB-S-4 were characterized as *Bacillus* sp. and ZSB-S-2 as *Pseudomonas* sp. Further the potential to correct the zinc deficiency was assessed and the results revealed that *Pseudomonas* sp. (ZSB-S-1) was able to correct the zinc deficiency when used along with 1%(w/w) of zinc oxide.

INTRODUCTION

Zinc is an important nutrient, which plays a major role in the metabolism of plants and animals. Though its requirements are small its role in the metabolism is significant. First evidence that zinc as an essential nutrient for the crops was given by Maze (1914). Zinc is stored predominantly in the seeds (Nikolaenko, 1971). Shuvalov and Kranovsky (1971) investigated the role of zinc ions in the biosynthesis of chlorophyll molecule precursors, which relates to the photosynthetic ability of a plant and more than 16 enzymes having metal complexes are activated by zinc (Parisi and Vallee, 1969). Thus zinc serves as predominant micronutrient in the plant system. Zinc forms an immobile zinc protein complexes under high nitrogen and low zinc conditions. The excessive application of high analysis fertilizers aggravate zinc deficiency in soil (Ozanne, 1955 and Chapman 1960).

Although plants take this metal in relatively very low concentration, over 50 per cent of Indian soils are deficient in the zinc status i.e. below the critical level of 1.5 ppm available zinc (Tiwari and Dwivedi, 1994) though the total zinc is very high 10-330 ppm in different soils (Swaine, 1956). Thus the problem of the zinc was confined only with its availability and not on its total quantity. Though zinc sulphate, a commercial fertilizer source, is applied to alleviate the zinc deficiency problem again the fixation may take place in soil and its use efficiency was also low (0.32-1.5 per cent). So a microbial approach was made, screening for certain microbes that is capable of solubilizing the insoluble zinc to soluble form. Thus the possibility of converting the total unavailable zinc which varied from 10-300 ppm in different soils, to available form, just to a maintain a critical level of 2.0 ppm is an easy task for microbes that do several transformations in soil. In the present investigation, Zinc solubilizing bacteria in the soil was investigated, isolated by enrichment technique and characterized.

MATERIALS AND METHODS

Zinc solubilizing bacteria was isolated from different sources like zinc ore (sphalerite) obtained form a private zinc sulphate manufacturing company and also from paddy soils. The medium used for isolation is the modified Bunt and Rovira medium (1955). The composition is as follows. Peptone–1.0g; yeast extract–1.0g; glucose–20.0g; ammonium sulphate–5.0g; dipotassium hydrogen phosphate–0.40g; magnesium chloride–0.10g; ferric chloride–0.01g; soil extract–250 ml; zinc oxide–1.00 g; tap water – 750 ml; agar – 20.00 g; pH– 6.6-7.0.

The organism was isolated by plating 1 ml aliquots of 10^{-3} dilution in the modified Bunt and Rovira medium containing 0.1 per cent insoluble zinc oxide. The plates were incubated at room temperature (30°±1°C). The colonies exhibiting clearing zone formation indicating dissolution of insoluble zinc (ZnO) was selected, purified and sub cultured in nutrient slants.

Isolation by enrichment technique

The zinc solubilizing bacteria was also isolated by enrichment technique. In this 100 g of paddy soil was collected to which insoluble ZnO was added @0.1 per cent, moistened to 40 per cent moisture content and incubated at room temperature for 3 days. After the incubation period, the sample was serially diluted and 1 ml of 10^3 dilution was plated in the Bunt and Rovira medium containing 0.1 per cent zinc oxide. The plates were incubated at room temperature and examined for colonies showing clear zone of zinc dissolution.

Identification and characteristics of isolated bacterial cultures

The bacterial cultures were identified based on their morphological structures like size, shape, motility and then based on nutritional characters like growth on different carbon sources like glucose, sucrose, lactose and mannitol. In these sources, growth was identified by acid and gas production and then Gram staining and endospore staining were also performed. The organisms were subjected to various biochemical tests like hydrolysis of starch, lipolysis liquefaction of gelatin, catalase and nitrate reductase activities. Based on the characteristics, the bacterial cultures were identified employing the key described in Bergey's Manual of Determinative Bacteriology (1994).

Alleviating zinc deficiency in soybean

The potentiality of ZSB to correct the Zn deficiency in plants upon inoculation was tested on soybean plants (soybean plants were taken because of its high zinc requirement) grown in acid washed sand. The sand was first sieved to remove clay and silt. Then it was subjected to acid treatment by saturating with a mixture of 10 per cent HCl and 1per cent oxalic acid to remove the macro and micronutrients. The saturated sand was left aside for 2 days, and after 2 days the sand was steamed at 20 psi in an autoclave, cooled, leached initially with rain water and then twice with piper nutrient solution. This ensured the displacement of adsorbed hydrogen ions, which would other wise cause severe injury to the plants. The tubular pots of 18 cm height and 12 cm diameter were filled with acid washed air dried sand (800 g/pot) moistened with sterile water and sterilized at 20 psi for 1 hr. The insoluble zinc compound, zinc oxide was added to each pot @ 0. 1per cent (w/w). These pots were inoculated with ZSB @1 g/pot. A separate control without zinc compound was maintained. Soybean seeds of var. Co-1 were sown and irrigated with Piper nutrient solution devoid of zinc. The following is the composition of the Piper nutrient solution (Piper, 1942) Potassium nitrate-1.5 g; potassium dihydrogen phosphate-0.5 g; sodium chloride-0.1g; calcium sulphate hydrated-0.5 g; magnesium sulphate hydrated-0.5 g; ferric citrate-0.02 g; boric acid-0.5 mg; sodium molybdate- 0.1mg; copper sulphate-0.003 mg; water-1000 ml.

RESULTS AND DISCUSSION

The total and zinc solubilizing bacterial population can be visualized from the (Table 1). The zinc ore (sphalerite) obtained from Hindustan Zinc Ltd. Rajasthan (stored for 6 months) showed $2x10^3 g^{-1}$ZSB compared to a total bacterial count of only $6 \times 10^6 g^{-1}$ on dry weight basis. The zinc ore sphalerite showed the ZSB population only in thousands compared to millions of total bacteria. The population of total bacteria was higher before the application of zinc compounds for enrichment ($60 \times 10^6 g^{-1}$ in clayey soil and $56x10^6 g^{-1}$ in black soil), but after enrichment a slight decrease was observed which may be due to toxic action of zinc on certain group of microbes. The ZSB population was less than thousand before enrichment, but recorded $3 \times 10^3 g^{-1}$ dry weight in clayey loam and $2 \times 10^3 g^{-1}$ in the black soil. The ZSB showing prominent clearing zones were purified and maintained in slants. The isolate from ore (sphalerite) was designated as ZSB-O-1 and other two isolates from paddy soil was designated as ZSB-S-2 and ZSB-S-4 and another isolate obtained after enrichment with zinc in paddy soil was designated as ZSB-S-3.

Table 1. Total and zinc solubilizing bacteria in soil and zinc ore

S. No.	Sample	Source	Total bacteria $1 \times 10^6 g^{-1}$	ZSB $1 \times 10^3 g^{-1}$
1.	Zinc ore	Rajasthan	6	2
2.	Clayey loam soil before soil enrichment	Central farm (Madurai)	60	1×10^2
3.	Clayey loam soil after soil enrichment	Central farm (Madurai)	56	3
4.	Black soil before enrichment	Perambalur (Trichy)	70	1×10^2
5.	Black soil after enrichment	Perambalur (Trichy)	60	2

The three isolates were identified based on their morphological and biochemical characteristics (Table 2). The ZSB-O-1 and ZSB-S-4 isolates are the Gram positive and very short to medium rods. The ZSB-O-1 produced colonies with white colour initially but turned to chocolate brown later, with creamy pigmentation and produced sub terminal spores. However, ZSB-S-4 produced milky white to creamy colony. The ZSB-S-2 produced light brown slimy colony, spreading quickly over the agar surface. Both the cultures (ZSB-O-1 and ZSB-S-4) showed positive reaction for starch hydrolysis, gelatin hydrolysis and lipolysis. They are also catalase positive and utilized citrate. They did not produce gas in glucose, lactose and mannitol and reduced NO_3. These two cultures were tentatively identified as *Bacillus* sp.

Table 2. Morphological, cultural and biochemical characteristics of the different zinc solubilizing isolates

Sl. No.	Tests	ZSB-O-1	ZSB-S-2	ZSB-S-4
1.	Shape	Short rods	Very small rods	Short rods
2.	Gram stain	+	-	+
3.	Endospore stain	+	-	+
4.	Colony character	White mucous but chocolate brown in older stage.	Light brown slimy spreading colony	Creamy white colony
5.	Pigments production	+	+	+
6.	Starch hydrolysis	+	+	+
7.	Lipolysis	+	+	+
8.	Acid production			
	glucose	+	+	+
	lactose	+	+	+
	mannitol	+	+	+
9.	Gas production			
	glucose	-	-	-

	lactose	-	-	-
	mannitol	-	-	-
10	Catalase	+	+	+
11	NO_3 reduction	+	-	+
12.	Gelatin hydrolysis	+	-	+
13.	Citrate utilization	+	+	+
14.	Tentative identification	*Bacillus* sp	*Pseudomonas* sp	*Bacillus* sp

The isolate ZSSB-S-2 is Gram negative, very small rods and non-spore forming bacterium, with yellowish green pigmentation in broth, which turned to brown colour later. It hydrolyzed starch and not gelatin, utilized citrate and lipid and was catalase positive. It however failed to reduce nitrate. It was tentatively identified as *Pseudomonas* sp. The results of the ZSB inoculation on correction of zinc deficiency are furnished in Table 3.

Table 3. Influence of ZSB inoculation on correction of zinc deficiency in soybean

Sl. No.	Treatments	Available zinc (ppm)	
		0 DAI	30 DAI
1.	Control (uninoculated)	-	-
2.	ZSB-S-1 @ 1g/pot	-	3.1
3.	ZnO @1 % (w/w)	4.3	6.1
4.	ZnO+ZSB-S-1 @ 1g/pot	5.0	13.1

The zinc deficiency was induced in soybean plants by growing them in acid washed sand and irrigated with sterile water. The typical deficiency symptoms appeared after 2 weeks of sowing with chlorotic interveinal areas, initially spreading and finally necrotic in appearance. After the appearance of deficiency symptom, a set of pots was treated with insoluble zinc oxide alone, another with both zinc oxide and ZSB-S-1. Within 15 days of inoculation, the deficiency symptoms disappeared gradually in the ZnO + ZSB-S-1 applied pots. Influence of ZSB inoculation on the available zinc are given in Table 3.

The available zinc increased from 5 ppm observed initially to 13.1 ppm after 30 DAI in the combined inoculation of ZnO+ZSB-S-1 treatment. The recovery from the deficiency was quick and rapid in the earlier stage of symptom development. The inoculation of either ZSB or application of ZnO alone did not correct the zinc deficiency rapidly, but their combination has corrected the deficiency.

The addition of insoluble ZnO alone did not correct the zinc deficiency, since it is insoluble in water. But addition of insoluble ZnO along with ZSB corrected the deficiency indicating the availability of zinc from the insoluble zinc compounds, is by the action of ZSB. Further ZSB may also help in the chlorophyll synthesis, which aided the faster recovery of plants. Importance of zinc in the chlorophyll synthesis was well established by Shuvalov and Kranovsky (1971). Zinc deficiency was found to be corrected within 48 hrs by supply of $ZnSO_4.7H_2O$ through spraying (Skoog, 1940) But in the present study, it took 20 days for full recovery as the insoluble zinc compound ZnO was applied to soil and also due to slow multiplication of ZSB in the sand poor in nutrients. The deficiency correction was quick when the symptoms were recognized early and diagnosed. The older plants with advanced deficiency symptom did not show full recovery form the deficiency, although there was some degree of correction observed.

Thus the zinc is an important nutrient and though it is not required in larger quantities like the major nutrients NPK, it is required in traces. At the same time, it is also surprising to note that even for this small requirements soil exhibit lack of available zinc despite the presence of higher total zinc.

Exogenous application of soluble zinc sources similar to fertilizer has been advocated to various crops. This causes transformations of about 96-99 per cent of applied available zinc to various unavailable forms. The zinc use efficiency from the soluble zinc compounds is also very low (0.32-1.5 per cent). Thus the water soluble zinc ($ZnSo_4$. $7H_2O$) advocated for crops and applied to the soil cannot be detected after 7 days (Rattan and Shukla, 1991). This makes the zinc nutrition to crops critical. These unavailable forms cannot be utilized by the plant system. Therefore, a system is evolved for dissolution of the zinc from the locked up unavailable system to a more available form.

In the soil system, the omnipotent microbes carry out several biological processes that are interdependent and cyclic one to maintain the balance of nature in the ecosystem. It is well known that in soil several mineral transformations occur through biological action and microbes do play a significant role in the mineralization of organic matter, transformations and solubilization of several nutrients and renders them available for absorption by plants. Thus, zinc may be solubilized by microbial action from the soil reserve to meet the crop requirement.

The total zinc content of soil ranges from10-300 ppm (Swaine, 1956) but its availability to crop is very low. Further continuous cropping and cultivation of high yielding varieties intensively lead to steady depletion of zinc in soil. Addition of high analysis fertilizers, which lack in Zn like nutrients renders soil unable to meet the trace element requirement of crops. For instance, a cropping system yielding 10 tonnes of food grains per hectare removes zinc form soil to an extent of 500 g. In fact a soil on the threshold of zinc contains that much of zinc (0.5 ppm which equivalent to 500 g of Zn in the top 15 cm of soil) only. Therefore, zinc nutrition is critical for analysis of the existing cropping system. Monoculturing or cultivation of same crop, season after season should be avoided. The roots, stubbles and other crop residue should be left in the field to replenish the soil with such of those trace elements. Further, a fallow period is also left for the soil to rest without feeding the crop from its nutrient reserve. Yet soils exhibit deficiency for zinc and to alleviate zinc deficiency, exogenous application of soluble compounds is advocated.

The present investigation has shown the natural occurrence of zinc solubilizing bacteria (ZSB) in the soil as well as zinc ores (sphalerite). But the population in the ore was low compared to total bacteria. When millions of bacteria occur in soil, only a few thousands of them have the ability to solubilize the insoluble zinc compounds. Zinc is an element required by the microbes but at the same time toxic at higher levels. If an organism by its action brings Zn into solution in larger quantity, it may become inimical to its growth. Only a limited number of microorganisms in soil have ability to attack insoluble zinc compounds and this depends on the tolerance to high levels. This variation might also due to formation of zinc chloro complexes in the soil, which causes toxicity to many microbes (Wong, 1980). Thus this ZSB can also used like phosphobacteria as a microbial inoculant for the zinc nutrition to crops.

REFERENCES

Bunt, J.S. and A. D. Rovira. 1955. Microbiological studies of some subantartic soils. **J. Soil Sci., 6:** 119-128.

Chapman, H. D. 1960. Leaf and soil analysis in citrus orchards. Univ. California Div. Agr. Sci. Ext. Serv. Manual p. 25.

Maze, P. 1914. Influences respective des elements de la solution minerale sur le development clo mais. **Ann. Inst. Pasteur., 28:** 11-46.

Nikolaenko, N.A. 1971. The accumulation of trace elements in the organs of spring wheat varieties cultivated in the Orenburg region. Abstract of dissertation Agricultural Institute, Orenburg, p.28-38.

Ozanne, P.G. 1955. The effect of nitrogen on zinc deficiency in subterranean clover. **Australian. J. Biol. Sci., 8:**47-55.

Parisi, B. and B.L. Vallee. 1969. Metal enzyme complexes activated by zinc. **J. Biol. Chem. 179:** 803-807.

Piper, C.S. 1942. The symptoms and diagnosis of minor-element deficiencies in agricultural and horticultural crops. II copper, zinc, molybdenum. **Emp. Jour. Expt. Agr., 8:** 199-206.

Rattan, R.K. and L. M. Shukla 1991. Influence of different carrier on the utilization micronutrients by rice. **J. Indian Soc. Soil Sci., 39:** 808-810.

Shuvalov, V.S. and A.A. Kransovsky. 1971. The luminescence of zinc protoporphyrins in microorganisms and plants. Phosphorescence and delayed fluorescence. **Molekulyasnaya Biologiya, 5:** 698-793.

Skoog, F. 1940. Relationship between zinc and auxin in the growth of higher plants. **Plant Physiol., 1:**231-249.

Swaine, D.J. 1955. The trace element content of soils Common Wealth. Bur. Soil Sci. Tech. Comm. No. 48: Herald printing works, Coney St. York (England).

Tiwari, K.N. and B.S. Dwivedi. 1994. Available zinc status of soil and delineation of the area of zinc deficiency in Uttar Pradesh. **Fert. News, 39(3):** 31-39.

Wong, M.H. 1980. Effect of inorganic salts in combination with heavy metals on microbes. **Microbios. 28:** 19-25.

The present investigation has shown that [illegible] occurred [illegible] soil [illegible] (25%) in the [illegible] as [illegible] population in the [illegible] low compared to [illegible]. When [illegible] bacteria [illegible] the [illegible] to solubilize the insoluble zinc compounds [illegible] element required by the microbes but at the same time toxic at higher levels. [illegible] action [illegible] in more solution [illegible] larger quantity, it may become inhibitory to its growth. [illegible] number of microorganisms [illegible] have ability to [illegible] compounds and that depends on the tolerance to high levels. The variation [illegible] due to formation of zinc chelate complexes in the soil, which [illegible] microbes (Wong 1980). Thus the ZSB can also [illegible] like phosphobacteria as a microbial inoculant for [illegible] and [illegible] regions.

REFERENCES

Bun[illegible] and A. D. Rovira. [illegible] Microbiological studies [illegible]. Soil Sci. [illegible]

Chapman, H. D. 1966. [illegible] soil analysis [illegible] California Div. Agric. [illegible] Manual [illegible]

Mezei, [illegible] influences [illegible] elements [illegible] Am. [illegible]

Mi[illegible] Wheat [illegible] in the Oriental [illegible] Abstract of dissertation [illegible] 25-33.

Ozaw[illegible] Ind. [illegible] Bot. [illegible]

[illegible] metal [illegible] activity [illegible] Chem. [illegible]

[illegible] The symptoms and [illegible] in higher plants [illegible] Exp. [illegible]

[illegible] Soil Sci. [illegible]

Sh[illegible] and A. [illegible] Mol. [illegible] Agric. [illegible] 108-[illegible]

[illegible] Relationship between zinc [illegible] growth [illegible] higher plants. Plant Physiol. [illegible]

[illegible] The trace element content of [illegible] Conference [illegible]

Tewari, [illegible] and [illegible] 1981. [illegible] of soil and [illegible] of the areas [illegible] deficient [illegible] Uttar Pradesh. [illegible]

Wong, [illegible] 1980. Effect of inorganic acids on [illegible] organisms [illegible]

The present investigation has shown that natural occurring zinc solubilizing bacteria (ZSB) in the soil as well as rhizosphere [illegible] population in the [illegible] low compared to [illegible]. When [illegible] tested [illegible] the ability to solubilize [illegible] insoluble zinc compounds. Zinc is an element required by the microbes but at the same time toxic at higher levels. [illegible] its action being in [illegible] larger quantity it may become inhibitory to its growth. Only limited number of microorganisms in soil have ability to [illegible] insoluble zinc compounds and this depends on the tolerance to high levels. The variation [illegible] due to formation of zinc chelate complexes in the soil which [illegible] (Wang 1959). Thus the ZSB can be used like phosphobacteria as a microbial inoculant for zinc nutrition to crops.

REFERENCES

Bunt, J.S. and A.D. Rovira. 1955. Microbiological studies of some subantarctic soils. J. Soil Sci. 6: [illegible].

Chapman, H.D. 1966. [illegible] soil analysis [illegible]. Univ. California Div. Agr. [illegible] Manual [illegible].

Mehta, [illegible] 1961. [illegible] influences [illegible] Am. [illegible] 28: [illegible].

[illegible] N. [illegible] 1984. [illegible] winter wheat varieties cultivated in the Orenburg region. Abstract of [illegible] International [illegible] 28-35.

[illegible] 1950. The effect of [illegible]. Aust. J. Biol. Sci. [illegible].

[illegible] and B.L. Vallee. 1960. Metal [illegible] complexes [illegible]. J. Biol. Chem. [illegible].

[illegible] 1949. The symptoms and diagnosis of [illegible] deficiency in [illegible] crops [illegible]. Imp. [illegible].

[illegible] and [illegible] 1939. [illegible] of different [illegible] by [illegible] inoculation. Soil Sci. [illegible].

[illegible] V.S. and A.A. [illegible]. 1972. The [illegible] in [illegible] and plants [illegible]. Mol. [illegible] Biology [illegible].

[illegible] 1940. [illegible] and [illegible] in the growth of higher plants. Plant Physiol. [illegible].

[illegible] 1970. The trace element content of [illegible]. Tech. Comm. [illegible] Commonwealth Bur. [illegible] England.

Tewari, [illegible] 1994. Available zinc status of soils and delineation of the areas of zinc deficiency in Uttar Pradesh. [illegible].

Wang, [illegible] 1959. Effect of [illegible] acids in combination with [illegible]. [illegible].

CHAPTER - 49

CLONING OF NEW *Bacillus thuringiensis* GENES FOR MANAGEMENT OF RICE PESTS

V. Udayasuriyan, K. Lenin, M. Asia Mariam and M. Bharathi

ABSTRACT

Insecticidal proteins from *Bacillus thuringiensis* (Bt) are becoming increasingly important for management of pests. A highly suitable strategy for delaying Bt resistance development in insects would be the simultaneous expression of multiple toxin (gene pyramiding) or expression of different toxins in rotation (gene rotation). In this context *cry2A* genes are potential candidates for resistance management strategies. In this study Cry2Aa and Cry2Ac proteins were purified from transformants of an acrystalliferous Bt strain 4Q7 and their toxicity against rice leaf folder larvae (RLF) was tested in comparison with Cry1Ac. The experiments resulted 86.7, 86.7 and 76.7 percent mortality of RLF after 5 days with Cry1Ac, Cry2Aa and Cry2Ac respectively at the concentration of 3200 ppm. In all the treatments (Cry1Ac, 2Ac, 2Aa) significantly less area of leaf (0.69 to 0.74 cm^2) was consumed by the larva in comparison with control (7.83 cm^2).

INTRODUCTION

Insecticidal proteins from *Bacillus thuringiensis* (Bt) are becoming increasingly important for management of pests. Bt proteins are active against many insects; particularly lepidopteran pests and they act by disrupting midgut cells of insects (Gill *et al.*, 1992). Bt toxins are widely used as insecticidal formulations. The molecular potency of Bt toxins is higher compared to that of other pesticides, 300 times higher than synthetic pyrethroids or 80,000 times higher than organophosphates (Feitelson *et al.*, 1992).

The advances in genetic engineering and molecular biology in the early eighties led to the cloning of Bt crystal protein (*cry*) gene for the first time in 1981 (Schnepf & Whiteley 1981). Since then, more than 100 *cry* genes have been successfully cloned and added to the growing list. Initially several *cry* genes were expressed in plant colonizing microorganisms to target the stem and root dwelling insect pest. Today, the most efficient way to deliver *cry* genes seems to be the development of transgenic plants expressing them.

The present classification of *cry* genes is mainly based on the homology of the amino acid sequence of their products. Currently, the *cry1* genes of *B. thuringiensis* are widely used in transgenic crop plants. The reliance on a single kind of Bt protein over the years can lead to the development of resistance in insects (Tabashnik *et al.*, 1997). A highly suitable strategy for delaying Bt resistance development in insects would be the simultaneous expression of multiple toxin (gene pyramiding) or expression of different toxins in rotation (gene rotation). In this context *cry2A* genes are potential candidates for resistance management strategies, owing to unique mode of action of the Cry2 proteins when compared to other proteins (Lee *et al.*, 1997). Moreover Cry2A proteins show toxicity towards insects belonging to the orders of lepidoptera, diptera and homoptera (Walters and English, 1995), showing a wide spectrum of control. However, the Cry2Aa is less toxic against lepidopteran larvae when compared to Cry1Ac (Chakrabarti *et al.*, 1998). Sasaki *et al.* (1997) reported significant difference in the level of toxicity between two Cry2Aa proteins. Variations of a single amino acid can significantly influence the level of toxicity in Cry proteins (Udayasuriyan *et al.*, 1994; Rajamohan *et al.*, 1996). Therefore, toxicity analysis of Bt proteins encoded by different Bt genes is being carried out to select potential Bt genes against target pests.

In this study Cry2Aa and Cry2Ac proteins were purified from transformants of an acrystalliferous Bt strain 4Q7 and their toxicity against rice leaf folder larvae was tested in comparison with Cry1Ac.

MATERIALS AND METHODS

Bacterial cultures were grown in T_3 medium (Martin and Travers, 1989) under aseptic conditions at 30°C. The bacterial sporulation was monitored through Phase Contrast Microscope. When more than 90% lysis was reached, the sporulated plates and broths were transferred to 4°C at least 30 min before harvesting. Harvesting was done by scraping and suspending the lysed cell mass in sterile, cold 10mM Tris HCl-1mM EDTA buffer (pH 8.0) containing 1mM PMSF, followed by 0.5M NaCl. The crude containing spore-crystal inclusion was centrifuged at 10,000 rpm for 10 min in 4°C. The supernatant was removed and the weight of pellet containing spore-crystal inclusion was measured. Known weight of spore-crystal mixture was mixed with known volume of alkali buffer with proteinase inhibitor, mixed and incubated at 37°C in an incubator-cum-shaker at a frequency of 200 rpm for different periods depending upon the Bt strain. (Table 1). After incubation period the tube containing spore-crystal mixture and alkali buffer was centrifuged at 10,000 rpm for 10 min in 4°C to remove the spores and undissolved inclusions from the supernatant. The soup, which contains purified form of Cry protein, was taken and the concentration of protein in the soup was estimated by Lowry's method (1951). The intactness and profile of the purified protein was checked by SDS-PAGE (Laemmli, 1970). To minimize the degradation of Cry proteins (to get intact form of protein) proteinase inhibitors ``complete cocktail tablets" – 50x concentration (Roche Gmbh, Germany) and 1mM PMSF (phenyl methyl sulfonyl fluoride) (Merck, Germany) were mixed with alkali buffers.

Table 1. Purification of Bt protein from spore-crystal mixture

Sl.No.	Cry protein	Size	Solubilizing Buffer	Incubation
1.	Cry 1Ac	135 kDa	50mM NaOH (pH 12.0)	30 min
			50 mM Na_2CO_3 +10 mM DTT (pH 10.0)	15 h
2.	Cry 2 Aa	68 kDa	50 mM NaOH (pH 12.0)	60 min
3.	Cry 2Ac	65 kDa	50 mM NaOH (pH 12.0)	30 min

Culturing of Rice leaf folder (*Cnaphalocrosis medinalis*) was performed as described by Waldbauer and Marciano (1970) at Paddy Breeding Station, Tamil Nadu Agricultural University, and Coimbatore. From these 3rd instar larvae were obtained and used for experiment. Paddy leaf bit of 5 cm x1cm sizes were prepared and waxy coat on the upper surface was removed by brushing with the triton 10 X solutions. Same triton acts as a surfactant also. Different concentrations of Cry1Ac, Cry2Ac and Cry2Aa were coated on the upper surface of leaf bit, 15 µl per leaf bit. One larva per leaf bit was allowed to feed. On 5th day the mortality was taken and the percentage was worked out.

RESULTS AND DISCUSSION

Solvent-solute ratio for the solubility of Cry protein is standardized. Different weight of pellets containing spore-crystal inclusions ranging from 10 mg to 50 mg were mixed with different volume of alkali buffer ranging from 100 µl to 500 µl and solubilized. As the result 25 mg of spore-crystal inclusions with 300 µl of alkali buffer gave maximum concentration of intact protein in the extracted soup. Solubilization period was also standardized by incubating the spore-crystal inclusions and alkali buffer with different incubation periods ranging from 30 min to overnight. The solubilizing buffer and incubation time for solubilization of each Cry protein to give maximum concentration of intact protein is listed in Table 1. After solubilization the quantity of protein in the soup was estimated by Lowry's method and the intactness was checked by SDS-PAGE. Bioassay carried out with the purified proteins showed higher level of toxicity in Cry1Ac, Cry2Ac and Cry2Aa against rice leaf folder (RLF). The experiments resulted 86.7, 76.7 and 86.7 percent mortality of RLF after 5 days with Cry1Ac, Cry2Ac and Cry2Aa respectively at the concentration of 3200 ppm. In all the treatments (Cry1Ac,

2Ac, 2Aa) less amount of leaf area (0.69 to 0.74 cm^2) was consumed by the larva in comparison with control (Table 2). There are numerous reports that describe rice transformation with Cry1A genes of Bt. (Maqbool and Christou, 1999; Bentur *et al.*, 2000a). Maqbool *et al.* (1998) transformed transgenic indica rice Basmati 370 and M7 with a novel *cry2A* gene from Bt, that showed considerable resistance to yellow stem borer and rice leaf folder. In the present investigation also, the Cry2Aa and Cry2Ac proteins showed effective control of rice leaf folder as in the case of the commercially used Cry1Ac protein. Hence the newly cloned Cry2A indigenous genes could be used to develop insect resistant transgenic rice plants and for the management of resistant development in insect pests.

Table 2. Toxicity of Cry 2A proteins against rice leaf folder larvae

Treatment	Protein Conc. (ppm)	Larval mortality (%) 5th Day	Leaf area damage per larva after 5 days (sq. cm)
Control	-	-	7.83 ± 0.89
Cry1Ac	3200	86.7 ± 9.4	0.69 ± 0.35
Cry2Ac	3200	76.7 ± 20.5	0.74 ± 0.32
Cry2Aa	3200	86.7 ± 9.4	0.74 ± 0.17

Another advantage of Cry2 is the small size of its protoxin (~65 kDa) compared with the Cry1 protoxins (~130-135 kDa). The Cry1 protoxins are processed in the gut juice of susceptible insect larvae to yield active toxins (~60 kDa). The activated toxins can be a limiting factor for optimal expression in plants. Through the use of protoxin the collateral damage to non-target insects could be minimized (Kota *et al.*, 1999). Therefore, the Cry2 transgenic plants containing the whole protoxin are preferable to the Cry1 transgenics containing the toxin fragment alone.

CONCLUSION

Proteins encoded by the new indigenous Bt genes (Cry2Aa and Cry2Ac) showed effective control of rice leaf folder larvae. These genes could be used for the development of next generation insect resistant transgenic crops.

REFERENCES

Bentur, J. S., D. A . Andow, M. B Cohen, A. M. Romena, F. Gould,. 2000a. Frequency of alleles conferring resistance to a *Bacillus thuringiensis* toxin in a Philippine population of *Scirpophaga incertulas* (Lepidoptera: Pyralidae). **J. Econ. Entomol., 93(5)**:1515-21

Chakrabarti, S. K., A.Mandaokar, P.A. Kumar, & Sharma, R.P. 1998. Efficacy of lepidopteran specific delta-endotoxins of *Bacillus thuringiensis* against *Helicoverpa armigera*. **J. Invertebrate Pathology, 72**: 336-337.

Feitelson, J. S., J. Payne and L. Kim.1992. *Bacillus thuringiensis* :Insects and beyond. **Bio/Technology, 10**: 271-275.

Gill, S.S., E.A. Cowles and P.V. Pietrantonio. 1992. The mode of action of *Bacillus thuringiensis* dotoxin. **Ann. Rev. Entomol., 37**:615-636.

Kota, M., H.Daniell, S.Va r a, S.F Garczynski,., Gould, F. & Moar, W.J. 1999 Over expression of the *Bacillus thuringiensis* (Bt) Cry2Aa2 protein in chloroplasts confers resistance to plants against susceptible and Bt-resistant insects. **Proc. Natl. Acad. Sci., USA, 96,** 1840-45.

Laemmli, U. K. 1970. Cleavage of structural proteins during the assembly of the head of bacteriophage T_4. **Nature, (London) 227**: 680-685.

Lee, M.K., Aguda, R.M., Cohen, M.B., Gould, F.L. & Dean, H.D. 1997. Determination of binding *Bacillus thuringiensis* δ-endotoxin receptors to rice stem borer midguts. **Appl. Environ. Microbiol., 63:** 1453-1459.

Lowry, O. H., N. J. Rosebrough, A. L. Farr and R.J.Randall.1951. Protein measurement with the Folin phenol reagent. **J. Biol. Chem., 193:** 265-275.

Maqbool, S. B. and P. Christou. 1999. Multiple traits of agronomic importance in transgenic *indica* rice plants: analysis of transgene integration patterns, expression levels andstability.**Mol.Breed., 5:** 471-480.

Maqbool, S. B., T. Husnain, S. Riazuddin, L. Masson and P. Christou. 1998. Effective control of yellow stem borer and rice leaffolder in transgenic rice *indica* varieties Basmati 370 and M7 using the novel delta-endotoxin *cry2A Bacillus thuringiensis* gene. **Mol. Breed., 4:** 1-7.

Martin. P. A. W and R. S. Travers. 1989.Worldwide abundance and distribution of *Bacillus thuringiensis* isolates, **Appl. Environ. Microbiol., 55**:2437-2442.

Rajamohan, F., Hussain, S.R.A., Cotrill, J.A., Gould, F. & Dean, H.D. 1996 Mutations in domain II, Loop 3 of *Bacillus thuringiensis* Cry1Aa and Cry1Ab delta-endotoxins suggested loop 3 is involved in initial binding of lepidopteran midguts. **J.Biol.Chem., 271**: 25220-25226

Sasaki, J., Asano, S., Hashimoto, N., Lay, B.W. & Hastowo, B.W. 1997 Characterization of a *cry2Aa* gene cloned from an isolate of *Bacillus thuringiensis* serovar *sotto.* **Curr Microbiol., 35:** 1-8.

Schnepf, H. E. and Whiteley, H. R.1981.Cloning and expression of the *Bacillus thuringiensis* crystal protein gene in *Escherichia coli.* **Proc. Natl. Acad. Sci., USA, 78:** 893-2897.

Tabashnik, B.E., Lill, Y.B., Finson, N., Masson, L. & Heckel, D.G. 1997 One gene in diamond back moth confers resistance to foliar *Bacillus thuringiensis* toxin. **Proc. Natl. Acad. Sci., USA, 94:** 1640-1649.

Udayasuriyan, V., Nakamura, A., Mori, H., Masaki, H. & Uozumi, T. 1994 Cloning of a new *cry1A(a),* gene from *Bacillus thuringiensis* strains Fu-2-7 and analysis of chimeric Cry1A(a) proteins of toxicity. **Bioscience, Biotechnology and Biochemistry, 58:** 830-835.

Waldbauer, G. P. and A. P. Marciano. 1979. Mass rearing of the rice leaf folder, *Cnaphalocrosis medinalis* Guene (Lepidoptera: Pyralidae) under greenhouse conditions. **J. Ent. Res.,** 3 - 8.

Walters, S.F. & English, L.H. 1995 Toxicity of *Bacillus thuringiensis* delta-endotoxins towards the potato aphid in an artificial diet bioassay. **Entomologia experimentalis et applicata, 77:** 211 – 216.

CHAPTER – 50

STUDIES ON THE EFFECT OF *Fusarium* CULTURE FILTRATE ON PADDY SEEDLING DEVELOPMENT

K. Jeyaram, K. Govindarajan and S. Kannaiyan

ABSTRACT

The gibberellins are an important group of phytohormones occurring in higher plants. Gibberellic acid (GA_3) is having potential scope for inducing panicle emergence in hybrid rice seed production. *Fusarium moniliforme* (*Gibberella fujikuroi*) a fungus, has been used widely in commercial production of GA_3. Hence, an attempt was made to isolate *Fusarium* sp. with the ability to produce GA_3 as a cheaper source for spraying during flowering stage of hybrid rice. Twenty seven strains of *Fusarium* sp. were isolated from soil and plant samples collected from different location. Based on GA_3 production under submerged fermentation, four *Fusarium* isolates (>500 µg/ml GA_3 production) capable of producing higher GA_3 than the type strains *G. fujikuroi* - 1019 (280 µg/ ml GA_3 production) were selected and their biological activity on rice seedling development was studied. Soil application of *Fusarium* culture filtrate in cup culture resulted in significant increases in shoot length. However, response to commercial GA_3 (100 µg / ml) was highest (2.7 times more than control) followed by *G. fujikuroi* - 1019 (2.01 times) and *Fusarium* isolate KJ-KG-23 (1.28 times). Apart from shoot elongation, root elongation was also noticed in all *Fusarium* culture filtrate treatment (*Fusarium* isolate KJ-KG-23 responded 2 times more than control) except *G. fujikuroi* - 1019 (clubbing of roots). It was concluded from the bioassay that paddy seedlings responded very well to *Fusarium* culture filtrate treatment. This could be exploited well for faster seedlings development in dry land farming and hybrid seed production in rice.

INTRODUCTION

Gibberellins are a classic example of the production of plant growth regulators by microorganisms (Kumar and Lonsone, 1989). Their biological activity involves both stimulation of cell division and elongation. Japanese scientists were the first to succeed in isolating gibberellic acid (GA_3) from the rice pathogen *Fusarium moniiiforme (Gibberella fujikuro)* during 1930s. Till date 99 different GA_3 have been isolated from microorganisms and plants, identified and chemically characterized (Mander *et al,* 1996). Among GA, GA_3 is considered as the most important gibberellin and had received the greatest attention. In recent years it has been tried along with seeds to improve seedling establishment in direct seeded rice (Carlson *et al,1992),* GA_3 is also having potential scope for inducing panicle emergence in hybrid rice seed production. Hence, an attempt was made to isolate *Fusarium* sp. with the ability to produce GA_3 as cheaper source for spraying at flowering stage of hybrid rice and their biological activity on rice seedling development was studied.

MATERIALS AND METHODS

GA_3 estimation

GA_3 production under submerged fermentation was estimated by UV spectrophotometer at 254 nm, a method standardized by Lu *et al,* (1995).

Bioassay

Germinated paddy (*Oryza sativa L)* seeds (variety CO43) were sown in plastic cups containing sterile soil. The seedlings were given soil application of 10 days old *Fusarium* culture filtrate at 2 days intervals. Growth pattern of the treated seedlings was compared with 100 μg/ ml commercial GA_3 treatment and control without any treatment. Seven days after seedling development, observation was recorded on plant height, root length, fresh root and shoot weight.

RESULTS

Twenty seven strains of *Fusarium* sp. were isolated from soil and plant samples collected from different location (Fig. 1 and 2). Based on GA_3 production under submerged fermentation (SmF), four isolates (>500 μg/ ml of GA_3 production) capable of producing higher GA_3 than the type strain *G. fugikuroi*-1019 (280 μg/ ml of GA_3 production) were selected and their biological activity on rice seedling development was studied. Soil application of most of the *Fusarium* culture filtrate in crop culture resulted significant increase in shoot length (Table l). However, commercial GA_3 (100 μg/ ml) responded very well (2.7 times more than control) followed by G. *fugikuroi*-l019 (2.01 times) and *Fusarium* isolate KJ-KG-23 (1.28 times) (Fig. 3). Apart from shoot elongation, root elongation was also noticed in all *Fusarium* culture filtrate treatment. *(Fusarium* isolate KJ-KG-23 responded 2 times more than control) except *G. fujikuroi* (clubbing of roots). No fresh weight gains in shoot was observed in *G. fujikuroi* treatment. But in *Fusarium* isolate KJ-KG- 7, 10 and 15 culture filtrate treatment, reduced fresh shoot weight and increased fresh root weight was noticed. In *Fusarium* culture KJ-KG-23 filtrate treatment was found to be more effective in both shoot and root growth promotion.

Fig. 1. Mycelial growth of the type stain *G.fugikuroi*-1019

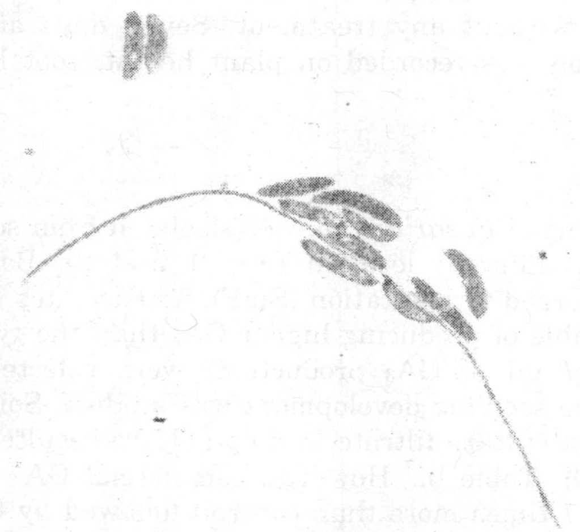

Fig. 2. Sickle shaped macroconidia of Fusarium sp. KJ-KG-7 (Magnification x450)

Table 1. Effect of *Fusarium* culture filtrate on paddy seedlings development

Treatments	Shoot length (cm)	Root length (cm)	Fresh shoot weight (g)	Fresh root weight (g)
Control	10.5	5.3	0.043	0.020
100 µg / ml commercial GA_3	27.2	6.1	0.060	0.026
G. fujikuroi culture filtrate (285 µg / ml GA_3)	20.1	5.7	0.042	0.016
Fusarium sp KJ-KG-7 culture filtrate (565 µg / ml GA_3)	13.7	8.3	0.030	0.021
Fusarium sp KJ-KG-10 culture filtrate (510 µg / ml GA_3)	9.7	7.1	0.031	0.026
Fusarium sp KJ-KG-15 culture filtrate (545 µg / ml GA_3)	9.7	5.5	0.029	0.022
Fusarium sp KJ-KG-23 culture filtrate (600 µg / ml GA_3)	12.8	10.9	0.051	0.045
SEd	1.16	0.75	0.0026	0.0059
CD (P =0.05)	2.48	1.61	0.0056	0.0127

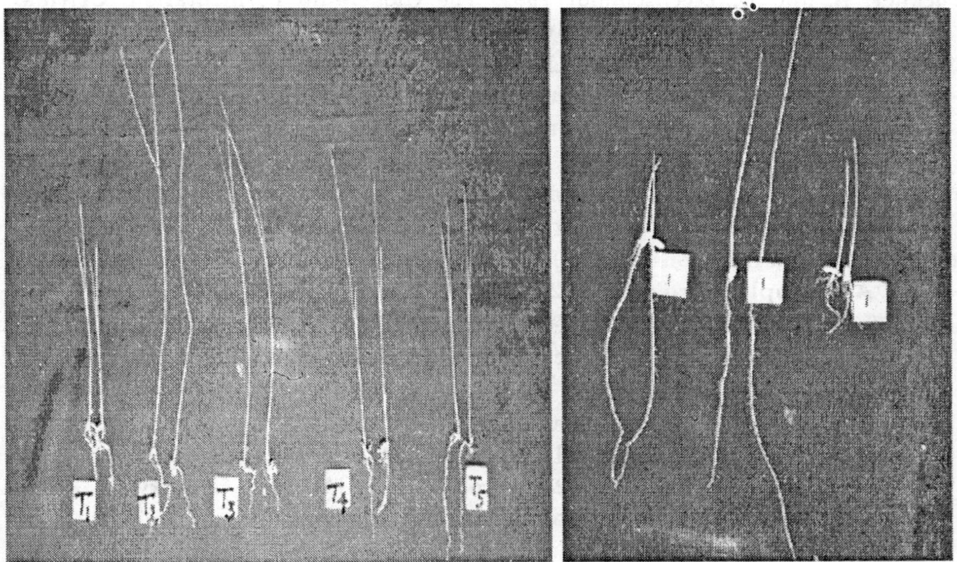

Fig. 3. Effect of *Fusarium* culture filtrate on Paddy seedling development

T1-Control

T2-100 µ/ml commercial GA_3

T3 -*Fusarium moniliformae* culture filtrate

T4-*Fusarium* isolate KJ-KG-7 culture filtrate

T5-*Fusarium* isolate KJ-KG-23 culture filtrate

DISCUSSION

Above plant bioassay revealed that GA activity of *Fusarium* culture filtrate on rice plant system was lower than the spectrophotometrically quantified values. This might be due to the specificity of different GA_3 in action. A large proportion of known gibberellin have little or no biological activity. They are either intermediates in the synthesis of active form or metabolic products that still retain the basic structure but are no longer biologically active. Structural requirements vary among plant systems and GA_3 that are highly active in one system may have low or zero activity in the another target tissue. Metabolic transformation of exogenous GA_3 vary from tissue to tissue and conjugation may interfere with the observed effects (Frankenbenger and Arshad, 1995). Even though *G. fugikuroi* culture filtrate responded very well in shoot elongation, reduced root length (clubbing of roots) was noticed. This might be due to fungal toxin already demonstrated in 'Bakane' disease (Bruckner and Blechschmidt, 1991). It was concluded from the bioassay that paddy seedlings responded very well for *Fusarium* culture filtrate treatment. This could be exploited well for faster seedling development in dry land farming and hybrid seed production in rice.

REFERENCES

Bruckner, B. and D. Blechschmidt 1991. The Gibberellin Fermentation. **Critical Reviews in Biotechnology. 11**(2): 163-192.

Carlson, R.D., N. Chang, C. Black-Schaefer and J.A. Fngiel 1992. Efficacy of release seed treatment in California rice production systems. In: Proceedings of the 4th Annual Conference of the Western Plant Growth Regulator Society, Sacramento, January 22-24, 109-129.

Frankenbenger, W.T. and Ir. M. Arshad 1995. **Phytohormones in soil microbial production and function;** Marcel Dekker Inc. New York. 137-225.

Kumar, P.K.R and B.K. Lonsone 1989. Microbial production of gibberellic acid: Sate of art. **Adv. Appl. Microbiol.,** 34:29-139.

Lu.Z.X, Z.C.Xie and M.Kumekura. 1995. Production of Gibberellic acid in *Gibberella fujikuroi* adhered onto polymeric fibrous carriers. **Process Biochemistry**, 30 (7) : 661-665.

Mander, L.N., D.J. Owen, S.J. Croker, P. Garkin, P. Hedden, M.J. Lewis, M. Talon, D.A. Gage, J.A.DD. Zeevant, M.L. Brenner and S.A. Chuxing 1996. Identification of three C_{20} gibberellin, GA_{97} (2β-hydroxy GA_{53}), GA_{98} (2 P -hydroxy GA_{44}) and GA_{99} (2P -hydroxy -GA_{19}).**Phyto cherm.,** 43(1) :23-28.

SECTION – VIII

QUALITY CONTROL OF BIOFERTILIZERS

CHAPTER – 51

QUALITY CONTROL AND CONSTRAINTS IN BIOFERTILIZER PRODUCTION TECHNOLOGY

D.J. Bagyaraj

ABSTRACT

Biofertilizers on application remain in soil, multiply and benefit crops. They do not get depleted as chemical fertilizers. They are cost effective, environment friendly and improve soil health. Their effect on plant growth may not be spectacular as that of chemical fertilizers, being biological material. Their use is of great significance in developing countries because even the marginal farmers can afford to buy and use them. Hence it is essential that the manufacturers of biofertilizers adhere to the quality control protocols in the production units ensuring a quality product. Care should also be taken to overcome other constraints coming in the way of biofertilizer technology. We should recognize that in adopting a rational approach to the use of management of biofertilizers in sustainable agriculture, the microbial labour force holds a vast potential for the future.

INTRODUCTION

Nitrogen, phosphorus and potassium are the three major plant nutrients. Fertilizers are chemicals supplying nutrients to plants. Fertilizer industry greatly depends on petroleum reserves and it is expected that in the next few decades, these reserves will be almost exhausted. Further use of chemicals is not only deleterious to environment, but injurious to human and soil health. Under these circumstances, alternate sources for fertilizers are looked for. The current population of India is nearly 100 million. The current gross cropped area is 190.6 million hectares and the food grain production is 190.3 million tonnes. With the current rate of population increase it is projected that India will have a population of 1500 million by 2025 A.D.

To feed this population we must produce 310 mt of food grains. The scope for increasing the gross cultivated area (horizontal expansion of production) is going to be marginal. At the most it may increase to 200 m ha (addition of 10 m ha from now) through reclamation of ravines, saline-alkaline, water-logged and other categories of cultivable waste land. That means with limited available gross cultivable area we have to produce more food per hectare. This points out that our food production per hectare (currently it is 1.3 t/ha), in the next 30 years must be doubled to feed the bulging population i.e. we should have vertical expansion of production. For this purpose use of external inputs, especially optimum use of fertilizer (off-farm) supplemented with farm manures, composts, city wastes, green manures and biofertilizers (on-farm) is indispensable.

It is estimated that annually 25.1 mt of nutrients are removed from the soil. Whereas, only 15.0 mt are supplied from soil sources including organics. The fertilizer (NPK) production and requirement, in million metric tonnes of our country is given in Table 1.

Table 1. Fertilizer requirement and production in India-Projections (Million tonnes)

Year	1996	2000	2011	2031	2051
Requirement	16.0	19.0	20.2	27.3	31.3
Production	12.8	14.9	15.8	20.9	23.9
Gap	3.2	4.1	4.4	6.4	7.2

Table 1 brings out that the fertilizer production in our country is less than the required amount. To fill this gap alternate sources of nutrients have to be looked for. Organic wastes and biofertilizers are the alternate sources to meet the nutrient requirement of crops and to bridge the future gaps. Further, knowing the deleterious effects of using only the chemical fertilizers on soil health, use of chemical fertilizers supplemented with organic wastes and biofertilizers will be an environmentally benign approach to nutrient management and ecosystem function. Such integrated approach will help to maintain soil health and productivity and thus should be advocated and popularized among the farming community (Bagyaraj and Suvarna, 1999).

Biofertilizers are preparations containing microorganisms in sufficient numbers helping plant growth and nutrition. Biofertilizers are widely accepted as low cost supplements to chemical fertilizers and have no deleterious effects either on soil health or environment.

Biofertilizers are known to make a number of positive contributions in agriculture (Motsara *et al.*, 1995).

For example :-

- They supplement fertilizer supplies for meeting the nutrient needs of crops.
- They can add 20-200 kg N/ha (by fixation) under optimum conditions and solubilise/mobilize, 30-50 kg P_2O_5/ha.

- They liberate growth promoting substances and vitamins and help to maintain soil fertility.
- They suppress the incidence of pathogens and control diseases.
- They increase crop yield by 10-50%, N-fixers reduce depletion of soil nutrients and provide sustainability to the farming system.
- They are cheaper, pollution free and based on renewable energy sources.
- They improve soil physical properties, tilth, and soil health in general.

Biofertilizers, which can supply nitrogen and phosphorus, two major plant nutrients, have been developed (Verma, 1993).

Bioferilizers, which can supply nitrogen and phosphorus, two major plant nutrients have been developed (Verma, 1993). Organisms capable of fixing atmospheric nitrogen like *Rhizobium*, *Azotobacter* and *Azospirillum* are commercially produced as biofertilizers to supply nitrogen and phosphate solubilising organisms are mass multiplied as biofertilizers aiding phosphorus nutrition (Tilak and Saxena, 1996).

BIOFERTILIZER DEMAND AND AVAILABILITY

The projected production of different biofertilizers in our country in given Table 2 (Singh, 1993).

Table 2. Target of production of different biofertilisers by 2000 AD

Biofertilizers	Quantity (tonnes)
Rhizobium	14,460
Azotobacter	7,230
Azospirillum	7,230
PSM	7,230
BGA	3,015
Total	39,165

Taking the cropped area to be around 200 mha, the demand for biofertilizers at present (2000 AD) should be:

For 1 mha biofertilizer required is = 500 t

For 200 mha, biofertilizer required = 200 x 500 = 1,00,000 tonnes.

It would be oversimplification to multiply the entire cropping area by 0.5 kg biofertilizer /ha to compute demand for biofertilizers of the country. From a realistic point of view, it is better to accept 25% of the theoretical figure (ie 1,00,000 tonnes) as the present annual biofertilizer requirement in India ie. 25,000 tonnes/annum. The present annual production of biofertilizer in the country is around 10,000 tonnes.

Thus the annual requirement and present production capacity for different biofertilizers which shows not only a tremendous gap, but also indicates the potential for organized production (Venkataraman and Tilak, 1990).

Constraints in Biofertilizer Technology

At present there are several constraints in biofertilizer production and its commercialization. They may be physical, chemical, biological, technological, infrastructural, financial, market related or concerned with human resource development.

Physical and Environmental Constraints

In summer months in some parts of India temperature is very high. Care must be taken to avoid transportation of biofertilizer packets in direct sunlight. Storage of biofertilizers should not be done in container/sheds where the temperature becomes very high. While making packets care must be taken to see that the carrier material has enough moisture (about 40%). If the carrier material is very dry, population of the organisms will decrease rapidly.

Chemical Constraints

Highly acidic or saline soils will adversely affect the population of introduced biofertilizer. In acidic soil, pelleting of seeds (after treating with biofertilizer) with lime and in saline soils with gypsum can be recommended. In soils less than 10 ppm available P, application of P fertilizer is essential without which the nitrogen fixing biofertilizers will not function effectively.

Biological Constraints

Soil flora and fauna can parasitize or devour biofertilizer added to soil. Bacteriophages called rhizophages can destroy rhizobia added to soil. Some protozoa like *Vorticella* and nematodes like *Meloidogyne* can act as predators of biofertilizer added to soil. *Azolla* is subjected to attack of pests, especially larvae of various lepidopterous and dipterous insects and aphids. Cyanobacteria can be attacked by phages, myxobacteria and grazers like snails and mosquito larvae.

Technical constraints

Raw material: Peat and lignite which are good sources of carrier materials are available only in specific places like Nilgiris and Neyveli, thus resulting in high cost of transportation to biofertilizer manufacturing units.

Strains: Suitable strains of biofertilizers for different crops have been identified through research. Many producers do not pay attention to this aspect and use whatever strains they get, resulting in poor quality biofertilizers.

Suitable Technology: In India biofertilizer production is generally done with nonsterile system which encourages considerable contamination. Suitable technology to minimize contamination should be developed. Eg.

Need for gamma irradiation of the carrier material, injection of liquid cultures into polythene bags containing sterile carrier material etc.

Lack of quality assurance: Substandard quality of inoculants is one of the most important factors resulting in failure at the field and lack of farmers confidence in the product. The Bureau of Indian Standards has prepared ISI specifications only for *Rhizobium* and *Azotobacter* inoculants. This was prepared in 1977. Revised specifications for *Rhizobium* and *Azotobacter* and ISI specifications for *Azospirillum* and phosphobacteria have been formulated recently. The standard specifications for *Rhizobium* and *Azotobacter* is given in Table 3. As per ISI (BIS) specification, the quality assessment tests required in India are summarized in Table 4a and 4b. The proposed BIS standards for *Azospirillum* and phosphobacteria are given in Tables 5 and 6.

Table 3. Indian Standard specifications for *Rhizobium* and *Azotobacter* biofertilizers

Parameters	Rhizobium inoculant (RI)	Azotobacter inoculant (AI)
1. Base	Carrier based	Carrier based
2. Cell number at the time of manufacture	10^8/g carrier within 15 days of manufacture	10^7/g carrier with in 15 days of manufacture
3. Cell number at the time of expiry	10^7/g carrier within 15 days before expiry date	10^6 /g carrier within 15 days before expiry date.
4. Expiry period	6 months from the date of manufacture	6 months from the date of manufacture
5. Permissible contamination	No contamination at 10^8 dilution.	No contamination at 10^5 dilution
6. pH	6.0-7.5	6.5-7.5
7. Strain	Should be checked serologically	Nothing specific. But *A. chroococcum* species is mentioned.
8. Carrier	Should pass through 150-212 micron, IS (72-100 mesh).	Should pass through 106 micron IS sieve.
9. Others	Nodulation test +ve Result in 50% or more dry matter than control	Minimum amount of N_2 fixation not less than 10 mg per gm of sucrose utilized.

Table 4a. Quality assessment tests for *Rhizobium*

Test		Observation	Expected result
1.	Streak-onto-Yeast Mannitol Agar (YMA) and Congored Yeast Mannitol Agar (CRYMA).	Colony character	Typical colony of Rhizobium should appear on YMA & it should not absorb congo red.
2.	Gram stain	Reaction	Cells must be Gram negative.
3.	Streak on Glucose peptone agar (GPA)	Whether growth is +ve or -ve	Growth should not come.
4.	Measure pH of the carrier	-	6.0-7.5

5.	Total plate count	No. of colonies	10^8/g carrier.
6.	Serological study	-	Agglutinate with the nominated antiserum.
7.	Nodulation study	-	Nodulation +ve.

Table 4b. Quality assessment tests for *Azotobacter*

1.	Streak onto Jensen's N-free medium	Colony character	*A. chroococcum* colonies are gummy, raised with or without striations, viscous and often sticky. The pigmentation varies from very light brown to black.
2.	Gram stain	Reaction	Cells must be Gram negative.
3.	Measure pH of the carrier.	-	6.5 to 7.5
4.	Measure N_2 fixation	-	The minimum amount of fixed nitrogen should not be less than 10 mg/g of sucrose utilized.
5.	Total plate count	-	10^7/g carrier.

Table 5. Quality control specification for *Azospirillum* inoculant

1. *Azospirillum* inoculant should contain a minimum of 10^9 viable *Azospirillum* cells/g of the dry carrier at the time of manufacture and 10^7 cells/g dry carrier at 15 days before the expiry date mentioned on the packets.
2. No contamination with other microorganisms at 10^6 dilution.
3. pH of inoculant should be 7.2-7.5
4. Carrier material should be neutral and pass through 100 micron IS sieve.
5. Each packet should be marked with strain number, date of manufacture, date of expiry, method of application etc.

Table 6. Quality control specification for Phosphobacteria

The following specification should be adopted by entrepreneurs to have a quality phosphobacteria inoculant.

1. The peat based phosphobacteria inoculant should contain a minimum of 10^9 viable cells/g of peat based inoculant, at the time of manufacture and 10^7 viable cells/g before the expiry date marked on the packet.
2. The pH of the peat based inoculant should be between 6.5 and 7.5.
3. The carrier material used for phosphobacteria inoculant preparation should not contain toxic material and should satisfy the required characters of high water holding capacity and high organic carbon content.
4. The carrier material prepared should pass through 100 micron sieve.
5. The polythene bag used for packing phosphobacteria should be low density (50-75 μm).

6. The front page of polythene bag should be printed with the name of biofertilizer, crops for which the inoculant to be used, application methods, methods of storage, date of manufacture, date of expiry, net weight and price.
7. Phosphobacteria inoculant should have no contamination with other microorganisms at 10^6 dilution.

Quality control consists of strain testing and selection service, and control of broth before incorporation into the carrier materials like lignite, charcoal, peat, vermiculite etc., for the carrier based culture after maturation and control of culture during the period of storage. Upto final broth stage, each individual manufacturer has to do his own quality control.

The BIS has no network to test biofertilizer samples. They send samples to Agricultural Universities or Research Institutes in order to make ISI mark. Once a manufacturer gets an ISI mark, they continue it without routine checking. This amounts to sub-standard product reaching the farmer. It is suggested that there should be separate testing laboratories with adequate infrastructure and manpower to check the quality of biofertilizers at various stages of production, marketing and application. Perhaps this can be done through strengthening National Biofertilizer Development Centre and its Regional Centers. More than anything, there should be honesty in the production units to ensure a quality product, especially as it is going to be used mostly by marginal farmers.

Infrastructural Constraints

Many production units do not have adequate space and other infrastructural facilities including electric supply needed for quality biofertilizers.

Human Constraints

Some biofertilizer production units do not have technically well qualified (Microbiologists) and trained persons who can produce high quality biofertilizers. It has been experienced that some organizations depend more on non-skilled labourers working on contract basis which leads to sub-standard biofertilizers.

Most of the farmers in the country do not have sufficient and clear knowledge on the use of biofertilizers. The farmers need to be educated. In order to educate farmers, educating and training extension workers is utmost important. This can be done through demonstration trials on the cultivators fields. The impact is greater and long lasting when farmers see the beneficial results for themselves on their lands.

An intense publicity programme is required for motivating the people, creating an assurance and disseminating knowledge amongst farming community. This can be done through media like T.V., cassettes, radio, advertisements in local newspapers, seminars, exhibitions and rural fairs.

Marketing Constraints

This includes weakness in the marketing network, retail outlet, storage facility etc. There is no standardization in packing, labelling and prices of biofertilizers. The size of the packet varies from 75 to 500 g/packet. Some manufacturers engage private dealers with lucrative trade discount. A uniform capacity of 200g packets is recommended so as to facilitate the designing of packing machine that can be manufactured and marketed all over the country. Some suggestions for distribution agencies are given in Table 7.

Table 7. Suggestions for Distribution Agencies

DO	DON'T
• Keep biofertiliser packets away from direct heat and sunlight. Store it in cool and dry place.	• Don't store biofertiliser packets under heat and sunlight.
• Sell only biofertiliser packets which contain batch number, the name of the crop, the date of manufacture and expiry period.	• Don't sell biofertiliser packets after their expiry period is over.
• If the expiry period of the packet is over, discard it. It will not be effective.	• Don't prick holes into the packets or puncture them
• Keep biofertiliser packets away from fertilizer or pesticide containers.	• Don't store heavy items above biofertiliser packets

REFERENCE

Bagyaraj, D.J. and C.V. Suvarna 1999. Biofertilizers and their application: **AGRI-EAST, 1:** 16-21.

Motsara, M.R., P. Bhattacharyya and Beena Srivastava 1995. **Biofertilizer-Technology, Marketing and Usage, Fertilizer.** Development and Consultation Organization, New Delhi.

Singh, T. 1993. Prospects and potential of biofertilizers. A mission mode approach. In: **Proceedings of the National Conference on Biofertilizers and Organic Farming**, Nov. 25-26, 1993. Madras. pp. 10-13.

Tilak, K.V.B.R. and A.K. Saxena 1996. Biofertilizers - Their role in crop production. **Fertl. Industry, XIX:** 187-194.

Venkataraman, G.S. and K.V.B.R. Tilak 1990. Biofertilizers in sustainable agriculture, In: **Soil Fertility and fertilizers. Vol. IV, Nutrient manage-ment and supply system for sustaining agriculture in 1990s,** (eds.), Kumar, V. Shortiya, G.C. and Kavre, S.V. India Farmers Fertilizers Co-operative Ltd., New Delhi, pp.132-148.

Verma, L.N. 1993. Biofertilizers in Agriculture. In: **Organics in Soil Health and Crop production** (eds.) Thampan, P.K. Peekay Tree Crops Development Foundation, Cochin. pp. 151-184.

CHAPTER – 52

CURRENT ADVANCES IN PRODUCTION AND QUALITY CONTROL OF RICE BASED BIOFERTILIZERS

R.N. Bisoyi

ABSTRACT

Although, India produces the largest quantity of biofertilizers in the globe the biofertilizers are produced using conventional production technology despite advanced technology available. The State of Tamil Nadu has been producing biofertilizers commercially in large scale of about 1400 tonnes per annum against production capacity of 3300 tonnes. The largely produced biofertilizers are in the order of *Azospirillum*> Phosphobacteria> *Rhizobium* and are used in different rice based cropping systems. This article reviews the conventional technology usually adopted by production units in India as against the recent technology known in terms of strain selection, strain improvement, fermentation technology, use of carrier materials in processing, inoculant formulations, use of semi-automatic or automatic production systems. Besides, the advanced systems of quality control protocols of biofertilizers are also discussed.

INTRODUCTION

Currently, rice is grown in about 44.6 million hectares in India with national unit area productivity of 1.9 tonnes per ha. On the other hand, rice is grown largely in the State of Tamil Nadu which produces about 3 times more unit area productivity of national average due to use of current available technology. The present day agricultural biotechnology has made replacement of traditional rice varieties with Hybrid rice and other improved varieties with characteristics of nitrogen responsiveness, pest tolerance, drought or salinity tolerance and also with the availability of recently known, the Golden Rice. The nutrients like nitrogen and phosphorus play a pivotal role in increasing the unit area productivity. Meeting the nutrient demand of N & P chemical fertilizers to increase unit area productivity has been a challenging task due to increased cost of fertilizers besides *its impact on soil health. However, the State of Tamil Nadu has been utilizing Biofertilizers (Microbial Inoculants) as an alternate or supplementary source of N or P nutrition since more than three decades. The biofertilizers (microbial inoculants) like *Azolla*, blue-green algae, legume green manures, *Azospirillum, Azotobacter, Rhizobium* and Phosphobacteria for rice based cropping systems have proved their immense potentialities in supplying crop nutrients. Current day technology of biofertilizers has witnessed a sea change due to development of biotechnological protocols and use of advanced instruments. The potentialities and limitations of different biofertilizers with technological developments are summerised (Table 1). Among the biofertilizers for paddy, legume green manures, *Azolla* and BGA although are very useful, have been under utilized. However, Tamil Nadu farmers have realized the significant contribution of use of stem nodulating *Sesbania rostrata* against the use of traditional green manures, *Azolla* hybrids against traditional *Azolla pinnata,* use of immobilized super strains of BGA cultures against conventional soil based BGA inoculum. However, farmers are always in search of an alternate technology which is economically sound, time saving, easily adaptable and sustainable. The biofertilizers like *Rhizobium, Azospirillum, Azotobacter*, Phosphobacteria have proved immense potentialities in rice based cropping systems and these carrier based biofertilizers have become very promising for adoption by the farmers. The State of Tamil Nadu has been largely promoting the use of biofertilizer technology in rice based cropping systems.

Biofertilizer production in the state of Tamil Nadu

A study of biofertilizer production scenario in the State of Tamil Nadu indicate that total production of biofertilizers like *Rhizobium, Azospirillum,* Phosphobacteria was about 410 tonnes during the year 1990-91 which has increased to 1410 tonnes during 1999-2000 (Table 2) (Ahmed *et. al.* 2000). Further, the first commercial biofertilizer production system was started in 1977 by the State Govt. of Tamil Nadu at Cuddalore and later 5 more production units were set up by the Tamil Nadu Govt. Among these 3 biofertilizers, *Azospirillum* was largely produced. About 16 major biofertilizer production units exist in the State of Tamil Nadu with production capacity of 3309 tonne per year, but biofertilizers are largely produced by Govt. Departments only (Table 3). A close look of biofertilizer production in the State of Tamil Nadu indicates that *Azospirillum* continues to be the largely produced biofertilizer followed by *Rhizobium* and Phosphobacteria. *Azospirillum* contributed about 54% of production followed by Phosphobacteria 26% and *Rhizobium* 20% (Table 4). The biofertilizers produced in the State have been assessed for its potential contribution. The Department of Evaluation and Applied Research, Chennai has assessed the impact of biofertilizers which contributed yield increase of 9.3% in paddy, 17% in groundnut, 12% in sugarcane and 6.8% in cotton (DEAR, 1999).

Table 1. Summarised Biofertilizer Technology Scenario for Rice Based Cropping System

Biofer-tilizer	Potentialities	Limitations	Conventional Technology	Current Technology
Legume Green Manures	Provide 50-80 kg N/ha on average	Bulkiness of green maures for incorporation in soil Labour intensive Economics often favour a food cash crop over green manure crop Non acceptance by Farmers	Use of local green manure species of *Sesbania*	Use of *Sesbania rostrata* with high N_2 fixing nodules in root and stem
Azolla	Provide 20 to 146 kg N/ha with average 70kg N/ha N_2-fixing activity per unit area is at par or higher than legume pastures	Needs water in the field Cannot withstand desiccation	Use of local *Azolla pinnata*	Use of improved strains of *Azolla filiculaides, A. caroliniana* and *Azolla* hybrids
	Easy to in corporate in soil and faster decomposition	Inoculum maintenance and distribution is difficult		Use of sporocarp technology for sporophyte regeneration
Blue Green Algae	Provide few Kg N/ha to 80 kg N/ha with average 27kg/ha	Inoculum establishment is sporadic	Use of soil based inoculum	Use of quality inoculum with efficient strains to suit paddy field conditions
Azospi-rillum	Provide few Kg N/ha to 40 kg N/ha with average 25kg N/ha	Availability of good quality inoculum	Use of unsterile carrier material and production of biofertilizer using traditional tech.	Use of sterile carrier material and use of advanced production protocols under aseptic conditions
Rhizo-bium	Provide 20 to 60 kg N/ha highly suitable for Rice follow up crop like pulses or leguminous oil yielding crop	Availability of good quality inoculum	Use of unsterile carrier material and production of biofertilizer using traditio-nal tech.	Use of sterile carrier material and use of advanced product-ion protocols under aseptic conditions
Phospho-bacteria	Provide 20 kg P_2O_5/ha on average	Availability of good quality inoculum	Use of unsterile carrier material and production of biofertilizer using traditional tech.	Use of sterile carrier material and use of advan-ced production protocols under aseptic conditions.

Table 2. Decadal Report of Production of Biofertilizers in Production Units of Govt. of Tamil Nadu

Years	Installed Capacity	*Rhizobium* (in tonnes)	*Azospirillum*	Phosphobacteria	Total
1990-91	400	165	244	1	410
1991-92	400	173	239	1	412
1992-93	600	209	308	17	534
1993-94	920	361	488	60	909
1994-95	1100	337	815	109	1261
1995-96	1300	310	821	349	1480
1996-97	1300	258	709	329	1296
1997-98	1400	246	521	298	1065
1998-99	1400	279	781	358	1480
1999-00	1400	260	786	364	1410

(Ahmed *et. al.* 2000)

Current knowledge of inoculum production

Although, commercial biofertilizer technology was pioneered in 1895 in USA, currently India continues to be the largest producer in the globe. Although there are large number of scientific papers available in dealing on principles and practices on inoculum production, most papers have not discussed the issue of presently followed inoculum production technology in most biofertilizer production units in India despite wide scope for their adaptation of currently available developed technology. Hence, this paper attempts to discuss the details of advances in biofertilizer technology and quality control.

Strain selection and improvement

In biofertilizer technology, strain selection occupies the foremost key area of technology, so as to explore its biological potentialities of the strain for the beneficiary crop. The selected strain used in biofertilizer production must be efficient to grow fast besides retain desirable potentialities of biological activities, which should be present in desired high number and must be stable.

Table 3. Biofertilizer production scenario in the state of Tamil Nadu

Agency	No. of Units	Capacity (tonnes)	Av. Capacity per unit
1. Government	6 (37.5)	1400 (42.3)	233
2. Fertilizer Industries	3 (18.8)	1075 (32.5)	358

3. Private	5 (31.3)	742 (22.4)	148
4. Seema Cotton Dev. Res.Assn.	1 (6.2)	75 (2.3)	75
5. TNAU	1 (6.2)	17 (9.5)	17
TOTAL	16(100.0)	3309(100.0)	207

% share of various agencies given in parenthesis

(Ahmed *et. al.* 2000)

Table 4. Production pattern of different biofertilizers

Biofertilizer	1997-98	1998-99	1999-2000	Average
Azospirillum	545 (52)	776 (55)	777 (55)	699 (54)
Rhizobium	237 (22)	285 (20)	267 (19)	263 (20)
Phosphobacteria	276 (26)	358 (25)	356 (26)	230 (26)
Total	1058 (100)	1419 (100)	1400 (100)	1292 (100)

% share given in parenthesis (Ahmed et. al. 2000)

In order to isolate the strain in the laboratory for the desired purpose, conventionally, microbiologists depend on mostly screening of hundreds of bacterial colonies to find a single strain possessing suitable combination of characters with commercial potential. The desirable potentialities of strain which are generally essential for biofertilizer production should have the characteristics like:

(i) Ability to infect/ability to fix nitrogen/ability to solubilize/mobilize phosphate

(ii) Cultural characteristics like ability to multiply in the broth

(iii) Survival in the carrier material

(iv) Stability during storage

(v) Competitive ability for survival in soil

(vi) Ability to migrate from initial site of inoculation

(vii) Ability to colonize soil in the absence of host

(viii) Ability to tolerate environmental stress

(ix) Ability to support nitrogen fixation or solubilize phosphates in a wide range of host genotypes

(x) Compatibility with agricultural chemicals

(xi) Ability to colonise early in the rhizosphere with host plant

Although to find a strain with all characters is extremely difficult, the strains which are used in biofertilizer production unit should have many of the desired characteristics. Conventionally, in order to improve the strain characteristics and to select a good desirable strain, the method employed was through mutation using chemical mutagens like NTG, spontaneous mutants through UV or X-ray irradiation etc. and after thorough screening, the strains are surveyed, isolated and selected.

Currently, due to rapid development of different protocols of genetic engineering, the bacterial strains could be improved through plasmid transformation, conjugation, electro-poration. The literature indicates the improvement of *Rhizobium* strains though transfer of plasmids, which carry important desirable characters like high nitrogen fixing ability, nodulation genes, tolerance to environmental stress or agro chemicals, which has been isolated and modified using Well lysis procedure (Denarie *et al.,* 1981). Similalry, using conjugation technique, the strain characteristics have been improved using a nonconjugative plasmid and superior strains of *Pseudomonas putida* have been developed (Chatterjee *et al.,* 1981). Using Electroporation technique, the desirable characters could be incorporated using Pulse Field and CHEF Gel Electrophoresis technique (Smith *et al.,* 1988). Using similar technology, the improved strain of *Rhizobium* with high nitrogen fixation and tolerance ability has been developed (Williams *et al.,* 1988). Using such technology, strain characteristics have been improved and many of the improved strains are available with Institutes of repute in India and abroad. However, biofertilizer production unit should use appropriate desirable strain to suit its own requirement, after procuring it from an Institute of repute and authenticating its desirable characteristics before use in the biofertilizer production unit. The quality of biofertilizer product could be better only when, the microbe used is efficient to do the desired activities. Hence, it is highly desirable to select a suitable microbial strain in the biofertilizer production unit with desired characteristics.

Fermentation

The second most important area in biofertilizer production is to retain the high microbial load of 10^8 or 10^9 populations in the inoculant, which usually is achieved only when the used strain could be cultured in a fermentor. Usually in a Production Unit, fermentation takes place under standard aseptic conditions with equipment designed specifically for this purpose. On the other hand, most biofertilizer production units in India are independent on stationery culturing system or orbital shake culturing system in glass bottles wherein, often the microbial load exceeds 10^7 despite fermentors are available. Commercial fermentors vary in size varying from 50 to 300 litre capacity so that, a particular strain could be allowed to grow fast, maintain requisite scale up and process control, temperature, agitation, pH and aeration in order to assess microbial population and density. Usually, the industrial fermentors use aerobic processes for production of *Rhizobium, Azospirillum, Azotobacter, Bacillus, Pseudomonas* etc., However, in order to maintain high population load, the fermentation needs adequate sterilizing systems, a clean working environment, sterile air supply, careful use of aseptic techniques in handling procedures, appropriate equipment, maintenance procedures so that, the desired microbe could grow fast and build population of 10^9 or even 10^{10} in the broth.

In order to support good fermentation, body construction of fermentor is most important so as to maintain sterility, supply sterile air, adequate temperature, process control and monitoring. A variety of fermenters are available in the market starting from 10 litres to enormous 50,000 litre industrial capacity. NifTAL designed 141 litre capacity which has been designed in stainless body at low cost, simple technology for operation, has appeared to be promising for Indian situations. On the other hand, modern fermentation technology has increased the efficiency of fermentation through automation and instrumentation. Recent available technology like automatic weighing, bulk storage, mass flow dispensor and large fermenter vessels have contributed much to reduce cost of biofertilizer production which show much promise.

Carrier Materials

The issue of quality inoculant production depends on use of good carrier materials in biofertilizer production unit. Although, peat has been regarded as one of the best carrier material. But, it is not available in plenty and so mostly lignite is used in India. On the other hand, alternate technologies like use of vermiculite has appeared recently which show good potentiality for its success for its use in biofertilizer production. Vermiculite could maintain sufficient water holding capacity which is a very useful character. Further, in order to maintain adequate nutrition in the product, addition of nutrient media to the broth before mixing of vermiculite, assists maintenance of high microbial load in the biofertilizer product. Further, the multilamellate structure of the vermiculite provides superior aeration and space for microbial proliferation besides, the finely powdered vermiculite could adhere readily to seeds. Studies conducted at RBDC, Bangalore indicate the high potentiality of vermiculite as compared to lignite (Bisoyi and Majumdar, 2000).

Processing of Carrier Materials

The foremost aspect of inoculant production in India is the development of technology of processing of carrier materials so as to increase the shelf life of biofertilizers. Conventionally, biofertilizers are being produced in India using unsterile carrier materials which often leads to sub-standard quality of biofertilizers. On the other hand, many of the countries in the globe use presterilized carrier materials which leads to production of quality biofertilizers since contamination is avoided. The technology to sterilize huge quantity of carrier materials, which are used in biofertilizer production, is the major problem and the technology rests with the use of conventional method like autoclaving of carrier materials in horizontal autoclave. However, horizontal autoclaving often consume high electricity, takes more time as well as bursts the polythene bags. On the other hand, gamma irradiation system of sterilizing bulk carrier materials are known, which is being used in many of the western countries like Australia. The gamma irradiated method of sterilizing offers wide scope for its use under Indian conditions. RBDC, Bangalore has conducted experiments with gamma irradiated carrier materials, which indicate production of quality product with increase of shelf life (Bisoyi and Majumdar, 2000).

Mixing of Broth with Carrier Material

The mixing operation serves the purpose of providing homogenous mixture of various ingredients primarily, *Rhizobium/Azospirillum/ Azotobacter*/PSB culture of the broth with the carrier material, so as to maintain the pH neutrality wherein, sufficient culture broth should not exceed the moisture level of 30-35% depending on the carrier material and to retain 10^8 or 10^9 cell count in the finished product. This could be done only when the mixing is done under aseptic conditions so that, the microbial contaminants are eliminated.

Conventionally, biofertilizers are mostly produced in India through open mixing system by which, even though the strain used has high potentialities, the quality of the biofertilizer produced often suffers to maintain standard. In order to mitigate the system, the biofertilizer could be produced through a closed system (Sterile System) of production like use of pre-sterilized and pre-packed carrier material, which could be injected with precalibrated volume of broth coming directly from the fermentor fitted through autosyringe or a dispensor aseptically under a laminar air flow. Using such systems, which is economical as well as involves innovative methods of modern technology, could deliver a quality product with no contamination. Among the technologies available for mixing, this sterilized system of injection appears to be more practicable by biofertilizer production units and useful under Indian conditions.

Alternate Inoculant Forms

Currently in India and most parts of the world, the carrier based biofertilizer formulations are being used. Although, different forms of biofertilizers like bottle or agar slants, lyophilized cultures, frozen concentrates particularly, dry cultures are available the carrier based cultures have been more common due to its easy adaptability and easy accessibility to different biofertilizer units. However, in India, due to the unavailability of good peat like quality carrier material, the product has often been regarded as spurious. However, currently, the recent technology of "Liquid form" of inoculants are known. The liquid form of inoculant of *Rhizobium* which was earlier patented by Nitragin Co., USA are being marketed with commercial name Biolidoz wherein, the liquid inoculant contained high populations and most importantly, retained shelflife of 30 months. Currently, many of the farmers of Australia and USA use liquid forms due to its accessibility as well as retention of high contaminant free microbial load and easy transportation. However, currently known use of liquid biofertilizers could serve the large Indian farming community under the present scenario of non availability of good carrier material, contamination problems as well as packaging and high labour intensive technology and this liquid biofertilizer technology is being currently pursued in India by many leading Microbiology Laboratories. Besides liquid formulations, granular based formulations which are currently used appears to show more promise. Although peat and clay based granular forms are available, in market for *Rhizobium* experiments has shown much promise in use of clay based granular forms. Current technology indicate the use of polymer capsules like alginate, polyacrylamide in the market retain high microbial load and maintain longer shelf life. But, the alginate and polymer gels are expensive which may not be economically feasible to use as a soil inoculant, but certainly has high scope for its promotion in some specified areas.

Use of Semi Automatic/Automatic System of Production

Conventionally, biofertilizers are produced/mixed using manual methods wherein, the manual methods often are dependent on manual labour and so, the biofertilizer produced retain high contaminants. On the other hand, currently, biofertilizer could be produced through use of fully or semi automatic packaging systems as outlined in the Flow Chart (Fig. 1). The closed systems of mixing and fully automatic system of packing using Form Fill and Seal (FFS) machines have major components of packing system such as Base Hopper, Screw Conveyor, Auger Feeder, Pouch maker and Sealer. In this method, the product is prepared by mixing carrier broth in a closed mixture, collected and transferred to Base Hopper for packing. On the other hand, Semi automatic packaging system show more promise. In the Semi automated packaging system, it is dependent on Elevated Main Feeder Hopper broth injection system. Here, the dry powdered carrier material is transferred to elevated main feeder Hopper. The carrier is passed on to Auger Feeder by horizontal screw conveyor. Preset quality is delivered in auger and collected in bag. The pre set quantity of broth is delivered into the packet directly from the fermentor through broth injection system. This has been possible by suitable integration and synchronization of both equipments i.e. Auger Feeder and Broth Injector. After collecting the material, the packet is transferred to sealing machine and sealed (Ramasami, 2000). These methods could be evaluated by the biofertilizer production units in India, for adopting either Semi Automatic Systems or Automated Systems. Keeping view of the modern facilities available, it is desirable to use the current technology of producing biofertilizer commercially so as to produce high quality biofertilizers at less cost which could assure quality product and in turn, increase promotion of good marketing. Although, biofertilizers produced in India suffers from challenges like use of ineffective strains, inadequate cell count, poor shelf life, high cost of production, non-availability of carrier materials, these constraints could be addressed through use of appropriate technology and good quality of biofertilizers could be produced (Bisoyi, 2000).

Quality Control

Evaluation of inoculant quality by enumeration of viable *Azospirillum*/*Azotobacter*/*Rhizobium*/Phosphobacteria is an accurate index of inoculating potential. Although basically, an inoculant producer should voluntarily check the quality of his own product in his own production unit before marketing most producers do not practice it strictly which ultimately fail to compete in the open market.

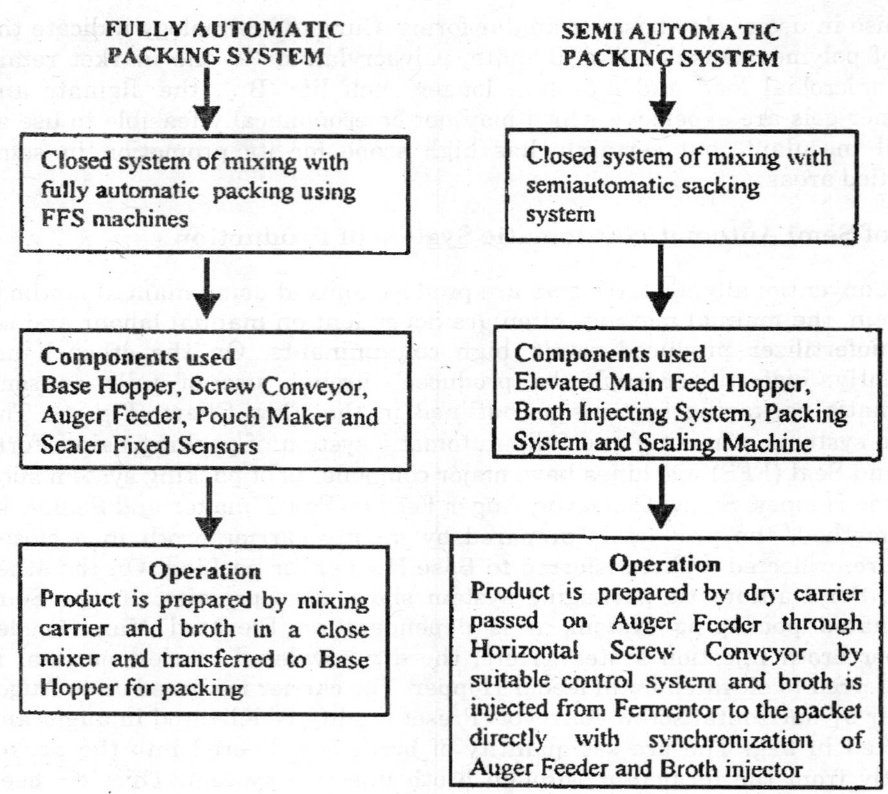

Fig. 1. Advanced Biofertilizer Production System

Although conventionally tests based on MPN counts and infection test (for *Rhizobium*) are used for quality analysis, they are time consuming and often is unreliable. On the other hand, current day agricultural biotechnology has developed the serological protocols for quality control of biofertilizers like agglutination, Fluorescent Antibody Test (FA Test), Membrane Immuno Blot Technique, ELISA etc. which could assess the quality of the biofertilizer product in large no of samples at less time (Somasegaran and Hoben, 1994). Further, using nucleic acid techniques like RFLP, PCR technology, the quality of the product could be assessed fast with authenticity. Further, use of reporter genes like gene encoding the enzyme β-glucuronidase helps to identify the inoculant based on the change of colour by the reporter gene in the rhizosphere of the beneficiary crop. This technology could be used to assess the inoculant quality in the farmer's field as well. This indicates the technological advances in inoculant quality assurance of the biofertilizer producer of the field level. Among the diagnostic tests, serological tests show much usefulness in the biofertilizer production unit for rapid identification. Such technologies could indeed improve quality so as to assist its commercial success. Recent years show the availability of new technologies which could be adopted in biofertilizer production units before its marketing, so as to assure supply of quality product to the consumers.

Despite the technologies available, the inoculant industry in India which is predominated by small scale production units are ill equipped to adopt current technology. Although a few proved technology could be adopted at less cost these technologies have not been adopted by small scale biofertilizer production units despite the fact that, Govt. of India is supporting nonrecurring grant of Rs. 20 lakh for infrastructure development under the Scheme "National Project on Development and Use of Biofertilizers" for strengthening/setting up of Biofertilizer Production Unit with annual capacity of 150 tonnes. Although replacement of with conventional technology presently adopted in biofertilizer production units, recent methods to not necessarily involves high cost if the motto of the biofertilizer producer is to produce quality product with use of recent technology, most of production units in India could produce good quality biofertilizers at less cost. Using current technology and proven strains, the quality biofertilizers at less cost could be produced and available for rice based cropping system for mass use of farming community, which inturn could ensure high marketing demand of biofertilizers besides could promote further growth of biofertilizer industry in India.

REFERENCES

Ahmed, S.A., T. Ramalinsgam and K. Arulmozhi. 2000. Current Status of Biofertilizer production technology in Tamil Nadu. In: **Proc. of Symposium on "Current Status of Biofertilizer Technology and Its Promotion in Karnataka,** Tamil Nadu, Kerala and Pondicherry" (eds. R.N. Bisoyi) RBDC, Bangalore, pp 110-115

Bisoyi, R.N. 2000. Constraints of Biofertilizer production and remedial measures. In: **Ibid**. PP 50-54.

Bisoyi, R.N. and P. Majumdar. 2000. Technology for Increasing Shelf life of biofertilizers. In: **Ibid**. PP 44-49.

Brockwell, J. and J. Bottemley.1995. Recent Advances in Inoculant Technology and Prospects for the future. **Soil Biol.Biochem., 47:** 683-697.

Brockwell, J., P.J Bottemley and J.E. Thies. 1995. Manipulation of Rhizobia microflora for improving legume productivity and soil fertility. A critical assessment. **Plant Soil. 174:** 143-180.

Chatterjee, D.K., S.T Kellog, S. Hamada and A.M. Chakraborthy, 1981. Plasmid specifying total degradation of 3-Chlorobenzoate by a modified pathway. **J. Bacteriol., 146:** 639-646.

DEAR, 1999. Evaluation Report of Department of Evaluation and Applied Research, Govt. of Tamil Nadu

Denarie, J., P.Boistard, F.Casse-Delbart, A.G., Altherly, J. O. Berry and P. Russel. 1981 Indigenous Plasmids of *Rhizobium*. In: **Biology of Rhizobiaceae** (eds.). K.L. Giles and A.G. Artherly, Academic Press. Orlando, PP 191-219.

Keyser, H.H., P. Somasegaran, and B.B. Bohlool 1993. Rhizobial Ecology and Technology. In: **Soil Microbial Ecology** (eds). F.B. Metting, Jr., Marcel Dekkar, Inc. Publ. New York, PP 205-226.

Ramasami, S. 2000. Automated Mechanized Systems in Biofertilizer Production: MFL's Experience In : **Proc. of Symposium on "Current Status of Biofertilizer Technology and Its Promotion in Karnataka,** Tamil Nadu, Kerala and Pondicherry", (Ed. R.N. Bisoyi) RBDC, Bangalore. PP 35-37.

Smith, C.L., S.R. Klco, and C.R.Cantor, 1988. Pulse-Field Gel Electrophoresis and technology of large DNA molecules. In : **Genome Analysis. A Practical Approach** (Ed). K.E. Davies, IRL Press, Oxford, PP 41-60.

Somasegaran, P. and H.J.Hoben, 1994. **Handbook for Rhizobia. Methods in Legume-*Rhizobium* Technology**, Springer-Verlag Publ. New York

Williams, M.K., F. Cannon, P. Mc Lean, and J. Beynon, 1988. Vector for the integration of genes into a defined site in the *Rhizobium* melilo genome. In: **Molecular Genetics of Plant Microbe Interactions** (eds.). R. Palacious and D. Verma. American Phyto Pathological Society Press. St. Paul, pp 198-199.

CHAPTER – 53

IMPORTANCE OF QUALITY CONTROL IN BIOFERTILIZERS

M.Thangaraju

INTRODUCTION

Biofertilizers are one of the important components of organic farming and sustainable agriculture. Awareness among farmers is increasing on the biological fertilizers. But, the success of biofertilizer inoculation depends on the quality of the inoculant used. Because farmers are unable to judge the quality of the inoculants at the time of purchase or even at use, the inoculant producers are negligent to institute quality control programmes (Thompson, 1991a). Unfortunately much of the inoculant produced in the world today is relatively of poor quality (Brockwell and Bottomley, 1995; Somasegaran, 1991). In such situation maintaining high quality inoculant products depend on the internal implementation of effective quality control systems by manufacturer. A major goal of a quality standard is to control the quality of inoculant to be used in any country. It is, therefore, important that each country should set-up its own standard based on research findings and their experience (Kannaiyan *et al.*, 2001).

INTERNATIONAL PERSPECTIVES ON QUALITY CONTROL

There are wide variations in the quality control programmes and parameters in developing countries and developed countries. The following are the perspectives on the quality control of microbial inoculants at international levels:

- Regulation on inoculant quality varies from country to country and no set of international standards exists. (Olsen *et al.*, 1996)
- Brazil (Thompson, 1991a), Canada (Olsen *et al.*, 1994), France (Wadoux, 1991) and Uruguay (Brockwell and Bottomley, 1995) have regulatory authorities supported by legislation.

- Australia, South Africa and New Zealand have quality control programs in which inoculant manufacturers participate voluntarily (Olsen *et al.*, 1996).
- In many countries (including USA, UK, India) product quality standards are left to the discretion of the manufacturers (Smith, 1992)
- A great disparity exists throughout the world today in the price, quality and effectiveness of microbial inoculants available to farmers (Joly, 1991; Day, 1991).

France currently has the highest standards world wide for inoculant quality, and also has a strict requirement that no contaminants be present in rhizobial inoculant products (Catroux, 1991). Australian standards are also high, requiring 1 x 10^9 rhizobia per g moist inoculant and fewer than 10^6 contaminants per g (Thompson, 1991b). England has no governmental standard, but Agricultural Genetics Company's self imposed minimum is 2 x 10^9 rhizobia per g with a maximum of 10^6 contaminants per g (Day, 1991). The Thai standard for its government program which produces inoculant is 5 x 10^7 rhizobia per g, but this minimum is sensibly tied to a requirement for a high inoculant application rate with the net result that 10^4-10^6 *Bradyrhizobium japonicum* per soybean seed are applied (Boonkerd, 1991). The Netherlands require from 4 to 25 x 10^9 rhizobia per g, and South Africa and New Zealand both require a minimum of 1 x 10^8 rhizobia per g (Smith, 1992). Rwanda prescribes 1 x 10^9 rhizobia per g and the contaminants should not be more than 0.001% of the rhizobial count (Scaglia, 1991). The United States applies no standards and inoculant quality is left to the manufacturer's discretion (Smith, 1992), but most US inoculants exported to Canada contain at least 1 x 10^8 rhizobia per g (Anonymous, 1992). The Canadian standard for minimum numbers of rhizobia in inoculants therefore varies with manufacturer, legume host, and application rate (Olsen *et al.,* 1994). For inoculants tested in Canada in 1992, the required number varied from 2.7 x 10^7 to 1.6 x 10^9 rhizobia per g (Anonymous, 1992).

BUREAU OF INDIAN STANDARDS (BIS) FOR CARRIER - BASED BIOFERTILIZERS

BIS have notified specifications for *Azotobacter* (ISI, 1979) and *Rhizobium* (ISI, 1986) long back for use of biofertilizer producers (Table 2) and suggested some modifications. The committee has also suggested specifications for phosphate solubilizing microorganisms and *Azospirillum*.

Table 1. Indian Standard Specifications for different biofertilizers

Parameters	Rhizobium	Azotobacter	Azospirillum	PSM
Base	Carrier based	Carrier based	Carrier based	Carrier based
Cell No. at the time of manufacture	10^7	10^6	10^7	10^8

Contamination	Nil at 10^5 dilution	Nil at 10^5 dilution	Nil at 10^5 dilution	Nil at 10^6 dilution
Expiry period from date of manufacture	6 months	6 months	6 months	6 months
PH	6.5 – 7.5	6.5 - 7.5	6.5 - 7.5	6.5 - 7.5
Strain	Should be checked serologically	Nothing specific *A. chroococcum* sp. is mentioned	Nothing specific	Nothing specific
Carrier should pass through	150-212 micron IS sieve	106 micron IS sieve	106 micron IS sieve	106 micron IS sieve
Others	Nodulation test +ve result in 50% or more dry matter than control	Minimum amount of N-fixation not less than 5 mg / g of sucrose utilized	Should produce white pellicle in NFB media at 10^7 dilution	Should show phosphate solubilizing zones of 12 mm in tricalcium phosphate medium at 10^5 dilution

(ISI, 1979; ISI, 1986)

QUALITY CONTROL METHODS

The general information based on the prescribed marking on the biofertilizer packets about the name of the biofertilizer, crop for which intended, batch number, date of manufacture, date of expiry, method of use, *etc* are to be checked first. Then the pH and physical characteristics of the carrier materials besides the moisture per cent of the product are to be assessed (ISI, 1979; ISI, 1986). The microbial population count of the concerned organism is to be assessed using appropriate methods and media. As soon as the colonies appear, the colony morphology like shapes of colony with margins are to be observed. Its morphological characteristics like cellular shape, size, flagella no., its attachment, *etc.* are studied under a phase contrast microscope and the contaminants if any are to be enumerated. Further, the strain is identified along with its confirmation test for strain authenticity using standard microbial protocols (Bisoyi, 1999).

Much research has been focused on refining and improving the sensitivity of existing techniques to develop quality assurance guidelines. The standard plant infection technique (Date and Vincent, 1962) is the most commonly used method for estimating the number of invasive/infective rhizobia in inoculants prepared with a non-sterile carrier. This standard technique has been modified (called pre-enriched plant-infection (PEPI) technique) to give most-probable number estimates of rhizobia in alfalfa inoculant (Biederbeck and Walker, 1981). Recent research has significantly improved the standard plant infection method for the enumeration of rhizobia in the presence of other microorganisms by making certain options available to allow greater flexibility to the user (Woomer *et al.*, 1990). They also have devised a computer programme called the Most Probable-Number Enumeration System (MPNES), which generates solutions and confidence limits for population estimates of MPN data. Currently, the MPNES is being adopted for the official standard procedure for inoculant quality in Canada.

TECHNIQUES FOR STRAIN IDENTIFICATION

Since, few strains are used in commercial production of biofertilizers and conventional microbial protocols take long time to authenticate, advanced protocols of strain authentication are available. Serological techniques have been successfully used in the quality control of inoculants. Serological methods like agglutination, fluorescent antibody technique (FA), enzyme linked immunosorbent assay (ELISA) and gel immunodiffusion techniques, *etc.* are being used to authenticate the strains of *Rhizobium.* (Somasekaran and Hoben, 1994) and *Azospirillum* (Suresh Babu and Thangaraju – unpublished data) The enzyme-linked-immunosorbent-assay (ELISA) has been used as a strain-specific serological technique for the identification of *R. meliloti* in commercial alfalfa inoculants (Olsen *et al.*, 1982). The same technique has been applied successfully for enumerating *Bradyrhizobium* sp. (peanut) in peat inoculant (Nambiar and Anjaiah, 1985). Serological methods have excellent potential in giving much quicker results for inoculant quality control if the necessary equipment, chemical and biological reagents are readily available. The membrane-filter immunofluorescence technique was evaluated for enumerating various rhizobial species in inoculants prepared with pre-sterilized peat gives good results (Somasegaran, 1985). Compared to the ELISA and immunofluorescence techniques, which give total cell counts (viable and non-viable), the immunoblot procedure (Olsen and Rice, 1989) makes a template of the resultant colonies (growing on an agar plate) on circular nitrocellulose membranes before identification of the rhizobia from contaminants. Use of nucleic acid fluorescent stains in the identification of *Rhizobium* was also attempted (Olsen and Rice, 1996).

Genetic markers, probes and PCR based techniques are recently employed for strain identification. Several PCR-fingerprinting techniques have been developed to identify bacteria at the strain level. Those methods are applied with cultured bacterial cells avoiding DNA-extraction and combining convenient analysis with universal applicability and the potential of automation (Versalovich *et al.*, 1994). Now, oligonucleotide primers were developed to study diazotrophic isolates and were applied for PCR-fingerprinting (Smida *et al.*, 1996). The electrophoretically separated band patterns were highly reproducible and strain specific. A variety of phylogenetic oligonucleotide probes was developed for *Azospirillum* spp. (Kirchhof and Hartman, 1992), *Herbaspirillum* spp. (Baldani *et al.*, 1996), *Azoarcus* (Hurek *et al.*, 1994), *Acetobacter diazotrophicus* (Kirchhof *et al.*, 1997). New whole-cell binding 16S rRNA-directed oligonucleotide probes have been developed and are available for *in situ* localization studies of

Herbaspirillum spp. and *Azospirillum* spp. (Hartmann *et al.*, 2000). Reports are available for the identification of lambda clones containing the *hupB* gene of *Anabaena* sp. using PCR technique (Gubili and Borthakur, 1996). Recently, strain specific identification of *Azospirillum* using *nifH-lacZ* fusions have been demonstrated in *Azospirillum*-wheat associations (Arsene *et al.*, 1994; Pereg Gerk *et al.*, 2000).

QUALITY ASSESSMENT IN CYANOBACTERIAL BIOFERTILIZER

The cyanobacteria (or) blue green algae are one of the important component of rice soil system contribute to the soil fertility through biological nitrogen fixation, secretion of growth promoting substances and also by the addition of organic matter. The beneficial effect of algal inoculation on the grain yield of rice has been well documented at a number of locations (Venkataraman, 1972; Kannaiyan, 1979, 1985). Blue green algal inoculation with composite cultures comprising 8 to 10 different culture was found to be more effective than single culture inoculation. The quality of the cultures used, whether composite genera or single genera, is very much necessary to achieve the benefit. The biofertilizer techniques of cyanobacteria are limited mainly due to the non-availability of good quality inoculants. Hence, the quality of the inoculum produced is paramount importance for ensuring the successful performance of the inoculant under field conditions.

QUALITY CRITERIA FOR CYANOBACTERIAL INOCULANTS

The quality parameters necessary for a quality cyanobacterial inoculants described by Kannaiyan (2000) is provided below.

1. Cyanobacterial inoculant shall be carrier based, the colour of the inoculant mainly depend on the colour of the carrier
2. Shall contain a minimum of 10^5 viable cell of the carrier on dry mass basis at the time of production and use when stored in dry state at room temperature and tested by the standard methods
3. Shall have minimum of two years of expiry periods from the date of production
4. Should not have contamination other than blue green algae, the limit of contamination with non-heteroystous BGA shall not exceed more than 10^2 when counted.
5. The pH of the cyanobacterial inoculant in carrier shall be between 7.0 and 7.5.
6. Shall show effective growth on the most surface of soil in the form of thin sheet or floating scum in / on water surface.
7. The nitrogenase activity shall be minimum of 0.158 ± 0.03 n moles C_2H_4 / mg / dry biomass / h (0.025 ± 0.04 n moles C_2H_4 / g chlorophyll a/ h)
8. The carrier material shall be in the form of soil capable of passing through 2 mm sine of finally ground.
9. The manufacture shall control the quality of final product and maintain records of the organism (records of isolation, location, identification of

strains *in vitro* and *in situ* nitrogenase activity, growth rate and level of tolerance to fixed nitrogen, agrochemical (insecticides herbicide etc) and pH.

10. The inoculant shall be packed in polythene lined gunny bags or thick polythene bags; the inoculant bags shall not be stored or packed in fertilizer bags or any pesticide bags.
11. Each packet shall be marked legibly to give the following information
    - Name of the product specifically as BGA inoculant
    - Crop for which intended - eg. Rice
    - Name and address of the manufacturer
    - Types of carrier
    - Batch or code number
    - Date of manufacture
    - Date of Expiry
    - Method of application
    - Moisture content on per cent basis
    - Net quality required for an acre or hectare
    - Storage instructions

Items like date of manufacture, date of expiry and moisture content on percent basis should be printed on a coloured ink background. Directions for the use of BGA biofertilizer shall be printed briefly on the packet. Each packet may also be marked with the ISI certification mark, if obtained.

BGA biofertilizer shall be stored by the manufacturer in a day place away from direct heat at room temperature. It shall not be allowed to become wet during storage, once dried. It shall also be the duty of the manufacturer to instruct the retailer and inturn, the users about the precautions during storage

METHODS OF CYANOBACTERIAL QUALITY ASSESSMENT

The quality of the cyanobacterial inoculant shall be determined by the cell count, chlorophyll content and nitrogenase activity. In addition, the strain identification of the different cyanobacterial genera may be carried out by the genetic analysis through polymerase chain reaction.

CONCLUSION

Quality of inoculant is created by variety of factors. These factors include incorporation of superior strain of the organism, the use of good quality carrier material, the survival of the organism in desired numbers in broth and carrier, free from excessive contamination, adequate shelf life, proper packing formulation that is easy and effective to supply, *etc.* A factor of primary concern is that the highest possible number of live cells of the specific strain of the organism concerned (Olsen *et al.*, 1996). A good policy and protocol including the availability of appropriate standards and establishment of proper testing and reporting agency is necessary for inoculant quality control (Thompson, 1980).

REFERENCES

Anonymous, 1992. Canadian legume inoculant and preinoculated seed product testing report. Food Production and Inspection Branch, Feed and Fertilizers Division, Government of Canada, Ottawa. p.30

Arsene, F., S. Katupitiya., I. R. Kennedy and C. Elmerich. 1994. Use of *lacZ* fusions to study the expression of *nif* genes of *Azospirillum brasilense* in association with plants. **Mol. Plant-Microbe Interactions 7(6),** 748-757.

Baldani, J. I., B. Pot, G. Kirchhof, E. Falsen Baldani, V. L. D. Olivares, F. L. B. Hoste, K. Kersters A. Harmann M. Gillis and J. Dobereiner. 1996. Emended description of *Herbaspirillum*; inclusion of (*Pseudomonas*) *rubrisubalbicans*, a mild plant pathogen, as *Herbaspirillum rubrisubalbicans* com. nov., and classification of a group of clinical isolates (EF group 1) as *Herbaspirillum* species. **Int. J. syst. Bacteriol. 46**, 802–810.

Biederbeck, V. O. and Walker, B. D. 1981 **Proc. 8th North American *Rhizobium* Conference**. (eds.) K W Clark and J H G Stephens. Printing Services, The University of Monitoba, Winnipeg, Canada. pp. 536-561.

Bisoyi, R. N. 1999. Role of regional biofertilizer development centers in quality control of biofertilizers. In: **Recent advances in microbial inoculants** – Training lecture notes Eds. M Thangaraju and P Santhanakrishnan. CAS, Tamil Nadu Agric. Univ., Coimbatore, India. pp 93-96.

Boonkerd, N. 1991. Inoculant quality control and standards in Thailand. In: **Report of the Expert Consultation on Legume Inoculant Production and Quality Control.** (ed.) J A Thompson. Food and Agriculture Organization of the United Nations, Rome, Italy. pp 121-129

Brockwell, J. and P. J. Bottomley. 1995 Recent advances in inoculant technology and prospects for the future. **Soil Biol. Biochem. 27**, 683-697.

Catroux, G. 1991. Inoculant quality standards and controls in France. In: **Report of the Expert Consultation on Legume Inoculant Production and Quality Control**. (ed.) J. A. Thompson. Food and Agriculture Organization of the United Nations, Rome, Italy. pp 113-120

Date, R. A. and J. M. Vincent 1962. Determination of the number of root nodule bacter ia in the presence of other organisms. **Austr. J. Exp. Agric. Anim. Husb. 2**,5-7

Day, J. M. 1991. Inoculant production in the UK. In: **Report of the Expert Consultation on Legume Inoculant Production and Quality Control**. Ed. J A Thompson. Food and Agriculture Organization of the United Nations, Rome, Italy. pp 75-85.

Gubili, J. and D. Borthakur. 1996. The use of a PCR cloning and screening strategy to identify lambda clones containing the *hupB* gene of *Anabaena* sp. strain PCC7120. **J. Microbiol. Methods 27**, 175-182.

Hartmann, A. M., B. Stoffels Eckert, G. Kirchhof and M. Schloter. 2000. Analysis of the presence and diversity of diazotrophic endophytes. In: **Prokaryotic Nitrogen Fixation : A Model System for Analysis of a Biological Process.** Horizon Scientific Press, Wymondham, UK. p.727

Hurek, T., B. Reinhold-Hurek, M. Van Montagu and E. Kellenberger. 1994. Root colonization and systemic spreading of *Azoarcus* sp. Strain BH72 in grasses. **J Bacteriol. 176,** 1913-1923.

ISI, 1979. Specification for *Azotobacter chroococcum* inoculants Standard IS:9138-1979 New Delhi, India. 14 p.

ISI, 1986. Indian standard specification for *Rhizobium* inoculants Standard IS: 8268-1986 New Delhi, India. 20 p.

Joly, C. 1991. Biological nitrogen fixation within FAO agricultural production programmes in the context of sustainable development. In: **Report of the Expert Consultation on Legume Inoculant Production and Quality Control.** Ed. J A Thompson. Food and Agriculture Organization of the United Nations, Rome, Italy. pp 9-14.

Kannaiyan, S. 1979. Blue-green algal biofertilizer for rice crop. **Farmers and Parliament 14(11),** 19-25.

Kannaiyan, S. 1985. **Studies on the algal application for low land rice crop.** Tech. Bulletin. Tamil Nadu Agrl. University, Coimbatore.

Kannaiyan, S. 2000. **Biofertilizer technology and quality control.** Tamil Nadu Agrl. University, Coimbatore. p 256.

Kannaiyan, S., K. Kumar, M. Thangaraju and K.Govindarajan. 2001. Microbial Inoculant Production in Developing Countries – Success, Problems and Potentials. Tamil Nadu Agricultural University, Coimbatore. p. 119.

Kirchhof, G. and Hartmann, A. 1992. Development of gene-probes for *Azospirillum* based on 23S-rRNA sequences. **Symbiosis 13**, 27-35.

Kirchhof, G., V.M. Reis, J.I. Baldani, B. Eckert, J. Dobereiner and A. Hartmann 1997. Occurrence, physiological and molecular analysis of endophytic diazotrophic bacteria in gramineous energy plants. **Plant and Soil 194**, 45-55.

Nambiar, P.T.C. and Anjaiah. 1985. Enumeration of rhizobia by enzyme-linked immusorbent assay (ELISA). **J. Appl. Microbiol. 58**, 187-193.

Olsen, P.E. and W. A. Rice. 1989. *Rhizobium* strain identification and quantification in commercial inoculants by immunoblot analysis. **Appl. Environ. Microbiol. 55**, 520-522.

Olsen, P. E. and W. A. Rice. 1996. Rapid evaluation of peat-based legume inoculant using immunomagnetic beads for cell retrieval and fluorescent nucleic acid probes for viability analysis. **Pl. Soil 186**, 75-79.

Olsen, P. E., W. A. Rice and M. M. Collins. 1994. Biological contaminants in North American inoculants. **Soil Biol. Biochem. 27**, 699-701.

Olsen, P. E., E. S. Sande and H. H. Keyser.1996. The enumeration and identification of rhizobial bacteria in legume inoculant quality control procedures. NifTAL, USA. 96 p.

Olsen, P. E., W. A. Rice, G. W. Stemke and W. J. Page 1982. Strain specific serological techniques for the identification of *Rhizobium meliloti* in commercial alfalfa inoculants. **Can. J. Microbiol. 29**, 225-230.

Pereg Gerk L, K. Gilchrist and I. R. Kennedy 2000. Mutants with enhanced nitrogenase activity in hydroponic *Azospirillum brasilense*-wheat associations. **Appl. Environ. Microbiol. 66(5),** 2175-2184.

Scaglia, J. A. 1991. Production and quality control of inoculants in Rwanda. In: **Report of the Expert Consultation on Legume Inoculant Production and Quality Control.** Ed. J A Thompson. Food and Agriculture Organization of the United Nations, Rome, Italy. pp 61-69.

Smida, J. S. Leibhard, A. M. Nickel, F. Eckardt-Schupp and L. Hieber 1996. Application of repetitive sequence-based PCR (Inter-LINE PCR) for the analysis of genomic rearrangements and for the genome characterization on different taxonomic levels. **Gen. Anal. (Biomolecular Engineering) 13,** 95-98.

Smith, R. S. 1992. Legume inoculant formulation and application. **Can. J. Microbiol. 38,** 485-492.

Somasegaran, P. 1985. Inoculant production with diluted liquid cultures of *Rhizobium* spp. and autoclaved peat: Evaluation of diluents, *Rhizobium* spp., peats, sterility requirements, storage and plant effectiveness. **Appl. Environ. Microbiol. 50,** 398-405.

Somasegaran, P. 1991. Inculant production with emphasis on choice of carriers, methods of production and reliability testing / quality assurance guidelines. *In* Report of the Expert Consultation on Legume Inoculant Production and Quality Control. Ed. J A Thompson. pp 15-32. Food and Agriculture Organization of the United Nations, Rome, Italy.

Somasegaran, P. and H.J. Hoben. 1994. **Handbook for rhizobia-Methods in legume-*Rhizobium* technology.** Springer-Verlag, New York, USA. 450 p.

Thompson, J.A. 1980. Production and quality control of legume inoculants. ***In* Methods for evaluating biological nitrogen fixation.** Ed. F J Bergersen. John Wiley Publications. pp 489-533.

Thompson, J. A. 1984. Production and quality control of carrier-based legume inoculants. **Information Bulletin No. 17, ICRISAT**, India.

Thompson, J. A. 1991a. Australian quality control and standards. In: **Report of the Expert Consultation on Legume Inoculant Production and Quality Control**. Ed. J A Thompson. Food and Agriculture Organization of the United Nations, Rome, Italy. pp 107-111.

Thompson, J. A. 1991b. Legume inoculant production and quality control. ***In* Report of the Expert Consultation on Legume Inoculant Production and Quality Control.** Ed. J A Thompson Food and Agriculture Organization of the United Nations, Rome, Italy. . pp 15-32.

Venkaratamam, G. S. 1972. Algal biofertilizers and rice cultivation. To-day and Tomorrow's Printers and Publishers. New Delhi. 75p.

Versalovich, J., M. Schneider, F. deBrujin and J. R. Lupski. 1994. Genomic fingerprinting of bacteria using repetitive sequence-based polymerase chain reaction. Meth. Mol. Cellul. Biol. 5, 25-40.

Wadoux, P. 1991. Inodculant production in industry using sterile carriers. ***In* Report of the Expert Consultation on Legume Inoculant Production and Quality Control.** Ed. J A Thompson. Food and Agriculture Organization of the United Nations, Rome, Italy. pp 33-42.

Woomer, P. J. Bennet and R. Yost. 1990. Overcoming in flexibility of most probable number procedures. **Agron. J. 82,** 349-353.

CHAPTER - 54

MOLECULAR TECHNIQUES FOR CHECKING GENETIC PURITY OF STRAINS IN THE CYANOBACTERIAL INOCULUM

K. Kumar

INTRODUCTION

Cyanobacteria (blue green algae or BGA) are unique among the prokaryotes in possessing the capacity of oxygenic photosynthesis. An important feature of many cyanobacteria is their ability to fix atmospheric nitrogen both under free - living and symbiotic conditions. The species of blue green algae which are known to fix atmospheric nitrogen are classified into three groups *viz.,* heterocystous – aerobic forms, aerobic unicellular forms and non heterocystous filamentous microaerophilic forms. Nitrogen fixing cyanobacteria could use sunlight as the sole energy source for the fixation of carbon and nitrogen and therefore have potential as biofertilizer (Kannaiyan, 1985). The composite inoculum consisting of cyanobacterial cultures *viz., Nostoc, Anabaena, Calothrix, Tolypothrix, Plectonema, Aphanotheca, Gleocapsa, Cylindrospermum, Oscillatoria, Aulosira* and *Scytonema* have been used for inoculation in rice (Kannaiyan, 1993).

Mass production of high quality inoculum of the desirable strains is an important stage when considering the inoculation of nitrogen fixing cyanobacteria as biofertilizer for rice crop (Watanabe and Roger, 1984). The biofertilizer techniques of cyanobacteria are limited mainly due to the non availability of good quality inoculants and the difficulty in re-establishing the inoculated strains varied in the rice field (Roger and Watanabe, 1986). Rapid and reliable quality control methods have not yet been developed for cyanobacterial biofertilizers which are considered as a major hurdle in the production of quality cyanobacterial inoculants.

Cyanobacteria traditionally have been distinguished on the basis of phenotypic properties, structure and physiology. Taxonomic characters change so drastically that reliable identification of species become difficult or even impossible (Kumar *et al.,* 2000). More importantly, the morphology of cyanobacteria in laboratory cultures is often considerably altered from the original morphology of environmental isolates and the diversity of strain within a culture may be reduced because of selective culturing condition (Ward *et al.,* 1990). Many techniques have been evolved to study the phylogenetic perspective which are more reliable in cyanobacterial strain identification. These techniques could be effectively utilized for checking genetic purity of the strains and ensuring supply of quality inoculants to the farmers

CHEMOTAXONOMIC METHODS

The fatty acid composition is one of the tools available for the taxonomic classification of cyanobacterial strains in culture. Kenyon *et al.* (1972) used fatty acid composition as a criterion, dividing cyanobacteria into several subgroups and observed uniform fatty acid composition for *Anabaena* and *Calothrix*. Li *et al.* (1998) investigated 28 axenic planktonic cyanobacterial strains including *Microcystis, Oscillatoria, Spirulina, Aphanizomenon* and *Anabaena* for their fatty acid composition. The results revealed a better chemical diversity among the different strains. Cohen and Vonshak (1991) distinguished many *Spirulina* strains from a variety of *Spirulina* like strains based on fatty acid composition. Lakshmanan *et al.* (2002) analyzed the fatty acid profile of filamentous cyanobacterial strains of the Composite culture inoculum used for algal biofertilization of rice crop using GC-MS. They found that the filamentous cyanobacterial cultures tested were highly diverse with respect to fatty acid composition, some have a fatty acid composition of bacterial type, some with chloroplast type and only a few possess the algal type.

Hertzberg *et al.* (1971) used carotenoid composition as a tool for species identification. In several morphologically very similar species, more or less stable modifications in the pigment content were found (Skulberg and Skulberg, 1985). The stability of pigment content can serve an intergeneric features (Anagnostidis and Komarek, 1988).

Isozymes provide a cheap and effective method of classifying the different varieties and populations according to their genetic composition. Isozymes are useful as genetic markers since they co-dominate and also exhibit low level of environmental interactions (Gottlieb, 1981).

Utilization of isozyme profiles has proven valuable for cultivar identification in many agricultural crops. Correlation between isozyme variation and the geographic origin of plant accessions is also possible (Soltis *et al.,* 1983; Vallejos, 1983 and Glaszmann, 1987). Glaszmann and Roger (1985) used electrophoretically detectable isozyme profiles for identifying cyanobacterial strains especially in soil inoculation studies.

Puppo and Rigaud (1986) have proposed that the enzyme superoxide dismutase (SOD) function as a protection mechanism of nitrogenase in diazotrophic system. They explained that SOD catalyzes the disputation of oxygen to H_2O_2 and thus protecting cells from the effect of oxygen that are likely to be largely due to primary production of oxygen. Bannister *et al.* (1987) reported the distribution of four different SOD isozymes containing copper, zinc, manganese, ferric or Fe-Mn as prosthetic metals among prokaryotic organisms. Proteins and isozyme patterns as taxonomic probes have been well documented (Aitken, 1988) and isozymes are useful as genetic marker due to their frequent polymorphisms, co-dominance, single gene Mendelian inheritance, rapid, simple and relatively inexpensive assay and their ubiquity in plant tissues and organs. Hagemann *et al.* (1990) investigated the salt induced changes in protein profile of cyanobacterium *Synechocystis* sp. and reported three new protein bands in salt shocked cells.

ANALYSIS OF rRNA SEQUENCE DIVERSITY

One area in which a significant application has been developed recently is the phylogenetic characterization of natural microbial communities by 16S rRNA sequence analysis (Pace *et al.*, 1980). For the analysis of natural microbial populations in which unknown diversity must be anticipated, the rRNA is being focussed because they are functionally and evolutionarily homologous in all organisms. (Stackebrandt and Woese, 1981). Moreover, the rRNAs are ancient molecules and extremely conserved in overall structure. Hence, homologous rRNAs are readily identifiable by their sizes (Lane *et al.*, 1988). Most of the studies of cyanobacterial phylogeny with 16S rRNA sequences (Giovannoni *et al.*, 1988 and Nelissen *et al.*, 1994) were based on isolates available as axenic cultures. A cyanobacteria specific oligonucleotide probe which distinguished, prior to cloning and sequencing, at least between the 16S rRNA genes from those of bacterial contaminants was constructed (Nelissen *et al.*, 1996). After PCR amplification of the 16S rRNA from the non-axenic cultures and their separation by agarose gel electrophoresis, the probe was used to identify cyanobacterial PCR products.

Dwivedi *et al.* (1996) developed a rapid method for DNA sequencing of rRNA genes amplified by PCR, without the need for cloning of the gene to be sequenced. Nubel *et al.* (1997) developed a set of oligonucleotide primers for the specific amplification of 16S rRNA segments from a variety of cyanobacteria and plasmids by PCR. The use of this specific PCR in combination with Denaturing Gradient Gel Electrophoresis (DGGE) is an extremely useful method to probe the diversity of cyanobacteria in culture, symbiotic associations with other life forms.

A diagnostic system using the sequence polymorphism within the 16S rRNA variable regions V_6, V_7 and V_8 for individual strain characterization and identification of toxin producing organisms belonging to the genera *Nostoc, Anabaena, Microcystis, Planktothrix* and some non toxin producing strains has been developed (Rudi *et al.*, 1997).

Felske *et al.* (1998) developed a novel approach to quantify rRNA in complex bacterial communities by amplification using reverse transcription RT- PCR with bacterium specific primers. This approach can be used in temperature gradient gel electrophoresis (TGGE) fingerprints of soil bacteria to estimate the ratios of bacterial 16S rRNAs. Moffitt *et al.* (2001) designed a specific 16S rRNA PCR for rapid detection of nodularin producing strains and found that *Nodularia spumigena* cluster was distinct from the benthic species *N. harveyana* and *N. sphaerocarpa*.

Rudi *et al.* (2000) developed a method to analyze genetically the cyanobacterial composition in natural water bodies, using multiple sequence specific labelling of oligonucleotide probes targeted to 16S rRNA and subsequent hybridization of the labelled probes to their respective complements spotted on to a solid support. This array make it possible to accurately analyze the whole microbial communities and to directly relate sequence information obtained from organisms in type culture collections to the biodiversity that exists in nature.

DNA/DNA HYBRIDIZATION

It is based on the feature that denatured single stranded DNA will renature under suitable conditions. The extent of renaturation and the stability of renatured hybrid molecule are proportional to the complementarity of the DNA strands involved. The relative binding and the differences in thermal stability can be used to assess the taxonomic status among organisms.

DNA/DNA hybridization have been used successfully to resolve taxonomic problems at the genus and species levels in cyanobacteria (Kelly and Cowie, 1972; Stam and Venema, 1977 and Wilmotte *et al.*, 1997). DNA/DNA hybridization allows the determination of the sequence similarity between genomes of different organisms without the determination of the extract nucleotide sequence (Stam and Stulp, 1984). Apte and Haselkorn (1990) demonstrated the potential of the hybridization method to identify differentially expressed genes. They reported that salinity induces expression of entirely different polypeptides in the two strains *Anabaena torulosa* and *Anabaena* L- 31. The DNA identified by *Anabaena*-31 probes exemplifies the osmotically inducible DNA which is common to both *Anabaena* strains.

RESTRICTION FRAGMENT LENGTH POLYMORPHISMS (RFLP)

RFLP markers have proven to be a reliable and highly informative tool for characterizing genetic diversity. Among the various molecular markers developed, RFLPs were the first to be used in human genome mapping and later they were adopted for plant genome mapping.

Restriction fragment length polymorphisms have been used extensively as an efficient DNA fingerprinting method to identify symbiotic cyanobacteria (Franche and Cohen - Bazire, 1985 and Nierzwircki - Bauer and Haselkorn, 1986). The widespread occurrence of the heptamer repeats in the *Nostoc* strain suggests that they may be useful diagnostics in the future. However the taxonomic value of southern hybridization by using the randomly repeated sequences as a phylogenetic tool may be restricted to the identification of identical strains (Mazel *et al.*, 1990).

Planktonic filamentous cyanobacterial strain belonging to the genera *Anabaena*, *Nodularia*, *Nostoc* and *Aphanizomenon* were characterized by SDS-PAGE of whole cell proteins and PCR/RFLP of the 16S rRNA gene (Lyra *et al.*, 1997). They reported that RFLP of 16S rRNA is advantageous because the sequencing of 16S rRNA gene is tedious and time consuming especially when one needs to classify a large number of strains.

DNA AMPLIFIED FINGER PRINTING OF CYANOBACTERIAL ISOLATES BY RAPD

By RAPD, polymorphisms can be easily analyzed by small amount of template DNA. Caetano Anolles *et al.* (1991) proposed a novel tool for DNA fingerprinting. The stratergy is based on amplification of characteristic DNA fragment by a thermostable DNA polymerase, directed by a single oligonucleotide primer of arbitrary sequence in a thermocycling reaction. DNA amplification fingerprinting has been very useful for distinguishing among closely related genotypes and it is a novel method for classification (Welsh and Mcclelland, 1990). The most commonly used technique for specific detection of DNA target segments are molecular hybridization (Southern, 1975) and DNA amplification by polymerase chain reaction (Williams *et al.*, 1990 and Mullis and Faloona, 1987). These techniques can be used to study variant or polymorphic sites for genotyping or to construct genetic maps.

Tandeau De Marsac (1990) reported the characterization of three distinct families of repeated sequences in the genome of the cyanobacterium *Calothrix* sp PCC 7601. These repeated sequences named short tandemly repeated repetitive sequences are present at a level of about 100 copies per genome and consisted of tandemly amplified hepta nucleotides. The RAPD technique in conjunction with PCR has been employed to identify many organisms to the strain level of classification (Welsh and McClelland, 1990). This technique is sensitive and specific because the entire genome of an organism is used as the basis for generating a DNA profile. The combination of PCR and random oligonucleotide primers has provided a method for the rapid and sensitive delineation of animal, plant, fungal, algal and bacterial strains (Akopyanz *et al.*, 1992; Kresoich *et al.*, 1992 and Scott *et al.*, 1992). Eskew *et al.* (1993) used DNA amplified fingerprinting by using single short nucleotide primers of arbitrary sequences to direct amplification of characteristic set of DNA products by Taq DNA polymerase. They used this technique to differentiate the isolates in *Azolla* - *Anabaena* symbiosis. Subhashini *et al.* (2000 a, b) characterized *Anabaena azollae* isolated from four different *Azolla* cultures based on their RAPD profile to study the variation existing among them. The RAPD profile generated the finger printing pattern characteristic of each strains based on the sequence of the primers used. Common band sharing observed between the strains *A. azollae* -RS-KK-SK- AM and *A. azollae* – RS-KK-SK-RP probably represents maternal inheritance of DNA to the progeny. The polymorphic bands were generated specifically for the isolates *A. azollae* - RS-KK-SK-RP and *A. azollae*- RS-KK-SK –AM with two primers which could be developed as possible markers for these isolates.

Current RAPD technology has been optimized for the identification of cyanobacterial cultures upto the strain level. The choice of primers with high G+C content provided good DNA profiles at a low concentration. Neilan

(1995) reported the use of RAPD-PCR to generate uniqueness in phylogenetic analysis of toxic cyanobacteria.

Nishihara *et al.* (1997) reported the use of RAPD analysis to discriminate genotypes in 5 species of the cyanobacterium *Microcystis* . The PCR amplification and sequencing of the *rpoc*1 (DNA dependent RNA polymerase) gene from nineteen Australian isolates of *Cylindrospermum raciborskii* revealed 99 to 100% nucleotide sequence identity among all strains, indicating all the isolates belong to one species, despite being morphologically distinct (Wilson *et al.*, 2000). Tamagnini *et al.* (2000) made an effort to elucidate the diversity of cyanobacterial hydrogenase using a molecular approach. All the tested nitrogen fixing strains showed hybridization with *hup*S and *hup* L (uptake hydrogenase) probes.

Kumar *et al.* (2002) analyzed the DNA profile of seventeen different cyanobacterial cultures including free- living and symbiotic forms using RAPD technique. The dendrogram, constructed based on the RAPD profile of cyanobacterial cultures, revealed 3 major clusters (Fig. 1). There was a similarity of 60 - 90 per cent within *Westiellopsis* cultures. *Nostoc* cultures shared 50 –80 per cent similarity with *Westiellopsis* cultures. *Anabaena* cultures were similar to *Westiellopsis* cultures by 60 –70 per cent. The dendrogram analysis clearly revealed that free- living cyanobacterial cultures are closely related with each other and are diverse from the symbiotic forms. Jeberlin Prabina *et al.* (2002) generated strain specific RAPD markers, for the standard cyanobacterial strains used in the preparation of cyanobacterial inoculants which could be exploited for checking genetic purity of strains in the cyanobacterial inoculum.

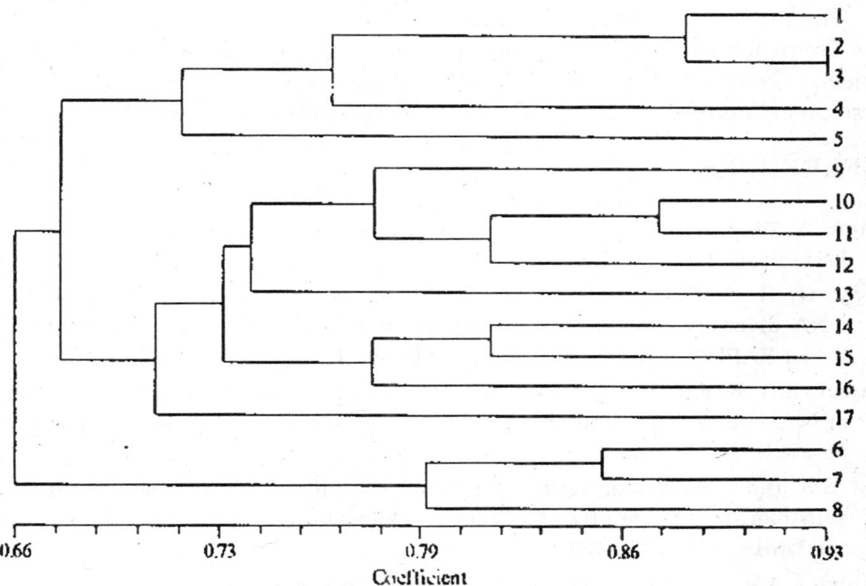

Fig 1. Phylogenetic relationship between the cyanobacterial cultures by RAPD analysis

1. *Westiellopsis* - ARM48
2. *Westiellopsis* sp.
3. *Westiellopsis* - C_{100} TR_5 ST_3 PA-SK
4. *Westiellopsis prolifica*
5. *Westiellopsis* $4A_2$
6. *Nostoc muscorum*
7. *Nostoc* sp.
8. *Anabaena variabilis*
9. *A. azollae* -SK-SLTNAU 1
10. *A. azollae* -MPK-SK-AM 24
11. *A. azollae* - MPK-SK-AM25
12. *A. azollae* - MPK SK-AF -38
13. *Anabaena* sp.
14. *Anabaena* -TR 52 ST 1
15. *Tolypothrix tenuis*
16. *Aulosira pseudoramosa*
17. *Fischerella* sp.

CONCLUSION

The concept of using nitrogen fixing cyanobacteria as nitrogen nourishing biofertilizer in rice fields is not yet fully explored and some major problems are still limiting the wide utilization of this biofertilization technique. Rapid and reliable quality control methods have not yet been developed for cyanobacterial biofertilizers which are considered as major hurdle in the production of quality cyanobacterial inoculants. In recent years, a number of new, valuable biochemical and molecular biological techniques for taxonomic purpose have been developed which could also be effectively utilized for checking the genetic purity of strains in the cyanobacterial inoculants

ACKNOWLEDGEMENT

The authors are thankful to the Indian Council of Agricultural Research, New Delhi for the financial support under National Agricultural Technology Project – Team of Excellence on Biofertilizers for rice based cropping system for some of the recent investigations reported in this review.

REFERENCES

Aitken, A. 1988. Protein sequences as taxonomic probes of cyanobacteria. **Methods. Enzymol., 167:** 145-154.

Akopyanz, N., N. O. Bukanov, T. U. Westblom, S. Kresovich and D. E. Berg. 1992. DNA diversity among clinical isolates of *Helicobacter pylori* detected by PCR-based RAPD fingerprinting. **Nucleic Acids. Res., 20:** 5137-5142.

Anagnostidis, K. and J. Komarek. 1988. Modern approach to the classification system of cyanophytes. 3. Oscillatoriales. **Arch. Hydrobiol. Suppl. 80. Algol. Stud., 50-53:** 327-472.

Apte, S.K and R. Haselkorn, 1990. Cloning of salinity stress-induced genes from the salt-tolerant nitrogen fixing cyanobacterium *Anabaena torulosa*. **Plant Molecular Biology, 15:** 723-733.

Bannister, J.V., W.H. Bannister and G. Rotilio 1987. Aspects of the structure function and applications of superoxide dismutase. **CRC Crit. Rev. Biochem, 28:**118-180.

Caetano-Anolles, G., B.J. Bassam and P.M. Gresshoff 1991. DNA amplification finger printing using very short arbitrary oligonucleotide primers. **Biotechnology, 9:** 553-557.

Cohen, Z. and A. Vonshak. 1991. Fatty acid composition of *Spirulina* and *Spirulina* - like cyanobacteria in relation to their chemotaxonomy. **Phytochem., 30:** 205-206.

Dwivedi, P.P., B. K. C. Patel, G.N. Rees and B. Ollivier. 1996. A rapid method for sequencing of rRNA genes amplified by polymerase chain reaction using an automated DNA sequences. **Indian J. Microbiol., 36:** 9-12.

Eskew, D.L., G. Caetano-Anolles, B. J. Bassam and P. M. Gresshoff. 1993. DNA amplification fingerprinting of the *Azolla-Anabaena* symbiosis. **Plant. Mol. Biol., 21:** 363-373.

Felske, A., A.D.L. Akkermans and W. M. Devos. 1998. Quantification of 16S rRNAs in complex bacterial communities by multiple competitive reverse transcription-PCR in Temperature Gradient Gel Electrophoresis, Fingerprints. **Appl. Environ. Microbiol., 16(11):** 4581-4587.

Franche, C. and G. Cohen-Bazire. 1985. The structural *nif* genes of four symbiotic *Anabaena azollae* show a highly conserved physical arrangement. **Plant Sci., 39:** 125-131.

Giovannoni, S.J., S. Turner, G. J. Olsen, S. Barns, D. J. Lane and N. R. Pace. 1988. Evolutionary relationship among cyanobacteria and green chloroplasts. **J. Bacteriol., 170:** 3584-3592.

Glaszmann, J.C. and P.A. Roger. 1985. Isozymes, possible markers for blue-green algae identification, **Intl. Rice Res. Newsletter, 10:** 4.

Gottlieb, L.D. 1981. Electrophoretic evidence and plant populations. In: **Progress in phytochemistry** Vol. 7, (eds.) L. Reinhold, J.B. Harburne and T. Swain. Royanme-Uni., Pergamon Press, Oxford, pp 1-46.

Hagemann, M., L. Wolfel and B. Kruger. 1990. Alternatives in protein synthesis in the cyanobacterium *Synechocystis* sp PCC 6803 after a salt shock. **J. Gen. Microbiol., 136:** 1393-1399.

Hertzberg, S., S. Liaaen-Jensen and H.W. Siegelman. 1971. The carotenoids of blue green algae. **Phyto Chem., 10:** 3121-3127.

Jeberlin Prabina, B., K. Kumar and S. Kannaiyan. 2002. DNA finger printing as a tool for checking genetic purity of strains in the cyanobacteril inoculum. **Plant and Soil** (Communicated).

Kannaiyan, S. 1985. **Studies on the algal application for low land rice crop.** Tamil Nadu Agric. University, Coimbatore, Tamil Nadu, India . p 24.

Kannaiyan, S. 1993. **Biofertilizers for rice.** Tamil Nadu Agricultural University, Coimbatore, Tamil Nadu, India. p 60.

Kelly, M. L. and D. B. Cowie. 1972. DNA-DNA hybridization studies of blue-green algae: **Carnegie Inst. Washington Yearbook, 71:** 276-287.

Kenyon, C.N., R. Rippka and R. Y. Stanier. 1972. Fatty acid composition and physiological properties of some filamentous blue-green algae. **Arch. Microbiol., 83:** 216-236.

Kumar, K., N. Annapoorna and S. Kannaiyan. 2000. Determination of genetic purity of the strains in the cyanobacterial inoculants by RAPD-PCR techniques. **Paper**

presented in the Asian Pacific Conference on Plant Tissue Culture and Agribiotechnology (APCPTCA), National University of Singapore, Singapore. 229 p (Abstr).

Kumar, K., B. Jeberlin Prabina and S. Kannaiyan. 2002. Phylogenetic analysis of symbiotic and free- living cyanobacterial cultures using DNA amplification fingerprinting. **Paper presented in the National Seminar and XII Southern Regional Conference on Microbial Inoculants** Annamalai University, Annamalainagar, March 21 & 22, 2002. (Abstr.) p.12.

Kresobich, S., J.G.K. Williams, J.R. McFerso, E.J. Routman and B.A. Schaal. 1992. Characterization of genetic identities and relationship of *Brassica oleracea* L. via a random amplified polymorphic DNA. **Theor. Appl. Genet., 85:** 190-196.

Lakshmanan, A., K. Kumar and S. Kannaiyan. 2002. Characterization of fatty acid profile of filamentous cyanobacteria for strain identification in the cyanobacterial inoculum. **Paper presented in the International Conference on Molecular Approaches for Improved Crop Productivity and Quality,** TNAU, Coimbatore May 22-24, 2002. (Abstr.) p.33.

Lane, D., K.G. Field, G.J. Olsen and N. R. Pace. 1988. Reverse transcripts sequencing of rRNA for phylogenetic analysis. **Methods Enzymol., 167:** 138-144.

Li, R., A. Yokota, J. Suglyania, M, Watanabe, M. Hiroki and M.M. Watanabe. 1998. Chemotaxonomy of planktonic cyanobacteria based on non-polar and hydroxy fatty acid composition. **Phycological Research, 46(1):** 210-28.

Lyra, G., J. Hantual, E. Vainio, J. Rapala, L. Rouhiainen and K. Sivonen. 1997. Characterization of cyanobacteria by SDS-PAGE of whole cell proteins and PCR/RFLP of the 16S rRNA gene. **Arch. Microbiol., 168:** 176-184.

Mazel, D., J. Houmard, A.M. Casets and N.T. deMarsac. 1990. Highly repetitive DNA sequences in cyanobacterial genomes. **J. Bacteriol., 172:** 2755-2761.

Moffitt, M.C., S.I. Blackburn and B.A. Neilan. 2001. rRNA sequences reflect the ecophysiology and define the toxic cyanobacteria of the genus *Nodularia* . **Int. J. Syst. Evol. Microbiol., 51:** 505-512.

Mullis, K.B. and F.A. Faloona. 1987. Polymerase chain reaction. **Methods Enzymol., 155:** 355-350.

Neilan, B.A. 1995. Identification and phylogenetic analysis of toxigenic cyanobacteria by multiplex randomly amplified polymorphic DNA-PCR. **Appl. Environ. Microbiol., 61(6):** 2286-2291.

Nelissen, B., A. Wilmotte, J..M. Neefs and R. DeWachter. 1994. Phylogenetic relationships among filamentous helical cyanobacteria investigated on the basis of 16S ribosomal RNA gene sequence analysis. **Syst. Appl. Microbiol., 17:** 206-210.

Nelissen, B., R. De Baere, A. Wilmotte and R. De Wachter. 1996. Phylogenetic relationships of non-axenic filamentous cyanobacteiral strains based on 16S rRNA sequence analysis. **J. Mol. Evol., 42:** 194-200.

Nierzwicki-Bauer, S.A and R. Haselkorn. 1986. Differences in mRNA levels in *Anabaena* living freely or in symbiotic association with *Azolla.* **EMBO. J., 5:** 29-35.

Nishihara, H., H. Miwa, M. Watanabe, M. Nagashima, O. Yagi and Y. Takamura. 1997. Random Amplified Polymorphic DNA (RAPD) analysis for discriminating genotypes of *Microcystis* cyanobacteria. **Biosci. Biotech. Biochem., 61:** 1067-1972.

Nubel, U., F. Garcia-Pichel and G. Muyzer. 1997. PCR primers to amplify 16S rRNA genes from cyanobacteria. **Appl. Environ. Microbiol., 63:** 3327-3332.

Pace, N.R., D. A. Stahl, D. J. Lane and C. J. Olsen. 1980. The analysis of natural populations by ribosomal RNA sequences. **Adv. Microbiol. Ecol., 9:** 1-55.

Puppo, A. and J. Rigaud. 1986. Superoxide dismutase: an essential role in the protection of the nitrogen fixation process. **FEBS Lett., 201:** 187-189.

Roger, P. A. and Watanabe, I. 1986. Technologies for utilizing biological nitrogen fixation in Wetland rice. Potentialities, current usage and limiting factors. In: **Nitrogen Economy and Flooded Rice Soils.** (Eds.) S K Dutta and N H Patrick Jr. Martinus Nijhoff Publishers, Dordrecht, The Netherlands. pp 37-77..

Rudi, K., O..M. Skulberg, F. Larsen and K. S. Jakobsen. 1997. Strain characterization and classification of oxyphotobacteria in clone cultures on the basis of 16S rRNA sequence from the variable regions V_6, V_7 and V_8. **Appl. Environ. Microbiol., 63:** 2593-2599.

Rudi, K., O.M. Skulberg, R. Skulberg and K.S. Jakobsen. 2000. Application of sequence specific labelled 16S rRNA gene oligonoculeotide probes for genetic profiling of cyanobacterial abundance and diversity by assay hybridization. **Appl. Environ. Microbiol., 66:** 4004-4011.

Scott, M.P., K.M Haymes. and S.M. Williams. 1992. Parentage analysis using RAPD PCR. **Nucleic Acids Res., 20:** 5493.

Skulberg, O.M. and R. Skulberg, 1985. Planktonic species of *Oscillatoria* from Norway: characterization and classification. **Arch. Hydrobiol. Suppl., 71: Algol. Stud. 38/39:** 157-174.

Soltis, D.E., C. H. Haufler, D. C. Parrow and G.J. Gastomy. 1983. Starch gel electrophoresis of ferns. A comparison of grinding buffers, gel and electrode buffers and staining schedules. **American. Fern. J., 73:** 9-27.

Southern, E.M. 1975. Detection of specific sequences among DNA fragments separated by gel electrophoresis. **J. Mol. Biol., 98:** 503-517.

Stackebradnt, E. and C.R. Woese. 1981. The evolution of prokaryotes. In: **Molecular and cellular aspects of microbial evolution** (eds.) M.J. Carlisle, J.R. Collins and B.E.B. Moseley, Cambridge Univ. Press, Cambridge, pp. 1-31.

Stam, W.T. and B. K. Stulp. 1984. New taxonomic methods DNA/DNA hybridization. **Methods Enzymol., 167:** 125-132.

Stam, W.T. and G. Venema, 1977. The use of DNA-DNA hybridization for determination of the relationship between some blue-green algae. **Acta. Bot. Neerl., 26:** 327-342.

Subhashini, R., Kumar, K., and Kannaiyan, S. 2000a. Variation revealed by RAPD – PCR Profile of Microsymbiotic *Anabaena azollae* Strains from Four N_2 fixing *Azolla* Cultures. **Acta Biotechnol. 20(2):** 169-174.

Subhashini, R., Kumar, K., and Kannaiyan, S. 2000b. Biodiversity of *Anabaena azollae* isolates from different *Azolla* cultures. **India J. Exp. Biog. 38:** 1168-1171.

Tamagnini, P., J. Costa, L. Almeidra, M.Olivera, R. Salema and P. Lindblad. 2000. Diversity of cyanobacterial hydrogenases, a molecular approach. **Current Microbiology, 40:** 356-361.

Tandeau De Marsac. 1990. Highly repetitive DNA sequences in cyanobacterial genomes. **J. Bacteriol., 172(5):** 2755-2761.

Vallejos, C.E. 1983. Enzyme activity staining. In: **Isozymes in plant genetics and breeding** Part A. (eds.) S.D. Tanksley and T.J. Orton. Elsevier Science Publishers, Amsterdam, pp. 469-516.

Ward, D.M., R. Weller and M. M. Bateson. 1990. 16S rRNA sequences reveal unclustured microorganisms in a natural community. **Nature, 345:** 63-65.

Watanabe, I. and P. A. Roger. 1984. Nitrogen fixation in wetland rice field. In: **Current Development in Biological Nitrogen Fixation.** (ed.) N S Subba Rao Oxford and IBH Publishing Co, New Delhi. pp. 237 – 276.

Welsh, J. and M. McClelland. 1990. Finger printing of genomes using PCR with arbitary primers. **Nucleic Acids Res., 18:** 7213-7218.

Williams, J.G.K., A.R. Kubelik, K.J. Livak, J.A. Raflaski and S.V. Tingey. 1990. DNA polymorphisms amplified by arbitrary primers are useful as genetic markers. **Nucleic Acids Res., 18:** 6531-6535.

Wilmotte, A., W.T. Stam and V. Demoulin. 1997. Taxonomic study of marine Oscillatoriacean strains (cyanophyceae, cyanobacteria) with narrow trichomes. III. DNA-DNA hybridization studies and taxonomic conclusions. **Algol. Stud., 87:** 11-28.

Wilson, K. M., M. .A. Schembri., P.D Baker, and C.P. Saint. 2000. Molecular characterization of the toxic cyanobacterium *Cylindrospermospsis raciborkskii* and design of a species-specific PCR. **Appl. Environ. Microbiol., 66:** 332-338.

CHAPTER – 55

EVALUATION OF NITROGEN FIXING POTENTIAL, AMMONIA EXCRETION AND PROTEIN CONTENT OF THE CYANOBACTERIAL ISOLATES UNDER LABORATORY CONDITIONS FOR DEVELOPING QUALITY CONTROL STANDARDS FOR CYANOBACTERIAL BIOFERTILIZERS

B. Jeberlin Prabina, K. Kumar and S. Kannaiyan

ABSTRACT

Diazotrophic cyanobacteria are recognized as the most potential nitrogen fixing biological system, under water logged condition of the rice field. The role of cyanobacteria as biofertilizer in relation to increased rice production as well as soil fertility is widely accepted. Quality control of cyanobacteria is important for ensuring the performance of the applied inoculum. An attempt was made to monitor the nitrogen fixing potential of nine different cyanobacterial cultures belonging to different genera *viz.*, *Anabaena, Aulosira, Fischerella, Nostoc, Tolypothrix* and *Westiellopsis* by estimating their nitrogenase activity, ammonia excretion and protein content. Tremendous variation was observed in the nitrogenase activity between different genera at different growth phases. *Nostoc muscorum* -DOH recorded the maximum nitrogenase activity to the level of 1654.95 n moles g^{-1} fresh weight of the culture per hour. In general the ammonia excretion was maximum on 20th day after inoculation. *Nostoc muscorum* -DOH released the maximum amount of ammonia (310 n moles ml^{-1} of the culture) followed by *Tolypothrix tenuis* (300 n moles ml^{-1}). Protein content of the cyanobacterial cultures ranged from 400-650 mg g^{-1} fresh weight of the cultures. The results indicated that these parameters will be useful for developing quality control standards for cyanobacterial biofertilizers.

INTRODUCTION

An attractive feature of cyanobacterial nitrogen fixation is its demand for energy and reductant are supplied by the photosynthetic activity of the cells. They possess multi enzyme system called nitrogenase which can break down the very stable dinitrogen molecule and reduce it to ammonia molecule with the aid of energy source and electrons (Moore, 1969). It is estimated that as much as 10^8 tonnes of nitrogen per annum is converted to ammonia by biological processes. The dinitrogen fixed by cyanobacteria in the form of ammonia is released either through exudation or through microbial decomposition after the death of the cells (Roger and Kulasooriya, 1980). It is observed that temporary immobilization of nitrogen by cyanobacteria conserve the nitrogen in their cell system and prevents the nitrogen loss in rice fields (Kannaiyan, 2000). In the present investigation attempts were made to evaluate the nitrogenase activity, ammonia excretion and protein content of the cyanobacterial cultures used in the inoculant preparation for their potential use in the quality control of cyanobacterial biofertilizers.

MATERIALS AND METHODS

Nine different cyanobacterial cultures viz., *Anabaena azollae* MPK-SK-AM-24, *A. azollae* MPK-SK-AM-25, *Anabaena* TR52-ST1, *A. variabilis* – SA_0, *Aulosira pseudoramosa, Fischerella* sp, *Nostoc muscorum*- DOH, *Tolypothrix tenuis* and *Westiellopsis prolifica* were evaluated for their nitrogen fixing potential, ammonia excretion and protein content. The nitrogenase activity was estimated by the standard Acetylene Reduction Assay (ARA) as described by Hardy *et al.* (1968). The release of ammonia by cyanobacterial cultures was estimated by Phenol – hypochlorite method (Solorzano, 1969). The protein was extracted by the method of Kamicker (1986) and quantified as per Lowry *et al.* (1951).

RESULTS ANS DISCUSSION

The nitrogen fixation by cyanobacteria plays a vital role in the built – up and maintenance of soil fertility. Nitrogen fixing cyanobacteria can give farmers benefits equivalent to the application of 25-30 kg ha^{-1}. In the present study, the nitrogenase activity of the cyanobacterial cultures markedly increased during different stages of growth *viz.,*10, 20 and 30th day after inoculation in BG-11 medium and was maximum on 30th day after inoculation (Table 1). Among the cyanobacterial cultures tested, *Nostoc muscorum* – DOH exhibited maximum nitrogenase activity (1654.95 n moles hr^{-1} g^{-1} fresh weight). Saha *et al.* (1982) showed that nitrogen fixed by cyanobacteria become available to rice plants only after it is released extracellularly into the environment, either as extracellular products or by mineralisation of their intracellular contents through microbial decomposition after death. The ammonia excretion increased till 20th day after inoculation and declined subsequently (Table 2). *Nostoc muscorum* – DOH excreted higher ammonia on 20th day after inoculation (310 n moles

ml^{-1}). The ammonia excretion varied widely with cultures. Even though, no direct correlation could be drawn between nitrogenase activity and ammonia excretion, the cyanobacterial cultures with higher nitrogenase activity excreted considerable amount of ammonia. Ammonia is lost from the cells in a process that is not energy dependent and leaks because of the inherent permeability of ammonia (Boussiba and Gibson, 1991) The protein content varied between 400-650 mg g^{-1} fresh weight (Table 3). The results of the present study suggested that with the available information it is possible to standardize the optimum levels of nitrogen fixing capacity, ammonia excretion and protein content for developing quality control standards for the efficient strains belonging to different genera of cyanobacteria used for the preparation of cyanobacterial inoculants in the near future.

Table 1. Nitrogenase activity of the cyanobacterial cultures during different stages of growth in BG - 11 medium under laboratory conditions

Sl. No	Cyanobacterial cultures	Amount of ethylene produced (n moles hr^{-1} g^{-1} fresh culture)		
		Days after inoculation		
		10	20	30
1	*A. azollae* - MPK-SK-AM-24	274.05	849.98	1292.73
2	*A. azollae* - MPK-SK-AM-25	240.19	682.77	760.72
3	*Anabaena* -TR-52-ST 1	276.37	844.36	1303.69
4	*A. variabilis*-SA_0	263.16	457.88	834.58
5	*Aulosira pseudoramosa*	457.88	703.04	1042.26
6	*Fischerella* sp	263.16	457.88	834.58
7	*Nostoc muscorum*-DOH	548.35	954.39	1654.95
8	*Tolypothrix tenuis*	361.36	781.61	1374.62
9	*Westiellopsis prolifica*	353.19	1061.84	1425.45
		SEd	CD (0.05)	
	Days	0.821	1.65	
	Culture	1.42	2.86	
	Days x Culture	2.47	4.95	

Table 2. Ammonia excretion by the cyanobacterial cultures during growth in BG-11 medium under laboratory conditions

Sl . No	Cyanobacterial cultures	Ammonia excretion (n moles ml^{-1})		
		Days after inoculation		
		10	20	30
1	*A. azollae* – MPK-SK-AM-24	130	210	80
2	*A. azollae* – MPK-SK-AM-25	80	230	30

3	*Anabaena* -TR-52-ST 1	80	160	140
4	*A. variabilis* -SA_0	150	230	70
5	*Aulosira pseudoramosa*	50	80	70
6	*Fischerella* sp	190	220	180
7	*Nostoc muscorum* –DOH	160	310	190
8	*Tolypothrix tenuis*	260	300	300
9	*Westiellopsis prolifica*	80	210	70
		SEd	CD (0.05)	
	Days	2.13	4.27	
	Cultures	3.69	7.39	
	Days x Cultures	6.39	12.81	

Table 3. Protein content of the cyanobacterial cultures grown in BG-11 medium under laboratory conditions

S1. No	Cyanobacterial cultures	Protein content (mg g^{-1} fresh weight)
1	*A. azollae* –MPK-SK-AM 24	590.91
2	*A. azollae* –MPK-SK-AM-25	636.36
3	*Anabaena* –TR-52 ST1	466.67
4	*Anabaena variabilis*-SA_0	675.00
5	*Aulosira pseudoramosa*	500.00
6	*Fischerella sp*	600.00
7	*Nostoc muscorum–DOH*	475.00
8	*Tolypothrix tenuis*	571.43
9	*Westiellopsis prolifica*	400.00
	SEd	2.45
	CD (0.05)	4.98

ACKNOWLEDGEMENT

The authors are thankful to the Indian Council of Agricultural Research, New Delhi, for the financial support extended to this work under National Agricultural Technology Project – Team of Excellence on Biofertilizers for Rice Based Cropping System.

REFERENCES

Boussiba, S. and J. Gibson 1991. Ammonia translocation in cyanobacteria. FEMS **Micorbiol. Rev., 88:** 1-74.

Hardy, R.W.F., R.D. Holsten, E.K. Jackson and R.C. Burns. 1968. The acetylene reduction assays for N_2 fixation: Laboratory and field evaluation. **Plant Physiol., 43:** 1185-1207.

Kamicker, B.J. and W.J. Brill. 1986. Identification of *Bradyrhizobium japonicum* nodule isolates from Wiscosin soybean farms. **Appl. Environ. Microbiol., 51(3):** 487-492.

Kannaiyan, S. 2000. **Biofertilizer technology and quality control,** Tamil Nadu Agricultural University Publication, Coimbatore, Tamil Nadu, India. p. 173.

Lowry, O.H., N.J. Rosebrought, A.I. Larr and R.J. Randall. 1951. Protein measurement with folin - phenol reagent. **J. Biol. Chem., 193 :** 265-275.

Moore, M.W. 1969. *Azolla* biology and agronomic significance. **Bot. Rev., 35:** 17-35..

Roger, P.A. and S.A. Kulasooriya. 1980. **Blue Green Algae and Rice.** International Rice Research Institute, Los Banos, The Philippines, p. 112

Saha, K.C., B.C. Panigrahi and P.K. Singh. 1982. Blue green algae-*Azolla* additions on nitrogen and phosphorus availability and redox potential of a flooded soil. **Soil. Biol. Biochem., 14:** 23-26.

Solorzano, L. 1969. Determination of ammonia in natural waters by the phenol-hypochlorite method. **Limnol. Oceanogr., 14:** 799-801.

CHAPTER - 56

GROWTH PATTERN AND CHLOROPHYLL CONTENT OF THE CYANOBACTERIAL STRAINS FOR THEIR UTILIZATION IN THE QUALITY CONTROL OF CYANOBACTERIAL BIOFERTILIZERS

B. Jeberlin Prabina, K. Kumar and S. Kannaiyan

ABSTRACT

Cyanobacteria (BGA) are a diverse group of prokoryotes possessing oxygen evolving photosynthetic system. The cyanobacterial contribution largely depends on their establishment and growth in rice soil. Hence, the growth pattern, biomass production potential and chlorophyll content of nine cyanobacterial strains belonging to six genera, *viz., Anabaena, Aulosira, Fischerella, Nostoc, Tolypothrix* and *Westiellopsis* which are being used for preparation of cyanobacterial inoculants, were examined at regular intervals. The peak of the growth phase was observed on 25 – 30 days after inoculation in all the cyanobacterial strains tested. With increasing time, the biomass of all the cyanobacterial strains tested increased whereas the total chlorophyll content decreased. In terms of dry weight, maximum biomass was produced by *Anabaena variabilis*-SA_0, *Nostoc muscorum*-DOH followed by *Tolypothrix tenuis*. The total chlorophyll content was maximum on 20 days after inoculation. *Westiellopsis prolifica* (91.47mg g^{-1} fresh weight) and *Anabaena variabilis*-SA_0 (84.99mg g^{-1} fresh weight) produced maximum chlorophyll. The results suggested the possibility of fixing minimum standards for biomass production and chlorophyll content of the cyanobacterial strains during mass multiplication under laboratory conditions.

INTRODUCTION

Cyanobacteria, a fascinating group of photosynthetic prokaryotic organisms, endowed with tremendous genome plasticity are distributed in all possible bio types of the world. Cyanobacteria synthesize cellular metabolites by drawing carbon from carbondioxide and energy from sunlight. These organisms synthesize compounds like amino acids, proteins, sugars, complex polysaccharides, vitamins and growth promoting substances for their own metabolism (Misra and Kaushik, 1989). Algal crusts on soil surface add organic matter, reduce evaporation from soil surface, remove compaction and increase nutrient utilization efficiency of rice crop. Adaptation to low light and effective nutrient uptake kinetics at low ambient concentration are important factors leading to cyanobacterial domination in rice fields (Kannaiyan, 1990). Cyanobacteria are characterized by their ability to form chlorophyll-*a*, phycocyanin and phycoerythrin. Hence the growth pattern, biomass production potential and chlorophyll content of different cyano-bacterial strains used in the inoculant preparation were examined for developing quality control standards for cyanobacterial biofertilizers.

MATERIALS AND METHODS

A total of nine cyanobacterial strains including both symbiotic and free - living forms belonging to six genera used in the inoculant preparation were studied. The cultures include *Anabaena azollae* MPK-SK-AM-24, *A. azollae* MPK-SK-AM-25 *Anabaena* TR52-ST1, *A. variabilis* – SA_0, *Aulosira pseudora-mosa, Fischerella* sp, *Nostoc muscorum*- DOH, *Tolypothrix tenuis* and *Westiellopsis prolifica*. The cyanobacterial cultures were grown in BG-11 medium and incubated at 3000 lux intensity at 25 ±1°C with a day and night cycle of 16/8 hrs. The different parameters were determined at every 10 days intervals for 30 days. The growth was measured by recording the optical density (OD) in a spectrophotometer at 750 nm. The biomass produced was calculated in terms of dry weight after oven drying at 65°C for 24 hours. The total chlorophyll content was estimated as suggested by Wintermans and Demots (1965). The total chlorophyll was extracted with 95 per cent acetone and absorbance was read at 665nm and 649 nm.

$$\text{Total chlorophyll} = \frac{6.10\,(A_{665}) + 20.04\,(A_{649}) \times V \times 100}{W}$$

A – Absorbance at specific wavelength

V – Final volume of chlorophyll content in 95 per cent acetone.

W – Fresh weight of the sample in mg.

RESULTS AND DISCUSSION

The role of diazotrophic cyanobacteria as biofertilizer in the nitrogen enrichment of the rice soil ecosystem and in the maintenance of soil fertility has been widely recognized. Earlier studies have shown a remarkable strain variation among the cyanobacterial cultures in their biomass production potential under field conditions (Kolte and Goyal, 1986). In the present investigation, all the cyanobacterial cultures recorded considerable growth in BG-11 liquid medium. In general, a gradual increase in the growth rate of cyanobacteria was observed till 30th day after inoculation (Table 1). Among the cultures tested, *Anabaena variabilis* – SA_0 recorded highest growth followed by *Nostoc muscorum* – DOH. Maximum dry weight was recorded on 30th day after inoculation. *Anabaena variabilis* – SA_0 registered the maximum dry weight (92.67 mg) on 30th day after inoculation. The biomass production of *A. pseudoramosa, Fischerella* sp and *W. prolifica* was markedly low compared to other cultures. The free living cultures *Anabaena variabilis* – SA_0, *Nostoc muscorum* – DOH showed higher growth and dry weight compared to symbiotic forms (Table 2). This may be due to the better adaptability of the free- living forms than the symbiotic ones in the natural ecosystem. Tremendous variations was observed in growth both at genus and species level which is in accordance with the results of Suseela and Goyal (1995). The cyanobacteria contain photosynthetic pigments like chlorophyll-*a*, phycobiliproteins and carotenoids in their filaments. The chlorophyll content was higher on 20th day after inoculation and started decreasing thereafter. The chlorophyll content was higher in *Westiellopsis prolifica* (91.47 mg g^{-1} fresh weight) followed by *Anabaena variabilis* – SA_0 (84.99) mg g^{-1} fresh weight (Table 3). The experimental results of Piorreck and Pohl (1984) with *Anacystis nidulans* and *Microcystis aeruginosa* revealed the decrease in chlorophyll content with increase in duration of growth. The data obtained in the present study showed that the chlorophyll content for different cyanobacterial cultures varied between 40-80 mg g^{-1} fresh weight and decreased with growth of the cultures which could be exploited for quality checking of the cyanobacterial biofertilizer.

Table 1. Growth of different cyanobacterial cultures in BG-11 medium under laboratory conditions

Sl. No	Cyanobacterial cultures	O.D. values (750 nm) Days after incubation					
		5	10	15	20	25	30
1	*A. azollae* – MPK-SK-AM-24	0.03	0.04	0.58	0.80	1.86	1.89
2	*A. azollae* – MPK-SK-AM-25	0.19	0.23	0.32	0.82	0.99	1.14
3	*Anabaena* -TR-52-ST1	0.04	0.19	0.34	0.95	1.21	1.23
4	*A. variabilis*-SA_0	0.22	0.67	1.28	1.64	2.18	2.30
5	*Aulosira pseudoramosa*	0.06	0.13	0.14	0.62	0.72	0.74
6	*Fischerella* sp.	0.46	0.59	1.27	1.61	1.38	1.91
7	*Nostoc muscorum*-DOH	0.24	0.38	1.00	1.42	2.05	2.21
8	*Tolypothrix tenuis*	0.05	0.10	0.22	0.62	1.86	1.89
9	*Westiellopsis prolifica*	0.07	0.42	0.60	1.01	1.07	1.18
		SEd	CD (0.05)				
	Days	0.02	0.03				
	Cultures	0.02	0.04				
	Days x Cultures	0.05	0.10				

Table 2. Biomass production by the cyanobacterial cultures in the BG-11 medium under laboratory conditions

Sl. No	Cyanobacterial cultures	Days after inoculation					
		10		20		30	
		Fresh weight (mg)	Dry weight (mg)	Fresh weight (mg)	Dry weight (mg)	Fresh weight (mg)	Dry weight (mg)
1	*A. azollae* - MPK-SK-AM-24	62.80	10.56	42.24	15.80	230.60	37.20
2	*A. azollae* - MPK-SK-AM-25	97.60	7.20	28.80	36.64	386.40	64.90
3	*Anabaena* -TR-52-ST1	81.20	16.00	92.80	22.30	219.20	43.60
4	*A. variabilis*-SA_0	41.20	12.70	284.20	52.00	406.89	92.67
5	*Aulosira pseudoramosa*	32.40	10.00	57.92	19.57	86.40	22.60
6	*Fischerella* sp	42.20	15.50	62.00	20.80	137.80	33.60
7	*Nostoc muscorum*-DOH	101.10	24.50	236.90	49.00	400.69	80.94
8	*Tolypothrix tenuis*	76.40	19.10	111.80	38.00	251.30	79.40
9	*Westiellopsis prolifica*	46.90	16.00	64.00	21.00	209.00	36.40

	SEd	CD (0.05)
Days	0.27	0.54
Cultures	0.47	0.94
Days x Cultures	0.81	1.63

Table 3. Chlorophyll content of the cyanobacterial cultures grown in BG-11 medium under laboratory conditions

S1. No	Cyanobacterial cultures	Chlorophyll content ($mg\ g^{-1}$ fresh weight)		
		Total Chlorophyll		
		Days after inoculation		
		10	20	30
1	*A. azollae* - MPK-SK-AM-24	33.62	49.98	46.82
2	*A. azollae* - MPK-SK-AM-25	39.48	52.66	39.41
3	*Anabaena* -TR-52-ST 1	46.20	67.40	60.78
4	*A. variabilis*-SA_0	58.69	84.99	73.62
5	*Aulosira pseudoramosa*	53.69	76.02	54.21
6	*Fischerella* sp	43.21	82.70	69.86
7	*Nostoc muscorum*-DOH	52.96	72.91	52.20

8	*Tolypothrix tenuis*	49.28	63.86	46.89
9	*Westiellopsis prolifica*	60.10	91.47	76.60

	Total chlorophyll	
	SEd	CD (0.05)
Days	0.69	1.22
Cultures	1.05	2.11
Days x Cultures	1.82	3.56

ACKNOWLEDGEMENT

The authors are thankful to the Indian Council of Agricultural Research, New Delhi, for the financial support extended to this work under National Agricultural Technology Project – Team of Excellence on Biofertilizers for Rice Based Cropping System.

REFERENCES

Kannaiyan, S. 1990. **Biotechnology of biofertilizer for rice.** Tamil Nadu Agric. Univ., Coimbatore, Tamil Nadu, India. p. 212.

Kolte, S.O and S.K. Goyal. 1986. Natural variation in the nitrogen fixing potential of cyanobacteria. **Phykos, 25:** 166.

Misra, S. and B.D. Kaushik. 1989. Growth promoting substances of cyanobacteria. I. Vitamins and their influence on rice plant. **Proc. Indian. Natn. Sci. Acad., 55:** 295-300.

Piorreck, M. and P. Pohl. 1984. Formation of biomass, total protein, Chlorophyll lipids and fatty acids in green and blue green algae during one growth phase. **Phytochemistry, 23(2):** 217-223.

Suseela, M.R. and S.K. Goyal. 1995. Effect of ammonium nitrogen on growth and nitrogen fixation by cyanobacteria. **Phykos, 34(1&2) :** 123-130.

Wintermans, J.F.G.M and A. Demots 1965. Spectrophotometric characteristic chlorophylls *a* and *b* and their phenophytins in ethanol. **Biochem. Biophys. Acta., 109:** 448-453.

Related Books

ISBN: 9789383692620

ISBN: 9789383692644

ISBN: 9789388043656

ISBN: 9789383692101

ISBN: 9789383692637

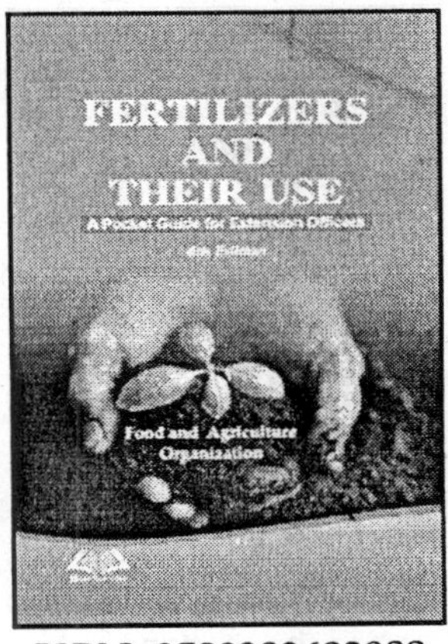

ISBN: 9789383692088

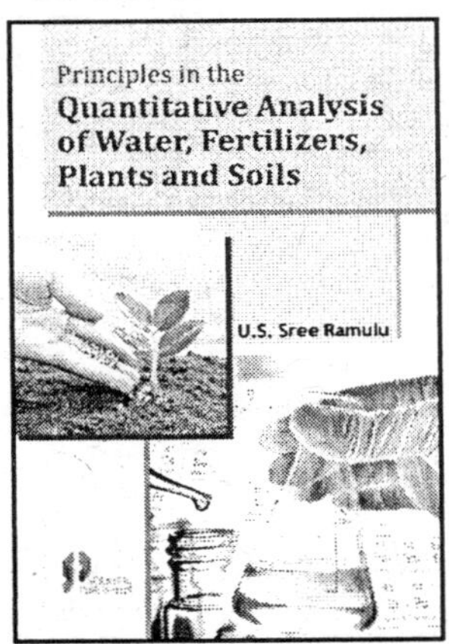

ISBN: 9788172333270

ISBN: 9789387307087

ISBN: 9789383692996

Related Books